Science, Technology, and Innovation for Sustainable Development Goals

T0323247

Science, Technology, and Innovation for Sustainable Development Goals

Insights from Agriculture, Health, Environment, and Energy

Edited by

Ademola A. Adenle
Marian R. Chertow
Ellen H.M. Moors
David J. Pannell

OXFORD
UNIVERSITY PRESS

OXFORD
UNIVERSITY PRESS

Oxford University Press is a department of the University of Oxford. It furthers
the University's objective of excellence in research, scholarship, and education
by publishing worldwide. Oxford is a registered trade mark of Oxford University
Press in the UK and certain other countries.

Published in the United States of America by Oxford University Press
198 Madison Avenue, New York, NY 10016, United States of America.

Library of Congress Cataloging-in-Publication Data
Names: Adenle, Ademola A., 1976– editor. | Chertow, Marian R., editor. |
Moors, Ellen, 1968– editor. | Pannell, David J., 1960– editor.
Title: Science, technology, and innovation for sustainable
development goals : insights from agriculture, health, environment, and energy /
[edited by] Ademola A. Adenle, Marian R. Chertow, Ellen H.M. Moors, David J. Pannell.
Description: New York, NY : Oxford University Press, [2020] |
Includes bibliographical references and index.
Identifiers: LCCN 2020003850 (print) | LCCN 2020003851 (ebook) |
ISBN 9780190949501 (hardback) | ISBN 9780190949518 (paperback) |
ISBN 9780190949532 (epub) | ISBN 9780190949525 (updf) | ISBN 9780197528907 (oso)
Subjects: LCSH: Sustainable Development Goals. | Sustainable development. |
Technological innovations—Environmental aspects. | Technological innovations—Economic aspects. |
Technological innovations—Government policy. |
Technological innovations—International cooperation.
Classification: LCC HC79.E5 S29236 2020 (print) | LCC HC79.E5 (ebook) |
DDC 338.9/27—dc23
LC record available at https://lccn.loc.gov/2020003850
LC ebook record available at https://lccn.loc.gov/2020003851

1 3 5 7 9 8 6 4 2

Paperback printed by LSC Communications, United States of America
Hardback printed by Bridgeport National Bindery, Inc., United States of America

Contents

Foreword by Mahmoud Mohieldin

In 2015, leaders from 193 nations made a commitment to achieve 17 Sustainable Development Goals (SDGs) by the year 2030, to end poverty, offer every person an opportunity for a better life, and protect our planet for future generations. This historic global agreement is one of the most ambitious ever conceived—and if it is to be successful, all of us will need to participate.

Each year, representatives from many of these signatory countries meet at the United Nations in New York as part of the High-Level Political Forum on Sustainable Development. Participants take stock of their nations' plans and progress toward the SDGs, with their 169 interlinked targets, measured by 232 indicators. I've attended all of these sessions as the World Bank Group's Senior Vice President for the 2030 Development Agenda, UN Relations and Partnerships. Watching these proceedings, I believe there is room for optimism that we can be successful. Yet it is also clear that we have set ourselves a daunting task that requires our best ideas, skillful implementation, and mobilization of resources on an unprecedented scale.

This book offers policymakers and practitioners a wealth of impactful ideas that could be instrumental in helping countries and the world reach the ambitious SDG targets, leveraging science, technology, and innovation (STI) to speed and improve the implementation of desired outcomes. It focuses on three key areas: environment and energy, health, and agriculture—areas which are critical to the SDGs and resource intensive. To reach the SDGs we will need STI solutions in areas such as these, displacing and disrupting existing systems, while protecting the people they are intended to help.

Indeed, we must carefully balance risks and rewards of these solutions so that progress is broad and fair to all people. Thus, our growth—increasingly driven by digital innovation—can and should be equitable and sustainable. Our planners and policymakers must also foster resilience against a variety of risks such as climate change and economic disruption. And we must invest in our people—especially in health, education, and workforce training. Successful interventions can help communities, including women, leapfrog to advanced technological solutions at an unprecedented pace and scale. For example:

- In healthcare, digital and web technologies have been shown to: expand health services in developing countries, increase health systems' efficiency, and lead to better patient outcomes through tele-medicine and improved access to information which is related to disease prevention, nutrition, and hygiene.
- In agriculture, breakthrough digital technologies have the potential to deliver significant positive impacts across food value chains. These range from innovations

that can make food systems more resource-efficient and climate-resilient such as precision agriculture, gene-editing, and biological-based crop protection; or technologies that improve traceability from farm to fork.

- Renewable energy is playing an increasingly vital role in helping countries develop modern, secure energy systems while disruptive technologies like smart grids, smart meters, and geospatial data systems have upturned energy planning and lowered carbon emissions.

These solutions are real and achievable, if we make the commitment to share and implement them at scale.

The authors who contributed their deep knowledge to this volume represent a diverse group of institutions and disciplines, from various regions of the world. As the authors note, progress on the SDGs will depend upon collaborative efforts between governments (at the national and local levels), civil society, the private sector, academia, multilateral organizations, and communities, each contributing their knowledge and resources to reach our shared objectives.

Of course, action on STI will also require adequate financing and policy tools to build institutional capacity to plan, implement, collect data, evaluate approaches, and to replicate and sustain successful interventions. The World Bank Group is working closely with the United Nations and many other partners to help mobilize resources and develop STI roadmaps to address specific challenges in STI solutions in country and subnational contexts and with diverse populations. This book is an important contribution toward making leaders aware of the financing challenges, and in sharing ideas for meeting revenue targets to implement STI solutions which can have a broad impact on people's lives.

The wise and regulated use of STI can boost productivity, raise global income levels, and improve quality of life for billions of people. Already, their aspirations are rising throughout the world, augmented by smart phones and social media. As STI redefines commerce and communities, we will have to adapt to a rapidly evolving workplace, requiring new skills and cultural shifts.

While there are risks and challenge on the road ahead, solutions enabled by science, technology, and innovation can and should be used to help us reach the SDGs—lifting up billions of people, protecting our planet, and achieving inclusive growth and shared prosperity.

Mahmoud Mohieldin
(Senior Vice President for the 2030
Development Agenda, United Nations
Relations and Partnerships
World Bank Group)

Foreword by Olusegun Obasanjo

The relative success of Millennium Development Goals (MDGs) over the period 2000–2015 encouraged the international community to venture into Sustainable Development Goals (SDGs). In a way, the SDGs are a continuation of the MDGs but deeper, wider, and more encompassing such that they capture major concerns of sustainable development around the world. From eight goals of the MDGs, SDGs consist of seventeen goals that not only involve developing countries but also bring developed countries and other relevant stakeholders on board. There are 169 targets and 232 indicators that are cross-cutting and multi-dimensional in nature, designed to monitor SDG progress, and provide accountability and performance assessment for broader implementation of SDGs.

The central issue of SDG's success, progress, achievements, and performance especially by developing countries hangs on the application of Science, Technology, and Innovation (STI) which are acknowledged as having a significant role in achieving the objectives of SDGs and therefore must be taken very seriously.

This book specifically focuses on using different types of science and technological innovation including digitization, artificial intelligence, modern biotechnology, agricultural technologies, information communication technologies, renewable energy and others to help achieve almost every target of SDGs. The authors' efforts are commendable given the fact that the book is unique in the context of STI for meeting SDGs, and touches on overarching issues that call for the indispensable role of STI in tackling sustainable development challenges around the world. Without hesitation, I agree with authors on the issues raised in the book because the key requirement for successful application of STI in the implementation of SDGs is partnership, collaboration, and cooperation at every level of government especially at the national, regional, and global levels. We have to swim together to survive together.

If anybody was in doubt that we are living on one world house, the recent COVID-19 global pandemic must have removed that doubt. SDGs are meant to make our world house more congenial for every occupant and to create harmony, wholesomeness, confidence, equity, comfort, availability of choices, freedom, peace, security, and human rights. And as a former United Nations Special Envoy, I strongly advocate all of these, where I am confident that STI can play a vital part in making them a reality. In this light, the authors believe and categorically state that a cardinal instrument or tool to achieve these goals and all that flow from them is STI. I unreservedly share their views.

I recommend this very readable and educative book to anyone who believes in SDGs as imperatives for environmental, social, and economic sustainability and stability of the world.

Olusegun Obasanjo, PhD
(Former President of Republic of Nigeria,
aka Father of Africa)

About the Editors

Ademola A. Adenle specializes in science and technology policy in addressing sustainable development challenges including climate change, food insecurity, energy, health innovation in developing countries, and social aspects of science and technology in sustainable agriculture and biodiversity conservation. He is the founder of a new initiative known as the Africa Sustainability Innovation Academy (ASI-Academy). He was a research fellow and principal investigator at the United Nations University Headquarters, Japan. Dr. Adenle was educated in Nigeria and the United Kingdom. He has a Bachelor in Natural Science from University of Lagos, a Master of Biotechnology from University of Sussex, and holds a PhD in applied toxicology with a focus on environment and health from Nottingham University, and a Master of Public Policy from the University of Oxford. He has over 20 years of combined experience in teaching and research at the international level. His publications straddle scientific assessment and stakeholder engagement for overcoming sustainable development challenges in developing countries. Most recently, he was lead editor of *Risk Analysis and Governance of GMOs in Developing Countries*, published in 2017 by Cambridge University Press.

Marian R. Chertow is a professor of industrial environmental management at the Yale University School of Forestry and Environmental Studies and Director of the Center for Industrial Ecology. She is also appointed at the National University of Singapore and the Yale School of Management. Her research and teaching focus on industrial ecology, business/environment issues, and reuse of waste and materials. Her primary research interest is thestudy of industrial symbiosis involving geographically based exchanges of materials, energy, water, and wastes within networks of businesses. Previously, Dr. Chertow spent 10 years in environmental business and state and local government. In 2019 she received the highest recognition of the International Society for Industrial Ecology, its Society Prize, for her "outstanding contributions to the field."

Ellen H.M. Moors is professor of innovation and sustainability at the Copernicus Institute of Sustainable Development of Utrecht University. She has more than 25 years of experience in science, technology, and innovation studies. Her research focuses on the dynamics and governance of technological innovations in science-based sectors in which emergent technology development occurs, such as in health and life sciences, and in the agri-food sector. Her research concentrates on studying changing institutional arrangements in health and food related innovations, using institutional, social entrepreneurship, and user-innovation theories. Dr. Moors publishes on the dynamics of sociotechnical innovations, sustainable innovation management, user-producer interactions, innovation-regulation issues, and responsible innovation, among other topics, in both life sciences and science, technology, innovation, and public policy journals.

David J. Pannell is professor of agricultural and resource economics, University of Western Australia; director of the Centre for Environmental Economics and Policy; an ARC Federation fellow (2007–2012); distinguished fellow and past president of the Australian Agricultural and Resource Economics Society; fellow of the Academy of Social Sciences in Australia; and a director of Natural Decisions Pty Ltd. He was a director on the Board of Land and Water Australia (2002–2005). Professor Pannell's research includes the economics of agricultural research and innovation, farmer adoption of new technologies and innovations, economics of land and water conservation, and environmental policy. His research has been recognized with awards from the United States, Australia, Canada, and the United Kingdom. He collaborates with a wide variety of environmental and natural resource management organizations.

Contributors

Ademola A. Adenle
Africa Sustainability Innovation Academy
(ASI-Academy), Nigeria and School of
Global Environmental Sustainability,
Colorado State University, United States

Robyn G. Alders
KYEEMA Foundation, Brisbane, Australia
and Maputo, Mozambique, Development
Policy Centre, Australian National
University, Australia and Centre for Global
Health Security, Chatham House, London,
United Kingdom

Brigitte Bagnol
International Rural Poultry Centre,
KYEEMA Foundation, Brisbane,
Australia and Maputo, Mozambique and
Department of Anthropology, University of
Witwatersrand, Johannesburg, South Africa

Catharina R. Bening
ETH Zurich, Group for Sustainability and
Technology, Switzerland

Peter Berrill
Center for Industrial Ecology, School of the
Environment, Yale University, United States

Nicola U. Blum
ETH Zurich, Group for Sustainability and
Technology, Switzerland

Janet Bouttell
Health Economics and Health Technology
Assessment, Institute of Health and
Wellbeing, University of Glasgow, United
Kingdom

Julia de Bruyn
Natural Resources Institute, University of
Greenwich, United Kingdom

Hilary H. Chan
Sydney School of Veterinary Science, Faculty
of Science, The University of Sydney, NSW
2006, Australia

Marian R. Chertow
Centre for Industrial Ecology, School of
Forestry and Environmental Studies, Yale
University, United States

Domenico Dentoni
Business Management & Organization
Group, Wageningen University, The
Netherlands

John Dixon
Australian Centre for International
Agricultural Research, Australia

Nora Engel
Department of Health, Ethics & Society,
Care and Public Health Research Institute,
Maastricht University, The Netherlands

Marcolino Estevão Fernandes E. Brito
Faculty of Agriculture, Universidade
Nacional de Timor Lorosa'e,
Timor-Leste

Tomer Fishman
School of Sustainability, Interdisciplinary
Center Herzliya, Israel

Cecilia G. Flocco
National Scientific and Technical Research
Council (CONICET) and University of
Buenos Aires, Argentina, and Leibniz
Institute DSMZ-German Collection of
Microorganisms and Cell Cultures,
Germany

Delia Grace
International Livestock Research Institute, Nairobi, Kenya

Eleanor Grieve
Health Economics and Health Technology Assessment, Institute of Health and Wellbeing, University of Glasgow, United Kingdom

Ruihua Guo
School of Computer Science, University of Sydney, Australia

Freek de Haan
Copernicus Institute of Sustainable Development, Utrecht University, The Netherlands

J. Brian Hardaker
Faculty of Science, Agriculture, Business and Law, University of New England, Australia

Kshama Harpankar
Lebanon Valley College, Annville-PA, United States

Neil Hawkins
Health Economics and Health Technology Assessment, Institute of Health and Wellbeing, University of Glasgow, United Kingdom

Niko Heeren
Center for Industrial Ecology, School of the Environment, Yale University, United States

Kathleen Hefferon
Department of Food Science and Technology, Cornell University, United States

Akira Homma
Bio-Manguinhos, Oswaldo Cruz Foundation (FIOCRUZ), Rio de Janeiro, Brazil

Koichi S. Kanaoka
Center for Industrial Ecology, School of the Environment, Yale University, United States

René Kemp
United Nations University-MERIT and Maastricht University, The Netherlands

Laurens Klerkx
Knowledge, Technology and Innovation Group, Wageningen University, The Netherlands

Anja Krumeich
Department of Health, Ethics & Society, Care and Public Health Research Institute, Maastricht University, The Netherlands

Felix Krussmann
International Development Studies, Wageningen University, The Netherlands

Cees Leeuwis
Knowledge, Technology and Innovation Group, Wageningen University, The Netherlands

Mu Li
School of Public Health, University of Sydney, Australia

Yiren Liu
Westmead Hospital, Western Sydney Local Health District, Australia

Mercy Lungaho
International Center for Tropical Agriculture (CIAT), Nairobi, Kenya

Mariana Machado
National Science and Technology Institute on Public Policies, Strategies, and Development (INCT-PPED), Federal University of Rio de Janeiro, Brazil

Tamar Makov
Department of Management, Guilford Glazer Faculty of Business and Management, Ben-Gurion University of the Negev, Israel

Reinaldo M. Martins
Bio-Manguinhos, Oswaldo Cruz Foundation (FIOCRUZ), Rio de Janeiro, Brazil

Agnes Meershoek
Department of Health, Ethics & Society, Care and Public Health Research Institute, Maastricht University, The Netherlands

T. Reed Miller
Center for Industrial Ecology, School of the
Environment, and Department of Chemical
and Environmental Engineering, School
of Engineering and Applied Science, Yale
University, United States

Marisa E. V. Mitchell
International Livestock Research Institute,
Hanoi, Vietnam

Ellen H.M. Moors
Copernicus Institute of Sustainable
Development, Utrecht University, The
Netherlands

Caroline Mwongera
International Center for Tropical Agriculture
(CIAT), Nairobi, Kenya

Chris M. Mwungu
International Center for Tropical Agriculture
(CIAT), Nairobi, Kenya

Babette Never
German Development Institute, Bonn,
Germany

Michael J. Nunn
Australian Centre for International
Agricultural Research, Canberra, Australia

Christopher Oster
Energy and Environmental Policy, University
of Delaware, United States

David J. Pannell
Centre for Environmental Economics and
Policy, University of Western Australia,
Australia

Simon K. Poon
School of Computer Science, University of
Sydney, Australia

Cristina Possas
Bio-Manguinhos, Oswaldo Cruz Foundation
(FIOCRUZ), Rio de Janeiro, Brazil

Maria Fay Rola-Rubzen
School of Agriculture and Environment,
University of Western Australia, Australia

Tobias S. Schmidt
ETH Zurich, Energy Politics Group,
Switzerland

Kalim U. Shah
Biden School of Public Policy and
Administration, University of Delaware,
United States

Hans De Steur
Department of Agricultural Economics,
Ghent University, Belgium

Casey Stevens
Providence College, Rhode Island,
United States

Govinda Timilsina
World Bank, Washington, DC,
United States

Bernhard Truffer
Eawag, Swiss Federal Institute of Aquatic
Science and Technology, Switzerland
Faculty of Geosciences, Utrecht University,
The Netherlands

Serdar Türkeli
United Nations University-MERIT and
Maastricht University, The Netherlands

Steve Twomlow
International Fund for Agriculture
Development (IFAD), Rome, Italy

Mara J. van Welie
Eawag, Swiss Federal Institute of Aquatic
Science and Technology, Switzerland
Faculty of Geosciences, Utrecht University,
The Netherlands

Renato Villano
Faculty of Science, Agriculture, Business
and Law, University of New England,
Australia

David Font Vivanco
2.-0 LCA consultants, Aalborg, Denmark

Rui Wang
Urban Planning and Design, Xi'an Jiaotong-
Liverpool University, China

Karin Wedig
Deutsche Gesellschaft für
Internationale Zusammenarbeit (GIZ)
GmbH, Germany

Justus Wesseler
Agricultural Economics and Rural Policy
Group, Wageningen University, The
Netherlands

Seerp Wigboldus
Wageningen Centre for
Development Innovation, Wageningen
University and Research, Wageningen,
The Netherlands

Kate Wingett
School of Life and Environmental Sciences,
University of Sydney, Australia

Paul Wolfram
Center for Industrial Ecology, School of the
Environment, Yale University, United States

Johanna T. Wong
School of Life and Environmental Sciences,
University of Sydney, Australia

Carlos Eduardo F. Young
Institute of Economics, Federal University of
Rio de Janeiro, Brazil

INTRODUCTION

1

What Can Science, Technology, and Innovation Offer in the Achievement of Sustainable Development Goals?

Ademola A. Adenle, Marian R. Chertow, Ellen H.M. Moors, and David J. Pannell

1. Introduction

Since 2000, significant achievements in global development have taken place, as evidenced by the lifting of one billion people out of extreme poverty and in the reduction of chronic hunger in many regions of the developing world (Word Bank 2018). Concerted international efforts aimed at meeting the millennium development goals (MDGs) in the period 2000–2015 have drawn the attention of many governments in developing countries, and helped shift their public policy and decision-making priorities (UN 2015). Despite these important achievements, much more needs to be done to bring people out of poverty, to improve public health, and to respond to environmental problems.

To expand and build on MDGs' successes, the United Nations' (UN) 2030 agenda for sustainable development (UN 2015) has included the establishment of a set of sustainable development goals (SDGs). SDGs break new ground in that they incorporate additional dimensions of socioeconomic and environmental concerns into the development agenda, within the setting of novel indicators of success across various sectors. The new United Nations 2030 Agenda for Sustainable Development has primarily been designed to end poverty, protect the planet, and ensure prosperity for all and nurture peaceful, inclusive societies (UN 2015). Associated with the 17 SDGs (Table 1.1) are 169 targets and 304 proposed indicators that are cross-cutting and multidimensional in nature, designed primarily to monitor SDG progress and to provide accountability for the implementation of the SDGs.

New policies that recognize the benefits of science, technology, and innovation (STI) and their potential risks are needed to implement the SDG agenda successfully by 2030. Recognizing this need, the Technology Facilitation Mechanism (TFM) was agreed to among UN member states in 2015. To support the achievement of the SDGs, "the TFM will facilitate multi-stakeholder collaboration and partnerships through the sharing of information, experiences, best practices and policy advice among Member

Table 1.1 The sustainable development goals

Goal 1. End poverty in all its forms everywhere

Goal 2. End hunger, achieve food security and improved nutrition, and promote sustainable agriculture

Goal 3. Ensure healthy lives and promote well-being for all at all ages

Goal 4. Ensure inclusive and equitable quality education and promote lifelong learning opportunities for all

Goal 5. Achieve gender equality and empower all women and girls

Goal 6. Ensure availability and sustainable management of water and sanitation for all

Goal 7. Ensure access to affordable, reliable, sustainable, and modern energy for all

Goal 8. Promote sustained, inclusive, and sustainable economic growth, full and productive employment, and decent work for all

Goal 9. Build resilient infrastructure, promote inclusive and sustainable industrialization, and foster innovation

Goal 10. Reduce inequality within and among countries

Goal 11. Make cities and human settlements inclusive, safe, resilient, and sustainable

Goal 12. Ensure sustainable consumption and production patterns

Goal 13. Take urgent action to combat climate change and its impacts

Goal 14. Conserve and sustainably use the oceans, seas, and marine resources for sustainable development

Goal 15. Protect, restore, and promote sustainable use of terrestrial ecosystems, sustainably manage forests, combat desertification, and halt and reverse land degradation and halt biodiversity loss

Goal 16. Promote peaceful and inclusive societies for sustainable development, provide access to justice for all, and build effective, accountable, and inclusive institutions at all levels

Goal 17. Strengthen the means of implementation and revitalize the Global Partnership for Sustainable Development

---

Source: United Nations (2015).

States, civil society, the private sector, the scientific community, United Nations entities and other stakeholders" (UN n.d.). The TFM includes three main elements: an interagency task team on STI for the SDGs, a multistakeholder forum on STI for the SDGs (occurring annually, starting in 2016), and an online platform that provides a gateway to information on STI initiatives, mechanisms, and programs (not yet operational, as of April 2020) (UN n.d.). The existence of the TFM, and the participation by numerous stakeholders, reflects a broad recognition of the essential role of STI in delivering SDGs, and of the potential for improving the capacity of many developing countries to undertake STI initiatives that will help them achieve SDGs.

STI applications can make multiple contributions to the achievement of SDGs. Developing countries in particular will need to harness STI, while managing resulting trade-offs, to deliver on the three pillars of sustainable development: environmental, social, and economic. The SDGs simultaneously touch upon all three aspects.

Integrating these aspects into the implementation of the SDGs is a key challenge for both policymakers and researchers who need to address them in interdisciplinary research and innovation projects (Biermann et al. 2017). There is a range of barriers to channeling STI toward accomplishing the SDGs. Therefore, to meet its SDG targets, the global community must mobilize STI across multiple sectors, new investments in innovation, and policy design that addresses the barriers.

A lack of clear vision and understanding among national governments about how STI can contribute to achieving the SDGs remains a significant challenge. Our aim in this book is to address the gap by raising understanding of STI among domestic and international organizations concerned with sustainable development in light of the SDGs.

This book contains 26 chapters and involves 74 authors from 55 institutions across 19 countries around the world. The chapters are not all inclusive across STI. Rather we target three themes in which STIs are crucial for sustainable development: environment and energy, health, and agriculture. Within each theme the chapters offer thoughtful background on particular issues concerning SDGs. We address each issue with analysis of a concept or theory, a set of tools or practices, or policy-relevant advice based on data collected to find paths forward to sustainability transitions. In this way our work is part textbook, part handbook, and part idea/concept book. Some chapters address SDGs in specific geographies and others are topical without a specific geographic focus. While intending to serve the broadest STI and policymaker audience, on balance the book tilts toward developing countries and regions more than developed ones for examples.

2. STI and Sustainable Development

Advances in science and technology can help to deliver basic human needs, enhance economic productivity, reduce environmental impacts, and improve the quality of products and services (Holdren 2008). The United Nations and other international organizations have long recognized the importance of STI to modern societies and the way the emergence of technological innovation has shaped the world and contributed to economic development (UN 2002).

As part of the efforts to achieve sustainable development, linkages between STI and the earlier set of MDGs were highlighted, particularly in the contexts of promoting industrialization, increasing productivity, achieving food security, promoting access to quality health systems, and creating decent jobs. Yet a limited emphasis was placed on the role of STI in meeting MDG targets both at the national and international levels, inhibiting the achievement of the MDGs, especially in developing regions (UN 2014).

The integration of STI into a broader agenda for achieving sustainable development remains complex, especially at the international level. According to the United Nations, despite efforts by the UN Conference on Trade and Development to conduct national STI policy reviews, such reviews are not compulsory so universal coverage was not achieved (UN 2016).

The World Bank (2010) argued that there has been a lack of partnerships between developed and developing countries around STI, meaning that systematic knowledge transfer between the Global South and Global North was lacking, thereby undermining the role of STI in fostering sustainable development and economic growth in the Global South. There are a number of reasons why international cooperation for STI activities has been problematic. In developing countries, lack of investment, poor institutional conditions, weak governance, and limited infrastructure, among other things, have been blamed for slow advances in scientific and technological development (Knack and Keefer 1995; Word Bank 2010). A case study by Ramón and Gaudin (2014), for example, argues that Central American countries fail to implement national STI policies owing to deficient funds and weak infrastructure, for research and development (R&D) and more generally. In addition, many developing countries lack the human capital, socioeconomic institutions, and technology systems to engage in international cooperation to advance STI's contributions to sustainable development (Adenle et al. 2015; Ramón and Gaudin 2014).

International regimes pertaining to intellectual property rights (IPR) are designed to protect the interests of (mostly developed-country) IP creators. While this has the advantage of incentivizing the creation of new IP that can be commercially exploited, it may not be the regime that would most effectively foster international cooperation to help drive sustainable economic growth in the developing world (Bozeman 2000; Roffe and Santa Cruz 2007). Nevertheless, some countries have tailored IPR to meet their specific needs by building national innovation systems that target human capital development, enterprise development, and STI policy development, all of which can contribute to economic transformation. Studies have shown that a number of countries have come out of poverty and built competitive economies through a growth trajectory aligned with strong STI capacities as underpinned by effective national innovation systems. For example, Chung (2002) argues that South Korea's industrialization is largely driven by STI policy that focuses on national R&D investment; entrepreneurship development; partnerships between academia, public research institutes, and industries; and a well-educated workforce. This experience reinforces the importance of STI in helping to achieve national level SDGs.

The role of frugal innovation in addressing sustainable development issues has also become more important in recent discussion (Khan 2016; UNCTAD 2017). Frugal innovation concerns the (re)development of products, services, and systems at the lowest possible cost, while retaining functionality. Here, the critical question raised in the UNCTAD report is how to harness STI policy to develop low-cost technologies that can service marginalized groups facing resource constraints as they try to advance sustainable development, and ultimately contribute to the SDGs. Unlike the STI activities driven by market-based incentives for R&D investments, frugal innovation is generally associated with untapped markets and very-low-income grassroots communities (Seyfang and Smith 2007). Yet STI policies to promote grassroots innovation for the achievement of SDGs are still largely absent at the global and national levels (Khan 2016; Levänen et al. 2016).

3. STI and the Framework of the Sustainable Development Goals

This book focuses on the 17 SDGs as guides and motivators to foster sustainable socioeconomic growth and improve quality of life around the world. Of 169 SDG targets, 48 targets are related to STI (GSDR 2016). Many of the remaining 121 targets also touch on STI, in that technological innovation has a role to play in reaching the targets. This underlines the critical role of STI in advancing economic performance and inclusive development, especially in light of the limited recognition of its contribution in the former MDG era (UN 2014). One could argue that weak coordination of STI at the global level contributed to the underperformance of STI in terms of its contributions toward the achievement of MDGs, especially in developing countries. Despite the 2004 Millennium Declaration promoting science and technology for MDGs (United Nations Economic and Social Council 2004), emphasizing the important role of human capital development and local capacity building to facilitate international technology transfer, the recognized role of new and emerging technologies as part of global STI activities was limited (see chapter 20).

We have structured this book around three science-based arenas in which it is clear that STI is important for achieving sustainable development: 1) environment and energy, 2) health, and 3) agriculture. While our themes tend to be studied by separate disciplines, we recognize that the SDG framework is cross-cutting, multidimensional, and interlinked (e.g., Biermann et al. 2017; Kanie et al. 2017). The framework covers some specific global thematic priorities, such as SDG2 (end hunger) or SDG5 (gender equality) and includes some with a broader scope, such as SDG11 (inclusive, safe cities) or SDG9 (sustainable industrialization).

As noted earlier, the lack of national STI policy frameworks has hampered the implementation of national sustainable development strategies. Such national policies can benefit from international frameworks providing guidance, especially the TFM (UN 2016). Through stakeholder participation in international forums and the sharing of knowledge and experiences related to STI initiatives, mechanisms, and programs, the TFM can make a substantial contribution to the delivery of SDGs. It is an important initiative, but only one element of many that will be needed if STI is to be fully effective in advancing SDGs.

Another initiative has been the clean development mechanism (CDM) created under the Kyoto Protocol by the UN Framework Convention on Climate Change (UNFCCC). To decrease greenhouse gas emissions, the CDM encouraged the transfer of low-carbon sustainable technologies. CDM technology transfer was relatively successful in China, Brazil, and India, and this was attributed to their strong technological capabilities (Dechezleprêtre et al. 2009), as well as to strong institutional support and high-quality infrastructure compared to other developing regions.

However, the CDM failed to achieve its aims in a number of developing regions, especially in Africa (Goldman 2010; van der Gaast et al. 2009). This can be partly attributed to the lack of STI capacity in developing countries. Moreover, CDM projects lacked an international STI framework that could sustain technology transfer and strengthen the capabilities of local firms to build strong and competitive global industries. As a result, the diffusion of low-carbon technology has been uneven. CDM projects have arguably been the largest market-based mechanism to facilitate low-carbon technology transfer to developing countries (Koch et al. 2014; Schneider et al. 2008). Nevertheless, the MDGs failed to emphasize the importance of the CDM projects.

As part of the 2030 Agenda, the SDGs and the recent Paris Agreement of UNFCCC aspire to transform the ways in which climate change and sustainable development issues are addressed. The Paris Agreement calls for a new STI policy framework to facilitate low-carbon technology transfer as part of the implementation of sustainable development programs. For both the Paris Agreement and the SDGs, the current absence of a broadly accepted international STI framework remains an impediment. Despite this limitation, donors including the international community and developing-country governments continue to promote the potential application of STI in addressing sustainable development challenges without a holistic approach to move the STI activities forward. A key question is how should stakeholder groups come together to prioritize investments, IPR reforms, and trade regulations that shape overall creation and deployment of STI to facilitate the implementation of the 2030 Agenda, including SDGs in developing countries.

Beyond the need for an international STI framework, the participation of a wider range of stakeholders including country governments, donors, nongovernmental organizations, academia, and the private sector is key to achieving global STI partnerships, especially where primary incentives to access innovation is driven by markets. The overall success of the SDGs will depend on various global socioinstitutional and governance factors, including the extent to which countries formalize their SDG commitments, strengthen global STI solutions and policy arrangements, and translate global SDGs into national contexts while integrating the environmental, economic, and social pillars of sustainable development.

4. Financing SDGs Requires Global Partnership

The pairing of SDGs with STI requires not only funding to carry out the needed research, development, and deployment to advance global goals, but also strong and coherent partnerships engaging diverse stakeholders to set priorities, evaluate plans, implement projects, and monitor the agreed upon programs at all levels. The recognition of the need for partnerships goes back to the 1992 Earth Summit in Rio and the creation of Agenda 21 which called for a global partnership that included important ideas such as science for sustainable development. By the time we get to the present day, recognition of the need for cooperation is so great that there is an entire SDG devoted to it, SDG17 (global partnership for sustainable development).

On the monetary side, the UN estimate is that meeting the global goals would require $5–7 trillion annually through until 2030. Roughly half of this amount is already being spent on infrastructure and other activities, so the estimate for additional "gap" funds is $2–3 trillion annually. To put this in perspective, annual global GDP is over $100 trillion on a purchasing power parity (PPP) basis, suggesting that providing $2–3 trillion could be in the realm of possibility (BSDC 2019). The three themes that we have chosen for this book, environment and energy, health, and agriculture, tend to be on the high end of expenditure given that all three engage the science to germinate ideas, the technology central to increasing productivity and well-being, and the flow of innovation needed to be able to adapt to a wide range of geographic and demographic contexts. Still, on a benefit-cost basis at the broadest level, it is reasonable to speculate that the health and welfare benefits of actually meeting the SDGs related to the three themes would greatly outweigh the costs.

Neither finance nor partnerships have been overlooked in the ambitious effort to create SDGs and move them forward. While the MDGs were more government-centered, the private sector has been called on many times to pave the way for creating the investment required to pursue the SDGs. The final report of the Business and Sustainable Development Commission, for example, is titled *Ideas for Action for a Long-term and Sustainable Financial System* (BSDC 2019). It is organized by focus areas including creating pools for long-term finance and getting infrastructure finance right. The Commission's flagship report, *Better Business, Better World*, describes how "sustainable business models" could attract $12 trillion in new market value and create as many as 380 million jobs by 2030 (BSDC 2017).

Regarding partnerships, Unilever, a company that has been active and effective in integrating SDGs, created a 2018 report on *How to ... Build Partnerships to Change the World*, based on the idea that SDGs require important work across business, civil society, and government. A key barrier confronting many of the suggestions in this book is that the vast majority of private capital is spent in developed countries and over half of the monetary estimate for SDG implementation is required in developing countries. One inspiring crossover example is a partnership by the UK Department for International Development and Unilever that created an innovation fund (£40 million), TRANSFORM, with the aim to enable 100 million people in sub-Saharan Africa and Asia to gain access to products and services related to health, livelihood, and environment or well-being (Unilever and Department for International Development 2019).

It is easy to observe that there is never enough money for STI. And failure to pay adequate attention to R&D investment for any of the three themes covered in this book may undermine the implementation of relevant SDGs at the national level and international levels. In chapter 3, Timilsina and Shah emphasize the need to increase R&D investment in renewable energy technologies especially in Africa and South

Asia, where billions of people still lack access to electricity and cooking fuels. This is reflected in another study which also emphasizes that investment in energy infrastructure is very limited in Africa as current annual spending is estimated at $8 billion compared to an estimate of investment need of $55 billion yearly until 2030 (Africa Progress Panel 2015). Also, in chapter 7, the authors Machado and Young analyze expenditure on R&D for environmental science in Brazil, finding that there is a substantial shortfall in government expenditures despite many improvements in the nation's STI programs. Beyond the funding itself, the fragmented practice of grant giving and the prolonged timetables of public funding surely inhibit the achievement of many good intentions.

Imagine expanding this single study and applying it globally. This suggests what is likely the most difficult aspect of managing SDGs—the sheer level of coordination needed to make the changes that SDGs demand for a better world. We have seen that some of the work on finance and partnerships has been thoughtful and, like the SDGs themselves, even visionary. The execution, however, is demanding—requiring wisdom, patience, and forbearance that can seem distant from our competitive, sped-up world.

One distinct challenge for the SDG agenda is how to actively engage various institutions, sectors, and actors especially at the national and regional levels in order to achieve synergy for mobilizing and accessing STI finance at the international level. Yet this problem persists without a coherent approach to accessing finance for the implementation of sustainable development projects, especially in developing regions (Adenle, Ford et al. 2017). Further, despite the increased demand for the mobilization of financial resources for achieving SDGs, previous evidence suggests that lack of transparency, potential interests of various actors, unevenly distributed finance, and limited capacity at the national level remain as obstacles particularly with regard to the implementation of relevant STI projects. For example, evidence indicates that mitigation projects such as renewable energy were only funded in Africa where institutional capacity was relatively strong (Adenle, Manning et al. 2017). These challenges call for global partnerships to increase access to finance especially at the UN level, where strong leadership, higher levels of commitment, and improved allocation of responsibilities can be coordinated.

5. The Structure of the Book

This book examines the relationship between STI and the 2030 agenda for sustainable development, providing examples, experiences, and case studies from around the world. It uses an interdisciplinary approach to examine the contributions of STI to the implementation of the SDGs across various continents including Asia, Africa, South America, and Europe. The inclusion of contributions from multiple disciplines provides for a broad range of perspectives and ideas on how to address the challenges at which the SDGs are targeted.

The book focuses on three human-development themes—environment and energy, health, and agriculture—as these sectors are major ones in which a STI approach can support SDG goals. The chapters highlight how STI initiatives have been applied to address each of the themes. They explore a range of STI solutions and governance arrangements. The themes are described further in the following sections.

5.1 Theme I: Environment and Energy

The environment and energy theme has nine chapters on SDG-related topics ranging from evaluation of biodiversity institutions at the global level to financing environmental STI at the country level (in Brazil), to the outlook on autonomous vehicles at the local, city level. There are four chapters on renewable energy, including two that are empirical and two that are relatively theoretical. All of these mention solar energy and, from an STI perspective, it is easy to see that better planning and implementation of solar energy could quickly change the status of millions of people in Africa and Asia who would achieve energy access. Altogether, the chapters unite many critical pieces that present a wider platter of possibilities for more rapid implementation of renewable energy.

Chapter 2 on biodiversity by Stevens proposes that innovation around wildlife conservation linked to SDGs is within our reach as a means of reducing biodiversity loss if we can learn from past efforts. In earlier days, a fragmented system of biodiversity governance came into being. Over time, this system has developed localized centers of innovation that would, passing on STI improvements from smaller groups to larger ones, raise the chances for acceptance and transfer of innovation across scales.

What can be learned from the four energy chapters (3, 4, 5, and 6) related to STI for SDGs? Overall, SDG7 stresses "universal access to affordable, reliable and modern energy services for all" by 2030. In chapter 3, Timilsina and Shah regard such energy as a "golden thread" that is linked to numerous other SDGs and is therefore critical to achieving many SDGs simultaneously. Adenle examines solar energy and SDGs in Kenya and South Africa in chapter 4. He emphasizes not only the educational, diet, wealth, and time-management benefits that are possible with increased dissemination of solar energy, but also the importance of African government policy enabling these benefits through investment in R&D programs and provision of appropriate subsidies. As a clean technology, solar energy has advantages over conventional sources because lower air emissions are accompanied by health and environmental benefits.

Schmidt and colleagues, in chapter 5, examine small, isolated renewable energy micro-grids that can serve a number of households independent of the main grid. The authors track these decentralized systems in three countries—Cambodia, Indonesia, and Laos—following the methodology of technological innovation systems that uses multilevel analysis to compare across these countries. They stress the importance of social and cultural differences and the need to consider each country's energy needs individually. Finally, Kemp and colleagues in chapter 6 explore three successful instances of the phasing in of solar energy and energy efficiency in China and India. They use an integrated framework that merges political economy elements of interests, ideas, and institutions with capabilities and policy delivery. Through this the authors find that there are positive and useful opportunities for developing countries to economize by advancing STI activities influencing several SDG targets simultaneously, given the interdependencies among them.

The theme of chapter 7, authored by Machado and Young, is whether countries provide sufficient resources for the environmental science and technology innovation that is needed to meet SDG-related goals in the detailed example of Brazil. By making reasonable and transparent assumptions about the level of financing that would be needed to fully fund R&D, the study reveals a significant financial gap and an urgent need to create alternative sources of funding or to find new sources from taxes and user fees.

The SDGs are highly cross-cutting and require expertise from numerous perspectives. Addressing these needs, chapter 8 by Chertow and colleagues presents the relatively new systems science of industrial ecology and outlines how it can contribute to the delivery of SDGs. Using tools such as life-cycle assessment and material flow analysis and approaches such as industrial symbiosis, industrial ecology provides many useful takes on physical resource use and efficiency in the quest to achieve SDGs.

As presented by Wang and Oster in chapter 9, transportation for passengers and freight plays a crucial role in achieving many SDGs. The core message of their work is that the revolution anticipated by the introduction of autonomous vehicles seems unlikely to radically change ground transportation in the near future given the challenges and uncertainties associated with highly automated ("driverless") vehicles.

Understanding how systems respond to technological change, including from STI interventions targeted at SDGs, it is important for policy leaders to know in advance about the rebound effect. Vivanco and Makov in chapter 10 present the idea of rebound effects, explaining how they can reduce the level of environmental benefit that a policy delivers when inherent conflicts arise based on interacting effects. For example, when the fuel efficiency of automobiles increases, drivers may decide to travel greater distances because it is cheaper. Rebound effects are a particular risk with the SDGs because they are so interlinked.

5.2 Theme II: Health

Chapters in the health theme examine how health is intrinsically linked to 16 targets across the 17 SDGs. The SDG framework provides an expansive approach to creating better health systems. The health theme has seven chapters on SDG targets, emphasizing the role of new and emerging technologies in solving key health problems in both developed and developing countries. It also covers a healthcare innovation approach and entrepreneurship programs targeted at health. The issues range from STI approaches to responsible innovation in health such as the development of vaccines, biotherapeutics, and antimalarial treatments, digital health, urban sanitation services and infrastructure, and regulatory innovations to tackle accessibility, affordability, and safety of healthcare.

Four chapters deal with empirical examples of STI to meet SDG goals, including vaccine innovations, antimalarial drug development and diffusion, sanitation innovations in informal settlements, and the role of animal-source food in healthy diets. One chapter focuses on the supportive role of digital health in meeting SDGs, and three chapters are more methodologically oriented, focusing on health technology assessment (HTA) methods to improve the contributions of STI to SDGs, a new sustainable innovation framework for meeting SDGs, and the role of responsive and responsible science and technology studies for global health. Overall, the chapters in the health theme cover crucial aspects of scientific and technological developments aiming for SDG targets. They discuss the socioeconomic, regulatory, and institutional challenges for sustainable innovations, and novel inclusive methodologies, frameworks, and supporting digital innovations to overcome these systemic barriers.

Chapter 11 by Possas and colleagues provides insights into the technological and regulatory challenges affecting access to vaccines in developing countries and recommendations for vaccine STI performance strategies to achieve relevant SDGs. From a global sustainability perspective, only SDG3.b.1 refers explicitly to vaccines. The authors, however, identify fourteen vaccine-related goals in the SDGs, of which SDG9 (sustainable industrialization) and SDG17 (global partnership for sustainable development) are specifically related to innovation and technological development of vaccines. The authors provide recommendations for specific vaccine STI performance indicators and strategies to achieve these fourteen vaccine-related SDGs.

Readers will also learn about how development-focused HTA can help support decisions about the introduction of new technologies for the achievement of SDG3 (healthy lives). Chapter 12 by Bouttell and colleagues sketches five different scenarios regarding biopharmaceutical innovations in low- and middle-income countries. It shows how HTA can improve the efficiency of research prioritization and development processes while ensuring that the needs of vulnerable populations are met.

A novel sustainable innovation framework based on availability, affordability, accessibility, and acceptability dimensions of the SDGs has been developed in chapter 13. This framework by De Haan and Moors can be applied in low-income and middle-income countries to innovations for communicable diseases, such as malaria, tuberculosis, and HIV/AIDS. The ending of these epidemics that are mostly prevalent in low- and middle-income countries (SDG3.3) cannot be separated from universal access to healthcare (SDG3.8). The same holds for malaria burden versus poverty (SDG1), education (SDG4), and equality between countries (SDG10). Additionally, water and sanitation conditions (SDG6) and global warming (SDG13) may affect living conditions of mosquitos and therefore also affect malaria infections.

Given this interrelated, multifaceted nature of SDGs, systemic approaches should be at the root of tackling global health challenges. For example, progress toward the sustainable development goal on water and sanitation (SDG6) is very slow and the lack of sanitation is especially persistent in rapidly growing cities in the Global South. Chapter 16 by Van Welie and Truffer shows how such a sociotechnical systems approach helps us understand the interaction between health and sanitation technologies and services, the infrastructure and institutions which need to be in place, and the user practices of new health technology.

There is also a need for responsive and responsible innovations for global health in context, as discussed in Engel and colleagues in chapter 15. For example, in the development and implementation of point-of-care diagnostics and development of cook stoves in low- and middle-income countries, there should be continuous interrogation and reflection on the localized consequences (intended and unintended) of new innovations rather than use of traditional technology transfer mechanisms. Chapter 14 by Poon and colleagues focuses on the benefits of digital health for the achievement of SDGs. This chapter evaluates the potential relations between digital and universal health (SDG3), inclusive and equitable education (SDG4), and reduction of inequality (SDG10). More specifically, the authors show how digital health could be articulated and evaluated in four dimensions—translation, education, transformation, and technology—to bridge the gap between digital health and SDGs.

The health section ends with a holistic approach for stakeholder engagement in animal-sourced foods in sustainable, ethical, and optimal human diets, as proposed by de Bruyn and colleagues in chapter 17. An ambition of the 2030 Agenda is to include health in development, and to recognize that good health depends on and contributes to other development goals, underpinning social justice, economic growth, and environmental protection (Dye 2018). As proposed by Dye (2018), it is important to expand the scope and enhance the effectiveness of the systems and the services that prevent and treat illness to advance health and development. In this way the 2030 SDG Agenda is also an expanded agenda for systems research, assuming that better systems can indeed deliver substantially better health and well-being for all.

5.3 Theme III: Agriculture

Agricultural STI has made enormous contributions to development in the past, and has great potential to do so again in future. Although the global percentage of people living in rural areas has fallen below 50%, it is still well over 50% in much of Asia and Africa. Particularly in these countries, agricultural issues are connected to multiple SDGs and often involve trade-offs between them. The eight chapters in this theme present a range of agriculture-related opportunities and challenges for achieving the SDGs.

Four of the chapters focus on particular technologies or sets of technologies: technologies for providing nitrogen to crops; integrated crop management for rice; crop biotechnology; and climate-smart agricultural technologies. In each case, the SDG-related issues raised range across food production (SDG2), poverty reduction

(SDG1), environmental protection (SDG6 and SDG15), and various others. The other four chapters are more cross-cutting, covering the nexus between production, supply chains, policy, and sustainability; transformation governance that strives for economic development while protecting the livelihoods of the poor who depend on the same resources; the method of value network analysis (VNA) that assists with reorganizing businesses in ways that can contribute to multiple SDGs; and a framework for responsible scaling of agricultural innovations, so that the delivery of SDGs is maximized with limited negative consequences.

The section starts with chapter 18 by Harpankar on nitrogen management. Ensuring that crops have adequate nitrogen to yield well is important for food production (SDG2, end hunger) and the economic welfare of farmers (SDG1, end poverty). In traditional systems, nitrogen largely comes from natural sources and is applied at modest rates, but as farmers adopt modern technologies in order to increase their productivity, they apply more nitrogen and rely increasingly on artificial fertilizers. While this helps farmers with their economic goals, it can result in serious problems of water pollution (SDG6, water and sanitation for all), and it involves higher emissions of greenhouse gases (SDG13, combat climate change). Harpankar discusses a range of technologies that may help to strike a better balance between production and pollution.

We then look at crop management for poverty reduction (SDG1) and food security (SDG2) for smallholder rice producers in Timor Leste. In chapter 19, Rola-Rubzen and colleagues investigate integrated crop management, a comprehensive package of measures including high-yielding varieties, high-quality seed, best-practice transplanting, and sound harvesting practices. They find that farmers who adopt these technologies have significantly higher yields and higher incomes. However, in common with some other favorable agricultural technologies, adoption across the population of farmers is disappointing. This throws the spotlight onto whether improved forms of agricultural "extension" (including information provision and training) can be devised and delivered to overcome adoption barriers.

While Rola-Rubzen mainly covers what might be termed "low-tech" solutions, Adenle and colleagues in chapter 20 discusses the potential of biotechnology to contribute to delivery of SDGs. The potential role of genetically modified organisms (GMOs) in addressing low agricultural productivity (SDG1), tackling malnutrition (SDG2), and adapting to climate change (SDG13) is considered to be very large. However, the persistent opposition to the application of GMOs in Europe and parts of developing regions (Africa, Asia, South America, Central America, and Latin America) is inhibiting the delivery of these SDGs. The authors suggest that an international GMO regulatory framework would assist in achieving acceptance of biotechnology solutions where appropriate. They also argue for use of risk-assessment models that suit local circumstances in developing countries rather than following the lead of certain developed countries that have chosen a highly conservative precautionary approach.

With an emphasis on climate change (SDG13), but with consequences for various other SDGs, Mwongera and colleagues explore the potential for "climate-smart agriculture" (CSA) in Africa (chapter 21). CSA encompasses measures for both adaptation to and mitigation of climate change, including minimum tillage, improved crop varieties, and integrated pest management. Mwongera presents a novel process for rapid appraisal of CSA, and applies it to case studies in Tanzania, Kenya, and Uganda. As with integrated crop management in Timor Leste, adoption of CSA in these three African case studies is found to improve farmers' incomes (SDG1), with some elements of the package being more beneficial than others.

In chapter 22, Flocco unpacks the soybean "production complex" in Brazil, with the aim of identifying levers that can help to deliver SDGs. She argues for a systems approach, including actions along the supply chain, from the adoption of best management practices in agricultural fields to development of supportive governance frameworks at the policy level. The approach is essentially market-based, but with an emphasis on conserving soils, with consequences for poverty (SDG1), hunger (SDG2), health (SDG3), water quality (SDG6), and others.

Another high-level perspective on SDGs in a primary industry is provided by Wedig (chapter 23), who presents the concept of "transformation governance" in the context of small-scale fishers in Lake Victoria, which contains Africa's largest inland small-scale fisheries sector, involving over three million people. Challenges are created by a growing commercial aquaculture industry operating in the lake, with risks to social and environmental sustainability. Wedig's three-part approach to governance is designed to manage these risks.

A different way of analyzing agricultural industries is provided by VNA, as explained by Dentoni and colleagues in chapter 24. They describe how VNA can be used as a diagnostic tool for actors seeking to reduce poverty (SDG1) and hunger (SDG2), enhance economic growth (SDG8), and other SDGs. Their case study of the Agricultural Commodity Exchange in Malawi reveals a number of options for building cross-sector partnerships that can utilize STI to help deliver SDGs.

Finally, we address the crucial issue of delivering SDGs at scale. Wigboldus and colleagues (chapter 25) identify the need for a "theory of scaling" to help make sure that efforts to deliver SDGs are successful at large scales rather than just locally. They highlight that the process of scaling up to large impact is often more complex than accounted for by policymakers and program managers. In unpacking what is inside the black box of unarticulated assumptions about scaling, they identify a broad range of relevant issues, including that scaling often involves multiple linked steps, the importance of understanding the characteristics of the innovation being scaled and how these characteristics fit with the needs of potential users, that scaling always involves a range of partners and stakeholders, and the need to link the theory of scaling to decision-making processes. This chapter has relevance to all of the SDGs, each of which needs to be delivered at large scale.

6. Conclusion

Readers of this book will come away with no doubt that STI can be a powerful tool for delivery of SDGs. Indeed, delivery of some SDG targets is likely to be impossible without major contributions from STI. However, readers will also come to appreciate that harnessing and applying STI effectively in the pursuit of development goals is often not straightforward.

The chapters of this book help to identify the challenges and complexities that must be grappled with in order to succeed in the delivery of SDGs through STI. The challenges and complexities described and analyzed are many and varied, ranging from the need to account for trade-offs between different SDGs when particular technologies or innovations are being considered for application, to the risk of unexpected consequences from application of technologies or innovations, to the existence of community opposition to certain types of science or certain technologies, removing them from the available toolkit via the political process, to the need for supportive policy environments, capable institutions, and good governance if STI is to deliver on its potential. These and many other issues are expanded on throughout the book. The book as a whole provides a wealth of analysis, experience, insights, and cautions that will be valuable to all who are involved in efforts to utilize STI in the delivery of SDGs.

References

Adenle, A. A., Azadi, H., and Arbiol, J. (2015). Global assessment of technological innovation for climate change adaptation and mitigation in developing world. *Journal of Environmental Management* 161, 261–275.

Adenle, A. A., Ford, J. D., Morton, J., Twomlow, S., Alverson, K., Cattaneo, A., Cervigni, R., Kurukulasuriya, P., Huq, S., Helfgott, A., and Ebinger, J. O. (2017). Managing climate change risks in Africa—a global perspective. *Ecological Economics* 141, 190–201.

Adenle, A. A., Manning, D. T., and Arbiol, J. (2017). Mitigating climate change in Africa: barriers to financing low-carbon development. *World Development* 100, 123–132.

Africa Progress Panel (2015). *Africa Progress Report 2015*. Africa Progress Panel. https://reliefweb.int/report/world/africa-progress-report-2015-power-people-planet-seizing-africas-energy-and-climate (accessed April 5, 2019).

Biermann, F., Kanie, N., and Kim, R. E. (2017). Global governance by goal-setting: the novel approach of the UN Sustainable Development Goals. *Current Opinion in Environmental Sustainability* 26–27, 26–31.

Bozeman, B. (2000). Technology transfer and public policy: a review of research and theory, *Research Policy* 29(4), 627–655.

Business and Sustainable Development Commission (BSDC) (2017). New paper. How the world can finance the SDGs. http://businesscommission.org/our-work/new-report-how-the-world-can-finance-the-sdgs (accessed April 8, 2019).

Business and Sustainable Development Commission (BSDC) (2019). *Better Business, Better World*. BSDC, London. http://report.businesscommission.org/uploads/BetterBiz-BetterWorld_170215_012417.pdf (accessed April 8, 2019).

Chung, S. (2002). Building a national innovation system through regional innovation systems. *Technovation* 22(8), 485–491.

Dechezleprêtre, A., Glachant, M., and Ménière, Y. (2009). Technology transfer by CDM projects: a comparison of Brazil, China, India and Mexico. *Energy Policy* 37(2), 703–711.

Dye, C. (2018). Expanded health systems for sustainable development. *Science* 359(6382), 1337.

Global Sustainable Development Report (GSDR 2016). *Perspective of Scientists on Technology and SDGs, Chapter 3, United Nations.* https://sustainabledevelopment.un.org/content/documents/10789Chapter3_GSDR2016.pdf (accessed April 5, 2019).

Goldman, M. (2010). Kuyasa CDM project: renewable energy efficient technology for the poor. *United Nations Development Programme (UNDP)* http://www.growinginclusivemarkets.org/media/cases/SouthAfrica_Kuyasa_2010.pdf (accessed April 5, 2019).

Holdren, J. P. (2008). Science and technology for sustainable well-being. *Science* 319(5862), 424–434.

Kanie, N., Bernstein, S., Biermann, F., and Haas, P. M. (2017). Introduction: Global governance through goal setting. In: Kanie, N., and Biermann, F. (Eds.), *Governing Through Goals: Sustainable Development Goals as Governance Innovation*, pp. 1–28. MIT Press, Cambridge, Massachusetts and London, England.

Khan, R. (2016). How frugal innovation promotes social sustainability. *Sustainability* 8, 1034.

Knack, S., and Keefer, P. (1995). Institutions and economic performance: cross-country tests using alternative institutional measures. *Economics and Politics* 7(3), 207–227.

Koch, N., Fuss, S., Grosjean, G., and Edenhofer, O. (2014). Causes of the EU ETS price drop: Recession, CDM, renewable policies or a bit of everything?—new evidence. *Energy Policy* 73, 676–685.

Levänen, J., Hossain, M., Lyytinen, T., Hyvarinen, A., Numminen, S., and Halme, M. (2016). Implications of frugal innovations on sustainable development: evaluating water and energy innovations. *Sustainability* 8, 4.

Roffe, P., and Santa Cruz, M. (2007). Intellectual property rights and sustainable development: a survey of major issues. Economic Commission for Latin America and the Caribbean (ECLAC). Sustainable Development and Human Settlements Division, the project ROA/49.

Schneider, M., Holzer, A., and Hoffmann V. H. (2008). Understanding the CDM's contribution to technology transfer. *Energy Policy* 36(8), 2930–2938.

Seyfang, G., and Smith, A. (2007). Grassroots innovations for sustainable development: Towards a new research and policy agenda. *Environmental Politics* 16(4), 584–603.

Unilever and Department for International Development (2019). Welcome to TRANSFORM. https://www.transform.global (accessed April 8, 2019).

United Nations (UN) (2002). *Implementation of the United Nations Millennium Declaration, Report of the Secretary-General.* United Nations General Assembly, New York.

United Nations (UN) (2014). Science, technology and innovation for the post-2015 development agenda. Economic and Social Council. Report of the Secretary-General. Seventeenth session, Geneva, May 12–16.

United Nations (UN) (2015). *Transforming Our World: the 2030 Agenda for Sustainable Development*, A/RES/70/1. United Nations, New York.

United Nations (UN) (2016). Science, technology, innovation and capacity-building: Chapter II.G. Addis Ababa Action Agenda-Monitoring commitments and actions. http://www.un.org/esa/ffd/wp-content/uploads/2016/03/2016-IATF-Chapter2G.pdf (accessed April 5, 2019).

United Nations (UN) (n.d.). Technology facilitation mechanism. https://sustainabledevelopment.un.org/tfm (accessed April 6, 2019).

United Nations Conference on Trade and Development (UNCTAD 2017). New innovation approaches to support the implementation of sustainable development goals. https://unctad.org/en/PublicationsLibrary/dtlstict2017d4_en.pdf (accessed April 5, 2019).

United Nations Economic and Social Council (2004). Promoting the application of science and technology to meet the Development Goals contained in the Millennium Declaration. Report by the Secretary-General, Commission on Science and Technology for Development, Seventh session, Geneva, May 24–28, 2004, Item 2 of the provisional agenda. https://unctad.org/en/Docs/ecn162004d2_en.pdf (accessed April 8, 2019).

van der Gaast, W., Begg, K., and Flamos, A. (2009). Promoting sustainable energy technology transfers to developing countries through the CDM. *Applied Energy* 86(2), 230–236.

World Bank (2010). *Innovation Policy: A Guide for Developing Countries*. World Bank, Washington, DC.

World Bank (2018). Poverty. https://www.worldbank.org/en/topic/poverty/overview (accessed April 5, 2019).

THEME I

ENERGY AND ENVIRONMENT

2

Learning to Innovate

The Global Institutions for Biodiversity Innovation in the Sustainable Development Goals

Casey Stevens

1. Introduction

Global biodiversity governance occurs in a highly fragmented institutional arrangement. The six major global biodiversity agreements capture a significant portion of the governance discussion, but significant amounts of global decision-making occur outside of these organizations. Market certification schemes have added another layer of governance. Nongovernmental organizations play a significant role inside and outside the treaties and market certification realm. And finally, government-to-government aid and assistance remains a bedrock of action and particularly shifts in action on biodiversity.

It is possible that such disparate efforts could be seen as an insurmountable weakness when it comes to developing coherent governance of science, technology, and innovation (STI). Fragmented governance allows countries and corporations to forum shop and find agreements that fit their policy preferences while ignoring the other governance regimes, creating a race to the bottom. In addition, fragmented action could cause low-level progress without ever coalescing into a large-scale global effort to reverse severe biodiversity loss. Finally, this fragmented institutional space could lead to technological uptake by the wealthy countries and further marginalization by poorer countries.

This chapter argues, in contrast, that the global biodiversity governance system has significant opportunities for innovation, particularly in connection to the sustainable development goals (SDGs). The argument is that while fragmentation may provide serious impediments to the development of coherent STI, there are contexts in which it can allow diverse ideas and technologies to develop and then spread throughout the system.

The global biodiversity governance system has not always been a system that could serve as a context for such global innovation, but currently it is. The global biodiversity governance system has constructed focused areas for innovative ideas to be created and the organizations have gotten strong enough to transmit innovative ideas throughout the system. Of central importance are the role of the National Biodiversity Strategies and Action Plans (NBSAPs) in the SDGs process and the development of stakeholder inclusion in the expert panels like the Intergovernmental Platform on Biodiversity and Ecosystem Services (IPBES). The architecture for global biodiversity governance has the key structure to facilitate innovation in the post-2015 agenda.

The focus of this chapter is on in situ wildlife conservation. In this field, three different scientific innovations could become core parts of the global system through the SDGs. First, integration of biodiversity and ecosystem services into wider discussions of development, housing, agriculture, climate change, and so on. Such connections and trade-offs have been significantly underexplored (McShane et al. 2011), but the specific targets of the SDGs emphasize some of these connections. Second, ecosystem-based management has been developing as a usable tool for conservation for a number of years now. The SDGs is the first major nonbiodiversity governance arrangement to explicitly focus on these as targets and indicators for progress. Third, the technology of remote sensing for both habitats and species has been used to effectively track changes in a number of different contexts. There are significant scientific and technical challenges to scaling up the use of such technology, but the SDGs, their targets, and the emphasis on monitoring could make significant gains in this respect.

What is the likelihood of innovation in global biodiversity governance through the SDGs? This chapter answers that it is very high, by exploring the architecture of the global biodiversity system. It proceeds through three sections. The first section explores the STI potential within the SDGs when it comes to in situ biodiversity conservation. The second section develops a framework for innovation that emphasizes the needs for developing science and technology in small groups and then spreading it throughout a larger system. The third section argues that global biodiversity governance has such a system in place. Therefore, the conclusion is that there are significant opportunities for innovation in biodiversity within the SDGs framework. Far from being guaranteed, changes that emphasize wider participation and which improve capacity of global biodiversity organizations and networks can make a significant difference.

2. Biodiversity Innovation and the SDGs

Goal 15 of the SDGs is the most relevant to the topic of in situ biodiversity conservation. The goal has the stated aim to "protect, restore and promote sustainable use of terrestrial ecosystems, sustainably manage forests, combat desertification, and halt and reverse land degradation and halt biodiversity loss." This goal includes nine specific biodiversity-related targets and three dealing with funding and resources.

In many ways, SDG15 builds on the earlier Aichi biodiversity targets developed in the Convention on Biological Diversity (CBD). For example, while most of the SDGs have a goal date of 2030, the biodiversity targets largely adopt 2020 as the goal date which is the same as the Aichi targets. In addition, some of the SDG targets are very similar to the Aichi targets. SDG target 15.5 commits actors to "take urgent and significant action to reduce the degradation of natural habitats, halt the loss of biodiversity and, by 2020, protect and prevent extinction of threatened species." Aichi target 5 says that "by 2020, the rate of loss of all natural habitats, including forests, is at least halved and where feasible brought close to zero, and degradation and fragmentation is significantly reduced." Similarly, the indicator for SDG15.9 is simply measured in terms of progress on the Aichi targets. An initial assessment may suggest that the SDGs add little additional benefits to these earlier targets.

There are three STI opportunities that the SDGs directly connect with that could be relevant for biodiversity conservation. First, SDG15 and its targets are more integrated into larger issues of sustainability than are the Aichi targets. SDG2.4 calls for sustainable food production which maintains ecosystems. SDG2.5 builds on this by explicitly committing states to "maintain the genetic diversity of seeds, cultivated plants and farmed and domesticated animals and their related wild species." Similarly, SDG6.6 states "by 2020, protect and restore water-related ecosystems, including mountains, forests, wetlands, rivers, aquifers and lakes." Goal 14 on conservation and sustainable use of oceans, seas and marine resources also connects to SDG15. While not as linked as other goals (Le Blanc 2015), this is an opportunity for new policy approaches and science policy that focuses precisely on these interactions. Importantly, these connections across various issues may prevent some of the myopia in the CBD that could restrain scientific progress (Prathapan et al. 2018).

Second, the SDGs overall and the specific goals and targets relevant to biodiversity conservation are a significant endorsement of the ecosystem approach to dealing with biodiversity loss. The millennium development goals and the Millennium Declaration notoriously adopted a very narrow and ill-defined framework toward biodiversity loss and ecological services. SDG14 and SDG15, in contrast, emphasize an ecosystem approach to biodiversity loss. This comes at a key point as the scientific base for ecosystem-based management is "moving—albeit slowly—from the 'what's, why's, and when's' to the 'how's' of operationalization and implementation" (Link and Browman 2017, p. 379). The goals dealing with ecosystems are vaguer than SDG6.5, which calls specifically for "integrated water resources management at all levels." However, the SDGs could provide the context in which applicable principles and approaches for ecosystem-based management can become an international focus.

Third, while much of the focus is on the goals and targets of the SDGs, the indicator discussion is a significant part of this agenda. Similarly to ecosystem-based management, the SDGs comes during a pivotal period for monitoring biodiversity loss, particularly through remote sensing. A recent review noted that remote sensing remains a "blunt tool" that is workable at the individual site scale, but not at larger or national scales (Corbane et al. 2015, p. 12). Lack of standardization and difficulty in operationalizing the Food and Agriculture Organization's land cover classification system both will be problems in scaling up this technological approach to monitoring habitats and species (Pause et al. 2016). Most of the biodiversity related targets and indicators will need significant conceptual progress in order to provide adequate assessments. SDG15.1, SDG15.3, and SDG15.4, in particular, rely upon basic land cover understandings, but could provide an impetus for developing better technological approaches for conservation. Crucially, the universal approach of the SDGs monitoring efforts could mean that more developing countries can contribute and benefit from such efforts. To date, such discussions have not figured predominantly in the discussions.

3. Architecture for Innovation

In addition to the available stock of scientific understandings and technological opportunities available for biodiversity conservation, this chapter argues that crucial is the ability of a global governance system to support and diffuse innovation throughout the world. This argument combines knowledge creation theory in management studies with the learning approach in international organizations. The basic argument is that the most effective innovation is that which starts in small sets of actors and is able to be broadcast from that small group of actors to a wider, connected network.

Knowledge creation theory begins with a distinction between tacit knowledge and explicit knowledge (Nonaka 1994). Explicit knowledge is the knowledge that is expressed, discussed, deliberated, and made apparent within an organization. The listing of species in biodiversity governance is a fantastic example of explicit knowledge where criteria are established, the threats to individual species are discussed, and standards of protection are developed and disseminated. Tacit knowledge, in contrast, is the personal knowledge of habit, design, and thinking from analogies. In relation to biodiversity governance, traditional and indigenous knowledge provides an excellent example because it is "inherently scattered and local in character, and gains its vitality from being deeply implicated in people's lives" (however, it is an example that should highlight the heterogeneity and limitations of scientific knowledge rather than reify some hierarchy of knowledge) (Agrawal 2014, p. 5).

From this approach, knowledge conversion or the processes by which tacit knowledge is made into explicit forms is the crucial process for innovation (Nonaka and von Krogh 2009). Tacit knowledge though does not easily lend itself to conversion and normal politics may tend to prevent knowledge conversion (Hannan and Freeman 1984). Knowledge creation theory argues that the key processes of knowledge conversion are 1) diversity of tacit knowledge, and 2) a process to facilitate discussion about tacit knowledge. If large groups of actors share a similar tacit knowledge (from organizational culture, similar training/education, etc.), then there will be limited contestation and disagreement to spur innovation. Similarly, if there is no forum for discussing tacit knowledge and transforming it into usable knowledge, then tacit knowledge is going to remain largely unexplored. Innovation happens from diverse epistemologies and processes of engagement that productively convert that diversity into new forms of explicit knowledge.

This conceptual framework resonates with on-the-ground examples of biodiversity management in local contexts. Many studies have found that "a great societal opportunity related to implementing voluntary biodiversity conservation initiatives is integration of various types of knowledge (social, ecological, scientific, and local) in the conservation planning processes for greater legitimacy and effectiveness" (Paloniemi et al. 2018, 7; see also Crona and Bodin 2006; Pretty and Smith 2004; Zerbe 2005). Similarly, an analysis of community-based adaptation policies found that wide participation and links to the larger development context were key to success (Reid 2016). Case studies in local environments have demonstrated that working together on projects can facilitate trust and effective biodiversity outcomes more than goal-based governance or working on preformed ideas together (Borg et al. 2015).

However, at the international level, these conditions present a number of challenges. First, while inclusion of diverse stakeholders is possible in local biodiversity efforts, scaling up to the international level adds additional logistical, representational, and accountability problems. This has been a central tension in the construction of IPBES. While IPBES developed a reflexive effort at stakeholder access (Esguerra et al. 2017), there remained pressures about including and developing connections with local level institutions (Soberón and Peterson 2015). Second, and relatedly, learning from diverse stakeholders is a difficult prospect in local biodiversity contexts, but developing systems for feedback from experience to future efforts is difficult at the international level. Those international negotiations that operate based on consensus, however, have found that including too many perspective becomes a barrier to social learning (Chasek and Wagner 2016). In addition, bureaucratic cultures and the battle for authority between different institutions present multiple ways for innovation to be blocked (Matthijs and Blyth 2018).

Haas and Haas (1995) provide an argument for how learning can occur in global governance. The crucial process in their approach is small-group consensus developing and influencing other actors. In particular, learning or innovation occur when small groups of experts (often scientists) develop arguments that resonate with a small number of important governments. These governments then form a coalition built around this consensus which then captures an international organization and uses it to transmit the argument to a wider group of states. The conveyor belt is built by different forms of small-group consensus along the way: expert consensus introduces novel ideas into the governance system, a coalition of states begins pushing for that consensus to be the basis of action, and finally consensus among member states or of the secretariat of an international organization turns that organization into a transmitter of novel ideas to a wider audience. Evidence from global health governance has indicated that legal obligations from the international organization are not necessary, and direct contacts between international organization staff and government bureaucrats and particularly technical assistance can be effective in this process (van Kerkhoff and Szlezák 2016).

4. Innovation Opportunities in Global Biodiversity Governance

The global biodiversity system has grown tremendously from its starting point in the 1970s. Since the 1990s, the CBD has been a primary international organization on biodiversity issues serving as a focal point for efforts, a negotiating forum on crucial issues, and an arena for the creation of global biodiversity efforts. Before the SDGs pursued multifaceted and interconnected goal-based governance, the Aichi biodiversity targets were created to focus efforts on slowing biodiversity loss by 2020. The Aichi targets and other efforts have created a rich terrain for innovative ideas to quickly develop, achieve critical support, and be transmitted across the globe. This section will highlight the national biodiversity planning process, international expert reporting, and the transmission capabilities of the international organizations.

4.1 National Biodiversity Strategies and Action Plans

SDG15.9 calls for integrating "ecosystem and biodiversity values into national and local planning, development processes, poverty reduction strategies and accounts." This connects the SDGs directly into the ongoing process of the CBD, most notably the Aichi targets. The CBD calls on all member states to develop National Biodiversity Strategies and Action Plans (NBSAPs) that approach biodiversity action in a manner integrated with the rest of government programs and efforts. These are crucial tools for global governance, linking international discussions to national and local level planning and policies. Ideally, these would be regular planning documents, agreed to in a wide participatory manner, that fully mainstream biodiversity into the overall policy approach being taken by a country and set new directions in terms of conservation, sustainable use, and equitable sharing of benefits (the three core goals of the CBD).

This ideal is rarely met and the NBSAPs have tended toward the technical, with minimal amounts of nonstate stakeholders involved, and are often light on specifics beyond declaring protected areas. The example of the Solomon Islands is instructive. The first time they attempted to develop an NBSAP they solicited individual plans from multiple different relevant ministries, but then when they met there was no common vision to bring everyone together. The Ministry of Environmental Conservation and Meteorology restarted the process in 2007, but with limited staffing and funding to pursue a broad, multisectoral plan (Carter 2007). The group completed an NBSAP in 2009 with the assistance of funding from multiple international organizations and nongovernmental organizations. A second NBSAP was completed in 2016 with direct reference to action to achieve the Aichi targets.

This example helps to understand the challenges which have to this point limited the innovative impact possible within the NBSAPs. A 2015 assessment found three relevant points throughout the submitted NBSAPs. First, few incorporated other biodiversity related conventions into their NBSAP. Second, conservation got more attention than either sustainable use or equitable sharing of benefits. And, finally, many NBSAPs appeared to be geared toward external funders and did not have effective domestic participation (Pisupati and Prip 2015). A 2016 analysis found that education campaigns were the primary mainstreaming effort involved in NBSAPs and that integration into decision-making beyond the environmental ministry was limited (UNDP 2016). The result of these problems led to Aichi biodiversity target 17, which specifically declared that "by 2015, each Party has developed, adopted as a policy instrument, and has commenced implementing an effective, participatory and updated national biodiversity strategy and action plan." By the end of June 2018, 146 states submitted NBSAPs that directly take into account the 2011–2020 Strategic Plan for Biodiversity.

While they have been limited to this point, the accumulated experience, focus, and commitment to NBSAPs makes them a potentially invaluable tool for innovation in global biodiversity governance. Earlier recommendations for improving NBSAPs have focused on the CBD providing clearer guidance, streamlining procedures, improving funding and support for NBSAPs, and widening stakeholder involvement (Adenle et al. 2015; Global Policy Unit-IUCN 2013; Pisupati 2007; Pisupati and Prip 2015; UNEP 2016). In addition, the process itself has gradually become more reflective of the NBSAP forum and their open peer review mechanism. SDG15.9 offers an opportunity for improving the NBSAP process and also possibly elevating these action plans to higher level discussions in domestic contexts.

NBSAPs and the supportive system they have built up offer unique opportunities for innovation in the SDGs. The SDGs, and particularly 15.1, 15.2, and 15.9, offer the opportunity to integrate specific, time-focused targets into each country's biodiversity plan and also to integrate it better across different ministries in government. Currently, both are limitations on NBSAPs (International Council for Science 2017). But, more importantly, the experience with prior NBSAPs and the existence of the NBSAP Forum both offer opportunities for significant efforts on biodiversity-relevant targets within the SDGs. In the SDGs process, the voluntary national reviews are being submitted by countries to demonstrate progress and the NBSAP process could be quickly combined, although it has not been in any of the national reviews to date. The peer review mechanism should become more institutionalized and develop clearer guidance on participatory inclusion. Institutionalizing the NBSAP peer review with other existing and nascent ones like the fossil fuel subsidy reform peer reviews, environmental management peer reviews, and so forth offers a unique opportunity for innovation. Domestic groups that have prepared NBSAPs should be well-positioned to participate actively in wider SDGs planning, and the community established to facilitate NBSAPs can be a guide to other areas of the SDGs.

Key recommendations for NBSAPs are to use the SDGs as an opportunity to 1) develop more participatory processes including more stakeholders; 2) develop, with the SDGs, time-bound targets that emphasize biodiversity outcomes; and

3) institutionalize and promote the NBSAP Forum and the peer review mechanism, particularly through greater synergy with the SDGs voluntary national reviews.

4.2 Expert Systems

Expert systems, those bodies tasked with developing robust information to guide decision-makers in international organizations, have traditionally been somewhat hindered in global biodiversity organizations. Some of the expert committees have been highly politicized, while others have been too weak. There are examples of times and forums which have worked effectively at developing robust understandings, including stakeholders, and delivering ideas to policymakers in timely and usable fashion. However, these have tended to be focused on very particular issues or individual species. For biodiversity innovation to occur or for SDG15.A to see progress, expert systems able to develop explicit knowledge would be necessary. The development of IPBES, established in 2012, has signaled a change from this prior political system. While the direct impacts from IPBES are still developing, the secondary impacts are equally important.

IPBES is the institutionalization of efforts that started in 2001 until 2005 with the millennium ecosystem assessment. Because of the scope of the problem, in both biological and socioeconomic terms, even during this period the construction of expertise was spread very widely. Each international organization had its own separate expertise procedures tailored to the specifics of the organization. In addition, a tremendous amount of expertise was constructed in locale-specific manner with only partial interconnections to other locales, for example in the man and biosphere agreement. Some of the expert bodies were quite effective, some were highly politicized, and some barely functioned at all (Gehring and Ruffing 2008; Haas and Stevens 2011; Koetz et al. 2008).

IPBES was designed with two particularly relevant features for the discussion here. First, it was designed to be multithematic and operate at multiple scales (Brooks et al. 2014). As opposed to the Intergovernmental Panel on Climate Change (IPCC), which is organized around focal reports every few years, the IPBES is allowed to bring together experts on various themes as decided by the IPBES plenary. To date, they have produced eight assessments on issues dealing with pollinators, land degradation, scenarios and models of biodiversity, regional reports for Europe, Asia, Africa, and the Americas, and a global report. At the 2018 plenary meeting, governments agreed to three new assessments on sustainable use of wild species, the multiple values of biodiversity, and invasive alien species. While some features of this are similar to the working group format of the IPCC, its scope is intentionally broader and more adaptive. Second, IPBES is "one of the first international expert organizations to have systematically developed a strategy for stakeholder engagement in its own right" (Esguerra et al. 2017, p. 60). While this is the case, in terms of initial institutional design and early efforts, stakeholder involvement has been focused and limited rather than broad and substantive. At the fifth plenary meeting in 2017, governments requested the secretariat to implement better participatory mechanisms particularly in dealing with traditional and indigenous knowledge and this appeared to show results by the sixth plenary in 2018 (Earth Negotiations Bulletin 2018).

Beyond the direct impact, IPBES has had a secondary impact that is potentially important for the topic of innovation. The focus on key integrative science topics to the IPBES has allowed much more of a focus on usable knowledge in the other science forums (Clark et al. 2016; Turnhout et al. 2016). What this means practically is that the science bodies of the various international organizations can focus more effort on the tractable issues of their governance efforts rather than on more fundamental issues under discussion. The clearest example of this is with the Subsidiary Body on Scientific, Technical and Technological Advice (SBSTTA) in the CBD. The SBSTTA has shifted its focus significantly from broad, largely bureaucratic recommendations to those focused significantly on technical issues and establishing scientifically-based frameworks for implementation. For example, at the sixth SBSTTA meeting in 2001, five of the nine recommendations dealt with basic organization of efforts and collaboration with other international organizations. By the 21st meeting in 2017, only one of the seven dealt with such topics and the rest were focused discussions.

The expert space is well-designed and offers a host of opportunities for consensual dialogue to occur. IPBES is one forum for such engagements, but it is not the exclusive organization and significant opportunities now exist in the science bodies of the other international organizations. This is a change from a decade ago where there was no clear institutionalized science forum dealing with biodiversity and where many of the bodies in the organizations were beset with significant challenges.

Key recommendations for expert systems are to 1) expand and institutionalize the participatory mechanism in IPBES; 2) encourage and foster stakeholder engagement in expert systems; 3) further strengthen the science bodies in the international biodiversity organizations; and 4) routinize feedbacks from IPBES into the international organizations and then back to IPBES.

4.3 Transmission-Capable Institutions

SDG17 deals with strengthening the means of implementation and focusing on a global partnership for sustainable development. The focus of the targets is at the national level, but improving the means of implementation at the global level may be equally important. As explored earlier, the biodiversity governance system may be able to develop diverse findings effectively, but the crucial aspect is whether they can diffuse those throughout the various organizations and at the local level. In this respect, the various international biodiversity organizations have developed such capacities and experiences. The examples are everywhere with the CBD-affiliated Global Partnership for Plant Conservation, Ramsar Advisory Missions, the Sustainable Ocean Initiative, the elephant efforts of the Convention on International Trade of

Endangered Species, and so forth. Working with national-level actors, both state and nonstate, through these efforts are essential to any transmission efforts because they ground the broad idea into the particular context of the state. In this respect, SDG15.9 and SDG17.9, which both focus on improving the capacity at the national level, offers a crucial opportunity for promoting innovation. In particular, operationalizing and standardizing ecosystem-based management and remote sensing could be essential aspects of allowing states to improve understanding of their ecosystems and take more meaningful steps to address them.

More crucially, though, is whether there is an effective system for learning that moves from one sector of the incredibly complex biodiversity system to others. If learning in one organization can move to other organizations, then that would be an architecture primed for innovation. This is crucial with the specific STI aspects highlighted in the first section of this chapter which have seen most of their development in the European Union context, but have seen only limited extension. This issue has been recognized for a number of years and led to the establishment of the Liaison Group of Biodiversity-Related Conventions in 2004. This brings together six biodiversity related organizations in regular meetings on key topics. Some ideas have moved between the organizations, for example the Addis Ababa Principles and Guidelines for Sustainable Use of Biodiversity developed largely within the CBD have had an impact on discussions in the other organizations (Secretariat of the Convention on Biological Diversity 2004). While such guidelines might be applicable across the various organizations, the evidence of specific innovative ideas moving across the different organizations is limited. The different discussions dealing with what types of access require a need for equitable sharing of benefits in the International Treaty of Plant Genetic Resources for Food and Agriculture and the CBD's Nagoya Protocol is illustrative. While largely supportive of each other, there are enough differences to make implementation complex.

The process to improve the ability of the regime complex to transmit across organizations is better interplay management by the various secretariats (Jinnah 2014). One option would be to establish one of the organizations as the primary coordinating agency, an approach the CBD wanted to take early in the 2000s, but there is no need for this coordination to happen only at the international level. Domestic coordination offers an opportunity for the implementing organizations to connect and transmit ideas at a focused level. Key in facilitating this would be the various national focal points of the agreement but also the officers of the United Nations Environment Programme and the United Nations Development Programme. Many forums have been organized recently which attempt to deal with these problems. Facilitating additional forums that bring in secretariat staff of the various agreements would be a positive step in both downward and horizontal ideal transmission.

Finally, there is an opportunity for temporal alignment and bringing various agendas together. The CBD has established a 2050 vision of biodiversity being "valued, conserved, restored and widely used." The plan at this point is for the 2011–2020 decade to be followed with a new strategic plan lasting until 2030. The ability to put all global biodiversity organizations on a similar timeframe would do much to facilitate the discussion and development of innovative ideas and their transmission across organizations. The correspondence between this strategic vision and the SDGs ending point in 2030 all provide an excellent opportunity for innovation of governance systems.

Specific recommendations are 1) continued use and strengthening of the Liaison Group; 2) strengthening the international biodiversity organizations to improve direct contacts and work with national level bureaucrats; and 3) temporal integration, bringing discussions in the various organizations together on a shared and overlapping time frame.

5. Conclusion

The temporal gap between the Aichi biodiversity targets and the SDGs offers an opportunity to reset the efforts after 2020 and really move forward on issues before 2030. How can the world make the most significant contribution possible for reversing biodiversity loss in the last decade of the SDGs? Three sets of scientific and technical ideas are primed to see significant progress forward: integration of biodiversity with other social spheres, ecosystem-based management, and remote monitoring of habitats and species. In order to do so, however, they will need to be spurred by the SDGs process to include more participants and be fostered by the global biodiversity organizations to spread to more settings. This chapter has argued that both of these are possible.

To achieve such innovation and scientific progress, however, would require planning right now for the post-2020 period. The urgency was reflected in both the High Level Political Forum meeting dealing with the SDGs in 2018 and in the regular meetings of the CBD. Maintaining a focus on widening participation, linking biodiversity with other spheres, and facilitating standards of remote sensing relevant to developing countries can help significantly. Innovation is not assured, but there are significant opportunities for turning the 2021–2030 period into one of significant scientific and technological progress on biodiversity issues.

References

Adenle, A. A., Stevens, C., and Bridgewater, P. (2015). Global conservation and management of biodiversity in developing countries: an opportunity for a new approach. *Environmental Science and Policy* 45, 104–108.

Agrawal, A. (2014). Indigenous and scientific knowledge: some critical comments. *Antropologi Indonesia* 55. DOI:10.7454/ai.v0i55.3331

Borg, R., Toikka, A., and Primmer, E. (2015). Social capital and governance: a social network analysis of forest biodiversity collaboration in Central Finland. *Forest Policy and Economics* 50, 90–97.

Brooks, T. M., Lamoreux, J. F., and Soberón, J. (2014). Intergovernmental Science-Policy Platform on Biodiversity and Ecosystem Services (IPBES) ≠ Intergovernmental Panel on Climate Change (IPCC). *Trends in Ecology and Evolution* 29(10), 543–545.

Carter, E. (2007). *National Biodiversity Strategies and Action Plans: Pacific Regional Review.* Commonwealth Secretariat, London.

Chasek, P. S., and Wagner, L. M. (2016). Breaking the mold: a new type of multilateral sustainable development negotiation. *International Environmental Agreements: Politics, Law and Economics* 16(3), 397–413.

Clark, W. C., van Kerkhoff, L., Lebel, L., and Gallopin, G. C. (2016). Crafting usable knowledge for sustainable development. *Proceedings of the National Academy of Sciences of the United States of America* 113(17), 4570–4578.

Corbane, C., Lang, S., Pipkins, K., Alleaume, S., Deshayes, M., García Millán, V. E., Strasser, T., Vanden Borre, J., Toon, S., and Michael, F. (2015). Remote sensing for mapping natural habitats and their conservation status: new opportunities and challenges. *International Journal of Applied Earth Observation and Geoinformation* 37, 7–16.

Crona, B. and Bodin, O. (2006). What you know is who you know? communication patterns among resource users as a prerequisite for co-management. *Ecology and Society* 11(2), 7–30.

Earth Negotiations Bulletin. (2018). Summary of the Sixth Session of the Plenary of the Intergovernmental Science-Policy Platform on Biodiversity and Ecosystem Services. International Institute for Sustainable Development. http://enb.iisd.org/vol31/enb3142e.html (accessed April 2, 2019).

Esguerra, A., Beck, S., and Lidskog, R. (2017). Stakeholder engagement in the making: IPBES Legitimization Politics. *Global Environmental Politics* 17(1), 59–76.

Gehring, T., and Ruffing, E. (2008). When arguments prevail over power: the CITES procedure for the listing of endangered species. *Global Environmental Politics* 8(2), 123–148.

Global Policy Unit: International Union for Conservation of Nature (IUCN) (2013). *An Engagement Strategy for IUCN in National Biodiversity Strategies and Action Plans (NBSAPs).* IUCN, Gland.

Haas, P. M., and Haas, E. B. (1995). Learning to learn. *Global Governance* 1(3), 255–284.

Haas, P. M., and Stevens, C. (2011). Organized Science, Usable Knowledge, and Multilateral Environmental Governance. In: Lidskog, R., and Sundqvist, G. (Eds.), *Governing the Air: The Dynamics of Science, Policy, and Citizen Interaction*, pp. 125–161. MIT Press, Cambridge, MA.

Hannan, M. T., and Freeman, J. (1984). Structural inertia and organizational change. *American Sociological Review* 49(2), 149–164.

International Council for Science (ICSU) (2017). *A Guide to SDG Interactions: From Science to Implementation.* ICS, Paris.

Jinnah, S. (2014). *Post-Treaty Politics: Secretariat Influence in Global Environmental Governance.* MIT Press, Cambridge, MA.

Koetz, T., Bridgewater, P., van den Hove, S., and Siebenhüner, B. (2008). The role of the subsidiary body on scientific, technical and technological advice to the Convention on Biological Diversity as science–policy interface. *Environmental Science and Policy* 11(6), 505–516.

Le Blanc, D. (2015). Towards integration at last? the sustainable development goals as a network of targets. *Sustainable Development* 23(3), 176–187.

Link, J. S., and Browman, H. I. (2017). Operationalizing and implementing ecosystem-based management. *ICES Journal of Marine Science: Journal du Conseil* 74(1), 379–381.

Matthijs, M., and Blyth, M. (2018). When is it rational to learn the wrong lessons? technocratic authority, social learning, and euro fragility. *Perspectives on Politics* 16(1), 110–126.

McShane, T. O., Hirsch, P. D., Trung, T. C., Songorwa, A. N., Kinzig, A., Monteferri, B., Mutekanga, D., Thang, H. V., Dammert, J. L., Pulgar-Vidal, M., Welch-Devine, M., Brosius, J. P., Coppolillo, P., and O'Connor, S. (2011). Hard choices: making trade-offs between biodiversity conservation and human well-being. *Biological Conservation* 144(3), 966–972.

Nonaka, I. (1994). A dynamic theory of organizational knowledge creation. *Organization Science* 5(1), 14–37.

Nonaka, I., and von Krogh, G. (2009). Perspective: tacit knowledge and knowledge conversion: controversy and advancement in organizational knowledge creation theory. *Organization Science* 20(3), 635–652.

Paloniemi, R., Hujala, T., Rantala, S., Harlio, A., Salomaa, A., Primmer, E., Pynnönen, S., and Arponen, A. (2018). Integrating social and ecological knowledge for targeting voluntary biodiversity conservation. *Conservation letters* 11(1): e12340. DOI:10.1111/conl.12340

Pause, M., Schweitzer, C., Rosenthal, M., Keuck, V., Bumberger, J., Dietrich, P., Heurich, M., Jung, A. and Lausch, A. (2016). In situ/remote sensing integration to assess forest health: a review. *Remote Sensing* 8(6), 471.

Pisupati, B. (2007). *Effective Implementation of NBSAPs: Using a Decentralized Approach: Guidelines for Developing Sub-National Biodiversity Action Plan.* UNU Institute of Advanced Studies, Yokohama, Japan.

Pisupati, B., and Prip, C. (2015). *Interim Assessment of Revised National Biodiversity Strategies and Action Plans (NBSAPs),* UNEP World Conservation Monitoring Centre, Cambridge.

Prathapan, K. D., Pethiyagoda, R., Bawa, K. S., Raven, P. H., Rajan, P. D., and 172 co-signatories from 35 countries. (2018). When the cure kills-CBD limits biodiversity research. *Science* 360(6396), 1405–1406.

Pretty, J., and Smith, D. (2004). Social capital in biodiversity conservation and management. *Conservation Biology* 18(3), 631–638.

Reid, H. (2016). Ecosystem- and community-based adaptation: learning from community-based natural resource management. *Climate and Development* 8(1), 4–9.

Secretariat of the Convention on Biological Diversity. (2004). *Addis Ababa Principles and Guidelines for the Sustainable Use of Biodiversity.* Convention on Biological Diversity, Montreal.

Soberón, J., and Peterson, A. T. (2015). Biodiversity governance: a Tower of Babel of scales and cultures. *PLoS Biology* 13(3), e1002108. doi: 10.1371/journal.pbio.1002108

Turnhout, E., Dewulf, A., and Hulme, M. (2016). What does policy-relevant global environmental knowledge do? the cases of climate and biodiversity. *Current Opinion in Environmental Sustainability* 18, 65–72.

van Kerkhoff, L., and Szlezák, N. A. (2016). The role of innovative global institutions in linking knowledge and action. *Proceedings of the National Academy of Sciences of the United States of America* 113(17), 4603–4608.

United Nations Development Program (UNDP). (2016). *National Biodiversity Strategies and Action Plans: Natural Catalysts for Accelerating Action on Sustainable Development Goals.* Interim Report. UNDP, New York.

United Nations Environment Program (UNEP). (2016). *Strengthening the National Biodiversity Strategies and Action Plans: Revision and Implementation.* UNEP, Nairobi.

Zerbe, N. (2005). Biodiversity, ownership, and indigenous knowledge: exploring legal frameworks for community, farmers, and intellectual property rights in Africa. *Ecological Economics* 53(4), 493–506.

3

Energy Technologies for Sustainable Development Goal 7

Govinda Timilsina and Kalim U. Shah

Energy is central to nearly every major challenge and opportunity the world faces today. Be it for jobs, security, climate change, food production or increasing incomes, access to energy for all is essential. Transitioning the global economy towards clean and sustainable sources of energy is one of our greatest challenges in the coming decades. Sustainable energy is an opportunity—it transforms lives, economies and the planet.

—United Nations SDG-7, 2008

1. Introduction

Globally more than one billion people do not have access to electricity and almost three billion people do not have access to modern fuels for cooking and home heating. More than two-thirds of the population without access to modern energy live in sub-Saharan Africa and South Asia. In sub-Saharan Africa alone, close to 90% of the rural population does not have access to electricity and 93% do not have access to modern energy for cooking (World Bank 2018). In 2015, the United Nations established the sustainable development goals (SDGs), a set of 17 global goals and 169 targets aiming to address a broad array of social and economic development challenges by 2030. These challenges include hunger, poverty, health, education, climate change, gender, clean drinking water and sanitation, modern energy, air pollution, and social justice (UN 2015). The theme of one goal, SDG7, centers on energy (ensure access to affordable, reliable, sustainable, and modern energy for all) and is composed of five targets, as listed in Table 3.1. Target SDG7.1 states the importance of access to affordable, reliable, and modern energy services for all by 2030. It is regarded as a "golden thread" because it is linked to other SDGs, namely SDG1 (end poverty), SDG2 (end hunger), SDG3 (healthy lives), SDG4 (equitable quality education), SDG6 (water and sanitation for all), SDG8 (economic growth, employment, and decent work), SDG9 (sustainable industrialization), SDG11 (inclusive, safe cities) and SDG13 (combat climate change) (de la Sota et al. 2017; Bhattacharya and Palit 2016). Therefore, meeting SDG7 is critical to achieving other SDGs as well.

Table 3.1 SDG7 goals and achievement indicators

Goals by 2030	Indicators
7.1: Universal access to affordable, reliable, and modern energy services	• Proportion of population with access to electricity • Proportion of population with primary reliance on clean fuels and technology
7.2: Substantial share of renewable energy in the global energy mix	• Renewable energy share in the total final energy consumption
7.3: The global rate of improvement in energy efficiency be doubled from 2015 level	• Energy intensity measured in terms of primary energy and GDP
7.a: Enhanced international cooperation to facilitate access to clean energy research and technology, including renewable energy, energy efficiency and advanced and cleaner fossil-fuel technology, and promotion of investment in energy infrastructure and clean energy technology	• Mobilization of $100 billion annually starting from 2020
7.b: Expanded infrastructure and upgraded technology for supplying modern and sustainable energy services for all in developing countries, particularly least developed countries, small island developing states, and land-locked developing countries	• Investments in energy efficiency as a percentage of GDP and the foreign direct investment in financial transfer for infrastructure and technology to sustainable development services

Source: Frankfurt School-United Nations Environment Programme (2015).

Since SDGs were launched in 2015, some progress has been made toward accomplishing SDG7. Global electricity access reached 85.7% in 2016 from 84.9% in 2014. An increase of 1% every year is needed on global energy access from 2016 to 2030 to achieve SDG7.1 (World Bank 2018). The share of renewables in the world's total final energy consumption (TFEC) increased 0.2% from 17.3% in 2014 to 17.5% in 2016. Still, more efforts are needed to achieve SDG7.

There are several challenges, however, in the quest to provide access to modern energy in access-deficit areas. Among the barriers to energy access, the main ones are lack of supply infrastructure, poor quality of supply, connection costs, and affordability. In many developing countries, especially in sub-Saharan Africa and South Asia, households and businesses do not have access to modern energy despite their affordability and willingness to pay because energy production capacity and supply infrastructure (e.g., transmission and distribution networks) do not exist. Per capita electric generating capacity in some poor countries, such as Chad, Nepal, and Cambodia, is a thousand times lower in comparison to developed countries (IEA 2018).

Even if supply is available, the quality is poor. Continuous disruption of electricity has become a characteristic of electricity supply systems in many developing countries. For example, unscheduled electricity disruptions occur more than 20 times in a typical month in several countries such as Yemen, Lebanon, Iraq, Pakistan, Guinea, Central African Republic, Democratic Republic of Congo, and Nigeria (IFC 2017). Scheduled power disruptions or load-shedding is common in many countries in South Asia and sub-Saharan Africa, where there is no electricity for hours and hours per day because of load-shedding. Therefore, reliable supply of modern energy is a key concern.

Affordability is another critical challenge for achieving SDG7 because households are not expected to spend more than 5% of their monthly income on energy in countries with tropical climates and 10% of spending in countries with temperate climates At present, however, 57% of the population living in energy access-deficit countries spend more than 5% of their gross nominal income on energy (World Bank 2017). A large number of poor people in South Asia and sub-Saharan Africa cannot afford electricity and modern cooking fuels such as liquefied petroleum gas (LPG) even if supply is available. In India, 22% of rural households and 6% of urban households in areas already electrified do not utilize electricity because they cannot afford connections or consumption (Ghosh-Benerjee et al. 2015).

In light of the challenges of reliability, affordability, and access to energy across the world, technology plays a pivotal role in achieving SDG-7. The transformation of the current energy system to a clean energy system requires new technologies to meet demand growth. This has catapulted the growth of renewable energy sectors such as solar, wind, and biomass (IEA 2017). Research, development, and innovation are making scalable renewable energy systems ready to be deployed to provide reliable and affordable power to people.

Renewable energy can be decentralized into off-grid systems thereby providing critical energy access to places with limited grid reach. Apart from the renewable energy sector, natural gas has grown into a comparatively cleaner way to generate electricity and heating, especially during peak periods. Along with renewable energy, energy efficiency technology has been touted to be one of the two foundational pillars of a sustainable energy system, but it has not taken off in developing countries owing to significant financing and technological challenges discussed later.

2. Conventional Technologies for SDG7

Various technologies and options are available to achieve SDG7. The main options are: (i) expanding the conventional energy supply systems; (ii) renewable and emerging technologies; (iii) modernization of traditional energy sources; and (iv) expansion of energy efficiency measures in residential and commercial buildings. Table 3.2 presents these key technologies required for meeting SDG7 and provides definitions

Table 3.2 Technologies and options for achieving SDG7

Technology/option	Definition	Examples
Expanding existing energy infrastructure	Rural electrification expanding electricity grids; adding infrastructure and supply chain for natural gas, LPG	Electricity grid expansion, natural gas distribution for home heating, cooking; LPG for cooking
Growth of renewable and clean energy technologies	Isolated single unit rooftop solar home systems; solar, hydro or wind power based mini or macro grid systems	Solar home systems, microgrids and mini grids for electricity, ethanol and biodiesel for cooking and transportation
Modernization of traditional energy resources	Conversion of traditional biomass to modern energy	Landfill gas or biogas for both lighting and heating, biomass fired electricity micro-mini grids
Expanding energy efficiency technologies	Reduction in amount of energy use for specific tasks	Energy efficient lighting, HVAC and appliances in residential and commercial buildings

Source: authors.

and examples. The critical challenge is that all of these options are relatively expensive and those without access to modern energy are mostly low-income populations who cannot afford these expensive energy services (Cecelski 2015). This implies that significant intervention is needed from governments and international development communities with respect to incentives and other policy supports for both producers and consumers of clean energy and energy efficiency technologies. (Dornan and Shah 2016; Ahlborg and Hammar 2014).

The number of people without access to electricity is striking (see Table 3.3). Based on 2016 data from the International Energy Agency, 1.06 billion people do have access to electricity in the world and over half of those defined as "energy poor" live in sub-Saharan Africa (Mohammed et al. 2013). North Africa, the Middle East and some Central and South American countries have experienced a remarkable increase in electrification levels by successfully supplying power to most of their residents. Haiti is the only country in the Americas with a 33% electrification rate, while other countries in the region have surpassed a 75% electrification level (Belt et al. 2018). Asia has experienced an increase in its electrification rate of nearly 22% for the 2000–2016 period, however. Yet 439 million people remain without electricity access.

Countries like Bangladesh, Nepal, Myanmar, and Laos have showed the highest increase in electrification for the 2000–2016 period, now supplying electricity to over 75% of their population from nearly 20% of their population in 2000. A similar scenario is seen in sub-Saharan Africa, which has experienced an increase of 20% for the same period (Mohammed et al. 2013; Brew-Hammond 2010). Less than 10% of the

Table 3.3 Electricity access

	Rate of access (%)						Population without access (million)
	National				Urban	Rural	
	2000	2005	2010	2016	2016	2016	2016
World	73	76	82	86	96	73	1060
Africa	34	39	43	52	77	32	588
• North Africa	90	96	99	100	100	99	<1
• Sub-Saharan Africa	23	27	32	43	71	23	588
Developing Asia	67	74	83	89	97	81	439
• China	99	99	99	100	100	100	—
• India	43	58	66	82	97	74	239
• Indonesia	53	56	67	91	99	82	23
• Other Southeast Asia	67	76	83	89	97	82	42
• Other Developing Asia	32	39	53	73	87	65	135
Central and South America	87	91	94	97	98	86	17
Middle East	91	80	91	93	98	79	17

Source: International Energy Agency (2017).

population in Chad, Central African Republic, South Sudan, and Sierra Leone have access to electricity, which is critically low and very alarming. The distribution of energy is also uneven in some regions—rural areas have a much lower electrification rate where the largest share of the population lives (Javadi et al. 2013; Khandker et al. 2012).

The main challenge to electricity grid expansion is that areas without electricity are often located far from the existing grids. Moreover, electricity load densities of those areas are too low. These factors cause the expansion of existing grids to be economically unattractive (Abdul-Salam and Phimister 2016; Akpan 2015; Nerini et al. 2016; and Javadi et al. 2013). Figure 3.1, reproduced from Nerini et al. (2016), indicates the importance of load density (in terms of number of households) and distance from the existing grids, on the cost of electricity supply through grid extension.

The figure also illustrates that the cost of electricity supplied through grid extension varies based on (i) distance from the existing grid, (ii) price of grid electricity, and (iii) load density. If the national grid electricity price is 0.15 USD/kWh, the electricity supply costs (or levelized cost of electricity supply or LCOE) increases up to 60% for each 10 km increase in distance to the grid.

An expansion of electricity grids would not be economically feasible without adequate demand or if the consumers cannot afford for it. These are typical characteristics of many areas not having electricity access around the world. The high costs of electricity extension and the low return coming from the extension in rural and remote areas helps to explain why electric utilities have little incentive have to pursue grid

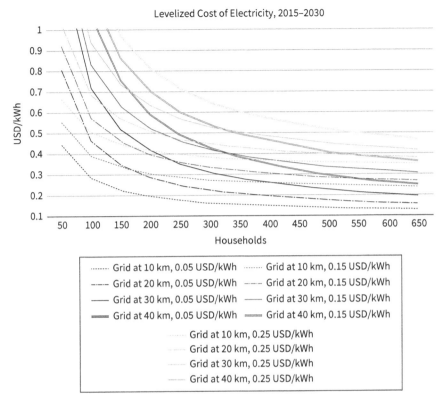

Figure 3.1 Cost of electricity through grid extension as a function of distance from existing grid and load density

Source: Nerini et al. (2016).

expansion or heavy investments on grid transmission and distribution assets (Dornan and Shah 2016; Williams et al. 2015). Therefore, governments should come forward with incentive packages to the utilities for grid expansion. Cross-subsidization schemes, where industrial consumers can pay more to generate the necessary funds, are the most common policy to promote grid extension (Picciariello 2015).

The situation is not different in the case of providing access to clean energy for cooking and heating. In many countries, modern fuels used for cooking (e.g., LPG, kerosene) are heavily subsidized. While conventional fuel subsidies are being phased out in many countries, it is unlikely that subsidies on clean cooking fuels, especially LPG, will be phased out any time soon Similarly, clean cooking technologies, such as improved biomass-fired cooking stoves, have been subsidized by almost every developing country around the world (Freeman and Zerriffi 2015). Landfill gas or biogas are also subsidized in virtually all countries where they are deployed (Siddiqui 2013).

Recent developments in smart grid technology have also provided technological options for expanding grid access. Such innovations have the potential to revolutionize the transmission, distribution, and conservation of energy in developed as well as developing countries. A smart grid is defined as "the electricity delivery system, from point of generation to point of consumption, integrated with communications and information technology for enhanced grid operations, customer services, and environmental benefits" (DOE 2008). A smart grid system in a region can provide useful insights in bottom-up estimation of electricity demand by (i) developing databases of electricity end-use equipment and services in different consumer segments; (ii) compiling market information and data on consumer demand; and (iii) encouraging the role of distributed generation options and including them in national plans.

Since smart grid actors are ushering in a more distributed and data-rich power system, the public sector in developing countries needs to step up research and development (R&D) efforts (in both the public and private sectors) tailored to their specific contexts and transfer mechanisms of the prerequisite knowledge and technical expertise to implement smart grids. This can only be possible through a nexus of public sector financing and international cooperation in technology transfer.

3. Renewable Energy Technologies

The need for widespread implementation of renewable energy technologies has become a global challenge as we transition to low carbon and green economy as portrayed in SDG-7 (Irsyad et al. 2017). Approximately 176 countries have stated their renewable energy targets, driven by a number of different motivations (REN21 2017). Renewable energy technologies are regarded as one of the pathways to decarbonize the energy sector and to reduce fossil fuel dependency (Dannenberg et al. 2008; Taylor et al. 2014; Ozcan 2017). While grid expansion may help bring affordable power to those who previously lacked such access due to poor connectivity, much remains to be done to ensure adequate generational capacity is also added to the grid to support the rapidly increasing consumer base (Pandey 2017). There are several sustainable and clean energy technologies available for expanding the generation base. Distributed generation (DG), which means the de-centralization of energy systems by placing them near to the point of consumption, can help in providing access to far-flung places where it is technically not feasible to provide grid-access DG systems, include technologies which tap into clean resources such as solar, wind, and biomass as well as energy storage technologies. The implementation of the technologies can be conducted in various configurations such as grid-tied systems and off-grid systems (mini or micro-grid), depending on the region and other specific factors.

Grid-tied systems, in the case of renewable energy technologies, can be either behind-the-meter (located at the customer site) or front-of-the meter (located at the producer site). For tackling the energy access issues with the help of renewable energy technologies, there are various technology options under the renewable resources mentioned earlier (solar, wind, and biomass). Large-scale hydro-powered energy has not been considered here owing to an ongoing debate over whether it is proper to include large-scale hydro to be a renewable energy source because of its negative impacts on water flow, fisheries, and displacement of communities. Also, biomass energy, although it is renewable, has been considered to be a traditional energy source and is discussed in the next section.

3.1 Solar Energy Technologies

Sunlight can be harnessed using sophisticated electronic chips to generate electricity in locations all across the world. It is especially profitable in places where there is a high level of solar radiation. According to the World Energy Council (2013), "the total energy from solar radiation falling on the earth was more than 7,500 times the world's total annual primary energy consumption of 450 EJ" (Urban and Mitchell 2011). This presents an opportunity to utilize technologies which can tap into solar energy to provide clean and sustainable energy access to people all over the world, and especially in developing countries.

The most prominent technology for harnessing solar energy is photovoltaics, commonly referred to as PV. It refers to a technology which uses modules (a collection of solar cells) to convert photons into usable electrical energy. The other technology that has reached commercial deployment is concentrated solar power (CSP). It uses mirrors or lenses to focus the rays of the sun to heat water, which can be used directly for household/industrial uses or be allowed to convert to steam, which can then drive a turbine to generate electricity. Grid-connected solar photovoltaic technology can offer significant opportunities for developing countries to increase energy access, create jobs, and produce revenue opportunities for people and communities. The potential for photovoltaic technology to support rural development arises from the fact that rooftop solar installations (<10 kW) can be used for household lighting, radios, and television sets, and to refrigerate medicines at rural clinics (Boccaletti et al. 2008). Another application of photovoltaic technology in a rural setting is solar water pumping (grid-tied and off-grid). This has several advantages over traditional systems; for example, diesel or propane engines require not only expensive fuels, they also create noise and air pollution in many remote areas (Bocacletti et al. 2008).

Community (>10 kW) and utility-scale PV systems (>5 MW) have also gained traction recently in the effort to electrify communities across the world. This is especially advantageous for regions with high solar irradiation and low grid access. Technological innovations such as introduction of single-axis tracking solar panels, use of thin film solar panels, and advanced cell manufacturing techniques have increased solar cell efficiencies leading to higher energy generation per unit area of the modules. In addition, battery storage systems are increasingly used in combination with community and utility-scale PV systems to create hybrid systems. Using a battery storage system (commonly Li-ion chemistry) with a PV system

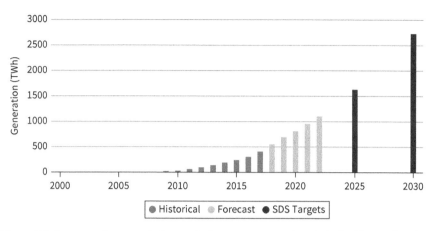

Figure 3.2 Increases in annual photovoltaic power generation globally, illustrating that solar is on track to meet the SDS target of 2700 TWh in 2030
Source: International Energy Agency (2017).

creates a semidispatchable system creating revenue opportunities for the utility/power producer.

Solar photovoltaic (PV) generation costs have been decreasing rapidly across the world owing to technological breakthroughs and production economies of scale. In several countries, the cost of PV power is already lower than coal and gas. China is one such country that has aggressively pursued PV module R&D, which has led to it become the largest producer in the world right now. According to SPV market research, PV development in China grew by 53% in 2017 to 52 GW, accounting for more than 50% of global solar capacity. This tremendous growth has been possible because of attractive incentives and policy supports provided by the government (Figure 3.2). It is also worth noting that low cost labor across Southeast Asia has significantly reduced the "soft costs," that is, the nonmaterial costs including labor and overhead (Figure 3.3). India is projected to reach utility-scale pricing of 65 US cents per watt based on reduction in soft costs as well.

Closely examining growth in solar energy, especially in the PV sector, is encouraging with renewables leading power-generation deployment globally, and solar leading renewables' deployment, with developing countries already representing more than half of global solar deployment (World Bank 2017).

3.2 Wind Energy Technologies

Wind energy (or wind power) refers to the process of creating electricity using the wind, or air flows that occur naturally in the earth's atmosphere due to uneven

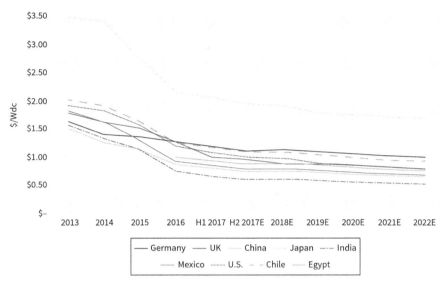

Figure 3.3 Historical and forecasted price of utility PV system pricing from 2013–2022
Source: Greentech Media Inc. (2017).

heating. The technologies to harness wind energy are called modern wind turbines, machines that can capture kinetic energy from the wind and generate electricity using a turbine-shaft connected to an electric generator. There are broadly three main types of wind energy installations: (i) on-shore utility-scale wind, (ii) off-shore wind, and (iii) small-scale or "distributed," also known as off-grid wind.

Over the year, wind turbines can generate electricity more than 90% of the time in suitable locations. On-shore as well as off-shore wind farms can provide a capacity factor of 40–45% which is comparable to fossil-fuel power plants. The generation profile depends on the location of the wind farm, the height of the tower, and the turbine technology. Wind turbines generate electricity from wind speeds ranging from around 15 km/h (4 m/s) to 90 km/h (25 m/s) (IEA 2008).

The cost of wind energy has been decreasing, which has been possible due to technological improvements that have compounded deployment in many regions in the world. The cost reductions have been driven by factors such as capital expenditure, operating expenditure, cost of financing, turbine performance, and project design life. Larger turbine size, larger rotor diameters, and higher hub heights are also key factors (Global Wind Energy Outlook 2017). These costs are projected to drop even further as the market matures and the technology improves even further (see Figure 3.5).

Another advantage of wind energy is that it is extremely independent of water resources in the region. Water shortage and stress are among the major impacts of worsening global climate change, and our profligacy with this precious resource is

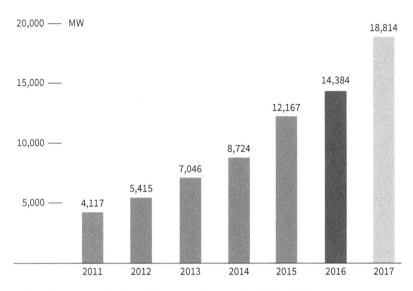

Figure 3.4 Global cumulative offshore wind capacity, 2011–2017
Source: Global Wind Energy Council (2018).

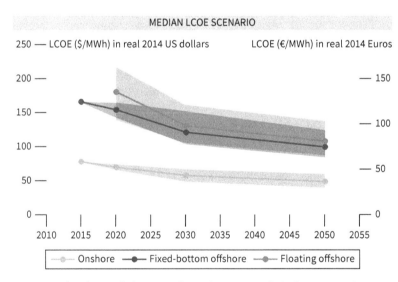

Figure 3.5 Levelized cost of electricity for various types of wind energy projects
Source: U.S. Department of Energy (2017).

a serious problem even without the threat of climate change (Wind Energy Outlook 2017). Wind energy systems are the least water-intensive of all the generation technologies (Mekonnen et al. 2015), making it an extremely valuable technology to achieve SDG-7.

3.3 Small-Scale Hydro Technologies

Hydropower has been seen, increasingly, as a two-fold solution to the provision of renewable energy and water storage, making it a pivotal technology for achieving SDG-7 targets (Mayer et al. 2017). Hydropower is broadly divided into two major categories: run of the river and reservoir-based. Usually, small-scale hydropower (SHP) systems tend to be run of the river, capturing the energy in flowing river water and converting it to usable energy. The capacity of SHP is usually defined to be less than 10 MW, although it depends on the region. Small-scale hydropower can be further subdivided into mini-hydro (<500kW) and micro-hydro (<100kW) (IEA-Hydro 2017).

Hydropower projects have efficiencies greater than 90%, making it the most efficient of energy conversion technologies in commercial deployment presently (Manzano et al. 2013). A small-scale hydroelectric facility requires a sizable flow of water and that a proper height of fall of water, called head, be obtained without building elaborate and expensive facilities (IEA Hydro 2017). SHPs can be developed at existing dams and have been constructed in connection with river and lake water-level control and irrigation schemes. By using existing structures, only minor new civil engineering works are required, reducing the cost of this component of a development.

Another advantage of hydropower technology is longevity. Hydropower plants usually have extremely long lifetimes and, depending on the particular component, are in the range of 30 to 80 years. There are many examples of hydropower plants that have been in operation for more than 100 years, with regular upgrading of electrical and mechanical systems but no major upgrades of the most expensive civil structures (dams, tunnels) (IPCC 2011). Although the potential for small hydro-electric systems depends on the availability of suitable water flow, where the resource exists it can provide cheap clean reliable electricity. A well-designed small hydropower system can blend with its surroundings and have minimal negative environmental impacts, unlike a large-scale hydropower system (>10 MW).

A 2010 report from the International Energy Agency (IEA) projected that global hydropower production might grow by nearly 75 % from 2007 to 2050 under a business-as-usual scenario and 85% under an aggressive climate action scenario. The report estimated that the majority of the remaining economic development potential is located in Africa, Asia, and Latin America (IEA 2008; IEA 2010c). The IEA notes that while small hydropower plants could provide as much as 150 GW to 200 GW of new generating capacity worldwide, only 5 % of the world's small-scale hydropower potential has been exploited (IEA 2008).

Figure 3.6 compares the cost of producing electricity from hydropower plants. The figure reveals that producing electricity from small-scale hydropower plants is expensive when compared to that from large hydropower plants, because the latter enjoys

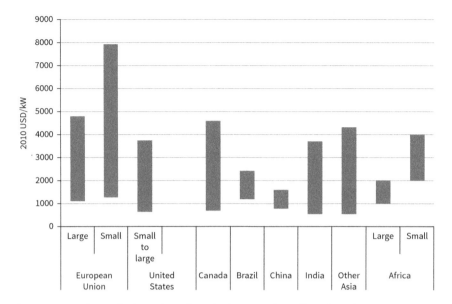

Figure 3.6 Total installed cost (USD/kW) for large- and small-scale hydro power across major markets in the world

Source: International Renewable Energy Agency (2012); International Energy Agency (2010); Black and Veach Corporation (2012).

economies of scale. Since the costs of electricity production are highly specific to the sites of hydropower plants or projects, there is a large difference in the costs across regions and countries.

3.4 Geothermal Energy Technology

Geothermal energy is derived from heat produced within hot water present beneath the surface of the earth. The heat can either be converted into electricity using turbines or utilized directly using heat pumps. This form of energy has been found in about 90 countries, and 79 of those have quantified records of geothermal utilization. Electricity is produced from geothermal sources in 24 countries, of which nine obtain 5–26% of their national electricity from geothermal. These include mostly high and moderate-income countries.

Geothermal energy can also be used directly through ground-source heat pumps (GHPs). The direct use of geothermal energy has become extremely important in many developing and transitional countries in Asia and Africa which can facilitate activities such as washing and bathing, and greenhouse cultivation and fish farming. In addition, tourism is often a substantial source of income at geothermal locations. The direct use of geothermal can even replace fossil fuels in densely populated areas where space heating and/or cooling is needed. The potential is very large, because space heating and water heating are significant parts of the energy budget in large parts of the world. Currently, the only commercially exploited geothermal systems for power generation and direct use are hydrothermal in nature (Dickson and Fanelli 2013). This means that until now only hydrothermal resources are equipped to provide fluid, heat, and permeability; the essential components required to produce electricity from geothermal energy.

3.5 Off-Grid Renewable Energy Systems

Off-grid, micro-grid, or mini-grids are the technologies promoted more recently to provide universal access to electricity under the SDG7. Off-grid renewable energy technologies offer a cost-effective, environmentally sustainable, rapidly deployable, and modular tool to accelerate the pace of electrification in developing countries. These technologies are also known as distributed generation (DG) technologies, because they are decentralized and located near the consumer of the energy.

These technologies have gained momentum for several reasons. First, they are modular (that can fit for a single household) and are suitable in areas where load density is low, where grid expansion is economically and sometime physically infeasible. Unlike grid extension, which could take many years including both expansion of generation and T&D systems, off-grid systems can be built within months, so households do not need to wait years to get electricity access. Moreover, the costs of off-grid systems (e.g., solar home systems) are rapidly declining. Although poor households still may not be able to afford these systems, the subsidy burden to the governments to provide electricity access through these technologies would be much cheaper than through the grid expansion. Thus, off-grid technologies have significantly contributed to enhance electrification rates (Tenenbaum et al. 2014; Bhattacharyya and Dow 2013). Decentralized electricity can also be supplied by expensive diesel-powered engines, but this leaves communities in need of power vulnerable to fuel price volatility and high transportation costs. In some cases, added transportation costs can exceed the original cost of fuel procurement leaving these communities in distress and in a state of energy deprivation again (Zahnd and Kimber 2009).

While fossil fuels, particularly diesel, can be used for off-grid, micro and mini grids; renewable energy, particularly solar and small hydropower suit the best of this approach.[1] The selection of off-grid, micro-grid and mini-grid depends on several

---

[1] Rising concerns on climate change persuading policymakers and international development communities that renewable energy should be pursued to provide electricity access so that infrastructure "lock-in" can be avoided. If power plants based on fossil fuels are built now to increase electricity access, these would turn to be stranded assets when they are replaced with renewable energy-based power plants in few years causing a loss to the society. In other word, these infrastructures are likely to stay until they retire to avoid such costs. This will cause an infrastructure lock-in situation. Moreover, in small island developing states, even though they contribute negligibly to global carbon dioxide emissions, shifting to renewables can be very desirable as this reduces their dependence on fossil fuels (mainly oil), increases resilience in light of energy security, and acts as a marketing strategy for their tourism sector, a very important pillar of their economy (Timilsina and Shah 2016; Roper 2005).

factors including characteristics of demand and supply sources (Tenenbaum et al. 2014; Bhattacharyya and Dow 2013). If households in a location to be electrified are scattered and their loads are small (lighting, battery charging, electronic devices), a solar home system would be appropriate. On the other hand, if household distribution is relatively dense, a micro- or mini-grid would be appropriate. Off-grid is very specific to a technology, solar PV; it is not appropriate for other sources of electricity, such as hydro and wind, because it is highly unlikely to develop hydropower or wind power for supplying a single home (Akikor et al. 2013). As mentioned, several studies have investigated economics of off-grid, micro-grid, and mini-grids. However, the results cannot be generalized because of differences in demand and supply characteristics across the locations to be electrified. Nevertheless, Figure 3.7, reproduced from Nerini et al. (2016), shows the cost effectiveness of an approach to provide electricity access depends on: (i) resource availability, which is measured in terms of capacity availability factor (CF); (ii) capital cost of technology; (iii) price of fuel (e.g., diesel), which would have been used if renewable technologies were not available or not appropriate; and (iv) load density (e.g., households per square kilometers).

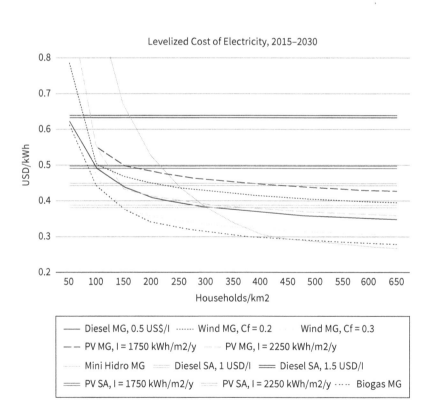

Figure 3.7 Cost comparison of off-grid and mini-grid systems with various types of electricity generation technologies
Source: Nerini et al. (2016).

4. Modernization of Traditional Energy Sources

Traditional energy sources still account for a majority of the energy needs globally. In particular, traditional biomass (fuel wood, agriculture residues, and animal wastes) are the main source of energy supply in many countries around the developing world. About 15% (or 55 exajoules) of the world's energy demand is covered by biomass resources, making biomass by far the most important renewable energy source used to date (Faaij 2005). This share is increasing even more owing to the growing population in developing countries.

Solid biomass, such as wood and garbage, can be burned directly to produce heat, which can then be used for cooking or heating purposes. Biomass can also be converted into a gas called biogas or into liquid biofuels such as ethanol and biodiesel. These fuels can then be burned to produce energy. Biomass, particularly when collected from natural forests, is one of the main sources of energy in many countries especially in rural sub-Saharan Africa and South Asia (see Figure 3.8). In countries such as the Democratic Republic of the Congo, Ethiopia, Tanzania, Nepal, and Nigeria, more than 80% of the total energy demand is met through use of biomass energy resources. Biomass is mostly used for cooking and to some extent in industry,

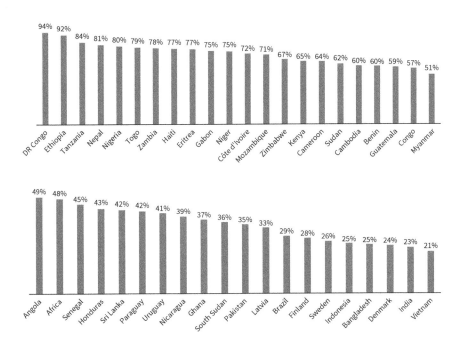

Figure 3.8 Local biomass to energy use in selected countries
Source: Compiled by Authors

such as brick manufacturing in developing countries, although recent technological advancements have opened up more avenues for its use.

One approach to meet SDG7 that has not been discussed much is converting local and traditional energy resources to modern energy services. Use of biomass for electricity generation, physical conversion of biomass to energy rich products such as briquettes, or pellets; chemical conversion of biomass to liquid biofuels (ethanol, biodiesel); and extraction of landfill gas are examples of techniques to generate modern energy commodities from traditional biomass. Modern biofuels including ethanol, biodiesel, and biogas derived from renewable biomass are being increasingly recognized as holding potential for offering a sustainable energy future and at the same time could have the potential for creating opportunities for achieving socioeconomic goals in the sustainable development paradigm (Sheelanere and Kulshreshtha 2013). Biomass resources that offer high yields, low cost, and low adverse environmental impacts are more suitable for use in modern energy systems (Larson 2000).

One argument against the use of traditional biomass to produce modern energy such as electricity and heat is that even though biomass energy is considered to be a renewable source, it is hard to claim it is clean or efficient power owing to the pollution caused by its combustion. It is also true that biomass has lower energy density as compared with fossil fuel resources (i.e., coal, oil, and natural gas). The cost of production and transportation might be higher if collected from natural forests. However, use of fuelwood and forest residues from sustainably grown forests to produce electricity and distribute it through a mini-grid to meet electricity demand in a village or suburban areas could be an attractive option compared to solar or wind power based mini-grid systems. Moreover, a mini-grid fed by a hybrid system where wind and solar are used at the time they are available and fuelwood or wood waste during the rest of the time could help provide a reliable solution to electricity access gaps in many areas.

According to various case studies conducted in Africa and Asia (Karlsson and Banda 2009), it seems that village-level projects on bioenergy have great potential in terms of sustainable fuel production and increased access to energy in rural areas of developing countries—if participatory processes are employed in the development and implementation of the projects. On a small scale, locally produced plant oils and biodiesel can successfully be used to power diesel engines and generators in rural villages for agricultural processing, new enterprises, and income generation (Karlsson and Banda 2009). Equity concerns under the sustainable development umbrella also need to be understood and addressed in the context of policies for rural energy provision. Gender impacts need to be analyzed carefully, since women in rural areas are responsible for collecting fuel and water for household needs, growing food, and gathering fodder, medicinal plants, and wild food.

Conversion of traditional biomass to produce modern energy for heating and cooking has been practiced for decades. Improved cooking stoves use much less fuel wood and do not emit harmful smoke, and have therefore been promoted in almost every developing country since the early 1980s. Agriculture residues have very low energy density; however, technology exists to improve it. One such technology, which involves biomass densification (i.e., conversion of sawdust into compressed blocks of dust, known as briquetting or the formation of briquettes) is extremely common in Southeast Asian countries like Bangladesh and India. This process improves the combustion performance and efficiency of the fuels, thereby aligning them to sustainable use of traditional sources of energy. Similar techniques could also be used to produce briquettes from corn, wheat, and cotton stoves. Production of biogas from animal wastes is also popular in many countries to supply modern energy from traditional biomass. Ethanol or vegetable oils produced from either first- or second-generation biofuel resources are another good option for providing modern energy for transportation and cooking. Traditional biomass has also been utilized for space and water heating, while kerosene has been predominantly used for lighting purposes historically. This causes indoor air pollution leading to health hazards for future generations. Government policies to improve heating energy access are required in such places by providing access to fuel sources such as natural gas and biogas for space and water heating, improving electric heating access in rural areas, and initiating technologies such as heat-only boilers (HOBs) in urban areas.

Scaling up the production of modern energy from traditional biomass is important. Yet it is not getting much attention in developing countries. The primary reason is again economics. In principle, the economics of bioenergy is not much worse than that of other sources to supply heating and cooking in rural areas. After all, energy access requires public subsidies no matter the type of energy carrier or option used. The question is whether the conversion of biomass to modern energy can be promoted successfully. In fact, this looks like the only viable option for now to supply modern energy for cooking and heating from nonfossil sources.

5. Technologies for Energy Efficiency

Energy efficient technologies are increasingly being recognized as a cost-effective and sustainable avenue in global energy discourse. The International Energy Agency projects that energy efficient technologies would contribute the most (34%) to reducing GHG emissions needed to meet the 2 egree target followed by carbon capture and sequestration (CCS) 32%, fuel switching 18%, and renewable energy 15% (IEA 2017c). McKinsey (2014) reports that energy efficiency accounts for 40% of GHG mitigation potential that can be realized at a cost of less than €60/tCO2. This indicates potential opportunities for public and private stakeholders in the national and international arena to collaborate and provide adequate financing to realize energy efficiency savings in electricity generation, buildings, transportation, and manufacturing, among other sectors.

It is important to understand the concept of energy intensity before measuring energy efficiency for an entity. Energy intensity is a measure of the amount of energy used to produce a unit of output. This indicator is usually calculated as primary energy demand per unit of GDP of a region/entity. Thus, reducing energy intensity can be seen as one of the indicators of successful energy efficiency technologies. A recent report suggests that energy efficiency gains will reduce the impact of growing levels of activity on energy demand in emerging economies, where energy intensity can be halved (IEA 2018).

As mentioned previously, energy efficiency technologies are being implemented in diverse sectors including residential and commercial buildings and appliances, transportation, and the industrial (manufacturing) sector. This creates a significant opportunity for investment into R&D to improve technology and scale it to deployment levels. A review of the technologies and their growth drivers in each sector is presented in Table 3.4.

Global investment in energy efficiency grew by 3% to $236 billion in 2017 (EIA 2018). Also, bonds issued in 2017 for energy efficiency measures rose from 18% to 29% of the total green bond market of $161 billion, which surprisingly equaled the issuance of bonds dedicated to renewable energy for the first time.

Supportive policies and regulations play an important part in adoption of energy efficient technologies and measures. The percentage of global energy use covered by mandatory policies and regulations increased from 32% in 2016 to 34% in 2017 (IEA 2018). As was the case in 2016, 99% of this increase was driven by replacements of

Table 3.4 A review of recent energy efficiency technologies/measures and their growth drivers

Sector	Technologies/Measures	Key Growth Drivers
Buildings and Appliances	Advanced cooling technology, efficient lighting (LEDs), demand response (DR), global building standards (LEED), global appliance standards (Energy Star)	Improvement in HVAC technology, predictable ROI for investors, advancement in building information management (BIM), increasing DR policies
Transportation	Fuel efficiency standards, enhancing internal combustion engines, improving tires and reducing vehicle weight	Complex and expensive technologies, price of crude oil, policies and regulations
Manufacturing	High efficiency industrial motors, energy management standards (ISO50001), material recycling (metals, plastics etc.), demand response (DR)	Stringent policies and standards for machines, improvement in motor technology, improved energy metering and monitoring

Source: authors.

vehicles, appliances, and equipment with new stock subject to existing energy efficiency policies; the remaining 1% was down to an extension of standards to new categories of energy-using equipment.

SDG7.3 seeks to double the global rate of improvement in energy efficiency by 2030, and is a crucial milestone bringing it into the public focus. Energy efficiency can be realized by targeted government policies and incentives, a variety of which have been implemented all over the world. EU member states lead the way through EU's Energy Efficiency Directive that established a binding target of 32.5% energy savings by 2030 and is required of all member states (EU 2018). These kinds of binding targets have been successful in pushing countries to adopt energy efficiency in their national plans and assess the yearly action initiated under their respective plans to visualize the trajectory being followed.

6. Challenges and Barriers to Technologies for SDG7

Technologies needed to meet SDG7 faces several barriers and challenges. These barriers, as highlighted in Table 3.5, include those that are technical, market-related, economic and financial, institutional, and political.

Although renewable energy based off-grid, micro-grid, and mini-grid options are better from a climate change mitigation perspective and also help to avoid infrastructure lock-in, they are still expensive and require government supports. They also face the same affordability challenges as grid extension. In fact, increasing the income level of the poor by providing economic or employment opportunities would be the best solution to address access to basic needs including modern energy.

Off-grid, micro-grid, and mini-grids also face several challenges, which explains their low penetration in the global energy market. The literature is very rich in terms of what are the challenges to the development of renewable energy focused electricity access.

Deployment of energy efficient technologies also face significant barriers that deter investments in energy efficiency. These include lack of proper knowledge of technologies available, access to finance due to high risks and uncertainty of the delivery of the technologies, and high discount rates or low value offered to households and industrial energy consumers for energy efficiency measures compared to other pressing priorities despite the benefits of energy efficiencies (Timilsina et al. 2013). The major research focus currently has been to improve the financing aspect of energy efficiency measures that are institutional in nature especially in emerging market countries such as China, India, and Brazil (Taylor et al. 2008). While multilateral international agencies have dedicated funds to address energy efficiency issues in developing and transition economies, it is the individual nation-states that must take the mantle by

Table 3.5 Barriers for renewable energy development

Type of Barrier	Specific Barriers	References
Market and economic barriers	1) Lack of financing mechanisms 2) High initial capital costs 3) Immature market for complementary technologies	Abdmouleh et al. (2015), Ahlborg & Hammar (2014), Dulal et al. (2013), Bhattacharyya & Palit (2016), Luthra et al. (2015), Richards et al. (2012), Sen & Ganguly (2017), Sovacool (2010), Sovacool (2011), Yaqoot et al. (2016)
Capacity barriers	1) Lack of technical knowledge and development skill 2) Lack of financial competencies	Abdmouleh et al. (2015), Dulal et al. (2013), Engelken et al. (2016), Dornan (2011), Luthra et al. (2015), Richards et al. (2012), Sovacool (2010), Sovacool (2011), Urmee et al. (2009), Yaqoot et al. (2016)
Policy, political, and regulatory barriers	1) Lack of subsidies 2) Lack of financial incentives like feed-in tariffs and net metering 3) Lack of government policies.	Abdmouleh et al. (2015), Dulal et al. (2013), Bhattacharyya & Palit (2016), Engelken et al. (2016), Luthra et al. (2015), Sovacool (2011), Richards et al. (2012), Yaqoot et al. (2016)

formulating targeted policies and market-based incentives to encourage energy efficiency in their respective economies.

7. R&D for Promoting Technologies for SDG7

R&D are ongoing on several fronts including technologies related to SDG-7. Modernizing energy is explicit in SDG7 and the science and technologies that advance new resources are important factors to consider. Renewable energy and energy efficiency technologies have seen tremendous growth in the last decade. This has been accelerated by rapid technological advancements in the fields of chemistry, electronics, engineering, and manufacturing, among others. Renewable energy and energy efficiency have been pushed further by increasing investments in R&D policies and economic incentives that have had the effect of reducing installed costs considerably.

In the specific case of solar energy, the rise in public and private sector investments benefitted from the substantial decline in the cost of the technology. While module prices have fallen to less than $1/watt in some markets, the target of new research is to reduce balance of system (BoS) costs, which include inverters, wirings, and all other components of a photovoltaic system other than the photovoltaic panels. Also, based on recent market trends, we see that module-level power electronics like microinverters are being preferred to large, centralized inverters. According to GTM Research, in 2016, utility-scale PV plants were outbidding coal and natural gas projects across the United States. This trend is occurring around the world with utility-scale solar applications and can be attributed to two factors: (i) falling installed costs of solar modules, and (ii) recognition of the emission free nature of installed solar with zero fuel costs. A significant addition to an already cost-effective PV system has been the addition of battery energy storage systems being paired with utility-scale PV to provide a semidispatchable energy source.

Energy storage technologies have been touted to be an extremely efficient and value-adding technology for achieving sustainability in the energy system. They have the ability to reduce the issues associated with high variable renewable energy penetration in the electric grid. Energy storage systems include pumped-hydro, compressed air energy storage (CAES), battery storage systems, and several more. The battery storage technology market has expanded considerably by providing multiple values to the electric grid, making it the technology that can combine most effectively with renewable sources such as solar and wind energy. Apart from being a backup power source, battery storage systems can also regulate frequency and voltage, and provide alternatives to significant investments in transmission and distribution systems (Denholm 2013). Thus, investments in R&D are focused on driving up efficiency, lifetime, and power density of batteries.

A majority of batteries being used for grid-scale application are of lithium-ion (Li-ion) chemistry or lead-acid chemistry. Recent trends show that Li-ion batteries have consolidated the market with 90% of the global installed market share in 2018 (GTM Research, 2018). Due to the exponential drop in installed costs ($/MWh) of the batteries, countries are building storage capacity all across the world (see Figure 3.9).

A significant increase in the number of deployed electric vehicles has also led to the market capture by Li-ion batteries leading to economies of scale in manufacturing. Battery costs for Li-ion chemistry are projected to fall even further from the present average cost of $273/kWh to $80/kWh in 2030 (BNEF 2017), with Germany leading the way in terms of upcoming installed capacity.

In the area of biomass, improved technologies to convert biomass feedstock into clean energy fuels via thermo-chemical and biochemical processes can potentially increase utilization of biomass-derived fuels, mitigate energy trade balance issues, and foster socioeconomic development (Ruiz et al. 2013). Biofuel industry interest in micro-algae primarily for transport fuels continues with clear recognition that achieving a positive energy balance requires technological advances and highly optimized production systems (Figure 3.9). To this end, commercialization aspects being addressed include energy required for pumping, the embodied energy required for

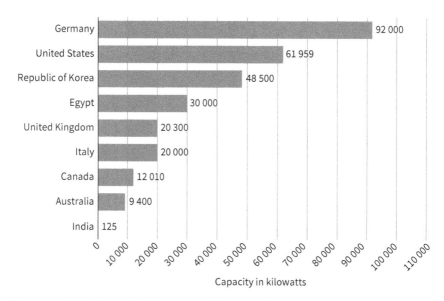

Figure 3.9 Global capacity of upcoming lithium-ion battery energy storage systems as of mid-2017, by country (in kilowatts)
Source: Energy Storage World Forum (2018).

construction, the embodied energy in fertilizer, and the energy required for drying and de-watering (Liew et al. 2014).

8. Conclusion

Universal access to modern energy is one of the main agendas (SDG7) of the SDGs established by the United Nations in 2015. Billions of people around the world, mainly from sub-Saharan Africa and South Asia, do not have access to electricity or modern cooking fuels or both. The problem is much more amplified in rural and remote areas than urban areas—around 90% of rural population in sub-Saharan Africa do not have access to electricity and 93% do not have access to modern energy for cooking.

To achieve SDG7, development and deployment of technologies that generate clean and affordable energy (i.e., supply-side energy technologies) and using energy more efficiently (i.e., demand side energy technologies) are crucial. The main supply side technological options to achieve SDG7 include:

1. expansion of conventional energy supply systems such as the electric grid, natural gas pipelines, and LPG supply infrastructure;
2. deployment of renewable and emerging technologies, such as residential solar systems or micro- and mini-hydropower; and

3. modernization of traditional energy sources, such as biomass for power generation, forest and agriculture residues for liquid fuels (ethanol), and animal wastes for heat and electricity (biogas).

Demand side energy technologies include technologies that utilize energy in a more efficient manner in all end-uses such as lighting, cooking, heating in buildings, electric motors, boilers and heaters in industrial facilities, and vehicles in the transportation sector.

On the supply side energy technology front, the development of renewable energy technologies and their economics have improved significantly bringing these technologies in the forefront of meeting the SDG7 goal. The most noticeable achievement is the reduction of the cost of solar photovoltaic technologies, which in turn are the most suitable technologies to provide electricity access in the most rural and remote areas on the earth, thereby helping meet the goal of universal access to electricity under SDG7. In addition to the improved economics of the technologies themselves, government policies have played an instrumental role toward scaling up their deployment. Almost every country in the world has set targets to increase its renewable energy deployment. One of the key options for providing access to modern energy is to convert traditional biomass, which is abundant and accessible in many rural and remote areas, to modern energy, particularly for heating and cooking. However, technological breakthroughs and necessary cost reductions of available technologies have yet to occur.

On the demand side, technological breakthroughs for energy efficient are noticeable. LED lighting is the best example, although affordability of energy efficient technologies is still an issue in areas inhabited by low income people, where the highest energy deficit in terms of modern energy access is likely to occur.

Despite the promising economic rollover and policy supports, technologies for SDG7 still face numerous challenges and barriers. Among them are the lack of supply infrastructure, poor quality of supply, connection costs, and affordability for the conventional technologies. High up-front costs, lack of information about the technologies and their applications, and lack of skilled local manpower for maintenance of the technologies and systems are the main barriers to renewable energy supply and energy efficiency technologies.

Further enhancement of R&D for lower cost technologies, policy support for their increased deployment, and political will and support for renewable energy and energy efficiency technologies in general are needed to address the remaining barriers and challenges to technologies for meeting SDG7.

Acknowledgment

The authors also thank Sashwat Roy, Doctoral Research Assistant at the Energy and Environmental Policy Program, University of Delaware.

References

Abdmouleh, Z., Alammari, R. A. M., and Gastli, A. (2015). Review of policies encouraging renewable energy integration and best practices. *Renewable and Sustainable Energy Reviews* 45, 249–262.

Abdul-Salam, Y., and Phimister, E. (2016). The politico-economics of electricity planning in developing countries: a case study of Ghana. *Energy Policy* 88, 299–309.

Ahlborg, H., and Hammar, L. (2014). Drivers and barriers to rural electrification in Tanzania and Mozambique: grid-extension, off-grid, and renewable energy technologies. *Renewable Energy* 61, 117–124.

Akikur, R. K., Saidur, R., Ping, H. W., and Ullah, K. R. (2013). Comparative study of stand-alone and hybrid solar energy systems suitable for off-grid rural electrification: a review. *Renewable and Sustainable Energy Reviews* 27, 738–752.

Akpan, U. (2015). Technology options for increasing electricity access in areas with low electricity access rate in Nigeria. *Socio-Economic Planning Sciences* 51, 1–12.

al Irsyad, M., Halog, A., Nepal, R., and Koesrindartoto, D. (2017). Selecting tools for renewable energy analysis in developing countries: an expanded review. *Frontiers in Energy Research* 5, 1–13.

Banerjee, S.G., Barnes, D., Singh, B., Mayer, K., and Samad, H. (2014). *Power for All: Electricity Access Challenge in India* (No. 92223). World Bank, Washington, DC.

Battaglini, A., Komendantova, N., Brtnik, P., and Patt, A. (2012). Perception of barriers for expansion of electricity grids in the European Union. *Energy Policy* 47, 254–259.

Belt, J., Allien, N., Mackinnon, J., and Kashi, B. (2018). Cost benefit analysis of power sector reform in Haiti. *The ICER Chronicle* 8, 23–38.

Bhattacharyya, S. (2013). *Rural Electrification Through Decentralised Off-Grid Systems in Developing Countries.* Springer-Verlag, London.

Bhattacharyya, S. C., and Palit, D. (2016). Mini-grid based off-grid electrification to enhance electricity access in developing countries: what policies may be required? *Energy Policy* 94, 166–178.

Black and Veach Corporation (2012). Cost and performance data for power generation technologies. Prepared for the National Renewable Energy Laboratory. https://refman.energy-transitionmodel.com/publications/1921/download (accessed May 3, 2020).

Boccaletti, C., Fabbri, G., Marco, J., and Santini, E. (2008). An overview on renewable energy technologies for developing countries: the case of Guinea Bissau. *Renewable Energy and Power Quality Journal* 1(6), 343–348.

Brew-Hammond, A. (2010). Energy access in Africa: challenges ahead. *Energy Policy* 38, 2291–2301.

Byrnes, L., Brown, C., Foster, J., and Wagner, L. D. (2013). Australian renewable energy policy: barriers and challenges. *Renewable Energy* 60, 711–721.

Cecelski, E., Dunkerley, J., Ramsay, W., Dunkerley, J., and Ramsay, W. (2015). *Household Energy and the Poor in the Third World.* Routledge, London.

Dannenberg, A., Mennel, T., and Moslener, U. (2007). What does Europe pay for clean energy? review of macroeconomic simulation studies. *Energy Policy* 36, 1318–1330.

Dickson, M. H., Fanelli, M., and Fanelli, M. (2013). *Geothermal Energy: Utilization and Technology.* Routledge, London.

Dornan, M. (2011). Solar-based rural electrification policy design: The Renewable Energy Service Company (RESCO) model in Fiji. *Renewable Energy* 36, 797–803.

Dornan, M. (2014). Access to electricity in small island developing states of the Pacific: issues and challenges. *Renewable and Sustainable Energy Reviews* 31, 726–735.

Dulal, H. B., Shah, K. U., Sapkota, C., Uma, G., and Kandel, B. R. (2013). Renewable energy diffusion in Asia: can it happen without government support? *Energy Policy* 59, 301–311.

Ellabban, O., Abu-Rub, H., and Blaabjerg, F. (2014). Renewable energy resources: current status, future prospects and their enabling technology. *Renewable and Sustainable Energy Reviews* 39, 748–764.

Energy Storage World Forum (2018). Energy Storage World Markets Report 2018, 296 p.

Engelken, M., Römer, B., Drescher, M., Welpe, I. M., and Picot, A. (2016). Comparing drivers, barriers, and opportunities of business models for renewable energies: a review. *Renewable and Sustainable Energy Reviews* 60, 795–809.

Frankfurt School-United Nations Environment Programme (2015). Global Trends in Renewable Energy Investment 2015. Frankfurt School-UNEP Collaborating Center, Frankfurt, Germany.

Freeman, O. E., and Zerriffi, H. (2015). Complexities and challenges in the emerging cookstove carbon market in India. *Energy for Sustainable Development* 24, 33–43.

Global Wind Energy Council (2018). *Global Wind Energy Outlook 2018*. Global Wind Energy Council, Brussels.

Greentech Media Inc. (2017). Market Research Report – PV Systems Pricing H1 2017: Breakdowns and Forecasts. Global Information Inc. https://www.giiresearch.com/report/gm515382-pv-system-pricing-h1-breakdowns-forecasts.html (accessed May 3, 2020).

He, Y., Xu, Y., Pang, Y., Tian, H., and Wu, R. (2016). A regulatory policy to promote renewable energy consumption in China: review and future evolutionary path. *Renewable Energy* 89, 695–705.

International Energy Agency (2017). World Energy Outlook 2017.

International Renewable Energy Agency (2012). Renewable energy technologies series: cost analysis series. Hydropower. Volume 1 Power Sector, Issue 3/5.

Islam, M. R., Mekhilef, S., and Saidur, R. (2013). Progress and recent trends of wind energy technology. *Renewable and Sustainable Energy Reviews* 21, 456–468.

Javadi, F. S., Rismanchi, B., Sarraf, M., Afshar, O., Saidur, R., Ping, H. W., and Rahim, N. A. (2013). Global policy of rural electrification. *Renewable and Sustainable Energy Reviews* 19, 402–416.

Karlsson, G., and Banda, K. (2009) *Biofuels for Sustainable Rural Development and Empowerment of Women: Case Studies from Africa and Asia*. Energia, Leusden, The Netherlands.

Khandker, S. R., Samad, H. A., Ali, R., and Barnes, D. F. (2012). Who Benefits Most from Rural Electrification? Evidence in India (No. WPS6095). World Bank, Washington, DC.

Luthra, S., Kumar, S., Garg, D., and Haleem, A. (2015). Barriers to renewable/sustainable energy technologies adoption: Indian perspective. *Renewable and Sustainable Energy Reviews* 41, 762–776.

Magagna, D., Uihlein, A. (2015). Ocean energy development in Europe: current status and future perspectives. *International Journal of Marine Energy* 11, 84–104.

Manzano-Agugliaro, F., Alcayde, A., Montoya, F.G., Zapata-Sierra, A., and Gil, C. (2013). Scientific production of renewable energies worldwide: an overview. *Renewable and Sustainable Energy Reviews* 18, 134–143.

Marques, A. C., and Fuinhas, J. A. (2012). Are public policies towards renewables successful? evidence from European countries. *Renewable Energy* 44, 109–118.

Mekonnen, M. M., Gerbens-Leenes, P. W., and Hoekstra, A. Y. (2015). The consumptive water footprint of electricity and heat: a global assessment. Environmental Science. *Water Research and Technology* 1, 285–297.

Mohammed, Y. S., Mustafa, M. W., and Bashir, N. (2013). Status of renewable energy consumption and developmental challenges in sub-Sahara Africa. *Renewable and Sustainable Energy Reviews* 27, 453–463.

Negro, S. O., Alkemade, F., Hekkert, M. P., 2012. Why does renewable energy diffuse so slowly? a review of innovation system problems. *Renewable and Sustainable Energy Reviews* 16, 3836–3846.

Nerini, F. F., Broad, O., Mentis, D., Welsch, M., Bazilian, M., and Howells, M. (2016). A cost comparison of technology approaches for improving access to electricity services. *Energy* 95, 255–265.

New Energy Outlook 2018. (2017). Bloomberg New Energy Finance, New York City. https://about.bnef.com/new-energy-outlook (accessed April 2, 2019).

Ozcan, M. (2018). The role of renewables in increasing Turkey's self-sufficiency in electrical energy. *Renewable and Sustainable Energy Reviews* 82, 2629–2639.

Pandey, R. (2017). Reinforcing energy grid in developing countries and role of energy storage. *Journal of Undergraduate Research at the University of Illinois at Chicago* 10(1), 1–6.

Picciariello, A., Vergara, C., Reneses, J., Frías, P., and Söder, L. (2015). Electricity distribution tariffs and distributed generation: quantifying cross-subsidies from consumers to prosumers. *Utilities Policy* 37, 23–33.

Radulovic, V. (2006). Are new institutional economics enough? promoting photovoltaics in India's agricultural sector. *Energy Policy* 33(14), 1883–1899.

REN21 (2017). *Renewables 2017: Global Status Report*. REN21, Paris.

Richards, G., Noble, B., and Belcher, K. (2012). Barriers to renewable energy development: a case study of large-scale wind energy in Saskatchewan, Canada. *Energy Policy* 42, 691–698.

Roper, T. (2005). Small island states: setting an example on green energy use. *Review of European Community and International Environmental Law* 14, 108–116.

Ruiz, J. A., Juárez, M. C., Morales, M. P., Muñoz, P., and Mendívil, M. A. (2013). Biomass gasification for electricity generation: review of current technology barriers. *Renewable and Sustainable Energy Reviews* 18, 174–183.

Sen, S., and Ganguly, S. (2017). Opportunities, barriers and issues with renewable energy development: a discussion. *Renewable and Sustainable Energy Reviews* 69, 1170–1181.

Shah, K. U., and Niles, K. (2016). Energy policy in the Caribbean green economy context and the Institutional Analysis and Design (IAD) framework as a proposed tool for its development. *Energy Policy* 98, 768–777.

Siddiqui, F. Z., Zaidi, S., Pandey, S., and Khan, M. E. (2013). Review of past research and proposed action plan for landfill gas-to-energy applications in India. *Waste Management and Research* 31, 3–22.

Sovacool, B. K. (2010). A comparative analysis of renewable electricity support mechanisms for Southeast Asia. *Energy* 35, 1779–1793.

Sovacool, B. K. (2011). An international comparison of four polycentric approaches to climate and energy governance. *Energy Policy* 39(6), 3832–3844.

Taylor, P. G., Upham, P., McDowall, W., and Christopherson, D. (2014). Energy model, boundary object and societal lens: 35 years of the MARKAL model in the UK. *Energy Research and Social Science* 4, 32–41.

Taylor, R. P., Govindarajalu, C., Levin, J., Meyer, A. S., and Ward, W. A. (2008). Financing Energy Efficiency: Lessons from Brazil, China, India, and Beyond (No. 42529). World Bank, Washington, DC.

Tenenbaum, B., Greacen, C., Siyambalapitiya, T., and Knuckles, J. (2014). From the Bottom Up: How Small Power Producers and Mini-Grids Can Deliver Electrification and Renewable Energy in Africa (No. 84042). World Bank, Washington, DC.

Timilsina, G. R., Hochman, G., and Fedets, I. (2016). Understanding energy efficiency barriers in Ukraine: insights from a survey of commercial and industrial firms. *Energy* 106, 203–211.

Timilsina, G. R., Shah, K. U. (2016). Filling the gaps: policy supports and interventions for scaling up renewable energy development in small island developing states. *Energy Policy* 98, 653–662.

Urban, F. and Mitchell, T. (2011) Climate change, disasters and electricity generation, strengthening climate resilience discussion paper No.8, IDS, University of Sussex, Brighton.

Urmee, T., Harries, D., and Schlapfer, A. (2009). Issues related to rural electrification using renewable energy in developing countries of Asia and Pacific. *Renewable Energy* 34, 354–357.

United States Department of Energy (2017). 2017 Wind technologies market Report, Washington DC.

Weng Hui, L., Hassim, M. H., and Ng, D. K. S. (2014). Review of evolution, technology and sustainability assessments of biofuel production. *Journal of Cleaner Production* 71, 11–29.

Williams, N. J., Jaramillo, P., Taneja, J., and Ustun, T. S. (2015). Enabling private sector investment in microgrid-based rural electrification in developing countries: a review. *Renewable and Sustainable Energy Reviews* 52, 1268–1281.

World Bank (2012). Addressing the Electricity Access Gap (No. 69062). World Bank, Washington, DC.

World Bank (2018). Tracking SDG7: The Energy Progress Report 2018 (No. 126026). World Bank, Washington, DC.

Yaqoot, M., Diwan, P., and Kandpal, T. C. (2016). Review of barriers to the dissemination of decentralized renewable energy systems. *Renewable and Sustainable Energy Reviews* 58, 477–490.

4

Linking Solar Energy Systems to Sustainable Development Goals in Africa

Recent Findings from Kenya and South Africa

Ademola A. Adenle

1. Introduction

Despite a 50% increase for energy demand in sub-Saharan Africa (SSA) since 2000, 620 million people still lack access to electricity and more than 700 million people still rely on traditional biomass (IEA 2014). Renewable sources including solar energy will play a significant role in improving energy security in Africa and diversifying the energy mix by reducing reliance on fossil fuels. Energy has been linked to tackling global challenges in sustainable development, including poverty reduction and climate change facing Africa in view of attaining millennium development goals (MDGs) (World Bank 2006). Energy was not stated as one of the MDGs but played an indirect role in helping meet the MDGs, especially in areas such as housing, health, education, and poverty reduction in Africa (UN 2015). However, these impacts were not quantified, in part because of a lack of focus on renewable energy technologies and the weak implementation of solar energy projects. Unlike MDGs, the 2030 agenda including SDGs devoted one of its 17 goals to energy. SDG7 specifically seeks to "ensure access to affordable, reliable, sustainable, and modern energy for all," thereby creating a vital role for the energy sector toward achieving SDGs.

Given the potential role of solar energy in tackling sustainable development challenges in Africa, this chapter examines some of the factors that could help facilitate the implementation of SDGs through the application of solar energy technologies. The chapter examines the advantages of solar energy technologies in the context of social, economic, and environment benefits using case studies from Kenya and South Africa. It also examines some of the key challenges that are associated with the application of solar technologies in these countries. Finally, the chapter discusses how solar technologies can help meet SDGs and summarizes policy and programs targeting the promotion of solar energy technologies for the implementation of SDGs.

In this chapter, a bibliometric approach and a qualitative analysis were undertaken to examine empirical research on solar energy technologies. The bibliometric approach was used to examine the impact of research and development (R&D) based on investment and solar energy patents awarded in the field of innovation studies across the African countries using appropriate database systems. Qualitative analysis focusing on household surveys and interviews with stakeholders was conducted through a literature review concerning the adoption of solar energy technologies including solar home systems (e.g., solar photovoltaic), solar lantern, solar heating water, and solar chargers. The chapter focuses on observations related to physical and practical components that influence the adoption of solar energy technologies at the household and provincial levels in Kenya and South Africa.

A number of sources (e.g., Google Scholar, Web of Science) with important keywords such as "solar energy and Kenya or South Africa" were used to identify relevant articles on this topic between 1990 and 2018 (up to June). The results were further refined through a focus on research related to solar energy technologies that was published in the social science research domain and the sample was reduced to 18 papers for in-depth analysis. Additional analysis generated the first set of codes used to identify the factors influencing the adoption of solar energy technologies. Following the previous observations, another set of codes were generated and grouped based on their relevance to benefits and challenges associated with the adoption of solar energy technologies as explained later in the text.

In the next section, I demonstrate the framework process for solar energy innovation. I also present the factors that could help promote solar energy innovation in Africa and the role of actors involved in developing solar energy industry.

2. A Framework for Harnessing Solar Energy Innovation

The application of science, technology, and innovation (STI) in developing solar energy cannot be overemphasized, as STI can play an important role toward achieving lasting sustainable energy development. One of the major driving forces behind STI in terms of knowledge production and technical capabilities is R&D. The R&D system is considered crucial for efficient innovation and its diffusion in the wider STI arena (Freeman and Soete 2009). According to Freeman and Soete (2009), the technological performance of countries, industries and sectors are most widely determined by R&D expenditures. They argued further that STI activities seem unrealisable without providing adequate support for R&D. Sagar and van der Zwaan (2006) point out that one of the key indicators to measure performance of innovation is financial investment in R&D. This reality exists when R&D expenditures are spearheaded by the national government, whereas private sector R&D tends to invest less in short-term and riskier projects. Empirical findings by (Garrone and Grilli 2010) indicate that publicly-funded energy R&D programs should be strengthened through the provision of adequate R&D policy measures, owing to the long-term and high risk nature of a short-term project. This is supported by the viewpoint that small and medium-sized energy projects will benefit more from public R&D expenditures because they are easy to set up and with limited risks (Fera et al. 2014).

Sagar and Vander Zwaan (2006) argued that public R&D may not be sufficient to drive fundamental changes in the energy sector, including the renewable energy industry, but will require important market-based elements which the private sector may bring into a partnership. While the public R&D funding in the energy sector may not be the only indicator of national capabilities, others factors including knowledge spillovers, strong institutions (e.g., fiscal policy, legal framework), structural change in economic growth, energy-supply sources, and domestic and international prices could determine the trajectory of a national energy sector (Caiazza and Volpe 2017; Sagar and van der Zwaan 2006). Empirical findings have recognized that knowledge spillovers can play an important role in new energy technology such as solar photovoltaic energy (Binz and Anadon 2018). These authors argued about the important role of public and private R&D in the context of the industrial sector, especially in creating incentives for market growth and cost reductions for solar PV products.

Achieving sustainable energy development in the context of social, economic, and environmental benefits being offered to society is a function of how various actors come together in identifying needs and priorities as well as providing resources and leadership required to implement solar energy projects (Binz and Anadon 2018; Caiazza and Volpe 2017). Part of this approach not only represents public and private R&D investment but also the stakeholder partnership required in the energy sector that can contribute toward achieving sustainable development. In this regard, it is important to understand the role of different actors including the national government, the international community (e.g., global R&D, development partners), and the private sector since all influence energy innovation in the policy system (Adenle et al. 2013; Stephens et al. 2008).

The individual or collective expectations of actors are very important and can influence the implementation of public policies as innovation systems gain maturity. In fact, public policies can be shaped by the institutional change within the innovation system and with respect to partnerships, learning, consumer preferences, and risk-taking (Solangi et al. 2011). Government policies including tax-incentives, subsidies, and feed-in tariffs have proven significant in facilitating the commercialization of renewable energy technology, in this case, of solar energy. UK universities, for example, have active R&D programs with a significant level of funding support from industry and government (Foxon et al. 2005). According to these authors, despite the limited presence of UK PV companies in the global PV market, market creation is guided by government policy support and incentives. Therefore, well-formulated public policies can bring about desired expectations to achieve sustainable development objectives including tackling poverty, providing access to clean energy, and reducing carbon emissions by investing in R&D, as discussed in the next section.

3. Assessment of Research and Development for Solar Energy Technologies

Prioritizing STI for solar and other renewable energy development in Africa is a key factor needed for the successful implementation of SDGs, particularly where energy usage is indispensable. While R&D investment in the global use of solar and other renewable energy continues to grow in important sectors such as cooling, drying, power, heating, and transport, only a few countries in Africa are taking advantage of investing in renewable energy. For example, in 2016, investment in solar PV represented $US 13.1 billion, $US 8.5 billion, and $US 3.5 billion in the United States, Japan, and China, respectively (Renewable 2017). Yet only one African country, South Africa, continues to dominate R&D investment in solar energy technologies (IRENA 2013).

The availability of STI data for almost two decades provides us with a vast amount of information on research expenditure in Africa. The data on gross domestic expenditure on research and development (GERD) as a percentage of gross domestic product (GDP) was obtained from United Nations Educational Scientific and Cultural Organization (UNESCO) and analyzed as shown in Figure 4.1, between 1996 and 2015. This analysis uses the ratio of GERD to GDP as the indicator or standard to measure the performance of STI investment. Figure 4.1 shows that South Africa has the highest average GERD: GDP ratio 0.78% followed by Tunisia (0.67%), Kenya (0.57%), Morocco (0.56%), Gabon (0.50%), and African countries such as Cape Verde, Gambia, Zambia, and Lesotho accounted for less than 0.1% each. In the past decade, and based on this analysis, only South Africa and Tunisia have maintained their consistency and financial commitment to R&D intensity. Also, over the past decade, these data indicate that Kenya (0.35%→0.57%), Morocco (0.27%→0.56%), and Gabon (0.41%→0.50%) have increased their financial commitments to R&D intensity. Overall, African countries perform below an Organisation for Economic Co-operation and Development (OECD) average of GERD: GDP of 2.3% (OECD 2018).

Similarly, using the World Intellectual Property Organization (WIPO) database to search for the number of solar energy patents in Africa, the majority of the patent activity has occurred in South Africa (84%), followed by Tunisia (10%), Egypt (4%), Morocco (1.75%), and Kenya (0.25%) (Figure 4.2). The largest number of solar energy patents in South Africa suggests higher investment in solar technology R&D. Without a doubt, investment in solar energy technology R&D could have significant implications for the achievement of SDGs in Africa. With the exception of South Africa, current trends in R&D investment are not shaping renewable industry and hence could present a significant challenge to the implementation of energy-related SDGs in many African countries.

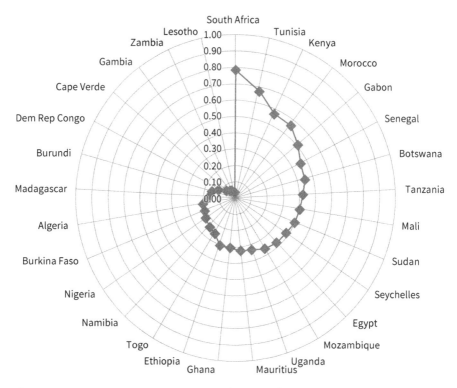

Figure 4.1 Average GERD as a percentage of GDP by African countries, 1996–2015

Source: Author's calculation based on UNESCO data (http://data.uis.unesco.org/index.aspx?queryid=74, accessed May 3, 2020).

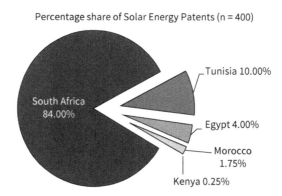

Percentage share of Solar Energy Patents (n = 400)

Tunisia 10.00%

South Africa 84.00%

Egypt 4.00%

Morocco 1.75%

Kenya 0.25%

Figure 4.2 Percentage share of solar energy patents by African countries, 2000–2017

Source: Author's calculation based on WIPO data (https://www.wipo.int/patentscope/en/, accessed May 3, 2020).

4. Adoption of Solar Energy Technologies in Kenya and South Africa

This section examines the reported or potential impacts of adoption of solar energy technologies in Kenya and South Africa, particularly with regard to social, economic, and environmental benefits and within the context of achieving SDGs. While solar energy projects reviewed for this chapter are not directly measured for meeting the SDGs, I show that most projects had positive impacts within the context of harnessing solar energy innovation for the implementation of SDGs across the two countries (Table 4.1), which of course is applicable to the rest of African countries in terms of achieving SDG targets. In the next section, the potential impacts of solar energy technologies are described and cover many of the priorities and aspirations for SDGs in the implementation of the 2030 agenda.

4.1 Social Benefits

The opportunity presented by a country's successful solar program not only touches energy, but is linked to several other social dimensions addressed by the SDGs such as end poverty (SDG1), end hunger (SDG2), healthy lives (SDG3), equitable quality education (SDG4), water and sanitation for all (SDG6), and, of course, energy for all (SDG7). The interaction among these SDGs, for example, SDG7 linked to SDG1, SDG2, SDG3, SDG4, and SDG6, can strengthen poverty and hunger responses, improve health and well-being, and enhance skill development and education. This section highlights cross-cutting issues and potential roles of solar energy technologies in reaching the interlinked SDG targets.

4.1.1 Provision of Electricity for Lighting and Powering Household Appliances

Access to regular electricity remains crucial to meeting the demand for household energy in Kenya and South Africa. However, a large number of people in these countries lack access to a regular power supply. In the reviewed studies (Figure 4.3), the need for electricity in Kenya (mostly among rural and urban poor) is higher than South Africa. Despite this, many poor households in South Africa are yet to be connected to the grid. The inability of households to access electricity reflects the high cost of the electric grid, the lack of a smart grid, and the unreliable grid extension, as indicated in the reviewed studies.

In view of this growing challenge, solar energy technology such as solar home systems (SHS) (made up of components such as solar panels, battery storage, inverters and charger controllers) can address the need for household services (Acker and Kammen 1996; Azimoh et al. 2015; Ulsrud et al. 2015). According to these studies,

Table 4.1 Impacts of solar energy technologies for the achievement of SDGs

Tiers of sustainable development	Indicator	Sustainable development goals	Reported/potential impacts	References
Social	Health	SDG3	Improvement of quality of health services	Acker and Kammen (1996), Azimoh et al. (2015), Biermann et al. (1999), Curry et al. (2017), Duke et al. (2002), Gustavsson (2007), Jacobson (2007), Kakaza and Folly (2015), Masekoameng et al. (2005), Ozdemir et al. (2012), Roche and Blanchard (2018), Ulsrud et al. (2015), Wlokas (2011).
	Poverty eradication; good livelihood; ensuring peace	SDG1, SDG2, SDG16	Poverty reduction	
	Access to affordable electricity; energy security	SDG7	Improvement of rural and urban electrification; improved energy security	
	Access to quality education	SDG4	Creating better learning environment and improved quality of education	
	Access to clean water and sanitation	SDG6	Improved access to adequate sanitation and safe drinking water	
	Food security (sustainable agriculture)	SDG2, SDG12	Adequate food available and reduced food waste	
	Capacity building	SDG4, SDG6	Increased level of capacity and knowledge sharing at the rural household level	
Economics	Creation of employment; women empowerment	SDG4, SDG5, SDG8	Improvement of employment rate among women and men; more women gainfully employed	Acker and Kammen (1996), Curry et al. (2017), Donev et al. (2012), Eskom (2013), Ulsrud et al. (2015), UNDP (2017), Winther et al. (2018), Wlokas (2011).
Environment	Reduced pollution (water, air, soil) in agricultural production; adapting to and mitigating against the impact of climate change; access to clean energy	SDG6, SDG7, SDG12, SDG13, SDG15	Improved level of energy efficiency and reduction of greenhouse gases; indoor pollution reduced	Acker and Kammen (1996), Biermann et al. (1999), Ozdemir et al. (2012), UNDP (2017).

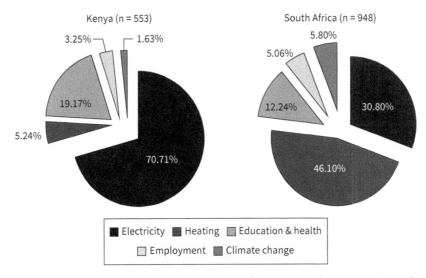

Figure 4.3 Application of solar energy technology for various end uses in Kenya and South Africa

Source: Author's calculations based on reviewed studies, including various sources of literature between 1990 and 2018 provided in the reference list.

solar PV or solar lanterns play a significant role in delivering lighting and powering radio, television, mobile phones, and other household appliances in rural homes.

In Kenya, solar PV is mostly used to power TV and light in the rural areas but only is available for households that can afford solar electricity (Jacobson 2007). Similarly, in Kenya, the Ikisiya Energy Centre, located in a village of 383 households, has a capacity 2.16 kilowatt (kW) solar PV system where services such as lantern renting and phone charging are provided only to people who can afford it (Ulsrud et al. 2015). In South Africa (Vhembe municipality), 50% of the households interviewed reported that their energy needs were met by SHS electricity, but not all households could afford it (Azimoh et al. 2015).

Unfortunately, the use of solar PV for TV and radio serves as the major source of new information on agricultural practices—information, therefore, that is unlikely to reach the poorest farmers on a consistent basis (Jacobson 2007). Also, solar-powered irrigation plays an important role in improving crop yields in rural Kenya. For example, in Western Kenya, smallholders have benefitted from the adoption of solar powered irrigation pumps to tackle challenges of frequent drought and unreliable rainfall (Schützeichel 2017).

In summary, using solar energy technologies to access mobile phone, radio, and TV can provide vital information and enhance the delivery of better agriculture extension services as well as improved agricultural productivity for the rural communities. All of these can play an important role in meeting the targets of SDG1 (target 1.5; build the resilience of the poor), SDG2 (target 2.4; increase resilient agricultural practices), and SDG13 (target 13.1; strengthen resilience and adaptive capacity), thereby contributing to the achievement of SDGs.

4.1.2 Provision of Energy Services, Time, and Energy Savings

The provision of energy services such as cooking or heating varies from village to village and even from urban area to urban area in Kenya and South Africa. The peer-reviewed survey shows that 5.24% of Kenyan households require heating compared to 46.10% of South African households (Figure 4.3). The fact that Kenya has a smaller heating percentage does not mean the demand for heating is very low in the country. As provided in the reviewed studies for both countries, the factors influencing the electricity grid or solar electricity use for heating depend on seasonality, location, income, and population density. For example, in South Africa, the Zanemvula Solar Water Heating (SWH) Project located in Port Elizabeth, an urban setting, will be different in terms of electricity consumption compared to the county of Nakuru in Kenya or rural South Africa (Roche and Blanchard 2018; Wlokas 2011). The reviewed studies emphasize the significance of hot water heating as a driver for energy consumption within the average household. For example, water heating accounts for more than half of energy consumption in South Africa (Kakaza and Folly 2015).

The adoption of solar energy technology does not only show that households benefit from solar energy services, but also encourages time and energy savings. A study by Biermann et al. (1999) analyzed fuel savings from a field test of different types of solar cookers in South Africa and found that the overall savings from solar cooking compared to three other fuel types (paraffin/kerosene, gas, and wood) are 38%. According to these authors, paraffin, gas, and wood represent 2000 litres, 2 tons, and 60 tons in terms of fuel savings, respectively (Biermann et al. 1999). They argued that by using solar cookers, fuel savings were considerable compared to use of kerosene, gas, and wood, which was costly (especially gas and kerosene), time-consuming, and a challenging task. The availability of solar energy saved women time, as they had to travel between 3km and 10km to collect firewood prior to the availability of solar energy (Masekoameng et al. 2005; Wlokas 2011).

These findings are corroborated by a similar study conducted in South Africa by (Curry et al. 2017), who reported that the use of SWH (63 out of 67 households) resulted in an average saving of 67% of electricity as compared to traditional electrical water heating by individual households. Another study in South Africa indicated that installation of SWH in Guateng reduced the peak electricity demand by 70% at the household level (Ozdemir et al. 2012). In Nakura, Meru, and Bungoma, Kenya, respondents interviewed mentioned that the use of solar PV made life much easier among the household members especially for women and children as solar energy saved them time and the hassle of travelling long distances every week to buy kerosene (Acker and Kammen 1996).

To summarize this section, energy efficiency, through the adoption of solar energy, can play an important role in achieving some of the SDG targets that are related to sustainable consumption and production (SDG12), terrestrial ecosystems and biodiversity (SDG15), and peaceful and inclusive societies (SDG16). Given the energy services provided by solar products, time savings remain very important, as family households can spend their time more productively due to the strong link between access to energy efficiency and firewood collection. Likewise, time savings can reduce different forms of violence and death (target 16.1) among women and children, who are more likely to experience violence related to rape and other sexual assaults, as there will be less long distance travel to fetch fuel woods. This can also reduce deforestation activities, thereby contributing to meeting the target 15.2 (restoring degraded forests and increasing afforestation).

4.1.3 Improved Access to Quality Education and Better Health Services
In addition to the provision of electricity and energy services, the improved access to quality education and better health services play an important role in the adoption process of SHS in Kenya and South Africa. Notably, access to light among school children and at health centers through the use of solar PV in the rural areas has been a very positive factor. Evidence from surveys showed that in communities where access to solar electricity has been provided, students have had more time at night to do their homework, and the availability of SHS has contributed to improved physical health and well-being of many households that participated in the solar programs.

In rural Kenya, relatively little solar energy has been available for students' education due to other domestic purposes (e.g., television viewing) (Jacobson 2007). This author found that of almost 80% of solar households with school age children (n = 76), solar lighting was used for studying in only 40% of them. Although the quality of light from solar lanterns is much higher than that of light from kerosene lamps or candles, the value of education appears not to be universal among solar home system user (Jacobson 2007). Further to education, social institutions such as health centers in rural areas have benefitted from government installation of solar PV in Kenya. The availability of solar electricity has helped to store vaccine and other medical products in addition to the provision of lights (Duke et al. 2002).

Azimoh et al. (2015) examined the impact of SHS electricity on children's education in four different solar households (Polokwane n = 23; Vhember Center n = 24; Tubatse n = 23 and Umkhanyakude n = 18) in South Africa. Their results showed that almost 90% of the respondents interviewed from these households mentioned that SHS electricity had a positive impact on their children's education. According to these authors, the availability of SHS electricity not only had positive impact in terms of extending study hours, but also contributed to better performance of their children at schools. This is consistent with a similar study that found improved performance for school children in rural Zambia where solar home systems were installed (Gustavsson 2007). The study reported by Wlokas (2011) found that there are linkages between well-being of households and solar water heater interventions. For example, the availability of warm water (due to adoption of SWH) helped improve the health of infants and children, thereby enabling them to attend school more regularly. In fact, this also contributed to improving mental health of the families, particularly women, by reducing depression caused by the burden of wood collection and processing.

Solar energy technology is strongly linked to quality education and improved health services, especially in the context of achieving SDG4 and SDG3. The provision of solar-powered products (SDG7, energy for all) to every household in Africa can play a significant role in meeting the top targets of SDG4 (target 4.1 and 4.2), particularly in ensuring that boys and girls have quality primary and secondary education with much more effective learning outcomes. And the availability of solar powered products can also contribute to meeting SDG3 (target 3.1), particularly in reducing the global maternal mortality ratio.

4.2 Economic Benefits

SDG1, SDG2, and SDG8 (economic growth, employment, and decent work) illustrate the strong links between economic dimensions and solar energy technologies.

4.2.1 Creation of Employment Opportunities

The use of local labor for installing, operating, and maintaining solar technologies creates jobs according to the reviewed surveys. For example, staff for a local energy center were recruited from the same location to deliver services including solar lantern renting, maintenance, and battery replacement (Ulsrud et al. 2015). In fact, the energy center encourages the employment of women, according to a respondent interviewed in one of the surveys: "Because of the lanterns which benefited us so much, and also when you opened the Energy Centre people got jobs because our girls are employed there, and they are good role models like [the woman manager]" (Winther et al. 2018, p. 68).

There is wide agreement that rapid expansion of the SWH market in South Africa has led to job creation. Over 42 companies were registered as SWH suppliers and manufacturers between 2007 and 2009 (Donev et al. 2012). According to these authors, the solar water industry is expected to increase employment by 40% in the country. This is consistent with Eskom's findings that the SWH market grew to 400 suppliers in 2011 from 20 suppliers in 1997 (Eskom 2013). The survey analyzed by Curry et al. (2017) looks into SWH as a source of employment in four townships (Mabopane View, Ga-Rankuwa View, Nelmapius and Soshanguve) where nearly 6000 SWHs were installed as of July 2011. Their results showed that 45 people constituting five different teams were trained and employed from these townships. The authors estimated that if the installation of 60,000 SWHs went ahead as planned, nearly 400 local jobs would have been created by 2015 in Tshwane alone. Still, temporary employment of the installers would mean loss of skills and training after they left their jobs, which is an important factor to consider for SWH projects (Wlokas 2011).

The need for quality employment and sustainable growth is highlighted in SDG8 (economic growth, employment, and decent work). The adoption of solar energy has been linked to job creation, an important driver of economic growth and one of the major tools for poverty eradication. In this context, there are strong links between the adoption of solar energy and decent job creation (target 8.3).

4.3 Monetary Savings

The uptake of solar technology not only reduces energy usage but also saves money based on the reviewed studies. Some years ago, Acker and Kammen (1996) found that solar PV was the most popular choice among the respondents interviewed in Kenya, who reported that solar PV products were the least expensive and saved money when compared to grid extension or other alternative energy sources such as diesel generators. Because of this, they spent less on electricity bills, kerosene, and paraffin and saved enough money to spend on other household items. More recent consumer surveys conducted by the United Nations Development Programme sought to find out why people wanted to purchase solar PV products in Kenya (UNDP 2017). Their study showed that 58% of those surveyed attributed cost savings as one specific factor, indeed one of the largest factors that attracted them to solar PV product adoption.

An analysis of savings on energy expenditures for low-income households (Tshwane, South Africa) was provided by Curry et al. (2017). According to this finding, more than 50% of households equipped with SWH were found to save 62kWh/month (R54/month) on electricity bills as compared to 92kwh/month used by the conventional electrical water heater. This study indicates that the SWH system lowers the cost of electricity in the surveyed households. The flip side is that some of the households interviewed experienced higher water bills from the usage of SWH, typically due to water leakage beyond their control. The authors concluded that water leakage has been solely responsible for the increased bills. But those households who experienced savings used the extra money to buy food for their children. The lowest annual expenditure and monetary savings with payback periods of three to four years were experienced among the middle-income households using SWH system in Guateng (Ozdemir et al. 2012).

Monetary savings through the uptake of solar energy technologies have been shown to reduce poverty as mentioned in these studies. Therefore, it is reasonable to conclude that encouraging the adoption of solar energy in Africa contributes to the implementation of the SDGs. There are strong links between the adoption of solar energy and income and monetary savings. The strong relationship between hunger and income suggests the important role solar energy is playing and can play through monetary savings because large parts of total income in many African countries are spent on food acquisition. The effects can be shown to directly or indirectly contribute to achieving SDG1 and SDG2.

4.4 Environmental Benefits

Access to clean energy (SDG7) is directly linked to SDG3 and SDG13 (combat climate change) and provides both health and environmental benefits to Kenya and South Africa.

4.4.1 Reduced Indoor Air Pollution

The use of wood, charcoal, and kerosene dominate the energy sector in both rural Kenya and South Africa. According to the World Health Organization (WHO), there has been a significant increase in the number of diseases (e.g., heart disease and lung cancer) largely attributed to household air pollution resulting from inefficient cooking and heating practices associated with the use of wood and kerosene in African countries, including Kenya and South Africa. While there is very limited research available, anecdotal evidence suggests important environmental benefits of solar energy. According to Acker and Kammen (1996), solar PV products offer environmental benefits in terms of decreased use of wood and kerosene in Kenya, thereby also providing air quality benefits. Reduced air pollution from firewood is one of the reported positive effects of adopting solar cookers in South Africa (Beirmann et al. 1999).

4.5 Reduced CO2 Emissions

In addition to reduced indoor air pollution, the adoption of solar energy has been linked to reduced emissions at the household level. A review of survey data suggests that the use of solar technology in reducing carbon emission is higher in South Africa than Kenya (Figure 4.3). Local promotion of solar energy technologies in Kenya, for example, solar lanterns and solar PV lighting can help reduce carbon emissions from kerosene and firewood use (UNDP 2017). The result of interviewed respondents in low-income households in South Africa shows that the installation of SWH systems is associated with the reduction of carbon emissions. According to Curry et al. (2017), 10,330 installed SWH systems implemented at the household level saved 610,500kg of CO_2 per month. The same authors argued that CO_2 emission savings of 0.96–1.20% could be achieved across the country just from SWH if targeted households have these systems installed. A further analysis of potential positive impact in South Africa from installed SWH systems in different types of households and commercial water heating revealed that greenhouse gas emissions were significantly reduced (Donev et al. 2012). These authors estimated that 186.4Mt CO_2 and 297.7Mt CO_2 emissions were reduced for moderate and accelerated SWH scenarios, respectively, compared to current emissions in South Africa of 434 Mt CO_2.

Taken together, it has been demonstrated that the adoption of solar energy technologies plays an important role in reducing greenhouse gases and contributes to the implementation of SDG7 and SDG13 in Africa.

5. Challenges to Application of Solar Energy Technology

A detailed review of survey-based research articles provides insight into current challenges associated with the implementation of solar energy projects in Kenya and South Africa, which are common in the majority of the African countries. Respondents interviewed in the selected studies showed that financial constraints, technical problems, and weak government policy are the major barriers to the implementation of solar energy projects in both countries as shown in Figure 4.4.

The largest economic barrier, financial constraints, is associated with the high capital, investment, installation, and operating costs of solar energy technologies, especially solar PV products. As noted in the review of many related studies, there are also inherent cost recovery problems, rising counterfeits and pirated products, and the inability to afford payment and inadequate access to credit, all of which represent significant challenges to the widespread use of solar energy in these countries and the rest of Africa (Donev et al. 2012; Duke et al. 2002; Roche and Blanchard 2018; World Bank 2010). In addition, inadequate financing schemes introduced by donors, development organizations, and nongovernmental organizations (NGOs), along with the lack of government subsidies, undermines the sales of solar home systems (SHS), leading to slow uptake of solar PV products, especially in rural Kenya (Duke et al. 2002). This problem has been attributed to high transaction costs involved in managing a large network of small-scale installations and arranging micro-credits with a highly dispersed customer base in the country.

Many low-income households would prefer to use other sources of energy such as kerosene and firewood because of their inability to afford solar PV products. For example, the cost of a 40 W solar PV is $US210 compared to $US5 for a kerosene stove (Karekezi and Kithyoma 2002). Similarly, limited installation of solar water heaters in areas of most need in South Africa is attributable to the high initial cost of SWH. For example, the average price of an installed solar water heater per square meter is 4191 rand, which was not affordable for the majority of low income households in the country (Donev et al. 2012). According to a World Bank report, lack of financing options remains a huge barrier to market penetration and growth of solar energy technologies in Africa (World Bank 2010).

Technical problems are one of the main factors affecting the uptake of solar energy technologies in African countries, including Kenya and South Africa. Other challenges associated with technical problems include limited knowledge and skills, poor installation and maintenance, and unreliable and low-quality components. According to the studies focusing on Kenyan solar energy development, absence of a solar research institute to support the manufacturing of the components (e.g., R&D), lack of education and a training center, and limited technical expertise were some of the largest limiting factors for Kenya's PV industry (Acker and Kammen 1996; Roche and Blanchard 2018; Simiyu et al. 2014). These particular challenges are not restricted to Kenya only. In South Africa, technical training for large-scale installation and maintenance of SWH was very limited, not well-guided, and uncoordinated (Donev et al. 2012). A study by Curry et al. (2017) assessed the large-scale installation of SWHs in the city of Tshwane. They found that both householders and the company in charge neglected the maintenance of SWH installations that were one and two years old. In fact, 57% of respondents interviewed attributed the abandonment of SWH to water leakages. Further, according to Donev et al. (2012), lack of standards for testing of solar collectors negatively affected performance measures and quality control of SWH products among local manufacturers.

Renewable energy policies of most African governments, especially those targeting the promotion of solar energy, appear very weak and fragmented (World Bank 2010). Their actions and policies continue to undermine the solar energy market and limit the diffusion of solar energy technologies across the continent. Kenya and South Africa are very good examples where solar markets continue to be affected by government policies. The results of reviewed studies indicate that policy intervention by the Kenyan government to increase solar market penetration in the country is absent. In fact, some authors point out that the lack of specific policies and regulations had serious impacts on solar PV markets (Ulsrud et al. 2015), thereby affecting the implementation of solar energy projects across the country. Based on the reviewed studies, the relative success of solar markets reported in Kenya has been attributed to the rural middle class where people are well-off, not because favorable policy was provided by the government.

The negative effect of policy fluctuations between the national government and local municipalities in introducing solar home systems in low-income households was also reported in South Africa. Lack of cooperation among municipalities concerning solar energy subsidies in low-income groups suggests that the weakness in national government policy could be an obstacle to the introduction of solar technologies in the most vulnerable communities (Lemaire 2011). According to Lemaire (2011), companies relying on subsidies helped keep the prices low, but withdrawal and reinstatement of subsidies disrupted the market thereby forcing out a lot of companies. This is corroborated by the lack of engagement and failure to specify the role of the key stakeholder groups in the institutional frameworks of government, causing delays or resulting in poor implementation of solar projects (Bikam and Mulaudzi 2006).

In addition to these challenges, lack of awareness and cultural problems were reported to be one of the main challenges to the introduction of solar technologies in African countries. According to Duke et al. (2002), lack of awareness and information on certification as it relates to the minimum quality standard of solar home systems, particularly in the rural areas, will create uncertainty in the market

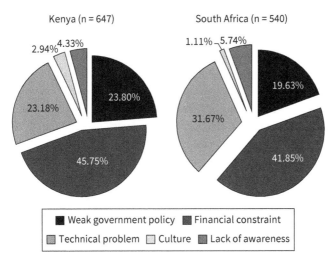

Figure 4.4 Observed barriers to implementation of solar energy activities by percent in Kenya and South Africa

Source: Author's calculation based on reviewed studies, including various sources of identified literature between 1990 and 2018, provided in the reference list.

as well as slow uptake of SHS in Kenya. A decade later this trend continues to be observed in Kenya's solar market, where lack of awareness and cultural differences contributed to slow adoption of SHS (Ondraczek 2013; Simiyu et al. 2014). In one case, the use of a solar cooker pilot program failed in part due to a culture that forbids cooking outside (van der Gaast et al. 2009). Similarly, in South Africa, lack of understanding of SWH installation resulted in limited application because of leakages. Lacking information, many low-income householders did not know who to approach or where to go for assistance (Curry et al. 2017). Moreover, a lack of community awareness in many parts of South Africa generated similar results (see Figure 4.4) (Donev et al. 2012).

6. Discussion and Policy Implications

There is an urgent need to harness science technology and innovation for sustainable development broadly and especially in meeting SDGs. Effective implementation of solar energy projects can help meet some SDGs in Africa, but African government policy also must be able to accommodate the use of solar energy technologies by increasing investment in solar R&D programs. Both basic and applied research are needed to create robust STI policy and to advance new knowledge about new and existing technologies (Foxon et al. 2005). But low investment in R&D programs (section 3) suggests that the technologies needed to improve solar energy may be lacking in many African countries, including the two case studies presented in this chapter.

The Kenyan national energy policy, for example, is absent in supporting solar R&D programs despite relatively notable solar markets (ERC-Kenya 2011).

With respect to policy, Kenya's Least Cost Power Development Plan (LCPDP) was released by the government in 2011, covering energy planning for electricity demand to the year 2031. It mostly relies, however, on low carbon technologies, but solar energy received little attention (ERC-Kenya 2011). In fact, in this report, there is no clear-cut strategy for solar technologies in the electricity supply between 2011 and 2031, suggesting that the Kenyan national innovation system does not prioritize solar energy technologies and therefore may play a limited role toward meeting the SDGs.

In South Africa, the disjointed energy policies between different levels of government created an unstable solar market as mentioned in Section 6. On the other hand, anecdotal evidence suggests that there has been some effort on the part of the South African government to support further development of solar energy technologies through the national research programs, given the rapid expansion of the solar market in the country (Jadhav et al. 2017). Still, the country has been criticized as being a latecomer in the solar market and for lacking a well-established industry for the manufacture of solar materials (Baker and Sovacool 2017), as Chinese companies currently dominate manufacturing of solar PV materials.

Given this challenge, the investment in R&D programs will continue to play an important role in creating new solar energy technologies that increase efficiency and facilitate adoption in Africa. Indeed, improving the efficiency of solar cells will require R&D that targets production of new materials, components, and devices conducted at the national level especially between private and public sectors (Sun et al. 2014). China and Germany have proven to be great world leaders in solar markets with a huge investment in R&D programs stimulated by STI policy (Solangi et al. 2011). Establishing a specialized solar research institute will be important in solving localized problems related to manufacture of solar local components. Therefore, harnessing STI through investment in solar R&D programs would not only pave the way for meeting the SDGs in Africa but also for a sustainable future on the continent.

Based on the analysis in this chapter, I have demonstrated that the adoption of solar energy technologies in African countries could play an important part in meeting many of the SDGs on the continent in view of its potential social, economic, and environmental benefits (Table 4.1).

From a social standpoint, the adoption of solar energy to generate electricity in many rural areas as explained in section 5 shows great potential to contribute to the implementation of the following SDGs: SDG1 (poverty), SDG2 (hunger), SDG3 (health), SDG4 (education), SDG5 (gender), SDG6 (water), and SDG7 (clean energy). The fact that these SDGs are interlinked in terms of services offered through solar energy products means that solar energy technologies could play a significant role in reaching some of the targets that are associated with the seven different SDGs.

The increased use of solar energy in rural areas could save more money and enhance farming productivity thereby helping to achieve target 2.1 of SDG2 (ending hunger through the provision of nutritious and sufficient food all year round). Given that the majority of the rural population uses firewood for cooking and heating, equal access to solar energy could help meet target 3.4 (promoting mental health and well-being) and target 3.9 (reducing the number of deaths and illnesses from air pollution) of SDG3. Similarly, in light of environmental benefits that are associated with the use of solar energy, the impact of climate change (SDG13) could be addressed through lowering greenhouse gas emissions especially in an industrialized country like South Africa.

SWH projects by the Kuyasa and Zanemvula (South African low-income communities) were implemented as a low-carbon development pathway to reduce poverty based on the clean development mechanism (CDM) under the Kyoto Protocol between 2008 and 2010 (Goldman 2010; Wlokas 2011). While several SWH projects were implemented during the MDG era, there was a lack of performance assessment partly because of the small size of the projects, thereby making it difficult to directly link it to the achievement of MDGs.

Given the nationally determined contributions of the Paris Agreement and SDG13, African governments and the international community must pay attention to performance assessment, especially by encouraging appropriate sets of indicators be used to monitor the progress of SDGs during the implementation of solar energy projects.

The economic benefits offered through the adoption of solar energy technologies should be embraced as more implementation of solar energy projects will continue to create jobs as described in this part of the chapter, thereby contributing toward meeting the targets of the interlinked SDGs. Promoting access to affordable and sustainable energy (SDG7) is connected to three targets of SDG8; target 8.3 (decent job creation, entrepreneurship-micro, small and medium enterprises), target 8.5 (employment and decent work for all women, men and youth), and target 8.6 (reduction of youth unemployment and training). The application of solar energy technologies is shown here to influence broader social and economic contexts that underlie the implementation of SDGs in Africa. More importantly, the use of solar energy is key to providing energy services that will serve the intersectoral linkage between the SDGs, thereby encouraging the implementation of many SDGs including SDG16 (peaceful and inclusive societies). In this regard the potential role of solar energy is crucial, particularly in poverty reduction and provision of good livelihoods, as indicated in the summary of benefits of solar energy. This could essentially contribute to the achievement of target 16.1 of SDG16, thereby encouraging peace in Africa.

In summary, the overall benefits of solar energy technologies toward meeting SDGs by 2030 in Africa cannot be overemphasized, but current challenges with respect to application and adoption of solar energy programs remain as important obstacles. There is a very little effort particularly on the part of the national governments to facilitate the uptake of solar energy. As mentioned in section 6, the high cost of solar PV products remains one of most important barriers and is largely due to weak government policy.

To overcome this challenge, the feed-in-tariff and subsidy policies targeting the promotion of solar energy technologies should be supported by a renewed national government energy policy framework, especially in partnership with the private sector.

It has been established that the application and diffusion of solar energy technologies in advanced countries owes much success to robust renewable energy policy that not only integrates R&D investment, but also provides a strong incentive structure to facilitate rapid growth and global expansion of the solar industry. In 2004, for example, effective implementation of a feed-in tariff and subsidy policies by the German government stimulated high investment in solar PV, thereby increasing country competitiveness in the solar global market (Hoppmann et al. 2014). As another example, in 2006, the Chinese government implemented new renewable energy policies that led to rapid growth of the solar PV industry in the country, which in turn played a large part in China's becoming a major player in international solar market (Sun et al. 2014).

Taken together, African governments need to overhaul their renewable energy policies so that R&D investment and a robust subsidy structure will remain strategic in national energy policy and encourage the development of the solar PV industry. In addition, other key stakeholders including development partners, the private sector, and NGOs must coordinate and align their strategies with national governments to implement solar energy projects across the continent and to facilitate, simultaneously, the implementation of the SDGs.

References

Acker, R. H. and Kammen, D. M. (1996). The quiet (energy) revolution: analysing the dissemination of photovoltaic power systems in Kenya. *Energy Policy* 24(1), 81–111.

Adenle, A. A., Haslam, G. E., and Lee, L. (2013). Global assessment of research and development for algae biofuel production and its potential role for sustainable development in developing countries. *Energy Policy* 61, 182–195.

Azimoh, C. L., Klintenberg, P., Wallin, F. and Karlsson, B. (2015). Illuminated but not electrified: an assessment of the impact of Solar Home System on rural households in South Africa. *Applied Energy* 155, 354–364.

Baker, L. and Sovacool, B. K. (2017). The political economy of technological capabilities and global production networks in South Africa's wind and solar photovoltaic (PV) industries. *Political Geography* 60, 1–12.

Biermann, E., Grupp, M. and Palmer, R. (1999). Solar cooker acceptance in South Africa: results of a comparative field-test. *Solar Energy* 66(6), 401–407.

Bikam, P. and Mulaudzi, D. J. (2006). Solar energy trial in Folovhodwe South Africa: lessons for policy and decision-makers. *Renewable Energy* 31(10), 1561–1571.

Binz, C. and Anadon, L. D. (2018). Unrelated diversification in latecomer context: the emergence of the Chinese solar photovoltaics industry. *Environmental Innovation and Societal Transitions* 28, 14–34.

Caiazza, R. and Volpe, T. (2017). Innovation and its diffusion: process, actors and actions. *Technology Analysis and Strategic Management* 29(2), 181–189.

Curry, C., Cherni, J. A. and Mapako, M. (2017). The potential and reality of the solar water heater programme in South African townships: lessons from the City of Tshwane. *Energy Policy* 106, 75–84.

Donev, G., Wilfried, G. J. H. M., Blok, K., and Dintchev, O. (2012). Solar water heating potential in South Africa in dynamic energy market conditions. *Renewable and Sustainable Energy Reviews* 16 (5), 3002–3013.

Duke, R. D., Jacobson, A., and Kammen, D. M. (2002). Photovoltaic module quality in the Kenyan solar home systems market. *Energy Policy* 30(6), 477–499.

Energy Regulatory Commission (ERC-Kenya 2011). The Least Cost Power Development Plan (LCPDP) 2011–2031. https://www.renewableenergy.go.ke/downloads/studies/LCPDP-2011-2030-Study.pdf (accessed April 5, 2019).

Eskom (2013). Solar Water Heating Rebate Programme. http://www.eskom.co.za/AboutElectricity/FactsFigures/Documents/OtherDocs/The_Solar_Water_Heating_SWH_Programme.pdf (accessed April 5, 2019).

Fera, M., Iannone, R., Macchiaroli, R., Miranda, S. and Schiraldi, M. M. (2014). Project appraisal for small and medium size wind energy installation: the Italian wind energy policy effects. *Energy Policy* 74, 621–631.

Foxon, T. J., Gross, R. Chase, A., Howes, J., Arnall, A. and Anderson, D. (2005). UK innovation systems for new and renewable energy technologies: drivers, barriers and systems failures. *Energy Policy* 33(16), 2123–2137.

Freeman, C. and Soete, L. (2009). Developing science, technology and innovation indicators: what we can learn from the past. *Research Policy* 38(4), 583–589.

Garrone, P. and Grilli, L. (2010). Is there a relationship between public expenditures in energy RandD and carbon emissions per GDP? an empirical investigation. *Energy Policy* 38(10), 5600–5613.

Goldman, M. (2010). *Kuyasa CDM Project: Renewable Energy Eefficient Ttechnology for the Poor.* http://www.growinginclusivemarkets.org/media/cases/SouthAfrica_Kuyasa_2010.pdf (accessed April 5, 2019).

Gustavsson, M. (2007). Educational benefits from solar technology: access to solar electric services and changes in children's study routines, experiences from eastern province Zambia. *Energy Policy* 35(2), 1292–1299.

Hoppmann, J., Huenteler, J. and Girod, B. (2014). Compulsive policy-making: the evolution of the German feed-in tariff system for solar photovoltaic power. *Research Policy* 43(8), 1422–1441.

International Energy Agency (IEA 2014). *Africa Energy Outlook: A Focus on Energy Prospects in Sub-Saharan Africa.* https://www.icafrica.org/en/knowledge-hub/article/africa-energy-outlook-a-focus-on-energy-prospects-in-sub-saharan-africa-263/ (accessed April 5, 2019).

International Renewable Energy Agency (IRENA 2013). Renewable Energy Auctions in Developing Countries. http://www.irena.org/publications/2013/Jun/Renewable-Energy-Auctions-in-Developing-Countries (accessed April 5, 2019).

Jacobson, A. (2007). Connective power: solar electrification and social change in Kenya. *World Development* 35(1), 144–162.

Jadhav, A. S., Chembe, D. K., Strauss, J. M., andVan Niekerk, J. L. (2017). Status of solar technology implementation in the Southern African Developing Community (SADC) region. *Renewable and Sustainable Energy Reviews* 73, 622–631.

Kakaza, M., and Folly, K. A. (2015). Effect of solar water heating system in reducing household energy consumption. *IFAC-PapersOnLine* 48(30), 468–472.

Karekezi, S., and Kithyoma, W. (2002). Renewable energy strategies for rural Africa: is a PV-led renewable energy strategy the right approach for providing modern energy to the rural poor of sub-Saharan Africa? *Energy Policy* 30(11), 1071–1086.

Lemaire, X. (2011). Off-grid electrification with solar home systems: the experience of a fee-for-service concession in South Africa. *Energy for Sustainable Development* 15(3), 277–283.

Masekoameng, K. E., Simalenga, T. E., and Saidi, T. (2005). Household energy needs and utilization patterns in the Giyani rural communities of Limpopo Province, South Africa. *Journal of Energy in Southern Africa* 16, 88–93.

Ondraczek, J., (2013). The sun rises in the east (of Africa): a comparison of the development and status of solar energy markets in Kenya and Tanzania. *Energy Policy* 56, 407–417.

Organisation for Economic Co-operation and Development (OECD (2018). *Main Science and Technology Indicators.* http://www.oecd.org/sti/msti.htm (accessed April 5, 2019).

Ozdemir, E. D., Marathe, S. D., Tomaschek, J., Dobbins, A., and Eltrop, L. (2012). Economic and environmental analysis of solar water heater utilisation in Gauteng Province, South Africa. *Journal of Energy in Southern Africa* 23, 2–19.

Renewable (2017). Advancing the Global Renewable Energy Transition: Highlights of the REN21 Renewables 2017 Global Status Report in Perspective. http://www.ren21.net/wp-content/uploads/2017/06/GSR2017_Highlights_FINAL.pdf (accessed April 5, 2019).

Roche, O. M., and Blanchard, R. E. (2018). Design of a solar energy centre for providing lighting and income-generating activities for off-grid rural communities in Kenya. *Renewable Energy* 118, 685–694.

Sagar, A. D., and van der Zwaan, B. (2006). Technological innovation in the energy sector: RandD, deployment, and learning-by-doing. *Energy Policy* 34(17), 2601–2608.

Schützeichel, H. (2017). Solar-Powered Irrigation: Food Security in Kenya's Drought Areas. Solar Energy Foundation. https://solar-energy-foundation.org/fileadmin/Dateien/Wasserpumpe/Waterpump_for_food_security_nf.pdf (accessed April 5, 2019).

Simiyu, J., Waita, S., Musembi, R., Ogacho, A., and Aduda, B. (2014). Promotion of PV uptake and sector growth in Kenya through value added training in PV sizing, installation and maintenance. *Energy Procedia* 57, 817–825.

Solangi, K. H., Islam, M. R., Saidur, R., Rahim, N. A., and Fayaz, H. (2011). A review on global solar energy policy. *Renewable and Sustainable Energy Reviews* 15(4), 2149–2163.

Stephens, J. C., Wilson, E. J., and Peterson, T. R. (2008). Socio-political evaluation of energy deployment (SPEED): an integrated research framework analyzing energy technology deployment. *Technological Forecasting and Social Change* 75(8), 1224–1246.

Sun, H., Zhi, Q., Wang, Y., Yao, Q., and Su, J. (2014). China's solar photovoltaic industry development: the status quo, problems and approaches. *Applied Energy* 118, 221–230.

Ulsrud, K., Winther, T., Palit, D., and Rohracher, H. (2015). Village-level solar power in Africa: accelerating access to electricity services through a socio-technical design in Kenya. *Energy Research and Social Science* 5, 34–44.

United Nations (UN) (2015). The Millennium Development Goal Report 2015. http://mdgs.un.org/unsd/mdg/Resources/Static/Products/Progress2015/English2015.pdf (accessed April 5, 2019).

United Nations Development Programme (UNDP) (2017). National Appropriate Mitigation Action on Access to Clean Energy in Rural Kenya Through Innovative Market Based Solutions. http://www.undp.org/content/dam/LECB/docs/pubs-namas/undp-lecb-Kenya_Clean-Energy-NAMA-2016.pdf (accessed April 5, 2019).

van der Gaast, W., Begg, K., and Flamos, A. (2009). Promoting sustainable energy technology transfers to developing countries through the CDM. *Applied Energy* 86(2), 230–236.

Winther, T., Ulsrud, K., and Saini, A. (2018). Solar powered electricity access: implications for women's empowerment in rural Kenya. *Energy Research and Social Science* 44, 61–74.

Wlokas, H. L. (2011). What contribution does the installation of solar water heaters make towards the alleviation of energy poverty in South Africa? *Journal of Energy in Southern Africa* 22(2), 27–39.

World Bank (2006). Energy and the Millennium Development Goals in Africa. The Forum for Energy Ministers in Africa (FEMA) for the United Nations World Summit. http://siteresources.worldbank.org/EXTAFRREGTOPENERGY/Resources/Energy_and_MilleniumFEMA_Report.pdf (accessed April 5, 2019).

World Bank (2010). Lighting Africa. Lighting Africa Progress Report. https://lightingafrica.org/wp-content/uploads/2014/01/Annual_Report_FY%202010_final.pdf (accessed April 5, 2019).

5
Comparing Renewable Energy Micro-Grids in Cambodia, Indonesia, and Laos

A Technological Innovation Systems Approach

Tobias S. Schmidt, Nicola U. Blum, and Catharina R. Bening

1. Introduction

Globally, of about 800 million people lacking access to electricity, the vast majority live in the rural areas of developing countries (World Bank 2018). Sustainable development goal 7 (SDG7) aims to ensure "access to affordable, reliable, and modern energy for all." SDG7.1 aims to achieve full electrification by 2030. SDG7.1-related activities are supported by the United Nations Sustainable Energy for All (SE4All) initiative and other global initiatives and institutions, which aim to facilitate large-scale investments in low-carbon electrification projects in developing countries (SE4All 2013). Low-carbon electrification projects typically utilize renewable energy (RE) sources, such as solar photovoltaic (PV), micro-hydro, or small biogas reactors. Regardless of the source of power, once generated, electricity is either stored or supplied directly to the end-user at the source or distributed through electricity grids of varying sizes and voltages.

Micro-grids (also called village-grids) provide energy access in a decentralized manner by generating power and then distributing it to multiple households independent of the main grid. This decentralized solution often is more economical and (technically) feasible than making extensions to an existing large grid (Szabó et al. 2011). At the same time, micro-grids typically offer higher poverty alleviation potentials than other completely independent, off-grid solutions that generate and provide power directly at the source to end-users, such as solar-home-systems (SHS) or solar lanterns (ARE 2011). In our study, we focus on decentralized micro-grids (MG) that are powered by renewable energy (RE) sources, or RE-MGs. A RE-MG can be defined as an isolated (i.e., off-grid), small (5 to 200kW) electricity grid based on renewable energy sources powering several customers, such as in a village (c.f., ESMAP 2007).

Despite the economic and practical advantages of micro-grid solutions, the global diffusion of renewable energy-based micro-grids (RE-MG) is rather low. In the past, the low use of RE-MGs were often related to poor reliability and maturity of critical components (such as batteries). Most of these types of bottlenecks have been alleviated with improved technological advances (Bhattacharyya 2013; Blum et al. 2013;

Schmidt et al. 2013). Recent research suggests that the current pattern of low diffusion is highly related to sociopolitical factors, including governance issues during the innovation, installation, operation, and maintenance of RE-MGs, as well as financing issues for these kinds of technologies (Blum et al. 2015; Comello et al. 2017; Malhotra et al. 2017; UNDP and ETH 2018).

In the case of Laos, Blum et al. (2015) have shown how governance and financial factors play out dynamically at the international, national, and local levels. Blum et al. then provide recommendations for improving RE-MG governance at all levels. However, the Blum et al. study is limited in its analysis to a single country, Laos. In order to support the diffusion of RE-MG globally and advance SDG7, it is important to understand whether the patterns found in Laos are similar to those found in other countries. Differences across countries would imply the need to adjust Blum et al. modes of RE-MG governance at relevant levels. There is, however, a lack of literature comparing the diffusion and governance dynamics of RE-MG across different countries.

To address this gap, we conduct a comparative study of RE-MG in Cambodia, Indonesia, and Laos, analyzing dynamics at international, national, and local levels. Our aim is to provide recommendations to decision-makers at all levels, but in particular to international initiatives supporting SDG7 (international coordination is explicitly aimed at under SDG7). Understanding country differences enables international organizations like SE4All to tailor their approach to countries' specificities and thereby make progress on SDG7 more effective.

In our study, we follow the same theoretical and methodological approaches used by Blum et al. (2015) in their study of Laos, also applying the functional framework of technological innovation systems (TIS) with interview data gathered in extensive field work. The TIS framework has been valuable in explaining innovation-diffusion dynamics of specific technologies in a general sense, as well as in developing country contexts,[1] see, for example, work by Blum et al. (2015), and Tigabu and colleagues (2015). In the case of Laos, Blum et al. (2015) identified that strong misalignments and the subsequent hampered interactions within and between institutions at local, national, and international levels were key characteristics of RE-MG diffusion bottlenecks. Consequently, we focus on these three levels and draw from the multilevel governance literature to derive policy recommendations.

The remainder of this chapter is structured as follows: section 2 introduces the analytical framework of TIS and relates it to the multilevel governance literature. In section 3 we describe our sampling strategy and the three case study countries (3.1) utilized in the comparison, as well as the methods used (3.2). We provide the results

---

[1] The TIS framework is specifically well suited for analyzing non–end-consumer technologies of a certain complexity, whose diffusion cannot be sufficiently explained with the diffusion of innovation framework (Rogers 2003). Such technologies, including RE-MG (ARE 2011), require technological experimentation and adaption to local circumstances in order to diffuse widely (Schmidt and Huenteler 2016).

of our comparison in section 4. A discussion of our results and their policy implications follows in section 5, and our chapter concludes in section 6.

2. Analytical Framework: TIS and Multilevel Governance

The innovation systems literature is rooted in evolutionary economic thinking and theories of interactive learning, highlighting the role of tacit knowledge (Edquist 1997; Nelson and Winter 1982; Rosenberg 1982). It gained relevance in innovation research and economics in the late 1980s (for an overview see Edquist 1997; Edquist 2005; Lundvall et al. 2009). The innovation systems literature regards technological innovation and diffusion as an evolutionary, systemic process, involving many feedback loops; insights are gained during the production and use of technology, which result in the development of tacit knowledge that is then iteratively fed back into the research and development (R&D) process (Arthur 1989; Carlsson and Stankiewicz 1991; Rosenberg 1994). Innovation systems' researchers use different system boundaries to study innovation systems: national (or, regional), sectoral (e.g., focusing on the chemical sector), and technological innovation systems (Binz et al. 2014; Carlsson et al. 2002; Malerba 2002; Stephan et al. 2017). A TIS focuses on a specific technology and is defined as a "dynamic network of agents interacting in a specific economic/industrial area under a particular institutional infrastructure and involved in the generation, diffusion, and utilization of technology" (Carlsson and Stankiewicz 1991, p. 93). In our study, we focus on the RE-MG TIS; that is, the dynamic network of agents engaged in the generation, diffusion, and utilization of renewable energy in micro-grids. As noted in the definition of TIS, these systems are characterized by "structural elements": *Technological artefacts* (e.g., solar panels), *actors* (i.e., those involved in the innovation and diffusion of technologies), the *networks* in which they interact, and the *institutions* that govern the actors' behavior and interactions. Studies of innovation systems derive policy recommendations to foster socially desired technological change (Edquist 1997).

Within the TIS literature, the concept of the *functions* of a technological innovation system (Bergek et al. 2008; Carlsson and Jacobsson 2004) has gained importance in the process of deriving informed policy recommendations (Bening et al. 2015). The functions approach goes beyond analyzing the building blocks mentioned here by attempting to make the system's performance "measureable" (Bergek et al. 2008; Hekkert et al. 2007). Bergek et al. (2008) describe the functional approach in their seminal paper as follows: "With an analysis of functions, we first desire to describe what is actually going on in the TIS ... [to] come up with a picture of an 'achieved' functional pattern, that is, a description of how each function is currently filled in the system." For an overview of the set of functions see Table 5.1.

The TIS literature assumes that informed policymakers can support these functions and thereby accelerate innovation and diffusion of the respective technology.

Table 5.1 Definitions of TIS functions

Function	Definitions
F1 Entrepre-neurial activities	Entrepreneurs play significant roles in innovation—they turn potentials (knowledge development, networks, markets, etc.) into concrete realities by acting in and on the system (e.g., taking advantage of business opportunities)
F2 Knowledge development (learning)	Mechanisms of learning are at the heart of any innovation process. For instance, according to Lundvall: "the most fundamental resource in the modern economy is knowledge and, accordingly, the most important process is learning." Therefore, R&D and knowledge development are prerequisites within the innovation system. This *function* encompasses learning by searching and learning by doing.
F3 Knowledge diffusion*	According to Carlsson and Stankiewicz the essential function of networks is the exchange of information. This is important in a strict R&D setting, but especially in a heterogeneous context where R&D meets government, competitors and market. Here policy decisions (standards, long term targets) should be consistent with the latest technological insights and, at the same time, R&D agendas are likely to be affected by changing norms and values. For example if there is a strong focus by society on renewable energy it is likely that a shift in R&D portfolios occurs toward a higher share of renewable energy projects. This way, network activity can be regarded as a precondition to learning by interacting. When user producer networks are concerned, it can also be regarded as learning by using.**
F4 Guidance of the search	The activities within the innovation system that can positively affect the visibility and clarity of specific wants among technology users fall under this system *function*. An example is the announcement of the policy goal to aim for a certain percentage of renewable energy in a future year. This grants a certain degree of legitimacy to the development of sustainable energy technologies and stimulates the mobilization of resources for this development. Expectations are also included, as occasionally expectations can converge on a specific topic and generate a momentum for change in a specific direction.
F5 Market formation	A new technology often has difficulties to compete with incumbent technologies, as is often the case for sustainable technologies. Therefore it is important to create protected spaces for new technologies. One possibility is the formation of temporary niche markets for specific applications of the technology.... This can be done by governments but also by other agents in the innovation system. Another possibility is to create a temporary competitive advantage by favorable tax regimes or minimal consumption quotas, activities in the sphere of public policy.
F6 Resource mobilization	Resources, both financial and human, are necessary as a basic input to all the activities within the innovation system. Specifically for biomass technologies, the abundant availability of the biomass resource itself is also an underlying factor determining the success or failure of a project.
F7 Creation of legitimacy	In order to develop well, a new technology has to become part of an incumbent regime, or has to even overthrow it. Parties with vested interests will often oppose this force of creative destruction. In that case, advocacy coalitions can function as a catalyst to create legitimacy for the new technology and to counteract resistance to change.

* The original definition by Hekkert et al. focuses exclusively on knowledge diffusion through networks. This definition falls short as knowledge is also diffused outside of networks; we therefore interpret the function slightly differently from Hekkert et al. (2007).

** For the analysis in this chapter, we use a broader definition of the function and include: "The function captures the breadth and depth of the knowledge ... and how that knowledge is diffused and combined in the system" (Jacobsson 2008, p. 1496), thereby going beyond the diffusion through networks, only.

Source: Based on Hekkert et al. (2007)

TIS have been analyzed at different spatial levels, for example, at the national level (e.g., Negro et al. 2007; Stephan et al. 2017), the global level (e.g., Binz et al. 2014), at two levels (Schmidt and Dabur 2014), or at various levels (e.g., Blum et al. 2015). In this chapter, we focus on the diffusion of one technology (namely RE-MG), that can substantially contribute to reaching SDG7. While the TIS and functions framework is not specific to SDGs, it is a well-proven framework to explain diffusion and diffusion bottlenecks of socially desired technologies, going beyond simple lists of drivers and barriers. The TIS approach has also been shown to be insightful when applied to low- and middle-income contexts (see e.g., Blum et al. 2015; Schmidt and Dabur 2014; Surana and Anadon 2015; Tigabu et al. 2013; Tigabu et al. 2015). For the case of RE-MG, Blum and colleagues (2015) have shown that in order to derive meaningful policy recommendations, at least three levels need to be considered: the *international* level, as many components used in a RE-MG are sourced internationally and often international development support is important; a *national* level, as many formal institutions (such as electricity sector regulation) as well as actors and networks (e.g., through associations) are very important at that level; and a *local* level, as many actors are locally constrained to a village in which a RE-MG is installed and local (informal) institutions that are relevant for the operation and maintenance can significantly differ from village to village.

In order to better explain how the support of RE-MG growth can respond to different sociopolitical contexts at multiple levels, we draw insight from the multilevel governance (MLG) literature (see e.g., Bache and Flinders 2004; Hooghe and Marks 2003) when formulating policy implications. In doing so, we follow the suggestion by Bening at al. (2015) to combine TIS with other bodies of literature in order to improve policy recommendations.[2] MLG literature is rooted in various streams of political science—such as international relations, European Union studies, or public policy—and goes beyond describing and analyzing the purely regulatory aspects of governance (Hooghe and Marks 2003).[3] MLG studies examine—among others—how responsibilities are shared (or overlapping) between different governance levels, and how these responsibilities shift over time and in response to which factors. In our analysis, we focus on how the various governance levels interact (Gupta 2007). Specifically, we analyze the RE-MG TIS building blocks and functions at the national

[2] Bening and colleagues also argue that the functions approach to TIS is a normative approach due to its focus on *desirable* technologies. Many institutions regard off-grid technology necessary in order to reach SDG7. We justify our technology focus on micro-grids by their higher poverty alleviation potentials than smaller off-grid solutions and the often-lower public costs compared to main grid extension (see section 1). The fact that we focus on renewables as energy source can be justified by the high social cost of carbon and other emissions from fossil fuels (van den Bergh and Botzen 2014) and the often lower costs of renewables over diesel-based alternatives (Blum et al. 2013).

[3] Gupta (2007) describes this in more detail: "Since the 1980s, there has been a general trend in the political science literature to focus more on governance than government, focusing on in addition to the state, hierarchical relations, the public sector and a command and control role for government to also analyzing the roles of civil society, networks and partnerships, private and community activities and the steering and enabling role of the state."

level in Cambodia, Indonesia, and Laos and their interactions with the international and local levels.

3. Case Study Countries, Method, and Data

In subsection 3.1, we describe our case selection and provide background information on each country. This is followed by subsection 3.2, detailing the method.

3.1 Cambodia, Indonesia, and Laos

We selected Cambodia, Indonesia, and Laos as our countries of study because all face massive rural electrification challenges, have high RE potentials in rural areas, have employed national electrification targets and policies, and are supported by international initiatives and institutions, such as SE4All. All three countries are furthermore characterized by rural poverty, low- to medium-level electrification rates, and have ambitious national targets and policies regarding electrification (i.e., SDG7) and economic development (compare Table 5.2). At the same time, important country differences exist concerning geography and topography: while Cambodia has a relatively flat topography with access to the South China Sea, Indonesia is an archipelago consisting of over 17,000 islands with high mountains (mostly volcanoes) and regularly hit by earthquakes and volcanic eruptions. Laos is landlocked and characterized by relatively high mountains (the foothills of the Himalayas).

Also, the political systems and the recent history differ strongly across the three countries, affecting the economy and its structure: Cambodia has been a constitutional monarchy since 1993 and operates as a parliamentary representative democracy. The current ruling party in Cambodia is the Cambodians People's Party, which was formerly Marxist-Leninist, but recently became more reformist. Generally, the political situation in Cambodia is still characterized by the country's recent history, including the Vietnam War, when the Red Khmers under their leader Pol Pot implemented an ultra-Marxist regime and social engineering policies, resulting in genocide.

Postcolonial Indonesia was ruled by Dictator Suharto for thirty years, but has been operating as a democratic republic since 2004, though the president continues to hold significant power. Indonesian economic policy is mostly characterized by market-based principles. Laos is a communist single-party socialist republic with the Lao People's Revolutionary Party (LPRP) as the only legal political party. Its economic policy is moving toward a more important role of the private sector, but still mostly based on Marxist planning principles.

In terms of business regulation, all three countries are characterized by the presence of critical barriers to private sector activities, especially to entrepreneurs through high levels of bureaucracy (reflected in the World Bank's Ease-of-doing-business rankings displayed in Table 5.2).

Table 5.2 Selected socioeconomic indicators of the three analyzed countries

	Cambodia	Indonesia	Laos
Electrification rates (national / rural) (International Energy Agency 2014)	34% / 18%	76% / 59%	78% / 70%
Population (Central Intelligence Agency 2014)	15.5 million	253.6 million	6.8 million
Gross National Income per capita (Purchase power parity) (United Nations Development Programme 2015)	USD2,805	USD8,970	USD4,351
Gini index (gauge of economic income inequality; 0% representing perfect equality, 100% representing perfect inequality) (World Bank 2015a)	34.7%	34.1%	35.5%
Share of people below 2$/day poverty line (International Fund for Agricultural Development 2011)	58%	76%	77%
Rank in the Human Development Index 2013 (of 187 countries; 1 being the highest level of human development) (United Nations Development Programme 2015)	136	108	139
Rank in the TI Corruption Perception Index (out of 175 countries; 1 being the lowest levels of perceived corruption) (Transparency International 2014)	156	145	107
Secondary/tertiary education: Enrolment rate (World Bank 2015b)	38% / 16%	76% / 32%	45% / 18%
Rank in the ease of doing business (World Bank Group 2015) (out of 189 countries; 1 being the most business-friendly environment)	135	114	148

With regard to the energy sector, major differences exist across countries (IEA 2014). While all three countries have high renewable energy potentials, Indonesia has very high coal and also significant gas and oil production. Laos and Cambodia satisfy their energy demands mostly by hydropower and fuel and electricity imports. Imports are particularly important for Cambodia. Due to the high share of imported electricity and the dependency of imported fossil fuels, Cambodia's electricity prices are among the highest in the region—up to 70 USD cents/kWh (van Mansvelt 2011). In Laos, where most grid-connected electricity is produced through hydro-power, the end-consumer tariffs are progressive (tariffs increase with consumption levels), start very low (around 4 USD cents/kWh for low consumer households), and are further cross-subsidized by the higher tariffs paid by business and industry consumers. In Indonesia, where over 90% of the grid-provided electricity is generated based on fossil fuels, household (and other consumer type) tariffs are highly subsidized and consequently very low (IEA 2014).

3.2 Method

We iteratively collected and analyzed data in a confirmatory qualitative case study design (Seawright and Gerring 2008), following the procedure proposed by Yin (2003). To this end we drew from primary data, mostly sourced through semistructured interviews and on-site visits of RE-MG conducted between 2010 and 2012, during extended stays in Laos, Cambodia, and Indonesia, as well as from reports, policy documentation, and other documentation (Eisenhardt and Graebner 2007).

We use the functional approach to TIS primarily to analyze the innovation and diffusion dynamics of MG within each country (as suggested by Tigabu et al. 2015), while at the same time taking into account that much of the technological innovation dynamics occurs outside of the three analyzed countries (Schmidt and Dabur 2014). We define the functions based on Hekkert et al. (2007) and Bergek et al. (2008) (see Table 5.1). To gain a multilayered picture of actors, institutions, networks, and the technology, as well as the processes in the TIS, we sampled interview partners from all relevant stakeholder groups. We account for a representative sample at the national level and include stakeholders from the local and international levels. Table 5.3 provides an overview of the interviewees, the stakeholder group they belong to, and the level at which they mainly operate.

We obtained a list of potential interview partners through web search and our international network. This list was extended when interviewees provided us with additional contacts. In sum, we conducted eight interviews with 10 people in Laos, 10 interviews with 14 people in Indonesia, and nine interviews with nine people in Cambodia, always including stakeholders from the local, national, and international level (compare Table 5.3). Additionally we interviewed four international experts working on electrification projects and policies in Southeast Asia. Interviews were prepared by reviewing related documents and tailoring the semistructured interview guideline to each interviewee. The interviews were conducted at a place of convenience for the interviewees in the capitals, provincial towns, or villages and lasted between 20 minutes and two hours. With the consent of the interviewee, interviews were recorded; otherwise, the interviewer took detailed notes. To triangulate information provided by the different interviewees, we included observations of site visits to micro-grids, one per country. These observations were documented in videotapes and the researchers' notes. Additional written data were provided by interviewees and considered of special value by the authors, as country-specific documents are often not available online. Interviewees, therefore, were an important source of recent policy drafts and other non-public documentation. For the analysis, interviews were transcribed and saved together with the written materials in an electronic case study database.[4]

---

[4] Note that due to the long publication process of the book in combination with the strong dynamics of the RE-MG sector, the situation in the three countries might have slightly changed by the time of publication.

Table 5.3 Overview of interviewees' positions and organizations

Stakeholder group	Cambodia	Indonesia	Laos
Private sector (MG-related companies)	– Director of a private Cambodian company – Project manager at a clean energy investor	– Director of a Swiss-Indonesian private company – Indonesian project manager at a private company – International project manager at a private company – Manager at a local investor company	– Director of a private Laotian company – Communications manager at a company
Public sector (national or regional government)	– Director at the Ministry of Mines and Energy – Head of the Rural Electrification Fund	– Secretary of a regional body – Senior manager at state-owned utility – Manager 1 at state-owned utility – Manager 2 at state-owned utility	– Head of the Rural Electrification Division at the Ministry of Energy and Mines
Development cooperation sector*	– Project manager 1 of a development agency – Project manager 2 of a development agency – Director of a Cambodian NGO – International renewable energy consultant	– Project manager 1 in a development agency – Project manager 2 in a development agency – Project manager 3 in a development agency – Director of a local NGO – Manager at an international research institute	– European project manager in a development agency based in Laos – European project manager in an NGO based in Laos – Laotian project manager in an NGO based in Laos
Local level stakeholders	– Local entrepreneur owning a biomass gasifier	– Head of a local village council	– Four villagers involved in RE-MG governance

* These include nongovernmental organizations, international organizations, and development agencies—all with focus on the respective country. Additionally, we interviewed four international experts with a focus on Southeast Asia, who are not explicitly listed in the table.

The data were then coded using the software Atlas.ti. The applied code list included all structural and dynamic elements of the TIS, and the three geographical levels. New code values were added during the analysis whenever a new peculiarity was qualitatively discovered which was not included in the original codes (Marshall and Rossman 1989). Based on the coding, all interview transcripts were qualitatively analyzed to identify differences and similarities between the RE-MG TIS and their functions at the three government levels in the three countries, as reported in the next section.

4. Results

This section describes the results of the TIS analysis first describing building blocks and then TIS functions.

4.1 Building Blocks

The TIS literature differentiates three types of building blocks: actors and their networks, institutions, and technology. In the following, we elaborate on these three blocks and how they differ from country to country and summarize our findings in Table 5.4.

4.1.1 Actors and Networks

In Cambodia, the number of RE-MG actors is quite low. Given the country's small size, almost all important actors interact with each other in one relatively small network. The four most important actors we identified are (i) the national Ministry of Industry, Mines, and Energy (MIME), which also co-owns the fully-integrated electricity utility Electricité du Cambodge (EdC) along with the Ministry of Economy and Finance; (ii) international donors operating at the national and/or local level; (iii) entrepreneurs operating at the national and/or local level; and (iv) actors involved in the Rural Electrification Fund (REF). One of our interviewees points to the important

Table 5.4 Summary of the RE-micro-grids innovation system building blocks in the three analyzed countries

	Cambodia	Indonesia	Laos
Actors and Networks	• Few actors • Small network	• Many actors • Large network but lacking density	• Few actors • Small but dense network
Formal institutions	• Lean RE-MG policy • Good cooperation among policy actors	• Unclear regulatory responsibilities and policy overlaps	• Very little RE-MG-related policy and institutions
Informal institutions	• High cultural homogeneity • Mistrust of community-based approaches • High perceived levels of corruption	• High cultural heterogeneity • High perceived levels of corruption	• Very high cultural heterogeneity • High perceived levels of corruption
Technology	• Focus mostly on solar-PV-based MG	• Focus on micro-hydro- and solar-PV-based approaches	• Focus on micro-hydro- and solar-PV-based approaches

Source: authors.

role of MIME: "They [MIME] are doing things like a renewable energy strategy, renewable energy policy, and a rural energy strategy."

Meanwhile, there are many more actors engaged in Indonesia's RE-MG network. For example, many of the country's ministries are involved in electrification, some having formulated individual energy and/or electrification goals. One exemplary interview quote pointed to the high level of fragmentation: "There are 36 Ministries here in Indonesia, several of them have RE programs, and there is no interest for cooperation." There are also many nongovernmental organizations (NGOs) (national and international) and international donors active in Indonesia, mostly at the national, but partly also at the local level. A dominant Indonesian actor is the national state-owned utility Perusahaan Listrik Negara (PLN; Indonesian for "State Electricity Company"). Due to the sheer number of actors, the size, and the geographical dispersion of the country, the RE-MG network is not very dense; nevertheless, denser clusters or subnetworks (often focusing on only one island) exist and often span the three levels of governance, from local to international.

In Laos, most actors are concentrated at the national level, including the fully integrated, state-owned utility (Electricité du Laos), the Ministry of Energy and Mines, a few NGOs, and a small research center. Not many actors can be found in Laos at the local level, with villagers principally playing the role of electricity consumers only. Due to the small country size, the Laotian RE-MG network is small and geographically dense. The few actors on the local level are not well connected owing to the mountainous topography and the scattered cultural/linguistic situation (for more details see Blum et al. 2015).

4.1.2 Institutions

In Cambodia, RE-MG-relevant regulation is relatively lean, and there is good cooperation among national regulatory institutions, such as the electricity regulator (Electricity Authority of Cambodia, EAC), or between national and local institutions. Independent power producers can get licenses from EAC and directly sell electricity to villagers. Indonesia has many national level institutional actors, but lacks manpower and clarity in regulatory responsibility. According to our interviewees, many electrification-related decisions are therefore made at the local level by entrepreneurs, local regulators, and NGOs, circumventing national level institutional engagement. In Laos, few policies exist to support RE-MG development. Licenses are administered by the national government, but the process is opaque and further complicated by the perceived high levels of corruption (compare Table 5.2).

Also culture (as an informal institution) matters for the diffusion of MG (Blum et al. 2015). Cambodia features a relatively homogeneous ethnos, with 90% of the population belonging to the Khmer group, which is also the name of the official national language. Specific to Cambodia, community-based approaches (playing out at the local level) that are planned centrally (at the national level) are not trusted by much of the public. This mistrust is related to the trauma of the Red Khmers era, who strongly promoted community-based approaches within a centralized governance structure. These Khmer top-down community-based approaches turned out to be disastrous from a socioeconomic perspective. Cambodia is also rather homogeneous in terms of religion with 97% of the population being Buddhist (Pew Research Center 2010).

Indonesia, on the other hand, is characterized by a much higher ethnic and cultural heterogeneity, with about 300 ethnic groups spread across the archipelago. While Bahasa Indonesia is the official language, many other languages are spoken (especially on the more eastern, less electrified islands). About 87% of the population is Muslim (Pew Research Center 2010). While freedom of religion is granted for most religions, the non-Muslim population—especially the economically successful ethnic Chinese population—regularly experiences discrimination (Marshall 2018).

Laos is at least equally culturally diverse as Indonesia featuring more than 100 different languages and dialects which are highly geographically scattered (see Messerli et al. 2008). The official language, Lao, is often not well known in remote villages. Buddhism is mixed with natural religions (Pew Research Center 2010).

All three countries have in common high levels of perceived corruption (compare their positions in the Transparency International Corruption Index provided in Table 5.2). However, there are some important differences regarding the understanding and treatment of corruption within the public sphere: in Indonesia, corruption recently became a more publicly salient topic; for example, corruption played a significant role in the latest presidential election campaigns. Yet corruption still seldom features in public discourse in Cambodia and Laos.

4.1.3 Technology

A RE-MG can be defined as an isolated, off-grid, small (5 to 200kW) electricity grid based on renewable energy sources, powering several customers such as in a village (compare ESMAP 2007). A MG schematic is provided in Figure 5.1. Power sources for RE-MGs differ across the three countries under study. While in all three countries solar photovoltaic- and biomass-powered MG are utilized, MGs in Laos and Indonesia also feature micro-hydro generated power. Cambodia's flat topography inhibits hydro-based power generation, but the country has good solar potential and PV technology is gaining in popularity. As one interviewee points out: "Cambodians even already have their own word for solar, everybody knows it" (Blum 2013).

4.2 Functions

In the following section, we summarize the key differences regarding each function of the RE-MG innovation system in the three countries. Figure 5.2 in the discussion section synthesizes these results and connects them to policy implications.

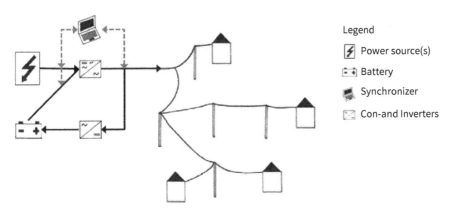

Figure 5.1 Schematic of a typical RE-MG setup as shown in Blum (2013); core components of a RE-MG are the power sources, synchronizers, transformers, a battery system, switchgear, inverters, the software balancing load and supply, and wiring (cf. figure legend)
Source: Blum 2013.

4.2.1 Function 1: Entrepreneurial Activities

In Cambodia, much entrepreneurial activity is observed at the village level, which is incentivized by national and international policy programs, most prominently by the World Bank. The state-owned national utility is responsible for electricity provision in provincial towns and the capital Phnom Penh, but not in villages, further opening up entrepreneurial opportunities in power generation and distribution within and between villages. Entrepreneurial activity in the energy sector is also high at the village level in Indonesia. In Indonesia, patterns of entrepreneurial activity are primarily driven by people seeing economic value and potential in the business of renewable energy. The low generation costs of micro-hydro (Blum et al. 2013) and the availability of international donor money motivates entrepreneurial activities.

At the national level in Indonesia, the provision of electricity is tightly controlled by a state-owned, vertically integrated monopolist utility PLN. The utility's electrification activities are primarily diesel-based MG. Diesel power generation comes at a higher cost, but is substantially offset by large subsidies (Blum et al. 2013; Schmidt et al. 2013) In Laos, entrepreneurial activity around electricity is very limited at the village level. Given the dominant communist paradigm, people expect the public sector to take charge of infrastructure developments (Blum et al. 2015). At the national level, state-owned utility Electricité du Laos (EDL), similar to the case of PLN in Indonesia, has a monopoly on electricity provision. EDL focuses mostly on extending its existing grid. The massive amount of cost-efficient electricity produced by EDL's large hydro dams contributes to EDL's focus on grid extension rather than MG development.

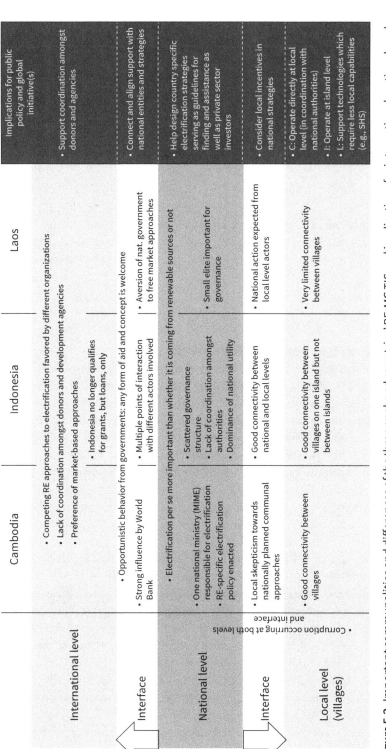

Figure 5.2 Important communalities and differences of the three analyzed countries' RE-MG TIS and implications for international, national, and local policymakers

Source: authors.

4.2.2 Function 2: Knowledge Development

The ability to share and create new knowledge about RE-MG is rather limited in Cambodia. This correlates with capacity issues: assembly and manufacturing activities of RE-MG (and industrial goods in general) are very limited in the country, and Cambodia's academic sector is small and focused on Phnom Penh. The general education level of Cambodians living and working in Cambodia is still low (Table 5.2). Knowledge-generation capacity in Indonesia is quite different. With the support of mostly Swiss and German development aid, local entrepreneurs can domestically manufacture hydro-powered RE-MG technology and compete in international markets.

Generally, the level of economic development and education is significantly higher in Indonesia than in the other two study countries (Table 5.2), facilitating local knowledge development, capacity for innovation, and access to international networks. Bandung University's engineering school, which offers several energy-related undergraduate and graduate engineering programs, has a strong, international reputation. In Laos, technology RE-MG-related knowledge is developed abroad with local activities focused mostly on the operations and maintenance of RE-MG. Knowledge development is therefore mostly related to and developed during these activities. While education levels are generally low (Table 5.2), there is a national Laotian Research Center (LIRE) that supports some technological advancement.

4.2.3 Function 3: Knowledge Diffusion

A number of entrepreneurs in Cambodia approach RE-MG development by applying RE-MG solutions from other, internationally sourced, biomass-based RE-MG projects. Often these sourced models offer electricity at significantly lower cost, but with lower reliability and quality. However, learning by doing enables such entrepreneurs to increase the reliability and quality through an iterative process. This development and improvement process is supported by international organizations that organize trips to other developing countries such as India that have a more advanced RE-MG sector. Knowledge diffusion in Indonesia seems to work relatively well at the international-national and national-local interfaces. Indonesia's rather mature industry of RE-MG component manufacturing strengthens knowledge flows, as technical information is easily generated and distributed within this industry. In Laos, the situation is different: while knowledge is transferred from the international level to the national level, it then hardly diffuses further within the country. This has to do with the mountainous topography as well as the scattered linguistic situation, both of which hamper information flow between villages. Due to the low level of exchange, knowledge is also more easily lost (Blum et al. 2015).

4.2.4 Function 4: Guidance of the Search

In all three countries, there seems to be very little guidance at the local level and villagers are not very involved in electrification strategies and technologies. Also, on the national level the situation is very similar across countries, in that a clear electrification strategy is missing and international grants are welcome. This is important because on the international level different electrification approaches (grid-extension, MGs, solar-home-systems, or solar lanterns) are favored by different electrification-related organizations. While in Cambodia the World Bank seems most relevant for this function of prioritizing technologies based on feasibility studies, in Indonesia and Laos there are many donors supporting different approaches. This means that no clear signal is provided to entrepreneurs and other actors of the innovation system in the respective countries, which would allow them to focus on preferred technologies or approaches.

4.2.5 Function 5: Market Formation

Concerning market formation, a highly critical function for the diffusion of RE-MG, we find major bottlenecks in all three countries, but they are rooted in different circumstances. While in Cambodia, the national electrification policy is favorable for RE-MG markets (see building block *institutions*, section 4.1.2), villages are often too small to be attractive for MG entrepreneurs. Therefore, many electrification entrepreneurs focus on few projects which include small industrial anchor loads. In Indonesia, the bottlenecks stem mostly from regulation. First, Indonesia's power sector is dominated by the state-owned utility PLN and generally not open to for-profit independent power producers. Also, power retail prices are regulated at the same level across the entire country, despite many independent grids which feature high generation cost differences and are highly subsidized, as is diesel fuel. This situation, together with the variable weather patterns, significantly penalizes RE-MG compared to diesel-based MG and grid-extension and thereby limits their diffusion. In Laos, RE-MG markets are in a very early phase and most projects have a demonstration character and heavily depend on international grant finance. The national private sector is hardly moving beyond the demonstration phase because investor money is scarce and flows into other investment opportunities instead because the villagers' willingness to pay is often lower than the generation costs. The few private sector-based projects are somewhat scattered owing to the topographic and cultural situation.

4.2.6 Function 6: Resource Mobilization

Regarding resources, we focus on financial and human resources. All three countries are receiving international development assistance. While Laos and Cambodia receive many grants, Indonesia typically only qualifies for concessional loans, because of its improved economic position. In all three countries, the international money is channeled to RE-MG projects through national government authorities. In Indonesia, however, with its 26 ministries, the money is split into many different pieces, which are often too small to help finance individual projects. Foreign direct investment (FDI) is almost nonexistent in the RE-MG sectors of all three countries. In Cambodia, financial resources, aside from international donations or impact investments, come mostly from the national private sector. This is not the case in Indonesia, due to regulation preventing profits in electrification projects. In Laos, a local financial sector is close to nonexistent. At the same time, international financiers support many different electrification approaches (from grid-extension, via MG and SHS, to solar lanterns), which, together with the lack of an electrification technology strategy, leads to an uncoordinated financing situation. All three countries are characterized by a high availability of labor resources but a lack of qualification and skills. The situation seems most severe in Laos, where training of personnel does not result in spillovers due to the topographic and cultural situation as discussed in section 4.2.3. In Indonesia, the growing manufacturing industry also results in more qualified workers being available.

4.2.7 Function 7: Legitimacy Creation

At the international level, RE-MGs enjoy a high level of legitimacy. SHS are also seen as good solutions for reaching SDG7 (UNDP 2011). The legitimacy of RE-MG is also high at both the national and the local level in Cambodia, as alternative means of electrification, especially diesel generators, have very high costs. Biomass-gasifiers and solar PV-powered MG are also popular, as they can provide electricity at lower cost. In Indonesia, RE-MG are also widely accepted. Owing to the high subsidies for diesel, however, fossil fuel-based off-grid power generation enjoys high popularity. In Laos, the legitimacy of RE-MG is rather low. The national government supports many forms of electrification depending on international support. At the local level, legitimacy is low, as villagers prefer grid extension based on the belief that electricity services from the main grid are much more reliable, despite evidence that the main grid is not very stable (cf. Blum et al. 2015).

5. Discussion and Policy Recommendations

After having explored the similarities and differences of the building blocks and functions of the RE-MG TIS in the three studied countries, we synthesize findings, identify important bottlenecks at all three governance levels, and provide recommendations to decision-makers at the three levels with a focus on international initiatives, such as the United Nations' SE4All initiative, targeting SDG7. We relate our findings to insights from the multilevel governance literature and specifically discuss which levels international initiatives should target and where coordination interventions by such initiatives are most helpful in order to tap the full potential in RE-MG technology in supporting SDG7. Figure 5.2 summarizes our main observations and implications for national and international decision-makers.

Organizations at the international level typically favor market-based approaches, but support different technologies to achieve SDG7. Coordination among these organizations seems limited. This situation trickles down to the national level (Figure 5.2), as none of the three countries has a clear electrification strategy that specifies or favors particular technologies/approaches. On the contrary, all three countries demonstrate conditions that favor opportunistic behavior, accepting international support regardless of the suitability of a technology or concept to meet national electrification targets and potentials. In the multilevel governance literature such situations are referred to as "socially perverse outcomes" due to spillovers from higher level dynamics and an overall lack of coordination across the levels (Hooghe and Marks 2003). This situation suggests an important future coordination role for global initiatives at the international level-national level interface.

Specifically, we suggest that global initiatives support countries in designing and implementing national electrification strategies, with specified roles for RE and MG that focus on those technologies most suitable to the country's unique characteristics for advancing SDG7. We also suggest that global initiatives can facilitate needed coordination among donors and development agencies to connect and align support among the approaches favored by national strategies and international donors. Coordination represents the main cost in multilevel governance, which increases with the number of actors and levels (Hooghe and Marks 2003). Hence, having one international initiative coordinating between the international and the national levels may limit costs while avoiding perverse social outcomes.

The national electrification strategy should also consider the current governance structure at the national level. In the case of fragmented government (such as that seen in Indonesia), establishing a coordination hub or similarly functioning entity between the institutional bodies is important to technology innovation and diffusion (Kanie et al. 2013). Hence, we suggest national electrification strategies with effective coordination entities. Implementing such suggestions implicates a political reform process at the national level. Meanwhile, in places like Laos and Cambodia, where energy and electrification governance is highly centralized, we suggest enhancing market completion (particularly in Laos), inclusivity, and stakeholder empowerment. While such strategies potentially improve RE-MG diffusion, higher fragmentation comes at increased coordination costs (Edler and Kuhlmann 2008).

National energy production and electrification strategies should also consider how national level dynamics play out at the local level. In the Cambodian example, there is general skepticism toward nationally planned communal approaches. Therefore, advancing that sort of national strategy is likely to be poorly received and ineffective. In Laos, however, where people have a much higher level of trust in the national government, top-down approaches may encounter fewer obstacles to implementation and public buy-in. The public expects the Laotian government to play an important role in any instrument supporting the national electrification strategy. International initiatives should be cognizant of such geopolitical, economic, and cultural differences to avoid recommending standardized policy strategies across different countries. In Cambodia, for example, international institutions have the opportunity to assume instrument-related responsibilities in order to address skepticism at the local level.

Our study reveals varying degrees of connectivity from the national to the local/village level and among villages at the local level. In Cambodia, good connectivity facilitates knowledge spillovers, an important feature of the dynamics and governance of diffusion of RE-MG (Bhattacharyya and Palit 2014; Blum et al. 2015). However, the small size of many Cambodian villages implies that strategies should include small-scale solutions. In Indonesia, village connectivity is high within, but not across, islands. We suggest strategies be tailored at the level of the island, representing an additional, unique governance layer, which, however, results in additional coordination costs (Scharpf 1997).

In Laos, actors responsible for the governance of RE-MG encounter challenges to learning about other actors' decisions and outcomes in other villages. The technological tacit knowledge developed within projects does not easily diffuse through the hilly and culturally fragmented country at the local level. As RE-MG depends on tacit technical knowledge for operation and maintenance, this is a major barrier for achieving SDG7 through RE-MG. Other, less knowledge-intense electrification technologies, such as grid-extension or SHS, might therefore be more promising for Laos.

Corruption at both the national and the local level (as well as at their interface) has implications for international initiatives. Energy and electrification strategies should comprehensively address corruption across the multiple scales and in their interfaces. Corruption is regarded as typical of multilevel governance (Hooghe and Marks 2003; Lowery and Lyons 1995), and therefore addressing and avoiding it becomes critical in schemes like RE-MG that favor complex institutional networks.

6. Conclusion and Avenues for Future Research

RE-MG can play an important role in reaching SDG7 and bringing electricity to the 800 million people without electricity access, while offering higher service levels often at lower cost than alternative electrification approaches. However, RE-MG diffusion in most developing countries is still relatively low (UNDP and ETH 2018). We highlight bottlenecks that hinder the diffusion of the technology and their contributions to the advancement of SDG7. Revealing consequential country differences, we underscore implications for international initiatives that aim to support electrification worldwide. Such consequential particularities should be considered in national policies and by international initiatives. Business models, technological approaches, instruments, and so forth might have to be tailored to individual countries. At the same time, international donors and development agencies favor different business models and approaches. Hence, coordinating between countries' individual needs and international offerings could be an important role of such initiatives.

Our chapter represents an initial attempt to improve our understanding of country variance with respect to the dynamics and governance challenges involved in reaching SDG7 and particularly in upscaling the diffusion of off-grid electrification technologies. Future research should include assessments of additional countries (spanning a variety of regions), and analysis of alternative technologies relevant for SDG7, such as solar-home-systems. Quantitative analyses would also enhance our qualitative assessment of RE-MG potentials and energy and electrification governance. For example, an economic assessment of the costs of coordination in light of a full spectrum of socioenvironmental benefits would be helpful and likely necessary to stimulate policy action effectively.

Acknowledgments

We thank Volker Hoffmann, Fah Yik Yong, and Marius Schwarz for their support. All errors remain our own. All authors declare no conflicts of interest in this chapter.

References

Alliance for Rural Electrification (ARE 2011). *Hybrid Mini-Grids for Rural Electrification: Lessons Learned. United States Agency for International Development.* USAID, Brussels.

Arthur, W. B. (1989). Competing technologies, increasing returns, and lock-in by historical events. *The Economic Journal* 99, 394, 116–131.

Bache, I., and Flinders, M. (2004). *Multi-Level Governance.* Oxford University Press, New York.

Bening, C. R., Blum, N. U., and Schmidt, T. S. (2015). The need to increase the policy relevance of the functional approach to technological innovation systems (TIS). *Environmental Innovation and Societal Transitions* 16, 73–75.

Bergek, A., Jacobsson, S., Carlsson, B., Lindmark, S. and Rickne, A. (2008). Analyzing the functional dynamics of technological innovation systems: a scheme of analysis. *Research Policy* 37(3), 407–429.

Bhattacharyya, S. (2013). *Rural rlectrification through decentralised off-grid systems in developing countries.* Springer, London.

Bhattacharyya, S. C., and Palit, D. (2014). *Mini-Grids for Rural Electrification of Developing Countries: Analysis and Case Studies from South Asia.* Springer, Cham.

Binz, C., Truffer, B., and Coenen, L. (2014). Why space matters in technological innovation systems: mapping global knowledge dynamics of membrane bioreactor technology. *Research Policy* 43(1), 138–155.

Blum, N. U. (2013). Fostering rural electrification: the case of renewable energy- based village grids in South East Asia. PhD thesis, Eidgenössische Technische Hochschule (ETH) Zurich.

Blum, N. U., Bening, C. R. and Schmidt, T. S. (2015). An analysis of remote electric mini-grids in Laos using the technological innovation systems approach. *Technological Forecasting and Social Change* 95, 218–233.

Blum, N. U., Sryantoro Wakeling, R. and Schmidt, T. S. (2013). Rural electrification through village grids: assessing the cost competitiveness of isolated renewable energy technologies in Indonesia. *Renewable and Sustainable Energy Reviews* 22, 482–496.

Carlsson, B., and Jacobsson, S. (2004). Dynamics of Innovation Systems: Policy-Making in a Complex and Non-deterministic World. International workshop on Functions of Innovation Systems' at the University of Utrecht, June 23–24, 2004.

Carlsson, B., Jacobsson, S., Holmén, M. and Rickne, A. (2002). Innovation systems: analytical and methodological issues. *Research Policy* 31(2), 233–245.

Carlsson, B., and Stankiewicz, R. (1991). On the nature, function and composition of technological systems. *Evolutionary Economics*, 93–118.

Comello, S. D., Reichelstein, S. J., Sahoo, A. and Schmidt, T. S. 2017. Enabling mini-grid development in rural India. *World Development* 93, 94–107.

Edler, J. and Kuhlmann, S. 2008. Coordination within fragmentation: governance in knowledge policy in the German federal system. *Science and Public Policy* 35 (4), 265–276.

Edquist, C. (Ed.) (1997). *Systems of Innovation: Technologies, Institutions and Organizations*. Routledge, Abingdon.

Edquist, C. (2005). Systems of Innovation: Perspectives and Challenges. In: Fagerberg, J., and Mowery, D. (Eds.), *The Oxford Handbook of Innovation*, pp. 181–208. Oxford University Press, Oxford.

Eisenhardt, K. M., and Graebner, M. E. (2007). Theory building from cases: opportunities and challenges. *Academy of Management Journal* 50 (1), 25–32.

Energy Sector Management Assistance Program (ESMAP 2007). Technical and Economic Assessment of Off-Grid, Mini-Grid and Grid Electrification Technologies. ESMAP, Washington, DC.

Gupta, J. (2007). The multi-level governance challenge of climate change. *Environmental Sciences* 4 (3), 131–137.

Hekkert, M. P., Suurs, R. a. a., Negro, S., Kuhlmann, S. and Smits, R. E. (2007). Functions of innovation systems: a new approach for analysing technological change. *Technological Forecasting and Social Change* 74(4), 413–432.

Hooghe, L. and Marks, G. 2003. Unraveling the central state, but how? types of multi-level governance. *American Political Science Review* 97(2), 233–243.

International Energy Agency (IEA 2014). *Energy Balances of Non-OECD Countries 2014*. World Energy Outlook 2014, Paris.

International Fund for Agricultural Development (IFAD 2011). Rural Poverty Report. IFAD, Rome.

Kanie, N., Suzuki, M. and Iguchi, M. (2013). Fragmentation of international low-carbon technology governance: an assessment in terms of barriers to technology development. *Global Environmental Research* 17(1), 61–70.

Lowery, D., and Lyons, W. 1995. The empirical evidence for citizen information and a local market for public goods. *American Political Science Review* 89(3), 705–707.

Lundvall, B.-Å., Joseph, K. J., Chaminade, C. and Vang, J. (Eds.) (2009). *Handbook of Innovation Systems and Developing Countries: Building Domestic Capabilities in a Global Setting*. Edwards Elgar Publishing Limited, Cheltenham, UK.

Malerba, F. (2002). Sectoral systems of innovation and production. *Research Policy* 31 (2), 247–264.

Malhotra, A., Schmidt, T. S., Haelg, L. and Waissbein, O. (2017). Scaling up finance for off-grid renewable energy: the role of aggregation and spatial diversification in derisking investments in mini-grids for rural electrification in India. *Energy Policy* 108, 657–672.

Marshall, C., and Rossman, G. B. (1989). *Designing Qualitative Research*. 4th ed. Sage Publications, Thousand Oaks, CA.

Marshall, P. (2018). The ambiguities of religious freedom in Indonesia. *Review of Faith and International Affairs* 16(1), 85–96.

Messerli, P., Heinimann, A., Epprecht, M., Phonesaly, S., Thiraka, C. and Minot, N. (Eds.) (2008). *Socio-Economic Atlas of the Lao PDR: An Analysis Based on the 2005 Population and Housing Census.* 1st ed. Geographica Bemensia, Swiss national Center of Competence in Research (NCCR) North-South, University of Bern.

Negro, S. O., Hekkert, M. P. and Smits, R. E. (2007). Explaining the failure of the Dutch innovation system for biomass digestion: a functional analysis. *Energy Policy* 35 (2), 925–938.

Nelson, R. R., and Winter, S. G. (1982). *An evolutionary theory of economic change.* Harvard University Press, Cambridge, MA.

Pew Research Center (2010). Global Religious Landscape 2010: Religious Composition By Country. Pew Research Center, Washington, DC.

Rogers, E. M. (2003). *Diffusion of Innovations.* 5th ed. Free Press, New York.

Rosenberg, N. (1982). *Inside the Black Box: Technology and Economics.* Cambridge University Press, Cambridge, UK.

Rosenberg, N. (1994). *Exploring the Black Box Technology, Economics, and History.* Cambridge University Press, Cambridge, UK.

Scharpf, F. W. (1997). *Games real actors play: actor-centered institutionalism in policy research.* Westview Press, Boulder, CO.

Schmidt, T. S., Blum, N. U. and Sryantoro Wakeling, R. (2013). Attracting private investments into rural electrification: a case study on renewable energy based village grids in Indonesia. *Energy for Sustainable Development* 17(6), 581–595.

Schmidt, T. S., and Dabur, S. 2014. Explaining the diffusion of biogas in India: a new functional approach considering national borders and technology transfer. *Environmental Economics & Policy Studies* 16 (2) 171–199.

Schmidt, T. S., and Huenteler, J. 2016. Anticipating industry localization effects of clean technology deployment policies in developing countries. *Global Environmental Change* 38, 8–20.

Seawright, J., and Gerring, J. (2008). Case selection techniques in a menu of qualitative and quantitative options. *Political Research Quarterly* 1975, 294–308.

Stephan, A., Schmidt, T. S., Bening, C. R. and Hoffmann, V. H. (2017). The sectoral configuration of technological innovation systems: patterns of knowledge development and diffusion in the lithium-ion battery technology in Japan. *Research Policy* 46, 709–723.

Surana, K. and Anadon, L. D. (2015). Public policy and financial resource mobilization for wind energy in developing countries: a comparison of approaches and outcomes in China and India. *Global Environmental Change* 35, 340–359.

Sustainable Energy for All Initiative (SE4All 2013). *Sustainable Energy for All Initiative.* United Nations, World Bank. https://www.seforall.org/ (accessed April 4, 2019).

Szabó, S., Bódis, K., Huld, T. and Moner-Girona, M. (2011). Energy solutions in rural Africa: mapping electrification costs of distributed solar and diesel generation versus grid extension. *Environmental Research Letters* 6(3), 034002.

Tigabu, A. D., Berkhout, F. and van Beukering, P. (2013a). Technology innovation systems and technology diffusion: adoption of bio-digestion in an emerging innovation system in Rwanda. *Technological Forecasting and Social Change* 90, 318–330.

Tigabu, A. D., Berkhout, F. and van Beukering, P. (2015b). The diffusion of a renewable energy technology and innovation system functioning: comparing biodigestion in Kenya and Rwanda. *Technological Forecasting and Social Change* 90, 331–345.

Transparency International (2014). Corruption Perceptions Index. Available online at http://www.transparency.org/cpi2014/ (accessed April 4, 2019).

United Nations Development Programme (UNDP 2011). Towards an 'Energy Plus' Approach for the Poor. United Nations Development Programme (UNDP), Bangkok.

United Nations Development Programme and Eidgenössische Technische Hochschule (UNDP and ETH 2018). *Derisking Renewable Energy Investment: Off-Grid Electrification.* UNDP and ETH, New York and Zurich.

van den Bergh, J. C. J. M., and Botzen, W. J. W. (2014). A lower bound to the social cost of CO_2 emissions. *Nature Climate Change* 4 (4), 253–258.

van Mansvelt, R. (2011). *The Solar Roadmap for Cambodia: How to Scale-Up Solar Diffusion in Cambodia? Strategy Formulated by Stakeholders.* Pico Sol Cambodia, Phnom Penh.

World Bank (2018). *Tracking SDG7: The Energy Progress Report 2018.* Washington, DC.

Yin, R. K. (2003). *Case Study Research: Design and Methods.* 3rd ed. Sage Publications, Thousand Oaks, CA.

6

Fostering Sustainable Development Goals Through an Integrated Approach

Phasing in Green Energy Technologies in India and China

René Kemp, Babette Never, and Serdar Türkeli

1. Introduction

Science, technology, and innovation (STI) present important levers to implement the sustainable development goals (SDGs). The United Nations has acknowledged this by creating a STI multistakeholder forum that calls for and draws together innovations to deal with the complexity of the many SDG targets in practice. The challenge for developing country governments now is to turn the broad SDG agenda into a manageable development agenda that fits their country's context, capacities, and interests. Developing country governments will need to find compromises between potentially differing local interests in, for example, the trade-off between economic growth and "classic" social protection, and donor interests in, say, investing a majority of funds and capacities into going green now.

Energy-intensive development paths may benefit the poor of today but can be detrimental to future generations (BMZ 2007, p. 3). Several conditions will influence which goals and targets receive priority, how they are interpreted, and how they are translated into practice.

In this chapter, we analyze the potential role of STI in achieving SDGs, and what governments in developing countries can do to phase in STI-based solutions, especially green energy technologies, for achieving SDGs. For this purpose, we examined three successful phase-in experiences in solar energy and energy efficiency in emerging economies (China, India). This is being done with the help of an integrated framework that merges political economy elements of *interests, ideas,* and *institutions* with *capabilities* and *policy delivery*. The experiences in dealing with observed difficulties (having to do with power, lock-in, weak institutions of governance, and information asymmetries) are used to outline what governments in developing countries can do to phase in green STI-based (energy) transitions by utilizing the opportunities already present that coincide with development objectives. Proposed solutions include proactive planning, rent management, use of adaptive policy mixes (informed by policy learning), and robust implementation mechanisms.

The structure of this chapter is as follows: in the next section, we discuss the analytical framework used, based on national systems of innovation, political economy, policy trajectories, and the economics of technology policy. After this, we document the relevance of these concepts in three successful cases of phase-in processes. The final section specifies key elements of phasing in green transitions for developing countries in the context of STI for SDGs by drawing on the lessons from these three cases. In this last section, we discuss what governments in developing countries can do to phase in STI-based solutions, especially green energy technologies in the context of SDGs in order to progress in achieving or to achieve SDGs.

2. Analytical Framework

Our analytical framework combines a framework of political economy with the framework of a national system of innovation (NSI). The NSI approach is focused on the firm or enterprise level and their interactions throughout the system (for a generic model of national innovation systems, please refer to Arnold and Kuhlman 2001). In NSI studies the focus is typically on activities of innovation, capacities for innovation, and policy instruments. Less attention is being given to the policymaking processes and the politics of innovation support policies.

From a political economy perspective, the role of ideas, interests, and institutions are central to understand, design, and implement policies relating to STI for any sectoral field in governing change. According to Smith et al. (2014, p. 334), ideas refer to the ways in which political actors and policy workers "make sense of the challenges that they face." This includes how problems are defined or defined away, what counts as evidence for such framing, and how possible solutions to policy issues are constructed (Mehta 2011). The notion of ideas also encompasses the very deeply and strongly held values and beliefs through which individuals make sense of the world—alternately referred to as ideologies or paradigms (Hall 1993). Ideas also take the form of public sentiments, "public assumptions [of the populace] that constrain the normative range of legitimate solutions available to policy makers" (Campbell 1998) (e.g., the attitude of the general public toward citizen-expert debates, participatory policymaking versus technocratic closed-circuit policymaking practices).

Interests are the material interests of key actors as conceived by them, as individuals or as groups (Hall 1997). Interests are often the proverbial elephant in the room that is only discussed indirectly. Interests are not purely fixed. Actors may see their interests in new ways, when sensing opportunities or threats. Institutions may be narrowly defined as organizations, which in our view is too restrictive. In a broader sense, institutions are "relatively enduring collection of rules and organized practices, embedded in structures of meaning and resources that are relatively invariant in the face of turnover of individuals and relatively resilient to the idiosyncratic preferences and expectations of individuals and changing external circumstances" (March and Olsen 2008, p. 3). Institutions as a collection of norms, rules, and bodies involved in policy processes have an important bearing on policy output and policy outcomes and thus performance at various levels (e.g., sectorial, national or temporal) (Hall 1997; Notermans 2000; Hall and Soskice 2001; Hall and Gingerich 2009).

Institutionalization processes play a significant role in the interplay of ideas and interests. Rodrik (2014) assigns a positive role to the embeddedness of policy in business, consisting of "strategic collaboration and coordination between the private sector and the government with the aim of learning where the most significant bottlenecks are and how best to pursue the opportunities that this interaction reveals." Others prefer an arm's-length relationship to avoid capture. There is widespread agreement that policy should evolve with experience and be adapted to new circumstances and take on board lessons from policy evaluation. Rigid policies are likely to be counterproductive and wasteful in terms of providing windfall gains (Kemp 2011).

Ideas, interests, and institutions affect policy goals and the choice of policy instruments and programs. They operate in complex ways, at multiple places, through all kinds of contingencies. Policymaking occurs within institutional fora, through instituted processes involving actors with material interests and viewpoints, discussing problems and policy solutions to the problems.

In relation to policymaking, implementation, and delivery it is useful to approach this as an issue of five contexts. The first context is the *context of influence*, which is about "where and how interest groups (interests) struggle over the construction of policy discourses, and where and how key policy concepts (ideas) are established" (Ball 1993; Vidovich 2003; Lall 2007). The second context is the *context of policy text production*. The first context is related to the second in terms of who is involved and the role of ideas. For example, the UN Intra-Agency Task Team on STI for SDGs (IATT) is mandated to work with the 10 representatives from civil society, private sector, and scientific community to prepare the STI Forum, where three spheres of sustainability (social, economic, environmental) and the interactions among them are being given consideration, in discussions about equitable, viable and effective solutions (UNEP 2015). The *context of practices*, where policy action confronts practical problems of implementation and enforcement and is subject to interpretation and recreation by implementers (Vidovich 2003; Lall 2007), is the third relevant context. The *context of outcomes,* the impact of policies on the issues of the policy recipients, necessitates monitoring and evaluation by policymakers; and the *context of political strategies* is where governments identify political activities to tackle these issues of the policy recipients while policymakers are concerned with devising calculated and/or sensemaking approaches to design policy solutions (Ball 1993; Vidovich 2003; Lall 2007).

Our framework is visualized in Figure 6.1. It contains the three elements of the political economy framework, namely ideas, interests, and institutions as three distinct, yet interacting elements, hence the acronym 3I. The interaction is visualized in a triple helix form. In our scheme, ideas, interests, and institutions are integral elements for the first plane, which provides a structured frame for ideas, interests, and institutions by 17 SDGs, and semistructured openings through interlinkages/overlapping boundaries among these 17

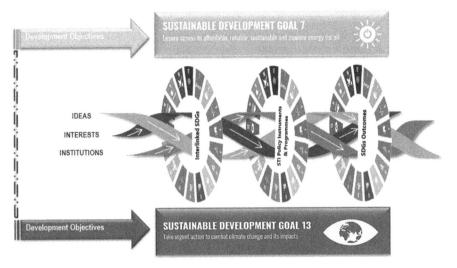

Figure 6.1 Helix model of ideas, interests, and institutions and SDGs
Source: authors

SDGs. In the process, the second plane consists of (STI-related) *policy instruments and programs*,[1] which are intentionally designed, adjusted, and implemented for achieving the targets relating to SDGs through deliberations and negotiations by relevant actors, some of which are resisting change and seeking delay and derogations. The final plane represents SDG outcomes which can be economic, social, environmental, or political outcomes, directly or indirectly interlinked through positive (or negative) externalities they can create over each other. SDGs outcomes (especially low-goal achievement) contribute to formation of the next cycle of policy deliberation, formation, and implementation, which again is shaped by ideas, interests, and institutions.

For the discussion of green energy transition, the most important goals among all SDGs represented in Figure 6.1 are especially SDG7 (ensure access to affordable, reliable, sustainable, and modern energy for all) and SDG13 (take urgent action to combat climate change and its impacts). Apart from mitigating the risk of climate change (SDG13), modern green energy also helps to limit the use of unsustainable firewood reducing deforestation and soil degradation (SDG15). In addition, modern energy is associated with health benefits (through the use of less polluting systems for cooking and heating (SDG2), promotes industrialization (SDG9) and the creation of jobs (SDG8), and is critical for the supply of safe drinking water (SDG6) and aiding the development of inclusive human settlements (SDG11) (Alloisio et al. 2016). SDG7 has five targets, for 2030, of which the first three targets are:[2]

---

[1] By policies and programs, we mean "programmatic ideas, which are precise, concrete, and policy-specific courses of action articulated consciously by policy makers and experts in the cognitive foreground" (Campbell 1998).
[2] UN SDGs, available at https://sustainabledevelopment.un.org/sdg7.

7.1: Ensure universal access to affordable, reliable, and modern energy services;
7.2: Increase substantially the share of renewable energy in the global energy mix;
7.3: Double the global rate of improvement in energy efficiency.

Achieving an energy transition requires policies of different types: supply-side policies, including policies for the supply of supplementary factors; demand-side policies in the form of adoption policies and information diffusion policies; and institutional change policies (Steinmueller 2010). In his discussion of technology policy, Steinmueller (2014) notes the significance of nonmarket relationships among actors, the role of expectations, the asymmetry of information among actors, the limits to knowledge exchange and transfer, and the need for policymaking capabilities. Formal and informal institutions are likely to be important. Mytelka et al. (2012) notes the importance of dialogues and confidence-building measures that recognize the legitimacy of local concerns, interests, and needs, with attention being given to the informal institutions shaping the behavior of actors involved in green energy transitions (the topic of this chapter).

3. Three Policy Approaches for Phasing in Technologies

3.1 Solar Photovoltaics (PV) in India

Although India has high solar irradiation and sufficient amounts of open space in several of its states, interests were organized around fossil fuels with a strong local coal industry for a long time. In the context of practices, there is a vast potential for both on- and off-grid solar energy, offering both energy and socioeconomic development opportunities. But lack of regulatory certainty and the lower costs of electricity from coal-fired plants discouraged *investments* and formation of interests in the solar sector. However, there was a clear idea that solar-based electricity would become much cheaper once it was produced at scale, leading the central government, as an institution, to come up with a solar PV program sought to change the situation. It offered a policy package to develop solar energy technology into a *viable* and competitive alternative by shielding it from the price competition from coal. This is an example of creating interests for a *programmatic idea*.

Interests in solar policies began to arise in the states of Gujarat and Rajasthan in the early 2000s. In 2008, the Jawaharlal Nehru National Solar Mission (JNNSM) was announced as part of the National Action Plan on Climate Change. It provided a long-term vision with clear targets to be achieved in each of its phases. The different phases included a comprehensive set of measures: preferential feed-in tariffs, renewable energy purchase obligations and certificates, tax incentives, and preferential loan schemes, as well as local content requirements to support the build-up of national manufacturing capabilities. In terms of targeted innovation support, government institutions receive 100% and the private sector 50% research and development (R&D) project subsidy for advancements in solar energy (Sahoo 2016; Quitzow 2015). Federal legislation is complemented by state level incentive programs in the context of policy text formation.

The most interesting element to nudge the investors' engagement and reduce the fiscal costs of the JNNSM is the reverse bidding approach that brought the required subsidies down at an unprecedented speed. The government of India auctioned licenses for solar power projects asking the bidders to hand in proposals below a reference price of €0.27/KWh that the government would guarantee for 25 years. In the first round, the government auctioned very small projects of just five MW. Already investors' engagement was overwhelming, and the lowest bidders offered to build such plants with a feed-in tariff of only €0.11/KWh. In subsequent rounds the government auctioned larger units, and from round to round, received lower bids. This way, the government managed to trigger an enormous growth of the solar energy market while at the same time bringing the price differential between conventional and solar energy down and greatly reducing the fiscal cost of the solar feed-in tariff (Altenburg and Engelmeier 2013).

Since 2009, the start of the National Solar Mission, India's installed solar capacity increased from virtually zero to 5,130 MW in January 2016. The solar energy market has been growing at an annual rate of more than 300% in the last five years. In 2014, around 125.000 people were employed in both on- and off-grid solar PV sector in India. The tariff that the government has to guarantee for new solar projects has been reduced to below €0,08/KWh on average in 2014–2015 (IRENA 2015), which is not too distant from the cost of a KWh from coal-fired power plants (Kemp and Never 2017). From a SDG perspective, this approach already contributed greatly to advancing the goal of sustainable energy systems (SDG7).

The JNNSM was less successful with regard to another objective that now falls under the new SDG12 on sustainable consumption and production, namely to boost domestic PV cell and module manufacturing capabilities. To achieve this objective, a local content requirement was included in the solar energy incentive package, as part of the context of political strategies. There are two types of PV cells: crystalline silicon and thin film. Before the JNNSM, India already had a decent production of crystalline silicon cells that was mainly export-oriented. The local content requirement made local sourcing mandatory for crystalline silicon cells but not for thin film cells, which were not produced locally. In the context of production practices, project developers then tried to avoid the mandatory local sourcing by almost exclusively offering projects on the basis of thin film technology (Johnson 2013). In the context of outcomes, a well-intentioned local content regulation resulted in a bias toward foreign thin film PV manufacturers rather than supporting the local crystalline silicon manufacturing industry (CEEW and NRDC 2012).

Overall, the solar policy was highly successful in terms of phasing in of solar power technologies from the beginning (contributing now to SDG7), but not necessarily for the promotion of local green industrial production (SDG12). Promotion attempts for local industries could be designed in the frame of SDG17, strengthening the means of implementation and revitalizing the global partnership for sustainable development, to be able to inform and coordinate foreign and local actors and project developers.

3.2 Energy Efficiency in India

Residential energy demand can be counted as a nonproductive form of energy consumption demand, as well as a considerable cost item for households. Forty-five percent of India's primary energy consumption is due to this type of residential energy demand, and 80% of it is attributed to only five technology appliances: ceiling fans, TVs, lighting, refrigerators, and air-conditioners (Jairaj et al. 2016). In terms of outcomes, the growth of the consumer appliance market between 2003 and 2013 at an annual average of 13% increased the overall energy demand (Jairaj et al. 2016). In the context of consumption practices, consumers typically choose inefficient appliances, which have a lower price on the market putting higher costs on their bills. In the context of production practices, manufacturers had no interest or incentive to invest in the development of more efficient, higher priced goods, which they consider that the consumers would not buy (Chaudhary et al. 2012). These are two market failures at the demand and supply side that had to be overcome.

In 2001, the government of India set up the Bureau of Energy Efficiency (BEE), as a new institution, with the mandate to develop an energy efficiency program for household appliances and lighting, buildings, and industries, including standards and labeling. This programmatic idea was to stimulate the use of more efficient appliances through energy efficiency *norms* for production and consumption. The program consists of a star-rating scheme based on minimum energy performance standards set by BEE (1 to 5 stars), as an institutional intervention. These labels were initially voluntary (norms) then made mandatory (rules) once labeled products reached 50% of overall sales. Facilities for testing the appliances were developed to allow for independent testing. *Demand side* market pull was created by mandating public procurement at the highest star ratings (Chaudhary et al. 2012). BEE as an institution seeks to stimulate innovation by tightening the performance standards for the stars (as an idea-based policy approach).

In the initial phase of the program, standards and labels were designed to include a majority of the market to support consumer understanding (demand side) with the clear communication (signaling strategies) that standards would be ratcheted up. Extensive consumer outreach and information was provided to build up the labels as a brand. In the context of production practices, manufacturers self-certified the products by adopting approved testing procedures and were responsible for the accuracy of the labels. Due to understaffing, BEE has outsourced the process of collection, verification, and processing of self-certificates to a consulting firm (Jairaj et al. 2016). In the context of new outcomes, BEE carries out market assessments regularly to understand market penetration, but these are not publicly available, making it difficult to judge how much the diffusion of efficiency has advanced in each product category (Khandari 2011).

Since the BEE has been functioning as both the designer of programs and facilitator to state governments and businesses at all levels, the reliance on outsourcing services and the development of an Energy Service Company (ESCO) market soon became inevitable (Kemp and Never 2017). While this ESCO market is evolving, it is still largely located in Delhi, Mumbai, and Pune (Harrison and Kostka 2014). Future challenges across India include establishment of trusted client-ESCO relationships (as a new institution) and increasing the attention of public agencies' interests as role models in electricity price and expenditures. Cobenefits, aligning multiple interests, are the key argument for energy efficiency in India: bundling interests of diverging and possibly opposing stakeholders (Harrison and Kostka 2014).

In the context of outcomes, in December 2012, the target of the standards and labeling program had already been surpassed: 7,766 MW of new generation capacity addition were avoided, greatly surpassing the estimated 3,000 MW (BEE 2014). By 2016, frost-free refrigerators, tubular fluorescent lamps, room air conditioners, and distribution transformers had progressed to mandatory labeling while 17 additional products were under the voluntary labeling scheme. Market uptake for appliances with higher levels of efficiency (4 and 5 stars) has generally been slower than for medium-level appliances (Jairaj et al. 2016). While the initial phase of the program had been highly participatory with a clear roadmap, the frequency of review and elevation of standards today is not sufficiently clear to all manufacturers. Consumer awareness programs have achieved growing brand recognition of the star label, but the share of inefficient devices is still significantly higher than the higher star-labeled ones. The absence of a targeted communication plan limits the full outreach potential in creating more energy efficient production and consumption culture, for instance via civil society organizations (Jairaj et al. 2016).

The factors responsible for the success of the program have been detailed understanding of the local needs (demand side), the sector (supply side), and the institutional context of practices. Examples include proactive establishment of the Bureau of Energy Efficiency and, in turn, BEE's proactive, dynamic engagement with producers and other institutions, outsourcing work with consultancies where necessary and the parallel development of the Energy Service Company sector. The success of this dynamic, smart rather than purely strong institutional engagement also clearly benefitted from extensive consultations with a range of stakeholders in the context of policy text production (Chaudhary et al. 2012, pp. 58–59). Embedded autonomy and communication worked well by ensuring inclusion of all relevant actors, including incumbents and innovators. The gradual tightening of standards and labels in line with the roadmap initially provided to stakeholders was also helpful, but currently needs to be improved so that the process is clear to all relevant stakeholders (Kemp and Never 2017).

3.3 Industrial Energy Efficiency in China

Like many other nations, the Chinese industry has been locked into an unsustainable, fossil fuel-based path, whereby 77% of all electricity is produced from coal. Industrial energy consumption is significantly increasing from 34% in 1990 to approximately 70% of total national energy consumption today (Kemp and Never 2017). Since energy prices for industry are regionally negotiated, the market does not send a consistent country-wide signal to invest in energy efficiency. In the context of production practices, before the Chinese government started its industrial energy efficiency programs, companies hardly invested in energy efficiency. In the context of political strategies, the government employs a "state-signaling approach" (Harrison and Kostka 2014, p. 1) where the central state provides targets and guidelines to local governments, but keeps policy goals broad enough to allow for local interest alignment and bundling of policies and incentives.

In the context of policy text formation, in 1995 the government published a guideline for energy management in industry, which already indicated to companies that future regulation was to come. Since the 1980s, China has been monitoring energy use in industry, but concrete interest in energy management systems in industry only started to rise in the early 2000s (Zhou et al. 2010). In the context of policy text production, following the introduction of the Energy Conservation Law in 2001, the government developed a comprehensive package of mandatory and voluntary policies and measures aimed at advancing energy efficiency and energy saving. The Medium and Long Term Energy Conservation Plan was published in 2004. In the 11th Five-Year Plan (2006–2010), command-control regulations were combined with taxes and subsidies—sticks, carrots, tambourine[3] measures (information campaigns), and prohibitions (Yang et al. 2015).

3.3.1 Sticks

The Differential Electricity Pricing Policy consists of four categories with different surcharges for industries that increase with consumption. More efficient enterprises pay less, which support development of cultural capital of energy efficiency in production. Between 2004 and 2010, these types of taxations were subsequently increased. However, initial charges were found to be inefficient in terms of increasing energy efficiency investments. This was due to the clashes in the context of political strategies: heterogeneous, counteracting local policies, fluctuating selling prices, and

---

[3] Tambourine measure in the energy efficiency field includes the provision of key information for decision-making and regular reminders of the benefits of adopting energy efficiency measures, and information disclosure requests that keep the topic on the agenda of the target group (e.g., companies).

other production costs (Yang et al. 2015). To phase out inefficient enterprises, the surcharge in this category was increased by a factor of four over time.

3.3.2 Carrots

In terms of financial incentives, companies can receive a reward of 250 Yuan per ton of coal equivalent saved through technical upgrading and engineering projects or the Ten Key-Energy Saving Projects. A number of R&D support strategies for different business sectors, financial compensation for technical retrofitting, and the phasing out of small and inefficient industrial plants are ongoing.

3.3.3 Tambourine

The Top 1000 energy savings agreements between key industry and government was voluntary when introduced in 2006, and gradually extended to provincial and local levels. In 2012, the program became mandatory, and was expanded to the Top 10,000. It requires companies to send their energy use statistics to government annually and to meet national and international standards. The Top 10,000 program and its combination with energy management systems helped to raise awareness among provincial authorities and top-level management. But the implementation is impeded by a lack of understanding in non–Top 10,000 companies, a lack of funding, and adoptable technical means (Goldberg et al. 2011). Proper adoption of energy management systems—either the national standard or ISO 50001—can actually make it easier for companies to comply with regulations and monitoring schemes. The electricity saved may then qualify the businesses for a different electricity price category (Kemp and Never 2017).

3.3.4 Prohibitions

The 2006 Plan on Energy Conservation and Emissions Reductions sets targets for the closure and phasing out of small and inefficient production facilities. However, in the context of political strategies, several provincial and local governments prioritize local economic development and jobs. They protect local factories from closure or let them operate unofficially, thus opposing national policies (Yang et al. 2015).

In the context of outcomes, China's approach to industrial energy efficiency has led to the envisioned reduction of the economy's energy intensity by 20% until 2010, but progress since has been rather slow. In the context of practices, implementation of national energy policies relies on creative bundling of interests, incentives, and policies by local government administration to minimize opposition from local players (Harrison and Kostka 2014). Thus, there is an interplay of politics and policies with a larger focus on experimentation than in the Indian approach to energy efficiency. In the context of political strategies, government struggled to achieve structural change that actually replaces less energy efficient firms and technologies with more efficient ones across all sectors during the 12th Five-Year Plan (2011–2015) (Ke et al. 2012). In the context of outcomes, official documents state that 340 million tons of coal equivalent have been saved under the Ten Key Projects program until 2010. A gradual, cautious introduction of a carbon trading system is now envisioned to complement existing measures. Diffusion of these measures to medium and small energy intensive companies presents a challenge in the next stage of the phase-in processes in China.

Concerning industrial policy effects on the supply side in the context of outcomes, the market for energy efficiency consultants and energy service companies (ESCOs) has been developing since 2006. Between 2010 and 2015, this market grew by 31.9% annually, employing 654,000 people in approximately 2,600 companies by 2015 (IbisWorld 2015). Yet ESCOs have been criticized for not working effectively enough due to imperfect business models, asymmetric information, high transaction costs, and lack of ability to build a relation of trust (Kostka and Shin 2011). Energy auditing capabilities vary greatly. While the market is developing well, it hasn't reached maturity in terms of utilization of its full potential across China (Kemp and Never 2017).

The Indian and Chinese examples provide important lessons for the implementation of the energy efficiency targets of SDG7 on sustainable energy. The cases show that both an explicit or implicit role of state institutions can be helpful, as long as it is proactive and allows for flexible engagement with stakeholders. While the Indian approach relied more on the market and consultancy work, the Chinese approach shows that explicit state-driven programs can work just as well.

For SDGs, China formulated a plan in 2016 outlining guiding thoughts and general principles (Government of the People's Republic of China 2016). India has thus far not done so, but the Research and Information System for Developing Countries (RIS) in collaboration with the United Nations in India pursued a rigorous research agenda to explore various facets of India's negotiations, adoption and implementation of SDGs. As part of a broader work program, "RIS launched a special paper series on each of the 17 SDGs and two cross cutting themes (technology and finance) authored by eminent experts in the related subjects." The special papers are combined in a report that lists achievements under the respective/related millennium development goal (MDG) targets, highlights remaining gaps in fulfilling targets under the respective/related MDG, articulates the philosophy and concept of the respective SDG and the targets, and formulates five pillars for an implementation framework for financing SDGs (RIS 2016). China's plan for working toward the SDGs and the recommendations of RIS are useful first steps for effectively working toward the SDGs. A big challenge for experts and policymakers is to consider multiple goals through coordinated policies. There is a big risk that the SDGs will be approached in a silo-based way.

4. Conclusions

In this chapter, we analyzed the experiences with phasing in green energy technologies in developing countries. We approached these experiences from analytical categories of political economy (ideas, interests, and institutions), the distinction between context of influence, policy production, practices, outcomes, and political strategies, and the economics of technology policy. As noted in the introduction and case descriptions, each situation requires its own approach, but we see the following six elements as useful elements for increasing the chances of success in achieving green and social-economic development benefits from modern green supply and demand technologies. The importance of these elements is brought out by the cases discussed and are in line with the innovation system dynamics literature (especially the literature on transition management, Rotmans et al. 2001; Kemp et al. 2007; Kemp and Never 2017). It is also in line with political economy studies of industrial policy that emphasize the importance of rent management and the need of a competent bureaucracy that is able to deal with information asymmetries (Altenburg and Engelmeier 2013; Schmitz et al. 2013; Rodrik 2014), and insights from economics of technology policy (Steinmueller 2010).

First, *proactive planning* on the part of government (in the form of a long-term vision, a clear roadmap with interim goals and steps) emerges useful starting points for any transition approach, in providing direction and guidance to innovation actors and investment decisions. The vision and roadmap needs to be attractive and communicated early and clearly to investors, innovators, and other stakeholders to identify technologies and innovations, and prepare the producers and consumers. Inclusion of relevant stakeholders such as manufacturers, business associations, and standardization bodies at an early development stage is advisable (embeddedness without political capture).

Second, the selection of options for support and the forms of support should be carefully done, and is best done with the help of independent experts. Auctions are a useful model for avoiding rents from being too high.

Third, a *sequential approach* helps to make use of contingencies and lessons, while maintaining a sense of direction. This can take the form of gradual tightening of regulations and standards or the testing in pilot projects before supporting a broader up-scaling.

This is connected to the fourth element, explicitly including *policy learning* in the phase-in process to achieve socioeconomically acceptable and successful implementation. These processes are complex and programs may turn out quite differently than planned. The Chinese and Indian cases have shown that allowing some policy space for strategic bundling of interests at local levels may be an important part of policy learning.

Fifth, designing a *policy package* has been helpful in many examples discussed here. This policy package can include both policy-push and market-pull policies as well as R&D, institutional capability, and skill and job creation measures, such as building up a consultancy and certification industry. But auctions based on costs may cause policymakers to overlook development benefits. Local content requirements are a way around this problem. In this respect, China has been more successful than India in the area of solar PV.

Sixth, *robust implementation control mechanisms* need to be put in place. Useful measures include the building and financing of technology testing facilities and the use of energy efficiency labels.

The experiences described show what can be done in terms of utilizing the opportunities already present that coincide with development objectives, through the use of proactive planning informed by identified opportunities, the use of special agencies tasked with energy efficiency promotion, and innovation and policy mixes that are fine-tuned to circumstances and informed by policy analysis and evaluation. The pursuit of multiple SDGs through STI may achieve more than the MDGs, which did not play much of a role in the policies pursued. It is relevant to note that in the context of practices (which includes the context of legislative/executive next to the context of production and consumption practices), developing countries need to develop their own set of capabilities, but in so doing they may want to make use of international platforms and agencies such as the STI forum.

The many links between the SDGs beg more attention than they are currently being given by policy agencies and politicians, but this is not an easy task. In our analysis, we came across few examples of effective collaboration across ministries. Of course, the programs predate the 17 SDGs that were officially introduced in 2015, the year of sustainable development, through Agenda 2030 of the United Nations (UN 2015), thus phasing in green STI-based transitions in the following years can benefit from utilizing the opportunities already present that coincide with development objectives and SDGs.

The influence of the STI forum and other initiatives to promote STI will critically depend on the extent to which they shape the triple helix of ideas, interests, and institutions at the national level and policies to deal with path-dependencies. Narrowing down the agenda to a set of relevant targets, interpreting and adapting them to local possibilities, is a process that is currently happening in many OECD countries as well (O'Conner et al. 2016; Weitz et al. 2015). Such a step has to be followed by active support of STI measures, identified as appropriate for support by capable agencies in stakeholder processes. For developing countries with limited resources and capacities, STI activities that have positive, reinforcing effects on several targets are most desirable. In general, it is easier to develop policies for support of STI than phasing out unsustainable practices, because of the interests that have grown into their continuation both at the supply side and demand side. The availability of green STI options may make this task easier, but it is still going to be a difficult task, in developing countries and developed countries alike.

Whereas most countries rely on market forces for any type of phase-out, China embarks on a planned closing down of energy-inefficient plants, with the help of a target measuring system that is linked to career prospects for local and regional authorities (Wu et al. 2017). While this approach is not without problems (Wu et al. 2017), it is an example of an effective approach. In our assessment, all countries should reconsider their policy frameworks for working toward the SDGs. For helping policymakers to achieve this difficult task, we have documented policy experiences in terms of context of influence, and the ideational element, practices, outcomes, and political strategies. SDG achievement requires frameworks for integrated thinking (across policy and SDG silos), evaluation by experts, processes of stakeholder interaction, capable agencies, and policy learning to drive policy change.

References

Alloisio, I., Zucca, A. and Carrara, S. (2016). SDG7 as an enabling factor for sustainable development: the role of technology innovation in the electricity sector, FEEM, draft copy. http://ic-sd.org/wp-content/uploads/sites/4/2017/01/AlloisioUpdate.pdf (accessed April 5, 2019).

Altenburg, T., and Engelmeier, T. (2013). Boosting solar investment with limited subsidies: rent management and policy learning in India. *Energy Policy* 59 (8), 866–874.

Arnold, E., and Kuhlman, S. (2001). RCN in the Norwegian Research and Innovation System, Background report No 12 of the Research Council of Norway, Oslo.

Ball, S. J. (1993). What is policy? texts, trajectories and toolboxes. *Australian Journal of Education Studies* 13(2), 10–17.

Béland, D. (2009). Ideas, institutions, and policy change. *Journal of European Public Policy* 16(5), 701–718.

Blyth, M. (2002). *Great Transformations: Economic Ideas and Institutional Change in the Twentieth Century.* Cambridge University Press, Cambridge.

Bundesministerium fur wirtschaftliche Zusammenarbeit und Entwicklung (BMZ) (2007). Sustainable Energy for Development Sector, Strategy Paper, Strategies 154 Online: https://www.bmz.de/en/publications/archiv/type_of_publication/strategies/konzept154.pdf (accessed April 5, 2019).

Bureau of Energy Efficiency (BEE) (2014). Energy Efficiency in India: Challenges and Lessons. Presentation by Baskar Serma, In-session Technical Expert Meeting on Energy Efficiency ADP, Bonn, March 13, 2014.

Campbell, J. L. (1997). Mechanisms of evolutionary change in economic governance: interaction, interpretation and bricolage. In: Magnusson, L. and Ottosson, J. (Eds.), *Evolutionary Economics and Path Dependence*, pp. 10–32. Edward Elgar, Cheltenham.

Campbell J. L. (1998). Institutional analysis and the role of ideas in political economy. *Theory and Society* 27(3), 377–409.

Chaudhary, A., Sagar, A. and Matur, A. (2012). Innovating for energy efficiency: a perspective from India. *Innovation and Development* 2(1), 45–66.

Council on Energy, Environment and Water (CEEW) and Natural Resources Defense Council (NRDC) (2012). Laying the foundations for a bright future: Assessing progress under phase 1 of India's national solar mission, Delhi (Interim Report): Council on Energy, Environment and Water and Natural Resources Defense Council.

Goldberg, A., Reinaud, J., and Taylor, R. (2011). Promotion Systems and Incentives for Adoption of Energy Management Systems in Industry. Institute for Industrial Productivity, Washington, DC.

Government of the People's Republic of China (2016). China's National Plan on Implementation of the 2030 Agenda for Sustainable Development. http://www.greengrowthknowledge.org/national-documents/chinas-national-plan-implementation-2030-agenda-sustainable-development (accessed April 5, 2019).

Hall, P. A. (1997). The role of interests, institutions, and ideas in the comparative political economy of the industrialized nations. In: Lichbach, M. I., and Zuckerman, A. S. (Eds.), *Comparative Politics: Rationality, Culture, and Structure*, pp. 174–207. Cambridge University Press, Cambridge.

Hall, P. A., and Gingerich, D. W. (2009). Varieties of capitalism and institutional complementarities in the political economy: an empirical analysis. *British Journal of Political Science* 39(03), 449–482.

Hall, P. A., and Soskice, D. (2001). An introduction to varieties of capitalism. In: Hall, P. A., and Soskice, D. (Eds.), *Varieties of Capitalism: The Institutional Foundations of Comparative Advantage,* pp. 1–68. Oxford University Press, Oxford.

Harrison, T., and Kostka, G. (2014). Balancing priorities, aligning interests: developing mitigation capacity in China and India. *Comparative Political Studies* 47 (3), 450–480.

IbisWorld (2015). Energy Efficiency Consultants in China: Market Research Report. http://www.ibisworld.com/industry/china/energy-efficiency-consultants.html (accessed April 5, 2019)

International Renewable Energy Agency (IRENA) (2015). *Renewable Energy and Jobs. Annual Review 2015.* IRENA, Abu Dhabi.

Jairaj, B., Agarwal, A., Parthasarathy, T. and Martin, S. (2016). *Strengthening Governance of India's Appliance Efficiency Standards and Labelling Program.* Issue Brief, World Resources Institute, Washington, DC.

Johnson, O. (2013). Exploring the effectiveness of local content requirements in promoting solar PV manufacturing in India, DIE Discussion Paper 11/2013 Deutsches Institut für Entwicklungspolitik, Bonn.

Ke, J., Price, L., Ohshita, S., Fridley, D., Khanna, N. Z., Zhou, N. and Levine, M. (2012). China's industrial energy consumption trends and impacts of the Top-1000 Enterprises Energy-Saving Program and the Ten Key Energy-Saving Projects. *Energy Policy 50,* 562–569.

Kemp, R. (2011). Ten themes of eco-innovation policies in Europe. *SAPIENS (Surveys and Perspectives Integrating Environment and Society)* 4(2) http://sapiens.revues.org/1169

Kemp, R., Loorbach, D. and Rotmans, J. (2007). Assessing the Dutch energy transition policy: how does it deal with dilemmas of managing transitions? *Journal of Environmental Policy and Planning* 9(3–4), 315–331.

Kemp, R. and Never, B. (2017). Green transition, industrial policy, and economic development. *Oxford Review of Economic Policy 33*(1), 66–84.

Khandari, R. (2011). Energy fixing. http://www.downtoearth.org.in/coverage/energyfixing-33562 (accessed April 5, 2019).

Kostka, G. and Shin, K. (2011). Energy conservation through energy service companies: Empirical analysis from China. *Energy Policy 52,* 748–759.

Lall, M. (2007). A Review of Concepts from Policy Studies Relevant to the Analysis of EFA in Developing Countries. Create Pathways to Access. Research Monograph No. 11, Institute of Education, University of London, London.

Letschert, V. E., McNeil, M. A., Kalavase, P., Fan, A. H. and Dreyfus, G. (2013). Energy Efficiency Appliance Standards: Where Do We Stand, How Far Can We Go and How Do We Get There? An Analysis Across Several Economies. (No. LBNL-6294E). Lawrence Berkeley National Laboratory (LBNL), Berkeley, CA.

Light I. (2001). Social Capital's Unique Accessibility. Paper presented at the Danish Building and Urban Research/EURA 2001 Conference in Copenhagen May 17–19, 2001.

Liu, F., Meyer, A., and Hogan, J. (2010). Mainstreaming Building Energy Efficiency Codes in Developing Countries: Global Experiences and Lessons From Early Adopters. World Bank Working Paper 204, World Bank, Washington, DC.

Loorbach, D. (2010). Transition management for sustainable development: a prescriptive, complexity-based governance framework. *Governance* 23 (1), 161–183.

March, J. G., and Olsen, J. P. (2008). Elaborating the new institutionalism. In: Rockman, B. A., Binder, S. A., and Rhodes, R. A. W. (Eds.), *The Oxford Handbook of Political Institutions,* pp. 3–20. Oxford University Press, Oxford.

Mytelka, L., Aguayo, F., Boyle, G., Breukers, S., de Scheemaker, G., Abdel Gelil, I., Kemp, R., Monkelbaan, J., Rossini, C., Watson, J., and Wolson, R. (2012). Policies for capacity development. In: Johansson, T., Patwardhan, A., Nakićenović, N., and Gomez-Echeverri, L. (Eds.), *Global Energy Assessment—Toward a Sustainable Future*, pp. 1745–1802. Cambridge University Press, Cambridge.

Nilsson, M., Griggs, D., and Visbeck, M. (2016). Map the interactions between sustainable development goals. *Nature* 534, 320–322.

Notermans T. (2000). *Money, Markets and the State: Social Democratic Policies Since 1918*, Cambridge University Press, Cambridge.

O'Connor, D., Mackie, J., Van Esveld, D., Kim, H., Scholz, I., and Weitz, N. (2016). *Universality, Integration, and Policy Coherence for Sustainable Development: Early SDG Implementation in Selected OECD Countries*. World Resources Institute, Washington, DC.

Quitzow, R. (2015). Assessing policy strategies for the promotion of environmental technologies: a review of India's National Solar Mission. *Research Policy* 44, 233–243.

Research and Information System for Developing Countries (RIS) (2016) India and Sustainable Development Goals: The Way Forward, a joint report of RIS and UN. http://ris.org.in/sdg/india-and-sustainable-development-goals-way-forward (accessed April 5, 2019).

Rodrik, D. (2014). Green industrial policy. *Oxford Review of Economic Policy* 30, 469–491.

Rotmans, J., Kemp, R., and van Asselt, M. (2001). More evolution than revolution: transition management in public policy. *Foresight* 3(1), 15–31.

Sahoo, S. (2016). Renewable and sustainable energy reviews solar photovoltaic progress in India: a review. *Renewable and Sustainable Energy Reviews* 59, 927–939.

Schmitz, H., Johnson, O. J., and Altenburg, T. (2013). Rent Management: The Heart of Green Industrial Policy, IDS Working Paper 418.

Smith, N., Mitton, C., Davidson, A., and Williams, I. (2014). A politics of priority setting: ideas, interests and institutions in healthcare resource allocation. *Public Policy and Administration* 29(4), 331–347.

Steinmueller, W. E. (2010). Economics of technology policy. *Handbook of the Economics of Innovation* 2, 1181–1218.

Surel, Y. (2000). The role of cognitive and normative frames in policy-making. *Journal of European public policy* 7(4), 495–512.

Swenson P. A. (2002). *Capitalists Against Markets: The Making of Labour Markets and Welfare States in the United States and Sweden*. Oxford University Press, New York.

United Nations (2015). *Transforming Our World: the 2030 Agenda for Sustainable Development*. United Nations, New York.

United Nations Environment Programme (UNEP) (2015). UNEP Environmental, Social and Economic Sustainability Framework (ESES), http://wedocs.unep.org/handle/20.500.11822/8718 (accessed April 5, 2019).

Vidovich, L. (2003). Methodological framings for a policy trajectory study. In: O'Donogue, T., and Punch, K. (Eds.), *Qualitative Educational Research in Action: Doing and Reflecting*, pp. 70–96. Routledge, London.

Waldstrøm, C., and Svendsen, G. L. H. (2008). On the capitalization and cultivation of social capital: towards a neo-capital general science? *Journal of Socio-Economics* 37(4), 1495–1514.

Weitz, N., Persson, Å., Nilsson, M., and Tenggren, S. (2015). Sustainable Development Goals for Sweden: Insights on Setting a National Agenda. SEI Working Paper No. 2015-10, Stockholm Environment Institute, Stockholm.

Wu, J., Zuidema, C., Gugerell, K., and de Roo, G. (2017). Mind the gap! barriers and implementation deficiencies of energy policies at the local scale in urban China. *Energy Policy*, 106: 201–211.

Yang, M., Patino-Echeverri, D., Yang, F., and Williams, E. (2015). Industrial energy efficiency in China: achievements, challenges and opportunities. *Energy Strategy Reviews*
6, 20–29.

Zhou, N., Levine, M., and Price, L. (2010). Overview of current energy-efficiency policies in
China. *Energy Policy* 38, 6439–6452.

7

Financing Environmental Science and Technological Innovation to Meet Sustainable Development Goals in Brazil

Mariana Machado and Carlos Eduardo F. Young

1. Introduction

In 2015, the United Nations launched the sustainable development goals (SDGs), a new round of global economic, environmental, and social goals to promote sustainable development for the entire world, continuing the worldwide efforts to implement the millennium development goals (MDGs) (Sachs 2012; UNDP 2018a).

In the MDGs, environmental issues were concentrated in goal 7, "ensuring quality of life and respect for the environment." Now, however, with the decline in environmental quality, there is a much more widespread understanding that worldwide environmental objectives play an important role alongside poverty reduction objectives (Sachs 2012). Thus, the 17 SDGs include new areas focused on environmental issues, such as climate change, biodiversity, conservation, and sustainable use of marine resources, innovation, and sustainable consumption (UNDP 2018b).

In the 2030 Agenda, science, technology, and innovation (STI) policies are also recognized as having a crucial role in the transition toward sustainable development. The United Nations (2015) suggests a technology facilitation mechanism (TFM) based on multisectoral cooperation among member states, civil society, the private sector, and the scientific community, with the aim of promoting the dissemination of information on initiatives, existing STI mechanisms, and programs to help facilitate the development, transfer, and diffusion of technologies relevant to the SDGs.

Brazil has already established a network of institutions and financial mechanisms with the great potential to perform this task of supporting the transition toward sustainable development (De Negri 2017). However, there are some barriers to overcome, such as fragile institutional arrangements (Bastos 2003; Melo 2009), legal framework gaps (Brito and Pozzetti 2017; Young et al. 2018), and lack of an investment orientation (De Negri 2017; Mazzucato and Penna 2016), to provide sufficient and stable funding for STI that helps the country to meet SDGs targets. In addition, recent budget cuts in STI funding jeopardize research and development (R&D) programs and undermine the ability to generate information on biodiversity and find solutions for the country's severe and increasing environmental problems (Gibney 2015; Oliveira et al. 2017; Overbeck et al. 2017).

The objective of this chapter is to discuss the financial conditions that promote scientific and technological development focused on environmentally-related SDGs, namely: SDG6 (water and sanitation for all), SDG13 (combat climate change), SDG14 (sustainable marine resources), and SDG15 (terrestrial ecosystems and biodiversity). The main questions addressed in this chapter are: 1) Do we have sufficient STI financing to help achieve the SDGs? 2) How does the current STI financial system work and what are the constraints? and 3) What are the possible alternatives to overcome the financial gap?

To answer these questions, we estimate demand for financing necessary to foster sustainable innovation in order to achieve SDGs targets and the financial gap to do so. Then we examine the current Brazilian STI financing system, highlighting its strengths and weaknesses. We dedicate special attention to the contribution of the sectoral funds, created with the objective of increasing the investments in STI within strategic economic sectors (oil and gas, mining, energy, health, and so forth) and fostering R&D for sustainable use of natural resources to help meet the selected SDGs. Finally, we discuss the possibilities to improve existing STI financing mechanisms and present proposals for new sources of funding to achieve sustainable development objectives.

In scientific and productive terms, Brazil is an emerging country in international terms, still far from the developed nations, but in a better position than countries with an equivalent level of development (MCTIC 2017). At the same time, it is a biodiversity superpower country and possessor of the largest tropical forest area in the world, but facing serious environmental issues such as high deforestation rates, with negative environmental, social, and economic consequences (Sant'Anna and Young 2010; Soares-Filho et al. 2010).

According to Sachs (2012, p. 2208), "middle-income emerging economies, like Brazil, are crucial leaders of the SDGs, and will have their own internal challenges of balancing growth and environmental sustainability; vulnerabilities to adverse trends such as climate change; and rising geopolitical roles, regionally and globally." Therefore, discussing the Brazilian case helps in understanding the challenges of establishing an efficient financing STI system oriented to develop innovation and technologies for the conservation and sustainable use of natural resources, and thus promoting socioeconomic development in line with the objectives of the 2030 agenda.

The chapter is structured as follows: section 2 presents the methodology adopted to carry out the analysis. Section 3 presents the estimated investment need in research, development, and innovation to promote sustainable use of natural resources that helps to achieve SDGs and the financial gap that exists. Section 4 focuses on the current STI structure of investments, its strengths, and the existing barriers. It is divided into two subsections: a brief description of the STI policies in Brazil, the expenditure devoted to the issue, and the results achieved; and the contribution of investment through sectoral funds to meet environmental issues in the selected SDGs. Section 5 presents a discussion of improvements in existing research, development, and innovation financing mechanisms. Finally, section 6 presents the main conclusion: to promote scientific knowledge and technological innovations that help Brazil to achieve environmental SDGs targets, it is necessary to make institutional adjustments to provide an adequate amount through strategically oriented funds and also to seek alternative forms of financing.

2. Methodology

The financial gap analysis adopted in this chapter follows the guidelines established in the Research Project "Analysis of SDGs and the Effectiveness of Public, Private and Mixed Funding Structures for the Promotion of Sustainable Development Improved," conducted by the Environmental Economics and Sustainability Research Group of the Federal University of Rio de Janeiro, cosponsored by the Ministry of the Environment of Brazil (Young et al. 2018).

The United Nations 2030 Agenda for Sustainable Development does not present numerical targets for many of the SDGs targets, including the targets related to enhancing STI (United Nations 2015). Therefore, to estimate the financing needs to foster R&D and innovation, we assumed the goal of investing 2% of the gross domestic product (GDP) in R&D by 2022, established in the STI National Strategy 2016–2022 document (MCTIC 2016). In addition, we estimated that the proportion of R&D investment for environmental issues would double in relation to its historical average. In this chapter, we extended the target of R&D investment as 2% of GDP to 2030, and we also considered that the proportion of R&D toward conservation and sustainable use of natural resources would reach of 2% of total R&D investment in the same period.

To estimate the available financial resources, we considered the current trend of fiscal austerity imposed by Constitutional Amendment n. 95/2016, which froze federal public spending at the 2016 level for the next 20 years. We also assumed that State and Municipal level public spending would follow the same pattern of frozen budgets established at the federal level.

In addition, we assumed that private sector investments will increase at a 2.5% annual growth rate. This growth rate is slightly higher than the Brazilian GDP average growth rate between 2000 and 2017 (2.3% per year). The monetary values, including the projections, were calculated at 2016 prices, using the implicit GDP deflator of the Brazilian Institute of Geography and Statistics (IBGE). The financial gap corresponds to the difference between the estimated value for the financial needs in the same period and the available resources projected up to 2030.

The analysis of the current STI scenario (funding and performance indicators) was based on official data provided by the Ministry of Science, Technology, Innovation, and Communication (MCTIC 2017a).

To be able to describe the contribution of sectoral funds in STI for biodiversity conservation and ecosystem services, we analyzed the data provided by MCTIC in the Aquarius Platform extracted from the module of the sectoral funds of the STI Management Information System (SIGCTI) of MCTIC (2017b). This platform presents data on the portfolio of projects supported, including information about the source of funding and the category of support.

The initial step was to identify among the lines of action inserted in the research, development, and innovation strategic areas, supported by the National Fund for Scientific and Technological Development (FNDCT), which are related to the SDGs selected for this analysis. Based on the objectives of the MCTIC lines of action (MCT 2007), we also made an effort to identify which specific targets could be benefited with R&D investments.

In Table 7.1 are presented the lines of action identified as related to specific SDGs.

It is important to emphasize that no individual analysis was carried out to verify the degree of adherence of each supported project to the action lines or to SDGs specific targets. Thus, we assumed that the projects supported by each specific line are adherent to the theme and help to achieve the selected SDGs as a whole. Based on the identification of the lines of action related to the selected SDGs, we performed an aggregated analysis presenting the number of projects supported, total value contracted and source of funds, development agency responsible for executing the resource, the average value of projects, financing instruments used, and supported project categories.

The discussion about the improvement of the financial mechanisms was based on a literature review about STI public policies and financing mechanisms in the current Brazilian context.

3. STI Financing Needs and the Financial Gap to Meet Selected SDGs

As presented in the methodology section, the estimation of the financing demand for STI considered the National Strategy for STI (2016–2022) (MCTIC 2016) and the historical proportion of the total R&D investment applied to environmental protection and control.

The projection of demand was based on the 2017 GDP (in 2016 prices) and the current R&D investment level (1.02%), considering an annual GDP growth of 2.5%

Table 7.1 Sustainable development goal and specific targets that could be benefited by adjustments in STI financing system in Brazil

Strategic areas defined as the PACT priorities and related SDG	Specific SDG targets (adapted names) that could be benefited from STI investments
STI for water resources SDG6	6.3 Improve water quality by reducing pollution, eliminating dumping and minimizing release of hazardous chemicals and materials, halving the proportion of untreated wastewater, and substantially increasing recycling and safe reuse 6.4 Increase water-use efficiency across all sectors and ensure sustainable withdrawals and supply of freshwater to address water scarcity and substantially reduce the number of people suffering from water scarcity
Climate Change National Program SDG13	13.1 Strengthen resilience and adaptive capacity to climate-related hazards and natural disasters in all countries 13.2 Integrate climate change measures into national policies, strategies, and planning 13.3 Improve institutional capacity on climate change mitigation, adaptation, impact reduction, and early warning 13.b Raising capacity for effective climate change-related planning and management in least developed countries
STI on Antarctica SDG14	14.3 Minimize and address the impacts of ocean acidification, including through enhanced scientific cooperation at all levels 14.a Increase scientific knowledge, develop research capacity on marine technology
STI for the explotation of seas resources SDG14	14.2 Sustainably manage and protect marine and coastal ecosystems to avoid significant adverse impacts, including by strengthening their resilience, and take action for their restoration in order to achieve healthy and productive oceans
RD&I in aquiculture and fishery SDG14	14.4 Implement science-based management plans, in order to restore fish stocks in the shortest time feasible, at least to levels that can produce maximum sustainable yield as determined by their biological characteristics 14.b Increase the economic benefits to least developed countries from the sustainable use of marine resources, including through sustainable management of fisheries and aquaculture
STI applied to biodiversity and natural resources SDG15	15.1 Ensure the conservation, restoration, and sustainable use of terrestrial and inland freshwater ecosystems and their services 15.2 Promote the implementation of sustainable management of all types of forests and restore degraded forests 15.4 Ensure the conservation of mountain ecosystems 15.5 Protect and prevent the extinction of threatened species 15.9 Integrate ecosystem and biodiversity values into national and local planning, development processes, poverty reduction strategies, and accounts

Continued

Table 7.1 *Continued*

Strategic areas defined as the PACT priorities and related SDG	Specific SDG targets (adapted names) that could be benefited from STI investments
STI integration program for conservation and sustainable development of Amazon region SDG15	15.2 Promote the implementation of sustainable management of all types of forests and restore degraded forests 15.9 Integrate ecosystem and biodiversity values into national and local planning, development processes, poverty reduction strategies, and accounts
STI for semi-arid sustainable development SDG15	15.2 Promote the implementation of sustainable management of all types of forests and restore degraded forests 15.3 Combat desertification, restore degraded land and soil, including land affected by desertification and drought 15.9 Integrate ecosystem and biodiversity values into national and local planning, development processes, poverty reduction strategies, and accounts

Source: Authors' own elaboration, based on MCT (2007), MCTIC (2017b), and United Nations (2015).

and the growth in R&D investment stipulated in the STI National Strategy (2% of GDP in 2030). In addition, the amount applied to R&D in the fields of conservation and sustainable use of natural resources would reach the proportion of 2% of the total R&D investment by 2030.

To meet the national STI goal of investing 2% of the GDP in R&D by 2030, a total of R$1.49 trillion for the 2018–2030 period would be necessary. Considering the austerity scenario for public investments, we estimated the available amount to be R$312 billion (2016 prices). The available private investment was estimated at around R$628 billion, totaling R$940 billion, with a financial gap of R$551 billion. This means that the R&D investments should increase at an annual rate of 8% in order to achieve the National STI goal of R$1.49 trillion.

The investment required to foster R&D related to selected SDGs (6, 13, 14, and 15) was estimated at a total amount of R$21.7 billion (2016 prices) for the 2018–2030 period. Considering the amount invested in the 2003–2017 period as a baseline, the amount available for environmentally-related R&D was estimated at R$7.5 billion for 2018–2030. Therefore, the estimated financial gap of R&D investment in the selected SDGs is R$14.2 billion (Figure 7.1).

As shown in Figure 7.1, the financial gap for R&D investments related to the environment is significant and should be addressed if Brazil, a mega-diverse country, wants to promote its development on a sustainable basis.

From this section, we conclude that overall funding of environmental-related R&D is insufficient to help meet the selected SDGs targets. Thus, in order to meet the Agenda 2030 commitments to invest in sustainable innovation, the country needs to

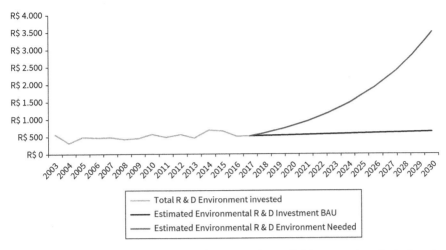

Figure 7.1 Environmental R&D invested amount, the available amount projected, and funding demand to meet SDGs targets (2016 prices)
Source: Authors' contribution.

make efforts to establish new funding sources, as well as to increase the efficiency of existing financing mechanisms. But in order to really assess this and to identify some of the opportunities to increase financial support, we need to look at the specifics of the current STI financial system, especially the major source of STI funding: the sectoral funds.

4. The Current Financial Scenario for Science, Technology, and Innovation

With the recent austerity scenario, intensified by the Constitutional Amendment n. 95/2016, Brazil is facing difficulties to invest an adequate amount on STI to promote truly sustainable development, as committed to in the 2030 Agenda. As estimated in the earlier section, the financial gap for environmentally-related R&D is significant, and there is an urgent need to seek alternative sources of funding and improvements of the existing financing mechanisms.

Thus, this section presents the evolution of the STI system in Brazil, its achievements, and the existing barriers that constrain the scientific and technological progress necessary for development on a sustainable basis. Specifically, we analyze the contribution of the sectoral funds, the main source of funding for fostering research, development, and innovation focused on the selected SDGs.

4.1   Financing Policies and Mechanisms for STI in Brazil and Results Achieved

The sectoral funds for STI were established in the late 1990s and early 2000s, based on the concept that a small percentage of the revenues of public utilities and specific private sector activities should be destined to R&D funds controlled by government-controlled STI agencies through the FNDCT. The STI legal and institutional framework has evolved since then, including:

- Law 11,196/2005 created incentives for exports, digital inclusion, regional development, and other mechanisms to encourage innovation in Brazil;
- Constitutional Amendment 85/2015 determined the role of the state in the promotion of scientific development, research, technical and scientific capacitation, and innovation, and institutionalized the National Science, Technology, and Innovation System (SNCTI), to be organized in collaboration between public and private entities;
- The Biodiversity Law (Law 13,123/2015) regulated access to the genetic heritage of Brazilian biodiversity and the traditional knowledge associated with it, defining rules for access to these resources by researchers and industry, regulating the right of traditional peoples to benefit sharing by using their knowledge of nature; and
- Law 13,243/2016, which amended Law 10,973/2004, provides for the reduction of bureaucratic obstacles and more freedom for scientific research, through the simplification of various processes and the encouragement of the integration of private companies to the public research system.

The MCTIC is the body responsible for coordinating the SNCTI and governance of the FNDCT.

The FNDCT is the largest funding source of STI in Brazil and its resources are operated by two development agencies linked to MCTIC, which are responsible for the allocation of public resources to support research, development and innovation activities. The National Council for Scientific and Technological Development (CNPq) fosters scientific and technological research, development, and innovation through partnerships with academic institutions, government agencies and the private sector. The Funding Authority for Studies and Projects (Finep) promotes the public support of STI in companies, universities, technological institutes, and other public or private institutions throughout the innovation chain operating reimbursable loans and specific nonrefundable programs (MCTIC 2016).

The implementation of STI activities depends on strong investments in the sector. De Negri (2017) recognizes that the advances brought by the recent innovation policies have built a relatively comprehensive and diverse framework, many of them used in most of the developed world to promote innovation, such as subsidized credit, fiscal incentives, grants to enterprises, and for research projects in universities and research centers. Even though subsidized credit and tax exemption mechanisms for private investment in R&D are important in some specific sectors, such as information and communication technologies (Kannebley Junior et al. 2016), the largest volume of investment in STI corresponds to direct public expenditures. This depends on the budgetary resource from tax revenues and earmarked revenues destined for specific STI sectoral funds (Zuniga et al. 2016).

This financing structure had resulted in improvements in national indicators for this sector. Between 2003 and 2015, gross domestic expenditure on R&D in Brazil increased from R$45.3 billion to R$82.9 billion in 2016 values (Figure 7.2). In 2015, public investments in R&D accounted for 50.17% of total expenditures, with the Federal government accounting for 35.6% and subnational governments accounting for 14.6%. The private sector accounted for 49.8% of National R&D expenditures (MCTIC 2017a).

The ratio between R&D expenditures and GDP increased from 1.1% to 1.3% between 2003 and 2015. The participation of the private sector in the R&D expenditures reached almost 50% of the total invested in 2015.

This level is similar to developed countries such as Italy and Spain and places Brazil in a prominent position among Latin American countries, where R&D investment is less than 1% of GDP. However, the country is still below the Organization for Economic Co-operation and Development average of 2.4% and higher-middle-income countries, which average 1.66% (UNESCO 2015).

The number of full-time researchers went from 51,600 to 180,000, an increase of 248% between 2000 and 2014 (MCTIC 2017a). Brazilian scientific production had also grown. From 1991 to 2013, national participation in international scientific

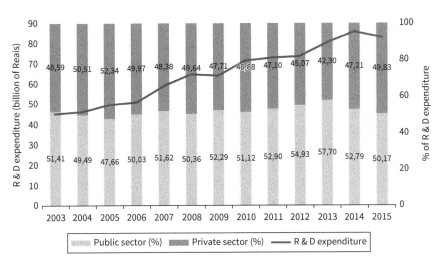

Figure 7.2 National expenditures in R&D by sectors, percentage, and total, in R$2016 prices (2003–2015)

Source: MCTIC (2017a).

production increased from less than 0.7% to almost 3%, especially in the field of life sciences and biomedicine which are related to biodiversity and natural resource use (Nascimento 2016).

Brazil also figures among the top 20 countries publishing in ocean science, specifically related to SDG14.A target (IOC-UNESCO 2017). However, in qualitative terms, progress was less significant. Between 2008 and 2012, the average citation rate (an indicator of the scientific impact of papers produced by a country relative to the world average) for Brazilian publications is 0.74, while that of the G20 countries was 1.02 (UNESCO 2015).

The number of patents granted by the US Patent and Trademarks Office (USPTO) is used as an indirect measure of the degree of the pursuit of an internationally competitive economy based on technology-driven innovation (UNESCO 2015). Between 2003 and 2014, the number of inventions granted to Brazilians by the USPTO increased by 74.6%. Although it registered significant growth in this field, the Brazilian proportion of total patents granted remained at 0.1%.

In 2014, the number of companies that implemented product and/or process innovations in the Technological Innovation Survey (PINTEC) of the Brazilian Institute of Geography and Statistics was 139 thousand. Of this total, 39.9% reported having received some public support to innovate, a significant growth over that declared in 2005 of 18.8% (MCTIC 2017a). The results show that in spite of the important advances in the Brazilian STI indicators, there remains a large gap when compared to the indicators of developed countries and other emerging nations.

Although the proportion of GDP invested in R&D had increased in the last decades, Brazil is now facing a scenario of a contingency of public expenditures. In the last couple of years, the STI budget (federal and state level) suffered severe cutbacks (Gibney 2015; Ribeiro 2017). Expenditures on R&D in relation to GDP in 2016 and 2017 were estimated at around 1.02% (Young et al. 2018). This austerity scenario makes it more challenging for the country to keep the gains made in the STI system and to increase its investment capacity in the sector to be able to reach the SDGs targets. In addition, if we consider the R&D investments in environmental protection and control, it is possible to verify that this agenda has not been a priority: the R$2.7 billion spent between 2003 and 2013 accounted for only around 1% of the total amount invested in R&D (Figure 7.3).

In this way, aside from increasing the investment in STI, it is also necessary to prioritize the research related to conservation and sustainable use of natural resources. In the National Strategies for STI from 2016 to 2022 (MICTIC 2016), Brazil established as research priority areas of investment: water, climate change, oceans, bioeconomics, and biotechnology. This fact demonstrates that at least in the country's planning instruments, the research related to the selected SDGs in this study remains a priority. In this sense, sectoral funds can be considered as one of the most important mechanisms to achieve these goals (Nascimento and Oliveira 2013).

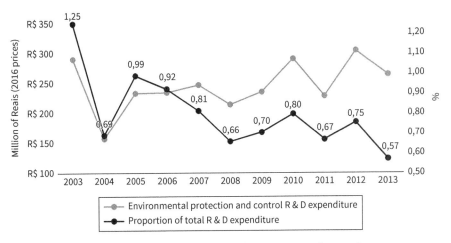

Figure 7.3 R&D expenditures on environmental protection and control, percentage, and total (2003–2013)

Source: Authors' own elaboration, based on MCTIC (2017c).

4.2 The Contribution of Sectoral Funds to Meet SDGs 6, 13, 14, and 15

There are currently 16 sectoral funds: 14 are sector-specific and two cross-sectoral (Table 7.2). Among the cross-sectoral funds, one is intended to support the improvement of infrastructure of science and technology institutes (CT-Infrastructure), and the other is aimed at university–company interaction (Green-Yellow Fund, FVA).

The committees controlling the sectoral funds have the autonomy to define the application of the resources, but the legislation defined the general guidelines for their operation:

1) To modernize and expand the STI infrastructure;
2) To promote greater synergy between universities, research centers, and the private sector;
3) To create new incentives for private investment in STI;
4) To encourage the generation of knowledge and innovation that contribute to the solution of major national problems; and
5) To stimulate the articulation between science and technological development, to include reduction of regional inequalities, and the interaction between universities and companies.

Between 1999 and 2014 the FNDCT contracted 36,123 projects, totaling an investment of R$26.76 billion. The sectoral funds were responsible for contracting 32,997 projects, with a total value of R$17.88 billion.

Table 7.2 Brazilian Sectoral Funds

Fund	Sector	Source of funds	Legislation
CT- AGRO	Agribusinesses	17.5% of the Cide[1] incurred on payments abroad for the purchase of technologies and technical services	Law 10,332/2001 Decree 4,157/2002
CT-AERO	Aeronautics	7,5% of the Cide incurred on payments abroad for the purchase of technologies and technical services	Law 10,332/2001 Decree 4,179/2002
CT-Amazon	R&D in the Amazon region	A minimum of 0.5% of the gross revenues of companies that produce IT goods and services in the Manaus Free Trade Zone	Law 10,176/2001 Decree 4,401/2002
CT-Waterway	Water Transportation and Shipbuilding	3% of the payment portion of the Merchant Navy Renewal Fee (AFRMM)	Law 10,893/2004 Decree 5,252/2004
CT-Bio	Biotechnology	7,5% of the Cide incurred on payments abroad for the purchase of technologies and technical services	Law 10,332/2001 Decree 4,154/2002
CT-Hydric	Water resources	4% of the financial compensation paid by electricity generating companies	Law 9,993/2000 Decree 3,874/2001
CT-Info	Computer Science and automation	0.5% of gross sales in the domestic market in the sale of products covered by the Informatics Law tax exemption or reduction	Lei 10,332/2001 Decree 5,906/ 2006
CT-Infra	R&D infrastructure	20% of each sectoral fund resources	Lei 10,197/2001 Decree 3,807/2001
CT-Energ	Energy	0.75% to 1% of net sales from generation, transmission, and distribution of electricity companies	Law 9,991/2000 Decree 3,867/2001
CT-Espace	Space	25% of the revenues from the use of orbital positions, 25% of the federal revenues from the commercial launch of satellites and sounding rockets from Brazil, and the total Union revenues related to the commercialization of data and images obtained from satellites and of the revenue received by the Brazilian Space Agency due to licenses and authorizations granting	Law 9,994/2000 Decree 3,800/2001

Table 7.2 *Continued*

Fund	Sector	Source of funds	Legislation
CT-Mineral	Mining	2% of the Financial Compensation of the Mineral Sector (CFEM) owed by the mining companies	Law 9,993/2000 Decree 3,866/2001
CT-Petro	Oil and natural gas	25% of the portion of the value of the royalties that exceed 5% of the production of oil and natural gas	Law 9,478/1997 Decree 3,318/1999
CT-Health	Health	17,5% of the Cide incurred on payments abroad for the purchase of technologies and technical services	Law 10,332/2001 Decree 4,143/2002
CT-Transport	Terrestrial transports	10% of the National Highway Department–DNER revenue from contracts signed with telephone operators, communications companies that use the Union's ground transportation infrastructure	Law 9,992/2000 Decree 4,324/2002
CT-Transversal[1]	Cross sector actions	At least 50% of each sectoral fund resources	Law 1,540/2007
CT-Green-Yellow	Green-Yellow Fund (Industry-University Cooperation)	10% of the Cide incurred on payments abroad for the purchase of technologies, technical services, and administrative assistance	Law 10,168/2001 Law 10,332/2001 Decree 3,949/2002

Source: Authors' own elaboration, based on Bastos (2003), Finep (2017).

[1] Contribution owed for intervention in the economic domain; [2] The cross sector actions do not constitute a fund in itself, but a budget line within the scope of FNDCT.

Considering the research themes related to the selected SDGs, FNDCT contributions totaled R\$953 million applied in 2,005 projects. Of these, 1,955 projects benefited from sectoral funds contributions totaling R\$694 million (Table 7.3).

The sectoral funds were responsible for 63.3% of the total investments made by the FNDCT. Only 3.6% of the amount applied by the FNDCT supported projects in the action lines related to the SDGs selected. Considering the number of projects, the sectoral funds were responsible for practically all the contracts (Table 7.3).

The sectoral funds were also the major source for fostering STI projects applied to the selected SDGs: 71.8% of the total applied to the selected SDGs (R\$498 million)

Table 7.3 Number of projects supported by FNDCT and sectoral funds and the invested value in 2016 prices (total and SDG related)

	FNDCT (Total)	FNDCT (SDG related)	Sectoral Funds (Total)	Sectoral Funds (SDG related)
Number of projects	36,123	2,005	32,997	1,995
Invested value (millions of Reais)	26,766	953	17,881	694

Source: Authors' own elaboration, based on MCTIC (2017b).

were contributions of the sectoral funds. The remaining contributions (28.2%) were from FNDCT—source 100, which consists of the fund's ordinary resources (Table 7.4).

In terms of the number of projects, CT-Hydric and CT-Transversal were the sectoral funds that supported the largest number of projects, 891 and 467, respectively. Followed by CT-Amazon, with 248 projects and CT-Agro, with 174 (Table 7.4).

Considering the two development agencies that operate the resources of the FNDCT, CNPq accounted for 83% of the supported projects related to SDGs and Finep for 17%. Although Finep had contracted about one-fifth of the projects, the agency was responsible for 61% of the contributions (R$581 million). The CNPq accounted for 39% (R$372 million).

The FNDCT operates five types of financing instruments to support the projects: public call, invitation letter, order (public bidding), grants, and events. The main financing instrument for contracting SDG- related projects was the public calls. Only 10% of the contracts were made through invitation letters, which deal with specific issues. In general, the public calls are not given for specific purposes, but result instead in supporting dispersed small projects not necessarily connected to each other.

In the Aquarius Platform, each project is classified within ten PACTI priority categories. Almost all (98%) of the projects related to the selected SDGs corresponded to the category Research and Development in Strategic Areas of PACTI, and they accounted for 84% of the investments. Only five projects corresponded to Infrastructure for Scientific and Technological Research, which accounted for 15% of the contributions. None of the projects were associated with the PACTI categories directly related to innovation processes in companies, one of the legally established guidelines for the sectoral funds.

A comparative analysis between the selected lines of action shows that the line that received the largest investment was STI for water resources (SDG6), with 936 supported projects, totaling R$154.06 million. The lines that received the lowest contributions from the sectoral funds were those related to SDG14, STI in Antarctica, and STI for exploration of the sea resources (Table 7.5).

Table 7.4 Number of projects related to SDG and the invested amount (2016 prices) by source

Source–Sectoral Fund	Number of projects	Invested value (Millions of Reais)
FNDCT–Source 100 (Ordinary budget)	50	195.89
CT-HYDRIC (Water resources)	891	188.94
CT-TRANSVERSAL (Cross sector actions)	467	166.24
CT-AGRO (Agrobusnisses)	174	53.65
CT-AMAZON (R&D in the Amazon region)	248	33.36
GREEN-YELLON (Industry-University Cooperation)	28	15.01
CT-BIOTEC (Biotechnology)	28	13.83
CT-WATERWAY (Water Transportation and Shipbuilding)	59	8.71
CT-INFRA (R&D infrastructure)	14	7.98
CT-ENERG (Energy sector)	21	5.02
CT-HEALTH (Health sector)	16	3.07
CT-MINERAL (Mining)	3	1.75
CT-PETRO (Oil and natural gas)	5	0.55
CT-INFO (Computer Science and automation)	1	0.01
Total	2,005	694

Source: Authors' own elaboration, based on MCTIC (2017b).

The sectoral funds' investments presented a growing trend in the value contracted until 2010, with a sharp decrease in 2011, a recovery between 2012 and 2014, and again a drastic reduction in 2014 (Figure 7.4).

From the data presented in this section, it is possible to describe the general profile of the projects related to SDGs supported by sectoral funds, as small grants contracted through broad public calls, and focused on scientific and technological research with

Table 7.5 Number of projects supported and invested value, by line of action selected

Line of action	FNDCT		Sectoral Funds	
	Number of projects	Invested value (R$ million)	Number of projects	Invested value (R$ million)
STI for water resources	936	274.95	913	276.54
Climate Change National Program	82	95.05	77	68.07
STI on Antarctica	5	142.19	3	0.18
STI for the exploration of seas resources	15	10.62	15	10.77
RDI in aquiculture and fisheries	152	88.57	150	87.73
STI applied to biodiversity and natural resources	222	126.11	220	114.06
STI integration program for conservation and sustainable development of Amazon region	389	166.62	378	101.75
STI for Semi-Arid sustainable development	204	48.90	199	34.89
Total	2,005	953	1,955	694

Source: Authors' own elaboration, based on MCTIC (2017b).

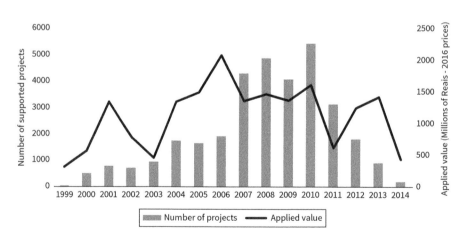

Figure 7.4 Number of projects supported by Sectoral Funds and the invested amount (R$ 2016 prices) in 1999–2014

Source: Authors' own elaboration, based on MCTIC (2017b).

little connection to private sector projects. This may explain the limited use of research by society in the form of new technologies or innovative products (De Negri et al. 2009; De Negri 2017; Melo 2009). Indeed, the tendency to invest in fewer robust projects reduces the chances of producing new knowledge, implementing disruptive technologies, and taking technological leaps that result in cutting-edge science and innovation (CNI 2016). Also, the STI has been suffering multiple public budget cuts, including to the FNDCT and sectoral funds, the most significant source of funding, which affects Brazil's ability to meet the commitments made under Agenda 2030.

In the following section, we will discuss some institutional issues to be overcome so that sectoral funds can foster STI activities with the adequate quantity of resources and results to help Brazil meet SDGs targets. These issues include the distant relationship between the academic sector and companies, lack of a systemic approach to sustainability, and the need to improve performance evaluation indicators.

5. Improvements in Funding Mechanisms for R&D

As presented in section 4, the sectoral funds' pattern of distribution of resources is characterized by excessive fragmentation, the low scale of projects, limited relations in the private sector, and a lack of strategic timing. This pattern of STI support results in more projects fragmentation and decreases the chances to achieve innovative processes and products, including those related to natural resource use.

To improve innovation policies, such as the FNDCT and the sectoral funds, public R&D investments should be strategically oriented (mission-oriented R&D) and new instruments for project selection and contracting should be adopted by public authorities, such as a specific type of bidding for contracting R&D (De Negri 2017; Mazzucato and Penna 2016). In this sense, the SDG targets could be institutionalized as missions to guide the investments of the sectoral funds, channeling resources to support projects capable of providing solutions to the most pressing problems of Brazilian society and helping to achieve the objectives of Agenda 2030.

Another FNDCT and sectoral funds institutional constraint to overcome is the budget contingency. Sectoral funds do not operate as funds as such, but as budgetary sources linked to the FNDCT, thus their resources are subject to the rules of the annual general budget (Bastos 2003). The FNDCT's budget execution series shows that between 2005 and 2010 the funds' expenditure were similar to the available budget. From 2011 onward, there was a gradual increase in the contingency of resources. In 2015, 89% of the available budget had not been released (Figure 7.5).

Adjusting the FNDCT legislation into a financial fund would allow for continuous and stable financing. There is currently a proposal to change the FNDCT and, if approved, the application of the resources of the fund can generate R$50 billion by 2030 (Castro 2017). Unfortunately, under the current circumstances of the Brazilian economy and politics, it is unlikely that this proposal will be accepted in the short run.

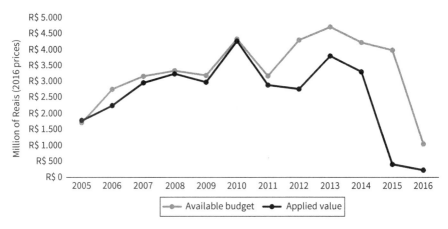

Figure 7.5 Budget expenditure of FNDCT (R$2016 prices) in 2005–2016
Source: Authors' own elaboration, based on MCTIC (2017b).

The compulsory R&D and innovation investment clause for the regulated sectors of oil and gas and energy could be a potential financing source for environmental sustainability R&D. These clauses consist of legal obligations for private investment in R&D defined by norms issued by competent regulatory agencies, which exercise the power of control and supervise the application of the resources. In the case of the oil and gas sector, the R&D clause establishes the application of a percentage of the gross revenue from production, according to specific conditions of each modality of contract. Among the types of projects and programs eligible for funding, as established in the National Agency for Petroleum, Natural Gas and Biofuels (ANP) Technical Regulation N. 03/2015, are basic research, applied research, and experimental development, including research in the environment and social, human, and life sciences, among others (ANP 2015).

Between 1998 and 2017, the total amount accumulated for investments in R&D established for oil and gas exploration and production contracts was R$13.3 billion (current values) (ANP 2018). However, from the 1,455 projects authorized by the ANP between 2005 and 2017, to receive funds under the RD&I clause, only 131 were related to environmental issues. These projects together accounted for only 4.7% of the total authorized project budget (ANP 2017). This result occurs partly because the current regulation on the subject does not establish criteria to prioritize research and scientific development in this area. Thus, one way to improve the regulated sectors R&D programs is to review the legislation in order to include criteria to direct the funds to projects with specific characteristics, such as environmental sustainability (Young et al. 2018).

Another potential financial mechanism to foster environmental R&D is the Biodiversity Law (N. 13123/2015). This law determines that if a company explores a product developed from existing Brazilian biodiversity material or from associated traditional knowledge, it will have to pay from 0.1% to 1% of the annual net revenue obtained from the economic exploitation as a benefit-sharing tax.

The benefit-sharing tax can be monetary or nonmonetary actions. The monetary payment goes to the National Benefit-Sharing Fund or directly to the communities that possess the associated traditional knowledge. The nonmonetary actions include investments in technology transfer, human resource training, conservation projects, and sustainable use of biodiversity.

The Brazilian benefit-sharing regulation is similar to that adopted in other countries in terms of the possibility of monetary or nonmonetary sharing, but the benefit sharing percentage varies significantly, with values that may relate to the survey, the gross revenue, or net income (Freire 2017). For example, in Australia, the recommended monetary benefit sharing is 5% on gross revenue (Medaglia et al. 2011). In the Philippines, the minimum percentage monetary benefit sharing is 2% of total gross revenue—25% of the amount goes to the federal government and 75% to the suppliers (Schmidt 2014).

Critical issues with the Brazilian law are the benefit-sharing payment exceptions (such as basic research), which violates the rights of the traditional populations who are the holders of this knowledge (Brito and Pozzetti 2017). It also has a negative impact on the SDG15.6 target (promote fair and equitable sharing of the benefits arising from the utilization of genetic resources and appropriate access to such resources).

Between 2002 and 2015, the total value of the benefit sharing agreements related to the economic exploitation of genetic heritage and associated traditional knowledge was R$8 million only (current value). In addition, 33% of the contracts provided for the development of socioenvironmental or conservation actions, and only 6% of the benefit-sharing contracts officially registered in Brazil provided for technology transfer (Souza et al. 2017).

Costa Rican regulation is a good example of an alternative resource allocation scheme. The National Biodiversity Institute, a nongovernmental, nonprofit association created in 1989, conducts most of the bioprospecting in the country through collaborative research agreements with industry and academia that involves technology transfer training, funding for publications and dissemination of scientific literature. The institute receives monetary benefits from direct payments, payment for samples supplied, and 50% of future royalties. In addition, the Ministry of Environment and Energy receive a fixed portion of 10% of the research budgets and 50% of the future royalties to be reinvested in conservation (Medaglia et al. 2011).

To be able to increase the funding potential through the Biodiversity Law, it will be necessary to readjust monetary benefit sharing percentages and to establish rules to allocate resources for scientific and technology development related to biodiversity. It is important to highlight that these adjustments could help to meet the SDG15.6, related to the equitable share of benefits. It could also raise funds to promote biodiversity conservation and sustainable use, which have a positive impact on SDG targets 14.A and 15.B.

6. Conclusions

As presented, Brazil has advanced in the consolidation of institutions, in the design of financing policies and in the adoption of broad legislation to foster STI. These advances had positive reflections in some important indicators, such as the proportion of the GDP invested in R&D, the number of researchers, and world publications. The R&D investments dedicated to environmental protection and sustainable use of natural resources, however, constitute a very small part of the total R&D expenditure.

The sectoral funds are a strategic source of funds for STI applied to the SDGs, but the fragmented pattern and the lack of orientation in resource application make it difficult to develop and apply new technologies to solve concrete environmental problems. Thus, some institutional adjustments will be necessary.

A possible strategy to improve the financing of sectoral funds could be the adoption of the SDG targets as mandatory for resource allocation, orienting the investments to produce the knowledge and technological innovations necessary to promote environmental conservation and sustainable use of natural resources. It is also possible to induce minimum targets for R&D investment in issues where Brazil has an important role in global terms, such as biodiversity conservation and greenhouse gas emissions. In addition, other financing mechanisms need some improvements in order to channel funds to environmental STI. Credit and fiscal benefits can be used to enhance private sector investment, while the public expenditures on R&D, particularly toward the academic institutions, can be oriented in the direction of the strategic objectives defined in terms of the SDG commitments.

In summary, to achieve the commitments with the 2030 Agenda, the Brazilian government must seek new sources of funds, as well as adjust existing institutional mechanisms and develop new ones to improve public policies that foster STI activities. This will not be possible, however, unless there is a significant shift away from the current trend of exacerbated austerity and frozen public expenditures in R&D and basic science.

References

Agência Nacional do Petróleo (ANP) (2015). Regulamento Técnico ANP N° 3/2015. http://www.anp.gov.br/images/Pesquisa_Desenvolvimento/Investimentos_PDI/Regulamentacao_tecnica/RT_ANP_03_2015_alterado.pdf (accessed April 5, 2019).

Agência Nacional do Petróleo (ANP) (2017). Boletim de Pesquisa, Desenvolvimento e Inovação—Circulação Externa n° 41 1° trimestre de 2017. http://www.anp.gov.br/images/publicacoes/boletins-anp/boletim_petroleo_p-e-d/Boletim_PD-e-I_Ed41_1trimestre2017.pdf (accessed April 5, 2019).

Agência Nacional do Petróleo (ANP) (2018). Regulamentação técnica de PD&I. http://www.anp.gov.br/pesquisa-desenvolvimento-e-inovacao/investimentos-em-p-d-i/regulamentacao-tecnica-relativa-aos-investimentos-em-p-d-i (accessed April 5, 2019).

Bastos, V. D. (2003). Fundos públicos para ciência e tecnologia. *Revista do BNDES* 10(20), 229–260.

Brito, A. C. L., and Pozzetti, V. C. (2017). Biodiversidade, conhecimentos tradicionais associados e repartição de benefícios. *Derecho y Cambio Social* 48, 1–13.

Castro, F. (2017). Medida Provisória muda fundo para dar verba à ciência. *Estadão*, August 7, 2017. https://ciencia.estadao.com.br/noticias/geral,medida-provisoria-muda-fundo-para-dar-verba-a-ciencia,70001926488 (accessed April 5, 2019).

Confederação Nacional da Indústria (CNI) (2016). *Financiamento à Inovação*. Confederação Nacional da Indústria, Instituto Euvaldo Lodi, Brasília, DF.

De Negri, F. (2017). Por uma nova geração de políticas de inovação no Brasil. In: Turchi, L. M., and Morais, J. M. (Eds.), *Políticas de Apoio à Inovação Tecnológica no Brasil: Avanços Recentes, Limitações e Propostas de Ações*, pp. 25–46. Instituto de Pesquisa Econômica Aplicada, Brasília, DF.

De Negri, F., Alves, P. F., Kubota, L. C., Cavalcante, L. R., and Damasceno, E. C. (2009). *Perfil das Empresas Integradas ao Sistema federal de CT&I no Brasil e aos Fundos Setoriais: uma Análise Exploratória*. Ministério da Ciência e Tecnologia/Financiadora de Estudos e Projetos/Instituto de Pesquisa Econômica Aplicada, Brasília, DF, and Universidade Federal de Minas Gerais, Belo Horizonte, MG.

Funding Authority for Studies and Projects (FINEP) (2017). *O que são os Fundos Setorais*. https://bit.ly/3e3cP7T (accessed December 16, 2017).

Freire, C. T. (2017). Pesquisa, inovação e a nova lei de biodiversidade no Brasil: avanços e desafios. In: Turchi, L. M., and Morais, J. M. (Eds.), *Políticas de Apoio à Inovação Tecnológica no Brasil: Avanços Recentes, Limitações e Propostas de Ações*, pp. 333–370. Instituto de Pesquisa Econômica Aplicada, Brasília, DF.

Gibney, E. (2015). Brazilian science paralyzed by economic slump. *Nature* 526, 16–17.

Intergovernmental Oceanographic Commission (IOC) and United Nations Educational, Scientific and Cultural Organization (UNESCO) (2017). *Global Ocean Science Report: The Current Status of Ocean Science Around the World*. UNESCO Publishing, Paris.

Kannebley Junior, S., Shimada, E., and De Negri, F. (2016). Efetividade da Lei do Bem no estímulo aos dispêndios em P&D: uma análise com dados em painel. *Pesquisa e Planejamento Econômico* 46(3), 111–145.

Mazzucato, M., and Penna, C. C. R. (2016). *The Brazilian Innovation System: A Mission-Oriented Policy Proposal*. Centro de Gestão e Estudos Estratégicos, Brasília, DF.

Medaglia, J. C., Perron-Welch, F. and Rukundo, O. (2011). *Overview of National and Regional Measures on Access to Genetic Resources and Benefit-sharing: Challenges and Opportunities in Implementing the Nagoya Protocol*. Centre for International Sustainable Development Law, Montreal.

Melo, L. M. (2009). Financiamento à inovação no Brasil: análise da aplicação dos recursos do Fundo Nacional de Desenvolvimento Científico e Tecnológico (FNDCT) e da Financiadora de Estudos e Projetos (FINEP) de 1967 a 2006. *Revista Brasileira de Inovação* 8(1), 87–120.

Ministry of Science and Technology (MCT) (2007). *Ciência, Tecnologia e Inovação para o Desenvolvimento Nacional. Plano de ação 2007–2010*. Ministério da Ciência e Tecnologia, Brasília, DF.

Ministry of Science, Technology, Innovations and Communications (MCTIC) (2016). *Estratégia Nacional para Ciência, Tecnologia e Inovação 2016–2022*. MCTIC, Brasília, DF.

Ministry of Science, Technology, Innovations and Communications (MCTIC) (2017a). *Indicadores Nacionais de Ciência, Tecnologia e Inovação*. MCTIC, Brasília, DF.

Ministry of Science, Technology, Innovations and Communications (MCTIC) (2017b). *Aquarium Plataform*. Accessed December 16, 2017. http://aquariusp.mcti.gov.br/

Ministry of Science, Technology, Innovations and Communications (MCTIC) (2017c). *Dispêndios públicos em P&D por objetivo socioeconômico, 2000-2013*. https://bit.ly/2UUKtoV (accessed December 15, 2017).

Nascimento, P. A. M. M. (2016). Áreas de maior especialização científica do Brasil e identificação de suas atuais instituições líderes. In: De Negri, F., and Squeff, F. H. S. (Eds.), *Sistemas Setoriais de Inovação e Infraestrutura de Pesquisa no Brasil*, pp. 617–637. Instituto de Pesquisa Econômica Aplicada, Brasília, DF.

Nascimento, P. A. M. M., and Oliveira, J. M. (2013). Papel das ações transversais no FNDCT: redirecionamento, redistribuição, indução ou nenhuma das alternativas? *Revista Brasileira de Inovação* 12(1), 73–104.

Oliveira, U., Soares-Filho, B. S., Paglia, A. P., Brescovit, A. D., Carvalho, C. J. B., Silva, D. P., Rezende, D. T., Leite, F. S. F., Batista, J. A. N., Barbosa, J. P. P. P., Stehmann, J. R., Ascher, J. S., Vasconcelos, M. F., De Marco, P., Löwenberg-Neto, P., Ferro, V. G., and Santos, A. J. (2017). Biodiversity conservation gaps in the Brazilian protected areas. *Scientific Reports* 7, 1–9.

Overbeck, G. E., Bergallo, H. G., Grelle, C. E. V., Akama, A., Bravo, F., Colli, G. R., Magnusson, W. E., Tomas, W. M., and Fernandes, G. W. (2017). Global biodiversity threatened by science budget cuts in Brazil. *BioScience* 68, 11–12.

Ribeiro, S. (2017). Ciência em retrocesso. *Revista Veja*, May 18. https://complemento.veja.abril.com.br/pagina-aberta/ciencia-em-retrocesso.html

Sachs, J. D. (2012). From millennium development goals to sustainable development goals. *The Lancet* 379(9832), 2206–2211.

Sant'Anna, A. A., and Young, C. E. F. (2010). Property rights, deforestation and rural conflicts in the Amazon. *Economia Aplicada* 14(3), 377–387.

Schmidt, L. (2014). *Avaliação sobre a Repartição de Benefícios no Brasil: Contratos Anuídos e em Tramitação no CGEN (Projeto BRA/11/001)*. Ministério do Meio Ambiente, Brasília, DF.

Soares-Filho, B., Moutinho, P., Nepstad, D., Anderson, A., Rodrigues, H., Garcia, R., Dietzsch, L., Merry, F., Bowman, M., Hissa, L., Silvestriniand, R., and Maretti, C. (2010). Role of Brazilian Amazon protected areas in climate change mitigation. *PNAS* 107(24), 10821–10826.

Souza, A. L. G., Santos Junior, A. A., and Silva, G. F. (2017). Os "royalties" das aplicações tecnológicas do patrimônio genético nacional e dos conhecimentos tradicionais associados: o estado brasileiro em questão. *Revista Gestão, Inovação e Tecnologias* 7(4), 4149–4158.

United Nations (2015). Transforming Our World: The 2030 Agenda for Sustainable Development. https://sustainabledevelopment.un.org/post2015/transformingourworld (accessed April 5, 2019).

United Nations Development Programme (UNDP) (2018a). Background on the goals. http://www.undp.org/content/undp/en/home/sustainable-development-goals/background/ (accessed April 5, 2019).

United Nations Development Programme (UNDP) (2018b). What are the sustainable development goals? http://www.undp.org/content/undp/en/home/sustainable-development-goals.html (accessed April 5, 2019).

United Nations Educational, Scientific and Cultural Organization (UNESCO) (2015). *Relatório de Ciência da Unesco: rumo a 2030. Visão geral e cenário brasileiro*. UNESCO, Paris, Representação da UNESCO no Brasil, Brasília, DF.

Young, C. E. F., Castro, B. S., Mathias, J. F. C. M., Penna, C. C. R., Ferraz, C. C. M., Pereira, G. S., Alvarenga Junior, M., Machado, M., Gatto, D. B., Batista A. K., Fontenelle, C. A., Duque, D. V. A., Arrellaga, M. M., Araújo, K. S., Jordão, C. S., Correa, M. G., Pessanha, A. L., Gonçalves, R., Fares, L. R., and Manzatto, L. H. R. (2018). *Análise conjuntural sobre ODS e efetividade das estruturas de financiamento públicas, privadas e mistas para a promoção do desenvolvimento sustentável aprimoradas.* Relatório de pesquisa. Grupo de Pesquisa de Economia do Meio Ambiente e Sustentabilidade—GEMA, Rio de Janeiro, RJ.

Zuniga, P., De Negri, F., Dutz, M. A., Pilat, D., and Rauen, A. (2016). Conditions for Innovation in Brazil: a Review of Key Issues and Policy Challenges. Discussion Paper. Ipea, Brasília, Rio de Janeiro.

8

The Systems Science of Industrial Ecology

Tools and Strategies Toward Meeting the Sustainable Development Goals

Marian R. Chertow, Koichi S. Kanaoka, T. Reed Miller, Peter Berrill, Paul Wolfram, Niko Heeren, and Tomer Fishman

1. Introduction

In the early 1970s at the dawning of the modern environmental movement, three eminent scientists battled over what became known as the IPAT identity. The work of Paul Ehrlich and John Holdren resulted in an equation to express the key drivers of environmental impact (I) as a product of population (P), affluence (A), and technology (T) (Chertow 2000). Each of these factors underlies aspects of the United Nations sustainable development goals (SDGs) with concern for demography, poverty and quality of life, and human development through technological and policy innovation (Table 8.1). The equation itself has an interesting and rich history that pitted Ehrlich's conviction that population (P) has the greatest impact against the involvement of a third prominent scientist, Barry Commoner, who embraced technology (T) as the most significant factor. It is important to note that the technology term was first depicted as a key driver of dramatic *increases* in environmental degradation (Chertow 2000). By the 1990s, however, the creation of a new systems science, industrial ecology—referred to early on as "the marriage of technology and ecology"—had transformed the more pessimistic views of technology toward an understanding that the T term could be the most influential in *reducing* environmental impact.

In 1995, the first textbook in the field declared an IPAT variant as the "Master Equation" of industrial ecology, operationalized as follows (Graedel and Allenby 1995; 2002):

$$\text{Environmental impact} = \text{Population} \times \frac{\text{GDP}}{\text{Person}} \times \frac{\text{Environmental impact}}{\text{Unit of GDP}}$$

$$I \quad = \quad P \quad \times \quad A \quad \times \quad T$$

Table 8.1 Sustainable development goals (SDGs) and relevant industrial ecology concepts discussed with case examples

Sustainable Development Goal	Targets	Relevant industrial ecology concepts discussed with case examples
2. Zero Hunger	2.4 By 2030, ensure sustainable food production systems and implement resilient agricultural practices.	E-LCA, EEIO
6. Clean Water and Sanitation	6.1 By 2030, achieve universal and equitable access to safe and affordable drinking water for all	E-LCA
	6.3 By 2030, improve water quality by reducing pollution, halving the proportion of untreated wastewater and increasing recycling and safe reuse globally	E-LCA, EEIO, IS
	6.4 By 2030, substantially increase water-use efficiency across all sectors and ensure sustainable withdrawals and supply of freshwater to address water scarcity.	MFA, IS
	6.5 By 2030, implement integrated water resource management at all levels, including through transboundary cooperation.	MFA, E-LCA, EEIO
7. Affordable and Clean Energy	7.3 By 2030, double the global rate of improvement in energy efficiency	E-LCA, IS
8. Decent Work and Economic Growth	8.3 Promote policies that support productive activities, decent job creation, and encourage the growth of micro-, small-, and medium-sized enterprises.	IS
	8.4 Improve global resource efficiency in consumption and production and endeavor to decouple economic growth from environmental degradation.	MFA, IS
	8.8 Protect labor rights and promote safe and secure working environments for all workers.	S-LCA
9. Industry, Innovation, and Infrastructure	9.4 By 2030, upgrade infrastructure to make them sustainable, with increased resource-use efficiency and greater adoption of clean technologies and industrial processes.	MFA, E-LCA, IS
10. Reducing Inequalities	10.4 Adopt policies, especially fiscal, wage, and social protection policies, and progressively achieve greater equality	SEIO
11. Sustainable Cities and Communities	11.2 By 2030, provide access to safe, affordable, accessible, and sustainable transport systems for all.	E-LCA
12. Responsible Consumption and Production	12.2 By 2030, achieve the sustainable management and efficient use of natural resources	MFA, E-LCA, EEIO, IS

Continued

Table 8.1 *Continued*

Sustainable Development Goal	Targets	Relevant industrial ecology concepts discussed with case examples
	12.3 By 2030, halve per capita global food waste at the retail and consumer levels and reduce food losses along production and supply chains.	MFA, E-LCA, EEIO
	12.4 By 2020, achieve the environmentally sound management of chemicals and all wastes throughout their life cycle and significantly reduce their release to air, water, and soil in order to minimize their adverse impacts on human health and the environment	MFA, E-LCA, IS
	12.5 By 2030, substantially reduce waste generation through prevention, reduction, recycling, and reuse	MFA, E-LCA, IS
13. Climate Action	13.2 Integrate climate change measures into national policies, strategies, and planning	MFA, E-LCA, EEIO, IS
	13.3 Improve education, awareness-raising, and human and institutional capacity on climate change mitigation, adaptation, impact reduction, and early warning	MFA, E-LCA, EEIO, IS

This table is not exhaustive and only includes SDGs and targets examined in this chapter. Abbreviations: MFA, material flow and stock analysis; E-LCA, environmental life-cycle assessment; S-LCA, social and socioeconomic life-cycle assessment; EEIO, environmentally extended input-output analysis; SEIO, socially and socioeconomically extended input-output analysis; IS, industrial symbiosis.

To offer an example of how the variables work, if global population increased 1.5 times by 2050 and GDP/person doubled by 2050, plugging 1.5 x 2 into the IPAT equation suggests that P × A will increase environmental impact (I) three-fold. While it sounds extraordinarily risky to triple environmental impact at this stage, it is also true that the SDGs aim to improve affluence (A)—thought of as a quality-of-life term concerned with improving people's standard of living. With only the technology (T) term remaining to balance the equation, we would need to decrease environmental impact/unit of GDP (the T term) at a tripled rate. As such, science and technology are essential counterweights to increases in population and affluence.

The oversimplification that we now understand about technology is captured in a polemical statement by Professor Melvin Kranzberg, an historian of technology, who wrote: "technology is neither good nor bad, nor is it neutral." Evaluating the trade-offs in assessing the benefits and burdens of new technologies in specific uses is a complicated matter, but accepting this complexity is crucial to getting the SDGs right. If one target (e.g., for air pollution), encourages capture of pollutants, but the specific technology achieves this simply by moving pollution to water rather than air, the problem is not solved, as negative impacts must be reduced rather than merely shifted.

Since the beginning of industrial ecology almost 30 years ago, the field has worked intensively to identify and hone assessment tools that apply a systems perspective to address whether one way of accomplishing a defined task is more environmentally beneficial than another way of accomplishing it. A commonplace example is whether paper bags or plastics bags for carrying home groceries have more environmental impact. Industrial ecology offers numerous tools for effectively evaluating trade-offs in decision-making and implementing strategies to help with meeting identified SDGs.

Though industrial ecology can be explained in many ways, Robert White, former president of the US National Academy of Engineering, offered a well-respected definition (White 1994, p. v):

> Industrial ecology is the study of the flows of materials and energy in industrial and consumer activities, of the effects of these flows on the environment, and of the influences of economic, political, regulatory, and social factors on the flow, use, and transformation of resources.

Industrial ecology takes a systems approach, aiming to reduce pollution from being created in the first instance in place of traditional strategies such as remediation and end-of-pipe control. Industrial ecology uses tools including material flow analysis and life-cycle assessment to holistically evaluate systems at the facility, regional and global levels, and across organizations. These tools are especially beneficial to rapidly industrializing countries, where improvements in economic performance and the environment must be carefully balanced (Erkman and Ramaswamy 2003; Smil 2004). By tracking flows of material and energy, industrial ecology places a premium on the study of physical resource flows through human systems. Industrial ecology also promotes resource efficiency, helping countries to secure supplies of resources over longer time horizons and uncovering opportunities for resource scarce nations to industrialize sustainably.

A later section of the chapter describes an additional concept from industrial ecology, a phenomenon termed *industrial symbiosis*, that is concerned with resource flows and local economic development. Diverse companies and other organizations, even those located in the same industrial cluster, often overlook the opportunity to reuse water, wastewater, energy, and materials such that the by-products of one enterprise become the feedstocks for another. Indeed, in the seminal paper of industrial ecology, "Strategies for Manufacturing," Frosch and Gallopoulos (1989, p. 144) stated:

> The traditional model of industrial activity ... should be transformed into a more integrated model: an industrial ecosystem. In such a system the consumption of

energy and materials is optimized, waste generation is minimized, and the efflu-
ents of one process … serve as the raw material for another process.

The next section of this chapter offers explanations of key assessment tools
and strategies used by industrial ecologists. These tools provide the SDG practi-
tioner, who is trying to compare technologies, products, or systems at different
scales, with information to identify where the largest environmental impacts are
occurring and with suggestions about how the life-cycle impacts can be improved,
substituted, or mitigated. Following the section on tools, we shift focus to industrial symbiosis, which could be employed as a collaborative strategy for improving
industrial resource efficiency in an economically beneficial manner. Policy implications of the assessment tools and industrial symbiosis are discussed in the
conclusion.

2. Industrial Ecology Methods, Tools, and Case Examples

The three key methods and tools of industrial ecology discussed in this section are
interrelated conceptually but differ in the scope of the assessment. These methods
and tools are: material flow and stock analysis (MFA),[1] life-cycle assessment (LCA),
and input-output (IO) analysis (Table 8.2). In the MFA represented in Figure 8.1,
a single material is tracked across some or all of its life cycle. The arrow thickness
represents the mass of the flow; while the material itself may find its way into multiple shorter-lasting products, the more durable products accumulate as stocks.
Environmental LCA (E-LCA) tracks natural resources, human-produced inputs,
and direct environmental impacts at some or all stages of the life cycle (Figure 8.2).
Social and socioeconomic LCA (S-LCA) parallels E-LCA, but instead examines the
social and socioeconomic problems, risks, and opportunities associated with labor.
Figure 8.3 depicts a simplified IO analysis, more specifically an environmentally extended IO (EEIO) analysis, showing how the tool identifies and brings together all
industries along the value chain needed to make one product. Multiple raw materials, products, and service sectors can be represented and assessed simultaneously
for environmental and labor impacts using national statistics.

In addition to explanations of the tools, this section presents case examples
of where industrial ecologists have made progress on systems-level analyses regarding food (SDG2), water (SDG6), energy (SDG7), economic development
(SDG8), industry, innovation, and infrastructure (SDG9), reducing inequalities
(SDG10), and cities and communities (SDG11). Responsible consumption and

---

[1] While MFA originally stood for material flow analysis, it has also been expanded to material and energy flow analysis and here we choose another variation, material flow and stock analysis, that takes account not only of the movement of *flows* but also of the accumulation of *stocks* that end up in one place.

Table 8.2 Overview of discussed industrial ecology tools

Industrial ecology tool	Description
Material flow and stock analysis (MFA)	Framework to account for the movement and accumulation of materials flowing in and out of, and within, a defined system. Based on the law of conservation of matter. Various material flow and stock accounts can be combined by daisy-chaining or nesting within each other, and dynamically tracked over time. MFA is a powerful and flexible tool to study the movement of materials and systems on various scales, from individual households and companies to entire countries (Schandl and Miatto 2018) and global supply chains (Graedel et al. 2015).
Life-cycle assessment (LCA)	Methodology to evaluate the environmental or social impacts of a product throughout its life cycle, from extraction of raw materials to manufacturing, consumer use, and disposal. Every LCA has four components: goal and scope definition, life-cycle inventory (LCI) analysis, life-cycle impact assessment (LCIA), and interpretation of results (ISO 2006; UNEP/SETAC 2009). The first component of an LCA requires a clear definition of the *goal and scope* of the overall project, the *functional unit* through which consistent comparisons can be made, and the overall *system boundary*.
Environmental LCA (E-LCA)	An LCA framework for assessing environmental impacts of a product system, also referred as process-based LCA. Guidelines for E-LCA have been standardized in ISO 14040 and 14044, and harmonized further in Environmental Product Declarations (EPDs, ISO 14025). The basis of an E-LCA is a life-cycle inventory (LCI), which quantifies exchanges between the environment (biosphere) and a specific process, product, or service in the technosphere across the entire life cycle. In the life-cycle impact assessment (LCIA), E-LCA translates these exchanges into environmental impact indicators, for example, cumulative energy demand (CED) or climate change impacts measured in carbon dioxide (CO_2)-equivalent emissions.
Social and socioeconomic LCA (S-LCA)	An LCA framework for evaluating the impacts of product systems on people. The steps of S-LCAs parallels that of E-LCAs but examines social rather than environmental aspects. Guidelines for conducting S-LCA are specified in UNEP/SETAC (2009). Assessments include five stakeholder categories: workers, consumers, local community, society, and value chain actors. Many S-LCA studies focus on a particular product in a particular location.
Working environment (WE) LCA	An S-LCA methodology developed specifically for examining impacts of the working environment on workers. WE-LCA studies may capture the risk of accidents or exposure to chemicals (Kim and Hur 2009).
Input-output (IO) accounts	Annual, national tables that account for the amount of inputs required from all economic sectors to produce commodities, in economic units.

Continued

Table 8.2 *Continued*

Industrial ecology tool	Description
Environmentally extended IO (EEIO) analysis	Methodology to quantify the environmental impacts of consuming commodities across their entire supply chain. Analyses focus on a single commodity or the annual consumption of an entire country. Accomplished by associating the annual environmental impacts of a commodity-producing sector with the corresponding sector in the IO table. EEIO results are less granular than process-based LCA results as specific commodities within a sector cannot be distinguished, but more comprehensive because they account for all sectors associated with the supply chain being studied, including, for instance, finance and retail. Common indicators produced by EEIO analyses include carbon, water, and energy footprints, while more recent indicators include nitrogen, biodiversity loss, mercury emissions, and metal ore extraction.
Single-country IO	The basic model for EEIO analysis, which focuses on a single country. An assumption is made that imports are produced in the same fashion as their domestically produced counterparts.
Multiregional IO (MRIO)	A popular extension of the single-country EEIO model, which accounts for different production recipes and environmental intensities in various countries or world regions, and can capture environmental impacts embodied in global trade (Lenzen et al. 2013; Stadler et al. 2018). MRIO analysis allows for tracking emissions associated with global production for consumption in another country.
Socially and socioeconomically extended IO (SEIO) analysis	A technique paralleling EEIO to evaluate the social impacts associated with a system being studied. An analysis of a purchase from a given sector and country provides an estimate of various risks and opportunities along the entire supply chain. For example, the Social Hotspots Database was created to streamline the social risk and opportunity analysis by using the Global Trade Analysis Project (GTAP)'s IO database which covers most countries (Norris et al. 2012).

production (SDG12) and climate action (SDG13) cut across these pertinent SDGs and inform many recent industrial ecology studies (see also Table 8.1 for targets identified).

2.1 Methods

Explanations are provided here for the methods of MFA, LCA, and IO analysis.

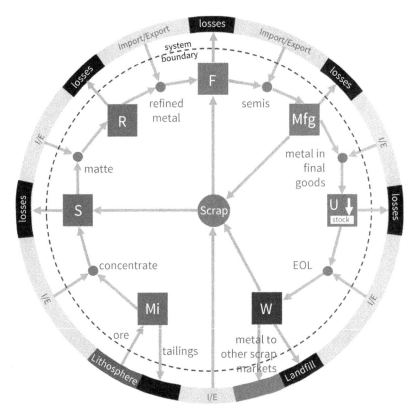

Figure 8.1 Generic circular diagram for metals, with the main processes mining/milling (Mi), smelting (S), refining (R), fabrication (F), manufacturing (Mfg), use (U), and waste management and recycling (W). The processes are connected through seven markets, each related to other regions through net import flows. EOL stands for end-of-life.
Source: Barbara Reck, Yale Center for Industrial Ecology (based on Reck et al. 2008).
Reproduced with permission from Barbara Reck.

2.1.1 Material Flow and Stock Analysis

MFA is a framework to account for the movement and accumulation of materials flowing into and out of, and within, a defined system. At its core is the law of conservation of mass which states that a material cannot simply disappear in ordinary chemical, physical, and economic processes. In other words, a material that enters a system must eventually come out, otherwise it must still be within the system. This understanding enables a full accounting of the movement and transformations of materials because it helps identify where to look for any material unaccounted for. If a factory purchased 10 tons of steel, for example, and produced 8 tons of widgets, then we can deduce that the missing 2 tons are either still on the factory floor or they left the factory as scrap or waste. The *flow* of 8 tons of steel products leaving the

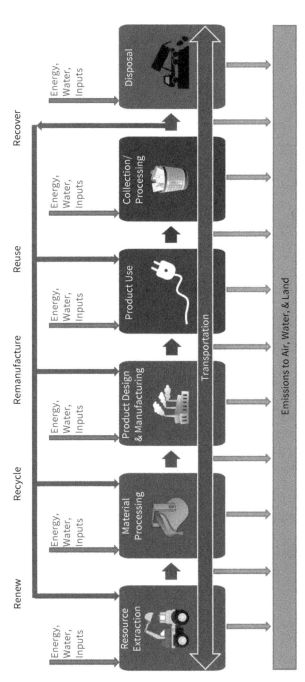

Figure 8.2 Generic representation of a product life cycle and resource flows studied in environmental life-cycle assessment

Source: authors.

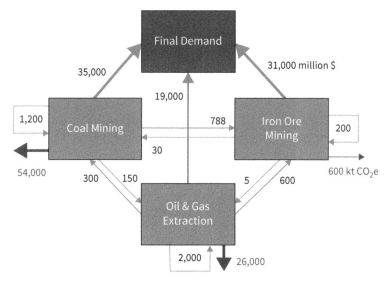

Figure 8.3 Generic representation of environmentally extended input-output analysis. Light gray arrows represent monetary units, while dark gray arrows represent climate change impacts of industries expressed in kilotons of carbon dioxide-equivalent (CO_2e) emissions.

Source: Adapted from Schmidt (2000).

factory now increases the *stock* of steel in the factory. If the missing 2 tons are used for scrap or are manufactured into something new, the flow of these materials will become additional stock elsewhere. While simple, this framework is a powerful and flexible tool to study the movement of materials and systems on wildly varying scales, from individual households and companies, through river basins and cities (Brunner and Rechberger 2016), to global supply chains (Graedel et al. 2015), whole countries (Schandl and Miatto 2018), and the entire planet (Krausmann et al. 2017; Schandl et al. 2018). Various material flow and stock accounts can be combined with one another and tracked over time.

2.1.2 Life-Cycle Assessment

LCA was initially developed as a method for quantifying environmental impacts across the entire life cycle of products, processes, or services from extraction of raw material to manufacturing and assembly, and end-of-life. Furthermore, it has the ambition to include indirect or upstream impacts that occur along the supply or disposal chain, since important environmental impacts can occur "off-site" or be "embodied" (i.e., not a consequence of the use phase). This conventional form of LCA is referred to as E-LCA or process-based LCA.

E-LCA has a long tradition as an assessment tool having first been applied to a study of beverage containers for the Coca-Cola Company in 1969 (Hunt and

Franklin 1996). The importance of E-LCA is that it allows us to identify the impact of individual life-cycle stages for different products or services. For example, many consumer electronics are manufactured efficiently but are difficult to dispose of at the end of their useful lives. Motor vehicles generate the majority of their greenhouse gas emissions not in the factory but in the use stage (e.g., when cars are being driven well after the vehicle has left the manufacturing site) (Chester and Horvath 2009). In such cases a detailed E-LCA study allows us to determine the savings owing to a replacement of existing products (vehicles) with more efficient ones. E-LCA has gathered a strong user community and is applied widely in industry, the consulting sector, and academia. Today it is one of the key tools used in the work of industrial ecology.

Every E-LCA must include four components: goal and scope definition, life-cycle inventory (LCI) analysis, life-cycle impact assessment (LCIA), and interpretation of results (ISO 2006). The goal and scope definition of an E-LCA typically requires a clear explanation of the *goal and scope* of the overall project, the *functional unit* through which consistent comparisons can be made, and the overall *system boundary*. The basis for an E-LCA is an LCI, which quantifies all exchanges between the environment (biosphere) and the process (technosphere) for the entire life cycle. The LCIA translates these exchanges into environmental impact indicators, for example, cumulative energy demand (CED) or climate change impacts measured in carbon dioxide (CO_2)-equivalent emissions (Matthews et al. 2014). E-LCA is a particularly useful tool to compare products or to gain system-level understanding and optimize the life cycle of a product.

Recently, methodologies to evaluate the social impacts of product systems have been developed, called S-LCA. The steps of an S-LCA parallel that of an E-LCA, but rather than focus on goods and services, S-LCA tracks the positive or negative impacts of a product system based on people—the human labor and communities along the supply chain (UNEP/SETAC 2009). A subset of methods was developed specifically to examine the impacts of the working environment (WE) on workers. WE-LCA studies may capture the risk of accidents or exposure to chemicals (Kim and Hur 2009). S-LCA is generally more expansive, attempting to include a set of five stakeholder categories: workers, consumers, local community, society, and value chain actors. Guidelines for S-LCA (UNEP/SETAC 2009) provide an overview of the assessment framework, and, more recently, have been complemented by methods for assessing 31 subcategories (Benoît-Norris et al. 2011), including child labor, consumer privacy, cultural heritage, and fair competition. Many S-LCA studies are thorough investigations into a particular product in a particular location.

2.1.3 Input-Output Analysis with Environmental and Social Extensions

By taking a detailed snapshot of a national economy in a given year, IO accounts show how much input from all sectors of the economy is required to produce a given amount of any commodity, in economic units. While the sectors of IO models tend to be less specific than products examined in process-based LCA models (e.g., combining all fruits and vegetables together or combining all iron and steel production together), their production recipes are systematically more comprehensive than those in LCA databases. For instance, IO models will incorporate design engineering and professional services as inputs to production of a bridge, while process LCA databases struggle to incorporate such nonphysical inputs of production.

Producing commodities typically involves resource extraction, land use, and pollution emissions. These environmental impacts can be associated with the economic output of sectors. By combining IO tables with environmental impacts of each economic sector, environmental footprints can be calculated for individual commodities, baskets of goods, or the entire consumption in a year (Kitzes 2013). This technique is referred to as environmentally extended input-output (EEIO) analysis. A popular extension of the basic single-country EEIO model is the multiregional IO model, which incorporates different production recipes and environmental intensities in different countries or world regions (Lenzen et al. 2013; Stadler et al. 2018), and can capture environmental impacts embodied in global trade.

Several authors have sought to address the social impacts of production and consumption with an approach similar to that of EEIO; here we refer to it as socially and socioeconomically extended IO (SEIO). For example, the Social Hotspots Database was created to streamline the social risk and opportunity analysis by using the IO database from the Global Trade Analysis Project (GTAP) which covers most countries (Norris et al. 2012). The result of the analysis of a purchase from a given sector and country provides an estimate of worker-hours associated with various risks and opportunities along the entire supply chain. The specific conditions of production will likely vary within a sector and country, but the analyses are particularly useful to guide more detailed investigations into areas of concern and to provide a general comparison of risks.

2.2 Cases

The following paragraphs demonstrate how MFA, LCA, and IO analysis can help identify opportunities for effectively pursuing several SDGs, drawing on existing studies.

2.2.1 Food + Footprint = Foodprint

Industrial ecology tools pose value in achieving progress toward SDG2, the ambitious and critical goal of zero hunger. E-LCA and EEIO methods can assess what is sometimes called the "foodprint" of scenarios of future food production systems, contributing directly to SDG2's fourth target: "By 2030, ensure sustainable food production systems and implement resilient agricultural practices." Foodprints of 100 cities were estimated based on 43 urban material and energy flow studies (Goldstein et al. 2017).

The carbon foodprints typically constitute 10–20% of a city's total carbon footprint. An exception is the island of Macao, which has a low carbon foodprint owing to low meat intake.

In another example, E-LCA was used to investigate the potential for actors along the Swedish milk supply chain to improve their collective environmental performance, identifying opportunities such as organic labeling, more efficient home refrigeration, and more efficient transport (Berlin et al. 2008). The carbon impact of historical and future scenarios of China's food system was also estimated using E-LCA, with researchers finding that the bulk of all food systems emissions derive from methane and nitrogen oxide from agricultural processes rather than energy, materials, and waste (Li et al. 2016). Once the authors knew where along the life-cycle chain such large environmental impacts were occurring, they explored pathways to reduce future emissions substantially, by increasing fertilizer efficiency, improving animal feeding technology, and shifting diets. The combined pathways could reduce the projected emissions from 2,510 to 950 $MtCO_2$-eq by 2050.

The third target of SDG12, responsible consumption and production, calls to halve per capita global food waste (see Table 8.1). MFA methods can track the flow of food along supply chains to identify loss reduction hotspots and opportunities. In one example, industrial ecology researchers analyzed government data to examine food losses in the United States and found that just over half of what is produced at farms is consumed (Heller and Keoleian 2015, Figure 8.4). They also used E-LCA to

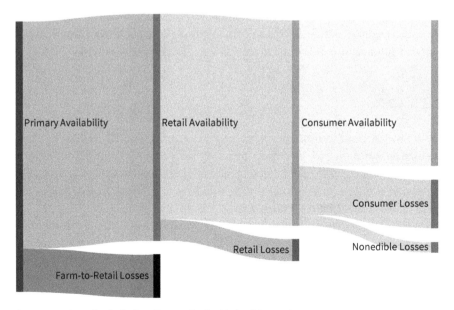

Figure 8.4 Supply chain food losses in the United States in 2010
Source: Adapted from Heller and Keoleian (2015).

examine the carbon emissions associated with certain foods and found that beef, veal, and lamb account for nearly a third of the impact from food losses. Another study using EEIO for consumer food waste in the EU found similar loss rates and similarly concluded that reducing waste of meat products is a priority as it drives the majority of potential environmental savings (Usubiaga et al. 2018).

2.2.2 Water

Industrial ecology methods can assist with several targets pertaining to SDG6, clean water and sanitation. Management of water supply, quality, and wastewater are multi-faceted sociotechnical issues that require extensive analyses to achieve sustainability.

As drinking water is fundamental to human life, the first target seeks to ensure universal and equitable access to safe and affordable drinking water for all. Technological innovations now enable use of relatively new sources of water, such as seawater and wastewater. A review of case studies demonstrated that E-LCA is a useful aid in decision-making around future drinking water systems (Godskesen et al. 2018). Several studies highlighted the intensity of energy in desalination systems versus traditional systems, but others found that bottled water, owing to the packaging, is still orders of magnitude more harmful to the environment than tap water from any system. Other comparisons noted that boiling or treating water with activated charcoal at home had higher impacts than improving water quality in centralized systems. Also, when exploring wastewater reuse, E-LCA practitioners can go beyond energy and carbon impacts to assess the toxicity impact of items such as discarded pharmaceuticals.

The third target aims to improve water quality, in part by addressing wastewater and reuse. For instance, a comparative E-LCA of four wastewater treatment scenarios for a plant in Spain resulted in the selection of tertiary treatment and a recommendation that substitution of potable water with reclaimed wastewater only take place in water scarce regions (Pasqualino et al. 2011). Another study applied EEIO methods to create a custom wastewater IO model for Tokyo and optimized combinations of alternative wastewater treatment technologies to minimize several types of environmental impacts simultaneously (Lin 2011).

Applying MFA to water systems can help investigate water-use efficiency and sustainable withdrawals as called for in SDG target 6.4 and integrated water resource management discussed in target 6.5. Researchers analyzed the water footprints for processing two kinds of copper ore in Chile and discovered that refining the sulfur ore required over twice the water as the oxide ore, and therefore targeted process stages for reduction in water loss (Peña and Huijbregts 2014). Similarly, an MFA of industrial cooling water systems identified opportunities for saving up to 24% of the water (Schlei-Peters et al. 2018).

2.2.3 Energy

Energy, the focus of SDG7, has been a primary focus of much industrial ecology research. In addition to footprint analyses of energy supply and consumption systems, industrial ecology methods have contributed greatly to the understanding of environmental trade-offs and resource requirements associated with "low carbon" energy options. In particular, the metal requirements of low-carbon energy transitions, a topic overlooked by the IPCC, have now been assessed in numerous studies. Metals that are actually by-products of commonly mined ores are increasingly required for low-carbon energy technologies such as wind turbines, solar panels, and batteries for electric vehicles or grid-scale energy storage. Because many of these by-product metals occur in the earth physically attached to the commonly mined ores, they cannot be extracted individually. Consequently, the *supply* of these materials is not assured, since their availability is tied to the *demand* for so-called parent metals (Graedel 2011). Low-carbon power generation was found to have considerably higher material and land requirements than fossil fuel generation (Berrill et al. 2016; Kleijn et al. 2011). Even for scenarios that result in excessive global temperature rise (in the range of 2 to 3°C by 2050), annual demand for metals required by low-carbon energy systems has been estimated at 3 to 4.5 times larger in 2050 than the total supply of all metals in 2000. Demand would be 5 to 7 times larger for the key rare earth elements neodymium and dysprosium (de Koning et al. 2018).

Another area of interaction between energy and SDGs studied in industrial ecology is in the relationship between buildings and energy use. Industrial ecologists highlight how the environmental impacts of manufacturing construction materials and constructing buildings, referred to as embodied impacts of buildings, can contribute significantly to life-cycle impacts of buildings. The relative importance of the embodied impacts to life-cycle impacts grows as buildings become more efficient and use cleaner energy, effectively reducing impacts of the operational phase (Sartori and Hestnes 2007). The quality and quantity of new buildings and building retrofits that take place in the coming years will play a large role in determining future energy demand.

Building lifetime is another attribute that is crucial in determining energy demand and related environmental impacts (Aktas and Bilec 2012; Miatto et al. 2017). Industrial ecology studies illustrate the effects that extending or decreasing building lifetimes can have on material requirements, debris, and associated environmental impacts. For example, if average Chinese building lifetimes were extended from 23 years to 50 years, Cai and colleagues (2015) showed considerable reductions possible in energy consumption (equivalent to Mexico's annual energy consumption in 2011), water withdrawals (equivalent to Belgium's annual withdrawals in 2011), and CO_2 emissions (equivalent to Italy's annual emissions in 2011).

Examining metal requirements for low carbon economies and the interplay between buildings and environmental impacts are central to guiding sustainable energy access and provision of shelter to the growing population throughout this century. These issues will be of utmost importance in developing economies, where energy demand and urban populations are expected to grow the most.

2.2.4 Industry, Innovation, Infrastructure, and Cities

Industrial ecology models can be used to support SDG9, which targets sustainable in-dustry and infrastructure, by analyzing shifts in industrial production practices. For example, Milford et al. (2013) showed that to reduce the steel sector's emissions 50% by 2050, no blast furnaces should be built after 2020. They did so using an MFA com-bined with E-LCA to forecast material dynamics and carbon emission trajectories for various efficiency strategies. Industrial ecology can also demonstrate the trade-offs between different infrastructure projects. For example, Banar and Özdemir (2015) applied E-LCA to railway passenger transport in Turkey and found that carbon emis-sions are significantly lower compared to private vehicles. The authors also noted, however, that the development of rail infrastructure causes significant environ-mental impacts and can even exceed the environmental loading of train operation. Larrea-Gallegos et al. (2017) studied life-cycle impacts of gravel road construction in a tropical rainforest environment in Peru and found that environmental impacts from material requirements are low. However, the deforestation necessary for road construction and the resulting removal of carbon stocks greatly exacerbates environ-mental impacts.

SDG11 addresses the sustainability of cities and communities, which can be addressed from a variety of angles. Case studies from an urban energy perspective in Malaysia and China illustrated that the electricity grids need stringent decar-bonization before electrification of the urban transport system yields any carbon emission reductions at all (Onn et al. 2017; Wang et al. 2013). Similar results can be expected for India or Indonesia, where the electricity mix is dominated by coal power.

2.2.5 Socioeconomic Metabolism, Economic Development, and Inequalities

The topics promoted by SDG8, decent work and economic growth, are often under the purview of economics rather than industrial ecology, but there is a clear ten-sion between meeting those goals and sustaining the environment (Machingura and Lally 2017). Economic development needs to be assured for impoverished de-veloping countries while minimizing environmental pressures. Long-term MFA studies showed that developed countries achieved their high socioeconomic levels through mass consumption of raw materials and accumulation of material stocks (Gierlinger and Krausmann 2012; Fishman et al. 2015), and rapidly developing countries like China are following suit (Huang et al. 2017). Martinico-Perez et al. (2018) showed that the Philippines achieved a higher growth rate of GDP compared to the growth rate of material requirements, which is referred to as relative decoup-ling. Absolute decoupling requires an absolute decrease in environmental impacts, which is difficult to achieve and would require highly coordinated strategic innova-tion to accomplish.

In addition, the social variants of industrial ecology tools can help identify areas needing attention in pursuing SDG target 8.8, to protect labor rights and promote safe and secure working environments for all workers, and SDG10, to adopt policies, especially fiscal, wage and social protection policies, and progressively achieve greater equality. Specifically, consumption in developed countries can undermine social development and aspects of global climate change mitigation in developing countries. Alsamawi et al. (2017) estimated an inequality footprint using SEIO to illustrate that consumption in developed countries is often supported through imports from developing economies with high inequalities. Simas et al. (2015) used SEIO to confirm that a significant share of Europe's imported products are produced in foreign low-skilled and labor-intensive sectors. These imports are also associated with low energy productivity, thereby making higher levels of energy efficiency gains and climate change mitigation efforts more difficult to achieve. Furthermore, Hosseinijou et al. (2014) identified key socioeconomic issues along the life cycle of building materials in Iran using S-LCA. The authors found that workers in the steel supply chain faced issues concerning fair salaries, ability to associate, and occupational health during the extraction phase. In the concrete and cement industry, the main concern identified was the poor living conditions of workers involved in cement production.

2.3 Discussion

This section has outlined the industrial ecology tools MFA, LCA, and IO analysis and demonstrated their application and relevance to SDGs 2 (food), 6 (water), 7 (energy), 8 (decent work and economic development), 9 (industry, innovation, and infrastructure), 10 (reducing inequalities), and 11 (cities and communities). In addition, SDGs 12 (responsible consumption and production) and 13 (climate action) have been indirectly addressed, as these are cross-cutting goals and therefore are strongly linked to the other SDGs. The presented case studies highlight various strengths of industrial ecology tools including capturing higher order supply chain effects (e.g., preconsumer food losses), impacts from indirect upstream operations (e.g., deforestation for construction purposes), trade-offs of different alternatives (e.g., public transport versus private vehicles), and tracing impacts beyond geographical boundaries (e.g., embodied emissions in trade). These examples demonstrate how industrial ecology tools offer a unique perspective through study of material and energy flows and their impact on the environmental and social spheres. Industrial ecology tools also have certain limitations and have been criticized as static and less comprehensive than other methods, such as integrated assessment models (Pauliuk and Hertwich 2016). Still, industrial ecology tools are valuable in guiding and evaluating progress toward reaching the SDGs.

3. Industrial Symbiosis

Emerging market countries place great emphasis on agendas for improving economic performance, often at the cost of environmental quality. Historically, many countries have followed the "grow first, clean up later" path of industrialization. The United Kingdom and the United States were no exceptions; reports were made of the Great Smog of London causing 400 fatalities and Cleveland's Cuyahoga River catching fire until the 1960s. Today, we see a similar situation in emerging market countries, as reflected in severe air pollution in Beijing, Delhi, and Jakarta. To meet the ambitious targets of SDGs, emerging economies could benefit from making use of an industrial ecology model for eco-industrial development, one based on resource sharing principles of industrial symbiosis (IS), to achieve economic development while limiting environmental impacts.

The concept of IS is inspired by mutualistic symbiosis in biological ecosystems, where organisms exchange resources and services for mutual benefit. In IS, firms exchange resources in the forms of 1) utilizing by-products from other firms as raw material inputs to industrial processes; 2) sharing utilities or infrastructure for steam, electricity, water, and wastewater; and 3) joint provision of services and materials not directly related to the core business of a company (Chertow et al. 2008). The application of IS is not limited to the manufacturing industry but applies broadly to industrial systems with large process material and waste flows, including agriculture, electric power generation, and industrial waste management. The first well-documented case of IS is the small Danish city of Kalundborg, where a by-product exchange that began

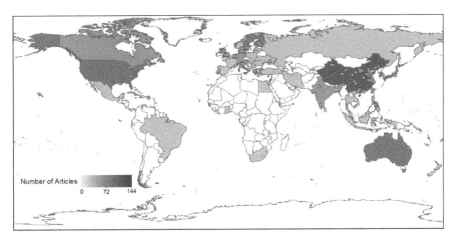

Figure 8.5 Countries and industrial symbiosis clusters featured in journal articles (1995–2017) from Web of Science and Scopus. Shades of gray represent the number of articles featuring each country. The dots denote locations of industrial clusters with material and energy exchanges reported in articles.

Source: Kanaoka and Chertow, Yale Center for Industrial Ecology (2018, unpublished data).

in 1972 has evolved into an IS network including a power plant, petroleum refinery, wallboard manufacturer, enzyme producer, insulin producer, local farms, and shared water and wastewater treatment projects, among others. As a global phenomenon, at least 82 *well-documented* IS clusters are operational in 20 countries today, totaling over 1,000 material and energy exchanges. IS in over 51 countries has been described in peer-reviewed journals (Figure 8.5).

In these industrial clusters, studies have repeatedly shown that IS led to economic or environmental benefits, or both. In the following sections, we present cases of three distinct models of IS to examine how interfirm synergies could support progress toward achieving the SDGs. In particular, we focus on SDG8, regarding decent work and economic growth, and SDG12 on responsible consumption and production. The first case, Ulsan Industrial District, illustrates how South Korea's national government stimulated Ulsan's regional economy while reducing pollution. The second case, Guitang Group and British Sugar, shows how IS could be applied as a strategy by businesses to generate additional revenue and jobs. Finally, we introduce and discuss policy implications of the industrial cluster in Nanjangud, India, where IS exchanges were initiated by firms in the absence of regulatory pressure (Bain et al. 2010). These cases offer insights into ways that governments and businesses could contribute to collective progress toward SDGs by promoting symbiosis as well as goals of individual stakeholders.

3.1 Cases

Several examples of implemented IS cases are presented to illustrate the potential of interfirm synergies in supporting progress toward meeting the SDGs.

3.1.1 Ulsan Industrial District

Industrial symbiosis in Ulsan is a prime example of how IS can help governments meet targets under SDGs 8 and 12, which pertain to decoupling economic growth from environmental degradation, through resource efficiency and waste reduction. The port city of Ulsan is South Korea's seventh most populous metropolis and its leading industrial powerhouse. Ulsan's industrial district is home to approximately 1,000 companies, including the world's largest automobile factory and shipyard, respectively operated by Hyundai Motor Company and Hyundai Heavy Industries (Behera et al. 2012). In 2005, the Korea National Cleaner Production Center launched the National Eco-Industrial Park Development Program, which aimed to reduce pollution, natural resource extraction, and operational costs of participating companies by implementing IS (Behera et al. 2012). IS networks were planned and implemented in Ulsan and seven other pilot regions by coordinating institutions (Park et al. 2018).

By 2012, 13 IS exchanges involving 41 companies were operational in Ulsan with each reporting significant economic and environmental benefits (Behera et al. 2012). Exchanges involved metallurgical plants, chemical plants, a paper mill, a petroleum refinery, waste management facilities, and Hyundai's automobile plant and shipyard. Investments in research and development and infrastructure for implementation of IS cost $66.8 million for the 41 companies, while economic benefits totaled $68.5 million per year (Behera et al. 2012). Payback periods were under one year for nine of the thirteen IS networks. Steam infrastructure sharing networks significantly improved the energy efficiency of the industrial area, reducing CO_2 emissions and air pollution (total of sulfur oxides, nitrogen oxides, and carbon monoxide) by 230,000 tons and 3,700 tons per year. In addition, the IS project in Ulsan created a significant number of jobs, which is an important component of SDG8, for implementing and managing the IS networks. For example, Hyosung Chemical Corporation's $150 million investment in a new manufacturing unit for utilizing excess steam was estimated to create 140 new jobs (Park and Park 2014). These benefits also contributed to the social benefit of improving the local community's perception of the industrial area (Behera et al. 2012).

Ironically, Korea's national IS program was discontinued in 2016, owing to its success (Park et al. 2018). The national government concluded that government funding was unnecessary because Ulsan and other pilot projects were generating significant profits for the involved companies. We have observed this occurrence before, when the United Kingdom's National Industrial Symbiosis Program (NISP) was dissolved for similar reasons. Ulsan's experience demonstrates how government initiatives and support for building IS networks could catalyze regional economic development while reducing environmental impacts. Other countries could also apply IS as a strategy for job creation or for improving the environmental performance of existing industrial areas to meet SDGs 8 and 12.

3.1.2 Guitang Group and British Sugar

Two instances involving sugar production, one in China (the Guitang Group) and the second in the United Kingdom (British Sugar), illustrate how IS could serve as a strategy to grow businesses and create jobs while pursuing the targets under SDG12 to achieve the sustainable management and efficient use of natural resources and to substantially reduce waste generation through prevention, reduction, recycling and reuse. The Guitang Group began in 1956 as a cane sugar and alcohol production facility in Guangxi Province, China. Between 1970 and 2000, the Guitang Group expanded its business to the production of pulp and paper, cement, fertilizer, caustic soda, and calcium carbonate through utilization of manufacturing by-products (Shi and Chertow 2017). The Guitang Group also formed IS networks with external organizations, including the purchase of molasses and bagasse from other sugar refineries and the sale of coal ash from Guitang Group's power plant to a cement mill and road material manufacturers. By 2017, Guitang Group had grown to a conglomerate with more than 3,000 employees, generating more than $150 million in annual revenue (Shi and Chertow 2017).

Similar to the Guitang Group, British Sugar began as a sugar refinery in 1925, using sugar beets instead of sugarcane, in Norfolk County, England. Between 1985 and

2010, British Sugar and its parent company, Associated British Foods, established four subsidiaries to sell products made from their sugar production by-products: Trident (animal feed from residual pulp), TOPSOIL (residual soil from sugar beet cleaning), LimeX (lime precipitate from sugar purification), and Cornerways Nursery (tomatoes grown with CO_2 supplied by British Sugar's power plant) (Short et al. 2014).

From other by-products, British Sugar produced betaine and raffinate, two high-value chemicals used in agricultural and cosmetics industries (Short et al. 2014). In 2007, British Sugar also established the first biorefinery in the United Kingdom, to produce bioethanol from sugar refining by-products. The by-product CO_2 from the biorefinery was sold to an industrial gas supplier (Short et al. 2014). These IS exchanges added considerable value to Associated British Foods' business by increasing revenue and decreasing risk by taking advantage of material flows that already existed.

These two examples of IS demonstrate how managing wastes as resources could benefit businesses and create jobs while contributing to the targets under SDG12 of improving resource efficiency and reducing waste generation. This approach was shown to be effective both in the contexts of developed and developing countries; however, it may be especially valuable to developing nations, where agriculture has a greater contribution to the economy. While the Guitang Group and British Sugar cases specifically focused on sugar producers, similar approaches could be applied to other agricultural by-products. Lessons from these cases could also be used to create policies that encourage IS for meeting SDG8's target to promote development-oriented policies that support productive activities, decent job creation, entrepreneurship, creativity, and innovation and encourage the formalization and growth of micro-, small-, and medium-sized enterprises.

3.1.3 Nanjangud Industrial Area

An examination of India's Nanjangud Industrial Area provides additional insights into ways that IS could help countries meet the targets of SDG12 to achieve efficient use of natural resources and substantially reduce waste generation. Nanjangud is a division of the Mysore district in the South Indian state of Karnataka and is home to over 60 industrial facilities of all sizes (Bain et al. 2010). In 2007, 42 facilities in Nanjangud generated a total of 897,000 tonnes of waste, of which a startling 99.5% was recovered through reuse and recycling (Bain et al. 2010). Of the recovered waste, 18.5% was utilized in 27 different IS exchanges. Unlike some other IS networks in industrialized regions that have been driven by regulatory restrictions on waste disposal, Nanjangud's IS emerged without such regulation as a direct driving factor (Bain et al. 2010). The IS networks in Nanjangud were previously undocumented but this symbiosis was "uncovered" by Bain and colleagues (2010) through a study of material flows for the industrial area.

Generally, implementing IS has been more successful in existing industrial areas rather than in new industrial areas designed from scratch (Chertow 2007). The case of Nanjangud suggests that other IS networks could be uncovered as sites for successfully pursuing SDG8's target to promote development-oriented policies based on IS, even in areas without policies that seem to promote IS.

3.2 Discussion

This section explored how industrial symbiosis offers a versatile approach that organizations can utilize to pursue SDGs 8 and 12. Several distinct IS cases were examined in terms of scale, types of industries, and local context, and they showed that individual firms can grow their businesses while creating jobs and reducing waste generation. Government bodies could create policies that promote IS to improve regional economic performance and reduce environmental impacts. In either case, resource efficiency would be improved for firms involved in IS exchanges. The improved efficiency of firms also contributes to meeting several other SDGs and targets, notably SDG13 on climate action, target 7.3 on energy efficiency, targets 6.3 and 6.4 on water (Chertow and Miyata 2011), and target 9.4 on improving industrial processes.

As beneficial as IS could be, developing and sustaining IS networks is a complex task that involves institutional capacity building, mutual trust, and information sharing across businesses, among other factors (Walls and Paquin 2015). Many envisioned IS networks have failed to get off the ground (Chertow 2007). In considering the future development of IS, significant economic and environmental benefits could be promoted through government, community, and business support for initial investment, data collection, establishment of coordinating institutions, facilitation, and waste regulation.

4. Conclusion

This chapter has considered how the interdisciplinary field of industrial ecology can help deliver on the SDGs. As a science, industrial ecology provides a source of knowledge that can guide sustainable production and consumption decisions, reduce waste and pollution, advance circular over linear material flows, and extend the life of physical resources. All of these objectives support the attainment of SDGs, as discussed in the chapter. Furthermore, this discussion of SDGs was not exhaustive, and industrial ecology approaches are useful to a broader array of SDGs and targets, including those pertaining to marine pollution and terrestrial ecosystems (SDGs 14 and 15).

Functionally, industrial ecology provides a quantitative means to approach the SDGs by introducing mathematical models for tracking material and energy flows, accounting for accumulating stocks, enumerating benefits and costs of eco-industrial development, and establishing benchmarks and target controls. That industrial ecology has, over time, developed many assessment tools suggests that it will not consider technological change as right or wrong, but will rather focus on analysis and complexities that must be considered for a long-term, comprehensive view. By looking at whole systems at different temporal and spatial scales, it becomes easier to reveal early warnings and reduce unintended consequences, which can help countries better prepare for rapid global and local change. Linking SDGs with industrial ecology frameworks can be a positive step to deeper and richer understanding that supports progress toward achieving the SDGs.

References

Aktas, C. B., and Bilec, M. M. (2012). Impact of lifetime on US residential building LCA results. *International Journal of Life Cycle Assessment* 17(3), 337–349.

Alsamawi, A., McBain, D., Murray, J., Lenzen, M., and Wiebe, K. S. (2017). The inequality footprints of nations: a novel approach to quantitative accounting of income inequality. In: Alsamawi, A., McBain, D., Murray, J., Lenzen, M., and Wiebe, K.S. (Eds.), *The Social Footprints of Global Trade*, pp. 69–91. Springer Nature, Singapore.

Bain, A., Shenoy, M., Ashton, W., and Chertow, M. R. (2010). Industrial symbiosis and waste recovery in an Indian industrial area. *Resources, Conservation and Recycling* 54(12), 1278–1287.

Banar, M. and Özdemir, A. (2015). An evaluation of railway passenger transport in Turkey using life cycle assessment and life cycle cost methods. *Transportation Research Part D: Transport and Environment* 41, 88–105.

Behera, S. K., Kim, J. -H., Lee, S. -Y., Suh, S., and Park, H. -S. (2012). Evolution of "designed" industrial symbiosis networks in the Ulsan Eco-industrial Park: "research and development into business" as the enabling framework. *Journal of Cleaner Production* 29–30, 103–112.

Benoît-Norris, C., Vickery-Niederman, G., Valdivia, S., Franze, J., Traverso, M., Ciroth, A., and Mazijn, B. (2011). Introducing the UNEP/SETAC methodological sheets for subcategories of social LCA. *International Journal of Life Cycle Assessment* 16(7), 682–690.

Berlin, J., Sonesson, U., and Tillman, A. -M. (2008). Product chain actors' potential for greening the product life cycle. *Journal of Industrial Ecology* 12(1), 95–110.

Berrill, P., Arvesen, A., Scholz, Y., Gils, H. C., and Hertwich, E. G. (2016). Environmental impacts of high penetration renewable energy scenarios for Europe. *Environmental Research Letters* 11(1), 014012.

Brunner, P. H., and Rechberger, H. (2016). *Handbook of Material Flow Analysis: For Environmental, Resource, and Waste Engineers* (2nd ed.). CRC Press, Boca Raton, FL.

Cai, W., Wan, L., Jiang, Y., Wang, C., and Lin, L. (2015). Short-lived buildings in China: impacts on water, energy, and carbon emissions. *Environmental Science and Technology* 49(24), 13921–13928.

Chertow, M. R. (2000). The IPAT equation and its variants. *Journal of Industrial Ecology* 4(4), 13–29.

Chertow, M. R. (2007). "Uncovering" industrial symbiosis. *Journal of Industrial Ecology* 11(1), 11–30.

Chertow, M. R., Ashton, W. S., and Espinosa, J. C. (2008). Industrial symbiosis in Puerto Rico: environmentally related agglomeration economies. *Regional Studies* 42(10), 1299–1312.

Chertow, M. R., and Miyata, Y. (2011). Assessing collective firm behavior: comparing industrial symbiosis with possible alternatives for individual companies in Oahu, HI. *Business Strategy and the Environment* 20(4), 266–280.

Chester, M. V., and Horvath, A. (2009). Environmental assessment of passenger transportation should include infrastructure and supply chains. *Environmental Research Letters* 4(2), 024008.

de Koning, A., Kleijn, R., Huppes, G., Sprecher, B., van Engelen, G., and Tukker, A. (2018). Metal supply constraints for a low-carbon economy? *Resources, Conservation and Recycling* 129, 202–208.

Erkman, S., and Ramaswamy, R. (2003). Industrial ecological solutions. In: Nemerow, N. L., and Agardy, F. J. (Eds.), *Environmental Solutions*, pp. 297–310. Elsevier Academic Press, Burlington, MA.

Fishman, T., Schandl, H., and Tanikawa, H. (2015). The socio-economic drivers of material stock accumulation in Japan's prefectures. *Ecological Economics* 113, 76–84.

Frosch, R. A., and Gallopoulos, N. E. (1989). Strategies for manufacturing. *Scientific American* 261(3), 144–152.

Gierlinger, S., and Krausmann, F. (2012). The physical economy of the United States of America. *Journal of Industrial Ecology* 16(3), 365–377.

Godskesen, B., Meron, N., and Rygaard, M. (2018). LCA of drinking water supply. In: Hauschild, M. Z., Rosenbaum, R. K., and Olsen, S. I. (Eds.), *Life Cycle Assessment: Theory and Practice*, pp. 835–860. Springer International Publishing, Cham.

Goldstein, B., Birkved, M., Fernández, J., and Hauschild, M. (2017). Surveying the environmental footprint of urban food consumption. *Journal of Industrial Ecology* 21(1), 151–165.

Graedel, T. E. (2011). On the future availability of the energy metals. *Annual Review of Materials Research* 41(1), 323–335.

Graedel, T. E., and Allenby, B. R. (1995). *Industrial Ecology*. 1st ed. Prentice Hall, Upper Saddle River, NJ.

Graedel, T. E., and Allenby, B. R. (2002). *Industrial Ecology*. 2nd ed. Prentice Hall, Upper Saddle River, NJ.

Graedel, T. E., Harper, E. M., Nassar, N. T., and Reck, B. K. (2015). On the materials basis of modern society. *Proceedings of the National Academy of Sciences* 112(20), 6295–6300.

Heller, M. C., and Keoleian, G. A. (2015). Greenhouse gas emission estimates of U.S. dietary choices and food loss. *Journal of Industrial Ecology* 19(3), 391–401.

Hosseinijou, S. A., Mansour, S., and Shirazi, M. A. (2014). Social life cycle assessment for material selection: a case study of building materials. *International Journal of Life Cycle Assessment* 19(3), 620–645.

Huang, C., Han, J., and Chen, W. -Q. (2017). Changing patterns and determinants of infrastructures' material stocks in Chinese cities. *Resources, Conservation and Recycling* 123, 47–53.

Hunt, R. G., and Franklin, W. E. (1996). LCA: how it came about. *International Journal of Life Cycle Assessment* 1(1), 4–7.

International Organization for Standardization (ISO). (2006). Environmental Management: Life Cycle Assessment: Principles and Framework (*ISO* 14040:2006). International Organization for Standardization, Geneva.

Kim, I., and Hur, T. (2009). Integration of working environment into life cycle assessment framework. *International Journal of Life Cycle Assessment* 14(4), 290–301.

Kitzes, J. (2013). An introduction to environmentally-extended input-output analysis. *Resources* 2(4), 489–503.

Kleijn, R., van der Voet, E., Kramer, G. J., van Oers, L., and van der Giesen, C. (2011). Metal requirements of low-carbon power generation. *Energy* 36(9), 5640–5648.

Krausmann, F., Wiedenhofer, D., Lauk, C., Haas, W., Tanikawa, H., Fishman, T., Miatto, A., Schandl, H., and Haberl, H. (2017). Global socioeconomic material stocks rise 23-fold over the 20th century and require half of annual resource use. *Proceedings of the National Academy of Sciences* 114(8), 1880–1885.

Larrea-Gallegos, G., Vázquez-Rowe, I., and Gallice, G. (2017). Life cycle assessment of the construction of an unpaved road in an undisturbed tropical rainforest area in the vicinity of Manu National Park, Peru. *International Journal of Life Cycle Assessment* 22(7), 1109–1124.

Lenzen, M., Moran, D., Kanemoto, K., and Geschke, A. (2013). Building Eora: a global multiregion input–output database at high country and sector resolution. *Economic Systems Research* 25(1), 20–49.

Li, H., Wu, T., Wang, X., and Qi, Y. (2016). The greenhouse gas footprint of China's food system: an analysis of recent trends and future scenarios. *Journal of Industrial Ecology* 20(4), 803–817.

Lin, C. (2011). Identifying lowest-emission choices and environmental pareto frontiers for wastewater treatment wastewater treatment input-output model based linear programming. *Journal of Industrial Ecology* 15(3), 367–380.

Machingura, F., and Lally, S. (2017). *The Sustainable Development Goals and Their Trade-Offs*. https://www.odi.org/sites/odi.org.uk/files/resource-documents/11329.pdf (accessed April 5, 2019).

Martinico-Perez, M. F. G., Schandl, H., Fishman, T., and Tanikawa, H. (2018). The socioeconomic metabolism of an emerging economy: monitoring progress of decoupling of economic growth and environmental pressures in the Philippines. *Ecological Economics* 147, 155–166.

Matthews, H. S., Hendrickson, C. T., and Matthews, D. H. (2014). *Life Cycle Assessment: Quantitative Approaches for Decisions that Matter*. https://www.lcatextbook.com (accessed April 5, 2019).

Miatto, A., Schandl, H., and Tanikawa, H. (2017). How important are realistic building lifespan assumptions for material stock and demolition waste accounts? *Resources, Conservation and Recycling* 122, 143–154.

Milford, R. L., Pauliuk, S., Allwood, J. M., and Müller, D. B. (2013). The roles of energy and material efficiency in meeting steel industry CO_2 targets. *Environmental Science and Technology* 47(7), 3455–3462.

Norris, C. B., Aulisio, D., and Norris, G. A. (2012). Working with the social hotspots database: methodology and findings from 7 social scoping assessments. In: Dornfeld, D. A., and Linke, B. S. (Eds.), *Leveraging Technology for a Sustainable World*, pp. 581–586. Springer-Verlag, Berlin.

Onn, C. C., Chai, C., Abd Rashid, A. F., Karim, M. R., and Yusoff, S. (2017). Vehicle electrification in a developing country: status and issue, from a well-to-wheel perspective. *Transportation Research Part D: Transport and Environment* 50, 192–201.

Park, J. Y., and Park, H. -S. (2014). Securing a competitive advantage through industrial symbiosis development. *Journal of Industrial Ecology* 18(5), 677–683.

Park, J. Y., Park, J. -M., and Park, H. -S. (2018). Scaling-up of industrial symbiosis in the Korean National Eco-Industrial Park Program: examining its evolution over the 10 years between 2005–2014. *Journal of Industrial Ecology* 23(1), 197–207.

Pasqualino, J. C., Meneses, M., and Castells, F. (2011). Life cycle assessment of urban wastewater reclamation and reuse alternatives. *Journal of Industrial Ecology* 15(1), 49–63.

Pauliuk, S., and Hertwich, E. G. (2016). Prospective models of society's future metabolism: what industrial ecology has to contribute. In: Clift, R., and Druckman, A. (Eds.), *Taking Stock of Industrial Ecology*, pp. 21–43. Springer International Publishing, Cham.

Peña, C. A., and Huijbregts, M. A. J. (2014). The blue water footprint of primary copper production in northern Chile. *Journal of Industrial Ecology* 18(1), 49–58.

Reck, B. K., Müller, D. B., Rostkowski, K., and Graedel, T. E. (2008). Anthropogenic nickel cycle: insights into use, trade, and recycling. *Environmental Science and Technology* 42(9), 3394–3400.

Sartori, I., and Hestnes, A. G. (2007). Energy use in the life cycle of conventional and low-energy buildings: a review article. *Energy and Buildings* 39, 249–257.

Schandl, H., Fischer-Kowalski, M., West, J., Giljum, S., Dittrich, M., Eisenmenger, N., Geschke, A., Lieber, M., Wieland, H., Schaffartzik, A., Krausmann, F., Gierlinger, S., Hosking, K., Lenzen, M., Tanikawa, H., Miatto, A., and Fishman, T. (2018). Global material flows and resource productivity: forty years of evidence. *Journal of Industrial Ecology* 22(4), 827–838.

Schandl, H., and Miatto, A. (2018). On the importance of linking inputs and outputs in material flow accounts: the Weight of Nations report revisited. *Journal of Cleaner Production* 204, 334–343.

Schlei-Peters, I., Wichmann, M. G., Matthes, I. -G., Gundlach, F. -W., and Spengler, T. S. (2018). Integrated material flow analysis and process modeling to increase energy and water efficiency of industrial cooling water systems. *Journal of Industrial Ecology* 22(1), 41–54.

Schmidt, K. D. (2000). *Mathematik*. Grundlagen für Wirtschaftswissenschaftler. Springer Berlin Heidelberg, Berlin and Heidelberg.

Shi, L., and Chertow, M. R. (2017). Organizational boundary change in industrial symbiosis: revisiting the Guitang Group in China. *Sustainability* 9(7), 1085.

Short, S. W., Bocken, N. M. P., Barlow, C. Y., and Chertow, M. R. (2014). From refining sugar to growing tomatoes. *Journal of Industrial Ecology* 18(5), 603–618.

Simas, M., Wood, R., and Hertwich, E. G. (2015). Labor embodied in trade. *Journal of Industrial Ecology* 19(3), 343–356.

Smil, V. (2004). *China's Past, China's Future: Energy, Food, Environment*. Routledge, Curzon, New York and London.

Stadler, K., Wood, R., Bulavskaya, T., Södersten, C.-J., Simas, M., Schmidt, S., Usubiaga, A., Acosta-Fernández, J., Kuenen, J., Bruckner, M., Giljum, S., Lutter, S., Merciai, S., Schmidt, J. H., Theurl, M. C., Plutzar, C., Kastner, T., Eisenmenger, N., Erb, K.-H., de Koning, A., and Tukker, A. (2018). EXIOBASE 3: developing a time series of detailed environmentally extended multi-regional input-output tables. *Journal of Industrial Ecology* 22(3), 502–515.

UNEP/SETAC. (2009). *Guidelines for social life cycle assessment of products*. United Nations Environment Programme, Paris.

Usubiaga, A., Butnar, I., and Schepelmann, P. (2018). Wasting food, wasting resources: potential environmental savings through food waste reductions. *Journal of Industrial Ecology* 22(3), 574–584.

Walls, J. L., and Paquin, R. L. (2015). Organizational perspectives of industrial symbiosis: a review and synthesis. *Organization and Environment* 28(1), 32–53.

Wang, D., Zamel, N., Jiao, K., Zhou, Y., Yu, S., Du, Q., and Yin, Y. (2013). Life cycle analysis of internal combustion engine, electric and fuel cell vehicles for China. *Energy* 59, 402–412.

White, R. (1994). Preface. In: Allenby, B. R., and Deanna, R. J. (Eds.), *The Greening of Industrial Ecosystems*, pp. v–vi. National Academy Press, Washington, DC.

9

Automated Vehicles and Sustainable Cities

A Realistic Outlook to 2030

Rui Wang and Christopher Oster

1. Introduction

Road transportation of passengers and freight plays a crucial role in achieving sustainable development goals (SDGs). Transportation activities generate crucial socioeconomic benefits by enabling exchanges of population, products and services, and information, while simultaneously imposing significant time, energy, environmental, and health/safety costs. In urban areas, according to the New Economic Geography theory (Krugman 1991), higher accessibility improves urban productivity by enabling the sharing of infrastructure and other investments, minimizing job market mismatch, and facilitating learning for technology and market innovation (Duranton and Puga 2003). Fast, safe, green, and inclusive road transportation is crucial for the sustainable development of cities.

Automated vehicle (AV) technology is a potentially disruptive innovation to transportation, especially at the urban and regional scale. The innovative characteristics and capabilities of AVs have direct and indirect implications for economic productivity, resource/environmental sustainability, and social equity. However, the role of AVs in sustainable cities and communities (SDG11), as well as interconnected SDGs including jobs and economic growth (SDG8), human health and safety (SDG3), ecosystem integrity (SDGs 14 and 15), and energy/climate (SDGs 7 and 13), depends not just on the development and deployment of AV technologies, but also on policy and planning measures in response to the potential impacts brought by vehicle automation.

This chapter first introduces AV technology's main innovations compared to today's unconnected human-operated road vehicles, and reviews how automated vehicles, fleets, and traffic can change the way we move passengers and freight, generating various direct and indirect effects on the economy, environment, and society. Compared to many observers of this sector, however, we are less optimistic about speedy adoption of "driverless" cars. We believe futuristic projections of the implementation of fully autonomous AV technologies, its effects, and potential policy responses (e.g., Cohen and Cavoli 2018; Fraedrich et al. 2018; Milakis et al. 2018;

Millard-Ball 2018; Shay et al. 2018; Taeihagh and Lim 2019) may be of limited value for policymaking given the huge degree of uncertainty about the potential of driverless cars, or highly automated vehicles (HAVs).

Why? Much advanced AV technology is still under development and the current experiments of HAVs are limited to small-scale trials in simple or restricted environments. We also know little about human interaction with HAVs and how rules should change to ensure humans use AVs in positive and cooperative ways. As a result, this chapter focuses on more realistic possibilities in the near future (corresponding to the United Nations' 2030 Agenda for Sustainable Development) guided by the maturity of technologies and the likelihood that today's urban transportation system will be able to accommodate or adapt to the new technologies. While vehicle automation in the near future is unlikely to revolutionize the way people travel in and around cities, the specific impacts of those foreseeable technologies may still lead to both great opportunities and significant challenges, thus calling for policy responses. This chapter concludes by discussing what cities and urban regions could do to maximize the benefits and minimize the costs for achieving SDGs with the most realistic AV technologies in the near future.

2. Vehicle Automation and Possible Impacts

To understand the rationale and potential pitfalls behind the enthusiasm and optimism about automated vehicles, it is necessary to first review major vehicle automation technologies and their possible impacts.

2.1 Vehicle Automation and Its Applications

The automobile industry has experienced significant changes, especially over the past decade, by incorporating computerization into driving. Recent models released to market have increasingly adopted features such as adaptive cruise control and front crash avoidance. Looking toward a very different future, both traditional automobile companies and information tech giants have invested heavily in the research of AVs to create transportation machines aimed at liberating humans completely from the labor intensive task of driving.

AVs refer to any vehicle that can perform at least some of the safety critical control functions (e.g., steering, throttle, or braking) without direct driver input (NHTSA 2013). The degree to which a vehicle can perform these essential functions is further refined in the levels of automation. The Society of Automotive Engineers (SAE 2014) have defined levels of automation, ranging from single-function driver assistance to full automation with no driver required, as listed in Table 9.1. According to Volpe Center (2014), possible applications of AV technologies include:

Table 9.1 Vehicle automation levels

Level 1	Driver Assistance	Vehicle is controlled by the driver, but some driving assist features may be included in the vehicle design.
Level 2	Partial Automation	Vehicle has combined automated functions, like acceleration and steering, but the driver must remain engaged with the driving task and monitor the environment at all times.
Level 3	Conditional Automation	Driver is a necessity, but is not required to monitor the environment. The driver must be ready to take control of the vehicle at all times with notice.
Level 4	High Automation	The vehicle is capable of performing all driving functions under certain conditions. The driver may have the option to control the vehicle.
Level 5	Full Automation	The vehicle is capable of performing all driving functions under all conditions. The driver may have the option to control the vehicle.

Source: SAE (2014)

- Collision avoidance (forward, backup, lane departure, and blind spot);
- Cooperative adaptive cruise control (CACC);
- Speed harmonization;
- Platooning;
- Intersection management;
- Lane change, merge, and demerge;
- Lane weaving;
- First and last mile mobility;
- Automated paratransit;
- Transit lane assist;
- Transit precision docking;
- Automated yard operations; and
- Full automation/driverless cars.

2.2 Possible Impacts of Vehicle Automation

By allowing for the implementation of these applications, AV technologies have the potential to improve road safety (SDG3.6), reduce travel time (SDG8), enhance driver comfort, and lower energy consumption and pollutant emissions (SDGs 3.9, 7.3, 8.4, 9.4, 11, 13, and 15).

Traffic accidents resulted in 1.25 million human casualties in 2013[1] and caused significant economic loss globally, up to 3% of GDP for low- and middle-income

---

[1] See http://www.who.int/gho/road_safety/mortality/traffic_deaths_number/en/.

countries (LMICs).[2] There is no doubt that human error is a major contributor to road accidents (NHTSA 2015). SDG target 3.6 calls for halving the number of global deaths and injuries from road traffic accidents by 2020. The socioeconomic benefits of vehicle automation is significant if it can meaningfully reduce or largely eliminate human error in road transportation. However, AV technologies capable of replacing human error with reliable automation are still under development. Also, it remains unknown how the interaction of human behavior and automation will influence safety.

Reducing travel time directly improves well-being (SDGs 3 and 11) and productivity (SDG8). Travel time reduction is likely to result from a combination of traffic smoothing, shockwave reduction, and road capacity increases (Lee and Park 2012; Shladover et al. 2012), as long as travel demand growth (as a result of faster travel) doesn't overwhelm infrastructure capacity improvement. Travel time reliability will also likely improve because of accident reduction and optimal routing.

In addition to improving safety and travel speed, vehicle automation will improve driver comfort by reducing the overall cognitive and physical burden associated with driving, especially under high-stress driving scenarios such as dense traffic or inclement weather. As a result, the cost of traveling by automobile will decrease due to a lower cost of travel time for drivers. In particular, full automation or driverless vehicles will turn everyone into a passenger and potentially replace private vehicle ownership with vehicle share (robot taxi service), which might further reduce the cost of travel by automobile.

Beyond cost reductions in accidents and travel time, vehicle automation can benefit energy efficiency (related to several SDGs) of transportation through multiple channels such as congestion reduction, traffic smoothing, better aerodynamics due to platooning (Fernandes and Nunes 2012), improved traffic through dynamic intersection management (Lee and Park 2012), and even smaller/lighter vehicle design enabled by full automation.

All of these things lead to a reduction in the generalized private cost per vehicle-kilometer traveled (VKT). As a direct result of reduced private cost per VKT, demand for vehicle travel in VKT will grow as individuals and businesses increase travel frequency and distance by road vehicles, thus likely reducing or even negating the safety, time, energy, and environmental benefits of vehicle automation (Stephens et al. 2016; Wadud et al. 2016). Furthermore, to maximize long term net private benefits, individuals and businesses may take advantage of per VKT cost reduction to move residences and businesses further away from central locations, suggested by historical evidence on urban sprawl (Glaeser and Kahn 2003; Nechyba and Walsh 2004). Finally, shifting urban passenger trips to AVs may result in lower demand and reduced revenue for conventional public transit with fixed routes and schedules, causing it to become more expensive to provide. Nevertheless, in the case of robot taxi service provided by

---

[2] See http://www.who.int/violence_injury_prevention/road_safety_status/2015/Executive_summary_GSRRS2015.pdf?ua=1.

level 5 HAVs, transit may be completely replaced by HAVs through vehicle and ride sharing. With sufficiently cheap HAV technology, driverless robot taxies can make door-to-door transit service affordable to everyone. In fact, even low-speed driverless neighborhood shuttles, a much simpler AV technology, can significantly enhance transit users' experience by providing high-quality first- and last-mile connection to transit stops. This could potentially enhance social equity in urban transportation by making accessible and convenient transit service to travelers who have physical or financial difficulties to use private motor vehicles.

2.3　Optimism

There are many economic, environmental, and social impacts of vehicle automation. The direct and intuitive impacts of AV technology breakthroughs seem overwhelmingly positive. The future of vehicle automation, especially driverless cars, is so tempting that many optimists, not only from private industry, but also from the government and even academia, predict or assume mass adoption of highly automated or driverless vehicles before 2030 (e.g., Wardrop 2009, Fagnant and Kockelman 2015, Chapin et al. 2017, Walker 2018). For example, Raj Nair, Ford Motor Co.'s product development chief, predicts that level 4 AVs will hit the market by 2020.[3] While the prevailing automation technologies available on the market have not exceeded level 2 automation, such as Tesla's Autopilot, the testing of higher levels of automation is currently underway as more and more state and local governments issue permits for the testing of HAVs on public roads, as long as a human driver is in the vehicle.

3.　Why Aren't HAVs Coming Soon?

Both the mass deployment of HAVs and the huge social benefits predicted by advocates are speculative given technological uncertainties, potential ethical and regulatory challenges, and public acceptance questions facing HAVs. The fact that we see so many optimistic projections and discussions of a near-future world of driverless cars may reflect both wishful visions by some and the vested interests of those who stand to benefit directly, at least in the short term, from the attention to and investment in the technologies and applications of vehicle automation.

A large share of AV technologies is still under development, ranging from sensors, artificial intelligence, and connection and communication to fleet and system management. Fully automated vehicles face many challenges and barriers owing to the multitude of possible scenarios the vehicle must be able to respond to, including the full range of other vehicles' maneuvers, pedestrian and bicyclist behaviors,

---

[3]　See http://www.autonews.com/article/20160227/OEM02/302299994/fords-nair%3A-sensors-software-are-self-driving-cars-main-obstacles.

animals and debris, work sites, inclement weather conditions, and poor lighting (US Department of Transportation 2015).

3.1 Evidence on Safety Record

The evaluation of HAV technology has focused on safety. It is, however, difficult to compare the safety performance of the test runs of driverless cars directly with the crash statistics of daily drivers for two reasons. First, ongoing testing happens mostly under "normal" weather and road conditions. Second, there is an actual driver monitoring any testing vehicle on public roads to take back control whenever s/he feels necessary.

One statistic might highlight the limitations of an "intelligent" machine's ability to handle itself in real-world tests. The California state government requires all companies testing driverless cars to record and report every disengagement; that is, every time an HAV is taken over by the test driver, either because of the vehicles' request ("system failure") or the decision of the human driver ("driver-initiated").[4] When automated driving fails or is limited by, for example, the inability of on-board computer algorithms to make a safe decision, the autonomous mode disengages and the drivers are expected to resume manual driving. In other words, these disengagements are "machine errors" that would have likely led to crashes or at least dangerous maneuvers.

On California's public roads, Waymo (a self-driving technology development company and a subsidiary of Alphabet Inc., also known as Google) has accumulated the longest testing distance and performed the best in terms of annual rate of disengagements from 2015 to 2017, by a significant margin compared to Mercedes, Nissan, GM, BMW, Ford, Volkswagen, and Tesla, as well as Bosch and Delphi. Figure 9.1 plots the monthly rate of disengagement by Waymo's test vehicles[5] on public roads in California since September 2014 (first available data point), until the latest month with available data. The test mileage per month ranges from about 4,200 to 83,700, with a mean of 36,231 and a standard deviation of 18,946. The most recent monthly rate, about 33 disengagements per million vehicle miles traveled (VMT) in November

[4] Disengagements can be broadly classified as automatic and manual ones. Automatic disengagements result from a vehicle system recognizing a (potential) failure in the ability to ensure safety under automated driving conditions. These generally occur because of failures in detection technology, communications breakdown, improper sensor readings, map or calibration issues, errors in data receptions, or some hardware issues. Manual disengagements occur when the driver suspects a precarious situation owing to discomfort with the autonomous mode, adverse weather conditions, construction activities, performing lane changing in heavy traffic, poor road infrastructure, and so forth.

[5] For example, in 2015, Waymo had 424,331 miles driven by 49 vehicles, followed by Delphi with 16,662 miles driven by 2 vehicles. In 2016, Waymo put on 635,868 miles driven by 60 vehicles, followed by 10,015 miles driven by 25 GM vehicles. In 2017, Waymo tested 75 vehicles for a total of 353,545 miles, followed by GM's 131,676 miles by 84 vehicles. Waymo had 0.8 disengagements per thousand automated miles in 2015, followed by a distant 17.4 by Volkswagen. In 2016, Waymo lowed the rate to 0.2, followed by 1.57 by BMW, which only tested 638 miles by one vehicle. In 2017, Waymo had a rate of 0.18, followed by GM's 0.8.

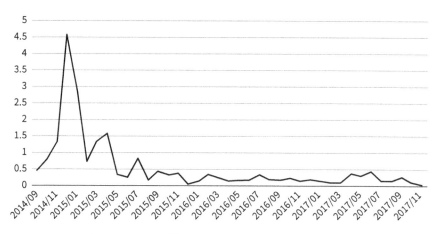

Figure 9.1 Monthly occurances of disengagements per million VMT by Waymo's test vehicles in California

2017, was the lowest recorded so far. The second lowest record, however, happened roughly two years ago in December 2015, at 51 disengagements per million VMT. The monthly rate of disengagements was fluctuating under 500 disengagements per million VMT since August 2015 without a clear trend of improvement.

In comparison, there were some 6.3 million fatal, injury, and property damage crashes that occurred in the United States in 2015,[6] and drivers in cars, trucks, minivans, and SUVs drove 3.1 trillion miles that year.[7] These translate into about two crashes per million VMT. The best monthly rate of disengagement so far (33 per million VMT) and the average crash rate of 2 per million VMT in the United States (or about 0.5 per million VMT in California in 2014, according to the California Highway Patrol Safety Database) are still different by an order of magnitude. Little improvement in disengagement rate has been achieved from 2015 to 2017 by Waymo, the leader in AV development.

There are caveats when comparing the AV disengagement rate directly with the human driver crash rate. On one hand, it is true that not every disengagement would have led to an accident without a responsive human driver and not every crash was reported. On the other hand, the disengagement statistics are from ongoing testing often under normal weather and road conditions (e.g., suburban boulevards in sunny California), thus a much more favorable measure of the "autonomous mobility" current technology can achieve. In fact, most experts would agree that it is relatively easy to build a driverless car that can handle 95% or even 99% of the situations, but extremely difficult to close the last gap of 5% or 1%.

---

[6] See https://www.statista.com/topics/3708/road-accidents-in-the-us/.
[7] See https://www.fhwa.dot.gov/pressroom/fhwa1704.cfm.

The challenge of automation is not the tedious task of routine driving. Instead, it is the exceptions to the normal situations that challenge the intelligence and reliability of machines. Some (Kalra and Paddock 2016) claim that 275 million miles of failure-free driving by AVs can statistically establish that AVs are as safe as human drivers with respect to the fatality rate. Depending on the number of test vehicles, it may take months to years and even decades to complete 275 million test miles. However, a crucial question is whether the test mileage resembles the full range of conditions of auto trips taken by human drivers in the real world. The ultimate challenge is: How long does it take to finish 275 million miles of failure-free driving given the current level and trend in the rate of disengagement?

3.2 Theoretical Challenges to Driverless Cars

The lack of continuous improvement in the safety record of HAVs is further supported by the theoretical discussions on the risks automation can introduce, such as hardware and software failures, malicious hacking, and increased risk-taking by humans (Fung 2015; Hsu 2017; Kockelman et al. 2016; Koopman and Wagner 2017; Ohnsman 2014).

Complex manmade systems are fragile. A modern luxury vehicle has a computer system that is more complex than an airplane because of the larger variety of interactions present on roadways than in the air (Government Accounting Office 2016). Producing such software is challenging and costly, and ensuring that it never fails is virtually impossible. There will almost certainly be system failures, which may lead to severe accidents and necessitate a less automated backup system. Experts acknowledge that significant technical progress is needed before level 5 automation is reliable, tested, and approved (Mervis 2017). Moreover, both human-driven cars and computer systems have been hacked since they existed. Few would expect that connected self-driven vehicles would be easy to guard against malicious hacking in order to protect not merely property and data, but human lives.

Not only is the readiness of HAV technology uncertain, but it is also difficult to predict the effects of high levels of automation. For example, the characteristics of AV traffic flow, the number of empty rides, and the impacts on travel mode choice and trip length are largely unclear (Meyer et al. 2017). In particular, perceived safety can invite risk-taking behavior. There have been a number of high-profile accidents and incidents involving vehicles either testing or using AV capabilities. From 2016 to 2018 AVs have been involved in at least four fatal crashes, three involving Tesla vehicles, largely attributable to driver overconfidence in technology.[8] This overconfidence is exemplified by a driver in the United Kingdom receiving a suspension of driving privileges for sitting in the passenger seat of his Tesla while in traffic.

---

[8] The fourth crash involved an Uber test vehicle with an operator in the cabin striking a killing a pedestrian, again indicating the limit of AV technology today.

We know very little about how rules should change to ensure positive and cooperative uses of HAVs in a fundamentally different transportation system. For example, there is the issue of liability for vehicle and systems manufacturers. Liability concerns may significantly slow the introduction of technologies, even if they are socially desirable (Kalra et al. 2009). Manufacturers may be held responsible under several theories of liability for systems that aid but do not replace the driver. Warnings and consumer education have to play a crucial role in managing manufacturer liability for these systems. Even after level 5 AV technology is mature, the behavior and interaction among in-vehicle and out-of-vehicle road users may bring numerous new questions over how to regulate the use of HAVs.

3.3 Summary

Road test performance casts a serious doubt on any optimistic forecast of driverless vehicles being significantly safer than human drivers in the near future. It seems not only impossible to predict when HAV technologies will become reality, but also highly uncertain how human behavior and transportation systems will react or adapt to HAVs. We believe it unrealistic to expect that HAVs would help us achieve or even surpass the target set by SDG3.6. As a result, it is premature to discuss policy response to the mass adoption of driverless vehicles.

4. Near-Future Technologies: What to Expect by 2030?

Despite the slim hope of HAVs in the near future, many current mass produced vehicles have level 1 and level 2 technologies such as adaptive cruise control, forward collision and blind spot warning, lane keeping, and automated parallel parking. Tesla's autopilot represents the most advanced of these systems, including some features that would be considered level 3. While many of these automated features are limited in terms of capacity and reliability, and by environmental and traffic conditions, these technologies are becoming increasingly more useful, more reliable, and less expensive.

Instead of making highly futuristic projections, we focus on more realistic possibilities in the near future (e.g., by 2030), based on relatively mature technologies indicated either by current market availability or by minimal technological breakthroughs and changes in infrastructure needed for deployment. Table 9.2 summarizes the major technologies that fit these criteria, based on data compiled by the US Department of Transportation (2015).

To illustrate the positive and negatives consequences of mass adoption of AV technologies in the near future, we discuss three groups of technologies: Advanced Driving Assistance System (ADAS), low-speed full automation in restricted

Table 9.2 Near-future AV technologies

Technology	Application	Availability	Level
Advanced Driver-Assistance Systems (ADAS)	Collision warning/avoidance	Currently available	L1
	Traffic jam assistance (low-speed car following, lane keeping, adaptive cruise control)	Currently available	L2
Low-speed full automation in restricted environments	Low-speed driverless shuttles/deliveries	Field demonstrations	L4 at low speed
Vehicle-to-Vehicle (V2V) and Vehicle-to-Infrastructure (V2I) communication	Cooperative adaptive cruise control (CACC), speed harmonization, intersection management, etc.	Field experiments	Connected L2

Source: the authors, based on US Department of Transportation (2015)

environments, and vehicles connected via Vehicle-to-Vehicle (V2V) and Vehicle-to-Infrastructure (V2I) communication. For each application we discuss the potential impacts of adopting the technology and how the impacts may be affected by penetration and infrastructure readiness.

4.1 Advanced Driving Assistance System (ADAS)

ADAS is best described as semiautomated features meant to enhance driver performance and comfort, with the goal of reducing crashes caused by human error. Since 2000, advanced driver assistance features have entered the market and helped improve safety. According to a six-country study (Fildes et al. 2015), autonomous emergency-braking system can prevent 38% of front to rear collisions. Lane-keeping systems were reported to have a 30–33% reduction in crashes related to lane changes (Sternlund 2017, Nodine et al. 2011). Still, further research is needed to establish the causal effect on casualty reduction (European Road Safety Observatory 2016).

Though current adoption rate of ADAS features are relatively low across major auto markets, with less than one fifth of recent car buyers in the United States adopting any ADAS feature, 89% of consumers have expressed willingness to purchase these systems in the future (McKinsey 2015). Price, instead of technological maturity or user acceptance, appears to be a major barrier to adopting ADAS. According to a consumer survey in 2015, car buyers are willing to pay up to $170 for the lane departure warning feature, while the cost is $400 per vehicle (BCG 2015). Currently, forward collision warning, automatic emergency braking, and lane departure warning are among the National Highway Traffic Safety Administration (NHTSA) recommended safety technologies that consumers should look for when shopping for a new vehicle.[9] The increasing regulatory interest in promoting safety is likely to lead to more required adoptions of ADAS features in the near future. As an industry-wide commitment in 2016, NHTSA and the Insurance Institute for Highway Safety announced that by 2022, virtually all new cars in the US market will have automatic emergency braking as a standard feature.[10] Similarly, in 2010, the EU adopted a legal framework to promote research and policy discussions in ADAS across Europe (European Road Safety Observatory 2016).

Lower levels of automation bring to us some of the benefits promised by HAVs but with fewer of the negative consequences. The direct benefits of ADAS are largely independent on the degree of adoption/penetration as ADAS only reduces a vehicle drivers' own errors in order to improve safety. Collectively, reduced human error leads to fewer accidents, smoother traffic, and associated energy and environmental benefits.

While level 1 and level 2 automation technologies produce fewer dramatic improvements or disruptions than HAV technologies, they can still induce a non-negligible amount of additional travel. This is because estimated long-run VKT is quite elastic with respect to generalized travel costs. Estimated generalized travel cost elasticity of VKT ranges from -1.0 to -2.0 (FHWA 2005) and -2.3 (Graham and Glaister 2004) for light duty vehicles (LDVs). For heavy duty freight vehicles (HDVs), the elasticity is estimated at about -1.0 (FHWA 2008, Cambridge Systematics 2009). That is, if generalized cost decreases by 10%, total VKT grows by 10–23% for LDVs and 10% for HDVs. The additional vehicle travel, while contributing to consumer and producer benefits, also generates more congestion, safety, and environmental costs. Unlike highly automated technologies, level 1 and level 2 automation does not dramatically alter vehicle or ride sharing behaviors, thus having little direct effect on the demand for parking spaces.

4.2 Low-Speed Driverless Neighborhood Shuttles/Deliveries

Low-speed driverless shuttles/deliveries require a high level of automation (L4), however they operate in low-speed conditions, often within a manicured and limited environment. A good example of such low-speed driverless shuttles/deliveries is the GATEway project funded by the UK government and industry.[11] Four driverless pods navigated a 3.4 km route around the Greenwich Peninsula in London to provide a shuttle service by picking up and dropping off passengers at four designated pod stops. The project also included automated urban deliveries.

---

[9] See https://www.nhtsa.gov/technology-innovation/automated-vehicles-safety.
[10] See https://www.nhtsa.gov/press-releases/us-dot-and-iihs-announce-historic-commitment-20-automakers-make-automatic-emergency.
[11] See https://gateway-project.org.uk.

Low-speed driverless shuttles reduce labor cost while providing last-mile access for transit users and urban logistics. User acceptance is less of a concern as low speeds and restricted environments make it much less likely to harm passengers in the event of a crash. Vehicles are often tailored to the specific needs of the application. By enhancing access to existing transit, automated low-speed shuttles can increase transit ridership and potentially decrease total VKT. While benefits are primarily internal to the users of such services, increased penetration of neighborhood shuttles serving public transit trunk lines can increase transit ridership and efficiency, which attracts more transit users, thus reducing driving and associated energy and environmental costs. Combining these shuttles with low-carbon electric vehicle infrastructure can further increase social benefit by avoiding fossil fuel-based trips and lowering local emissions.

4.3 Connected Vehicles

Future vehicles are expected to be connected to and interact with other vehicles and road infrastructures by wireless communication in order to enhance safety, efficiency, and sustainability. Connected vehicles use any of a number of different communication technologies to communicate with the driver (in-car content and services), other vehicles on the road (V2V), infrastructure (V2I), and the "cloud" (Science and Technology Select Committee 2017). V2V connection may cost only about $350 per vehicle in 2020 and V2I technologies such as Audi's "time to green" feature are already under pilot implementation.

There are many possible applications of connected vehicle technology. The most important application seems to be cooperative adaptive cruise control (CACC). CACC realizes automated speed control based on V2V and V2I communication of the information about both the forward vehicles and the traffic further ahead, as well as local speed recommendations (Shladover et al. 2015). Through keeping short but safe headway between vehicles and attenuating traffic flow disturbances while ensuring driving comfort, CACC can reduce traffic congestion by improving highway capacity. Some of the largest benefits of CACC can be realized in the freight sector through platooning. Platooning utilizes V2V communication to connect multiple vehicles to minimize following distance, creating an aerodynamic effect that decreases fuel consumption.

Large-scale penetration is important to fully realizing the benefits of CACC as it only works well when vehicles can communicate with each other to share information and coordinate platooning. With a 100% penetration level of CACC in the traffic stream, traffic throughput could reach more than 4,200 veh/hr/ln, more than doubling the maximum throughput in manual driving conditions of around 2,000 veh/hr/ln (Dey et al. 2016).

Compared to ADAS and low-speed driverless shuttles/deliveries, major applications of connected vehicles (e.g., CACC/platooning) are further away from market reality. There remain unresolved issues such as driver reengagement (driver may not be able to restore situational awareness and respond quickly enough as a platoon dissolves), user acceptance (challenge to comfort level when headway is too short), platoon entry/exit, and maximum platoon length.

4.4 Summary

Overall, much of the near term AV technologies generate benefits such as driver comfort and road safety. However, both ADAS and connected vehicles will still require a licensed operator to perform the majority of driving functions. Therefore, these technologies are unlikely to bring about revolutionary changes toward mass vehicle sharing, parking reduction, the enabling of unlicensed travelers, the generation of empty rides (relocation trips), and the self-fueling/charging of vehicles. Worries about increased energy consumption and environmental pollution due to induced demand and new user mobility appear to be more of a concern at full automation (Wadud et al. 2016).

5. What Can Cities Do?

More capable vehicles reduce the private cost of moving passengers and freight, benefiting the society as long as users pay for the full cost of using such vehicles—otherwise increased private benefits may be partially or fully offset by additional negative externalities. Internalizing vehicle users' social costs is not a new challenge to transportation and urban policymakers—policies such as congestion pricing and subsidizing public transit are well known tools to promote sustainable urban transportation. However, to maximize the benefits and minimize the costs of the AV technologies expected in the near future, policymakers should pay attention to the direct and indirect impacts of these technologies, their dependency on infrastructure and regulations, and potential for further improvements.

AV technologies are unlikely to revolutionize ground transportation in the near future given the challenges and uncertainties associated with HAVs. Nevertheless, the adoption of realistic future AV technologies could still trigger a shift toward a more efficient, environmentally friendly, and socially inclusive urban transport system if appropriate policies are implemented. Without such policies, connected and automated vehicle technologies can exacerbate policy challenges such as energy consumption and emissions, space occupied by parking, and infrastructure maintenance and investment, all caused by a higher travel demand induced by the decrease in travel cost per kilometer. Without going into the detailed context of specific cities, policymakers may nevertheless benefit from the four suggestions in the following sections in order to best achieve SDG11 and associated other SDGs.

5.1 Integrate Driverless Shuttles with Transit

Public transport is crucial to urban sustainability and social inclusivity (SDG11.2). Low-speed driverless shuttles can connect mass rapid transit (e.g., subway) with all users, including those who cannot drive or bike and those who cannot afford to use a private automobile. Such shuttles can potentially be much affordable than conventional feeder buses even at the current cost of low-speed automation. With a sufficiently low cost, a large number of reliable, low-speed driverless shuttles can be deployed to ensure a high level of convenience. This attracts more travelers to use public transit, which will then be able to offer service at a higher frequency (Mohring 1972), further incentivizing people to use transit instead of private vehicles.

5.2 Coordinate Infrastructure Provision

While the near future AV technologies may not require any major change to infrastructure, low-speed neighborhood shuttles/deliveries and truck platoons should be easier to adopt with friendly infrastructure. Neighborhood shuttles would need curbside passenger boarding/dropping space, while truck platoons would be much safer to operate if dedicated lanes are used where there is busy traffic of both passenger and freight vehicles. By implementing low-speed neighborhood shuttle facilities and automated freight lanes, cities can reduce private vehicle use, mitigate safety risks, and maximize platoon benefits, improving energy and economic efficiency (SDGs 7 and 8), environmental quality (SDGs 13, 14, and 15), and urban livability (SDG11).

5.3 Adopt Optimal Pricing Through V2I Connectivity

While political opposition is often cited as the foremost barrier to congestion charging, we cannot deny that setting up a city-wide system to charge road users can be costly. For example, the London congestion scheme costs the city one-third of its congestion charge revenue (Prud'homme and Bocarejo 2005), which could have been used toward general tax abatement and/or transit improvement. Built-in V2I connectivity will allow for more efficient travel management such as dynamic road pricing. While privacy protection remains an industry-wide challenge, as vehicles become smarter and routinely exchange information with other vehicles and infrastructure such as signalized intersections, the benefit-cost calculation will be more favorable toward making users pay for their use of infrastructure and external impacts on others. The private sector, such as the insurance industry, can also benefit from V2I connectivity when implementing mileage-based insurance. In the longer term, optimal pricing of infrastructure use removes market distortions (SDG12.c) and discourages inefficient, unsustainable, and often socially segregating urban sprawl, a widely known challenge to achieving SDG11.

5.4 Share Local Experience Across Cities and Governments

Real-world experience provides crucial lessons for improving AV technologies and transportation system. Cities have a unique advantage in acquiring detailed information on the local impacts of specific urban transportation technologies and implementation strategies. Connected vehicle and infrastructure technologies open new opportunities for cities to obtain high-quality, timely, and reliable data (SDG17.18), potentially with the help of nongovernmental knowledge partners, such as a university or research organization (SDG17.17). Sharing data and learning from each other's experience (SDG17.6–8) can greatly speed up the diffusion of best practice and innovations in technology and management, from which late adopters, such as cities of the developing world, would benefit. With the continuous advancement of AV technologies, new standards of vehicles and infrastructure and new regulations will be increasingly necessary. Sharing local experience with central governments and across countries could help formulate the most informed industrial standards and most sensible regulations nationally and globally.

References

Boston Consulting Group (BCG) (2015). A Road Map to Safer Driving through Advanced Driver Assistance Systems. http://image-src.bcg.com/Images/MEMA-BCG-A-Roadmap-to-Safer-Driving-Sep-2015_tcm9-63787.pdf (accessed April 5, 2019).

Cambridge Systematics (2009). Assessment of Fuel Economy Technologies for Medium and Heavy Duty Vehicles: Commissioned Paper on Indirect Costs and Alternative Approaches, Draft final paper. Cambridge Systematics Inc., Medford, MA.

Chapin, T., Stevens, L., Crute, J. (2017). Here Come the Robot Cars. *Planning*, April. https://www.planning.org/planning/2017/apr/robotcars/ (accessed April 5, 2019).

Cohen, T., and Cavoli, C. (2018). Automated vehicles: exploring possible consequences of government (non) intervention for congestion and accessibility. *Transport Reviews* 39(1), 129–151.

Dey, K. C., Yan, L., Wang, X., Wang, Y., Shen, H., Chowdhury, M., Yu, L., Qiu, C., and Soundararaj, V. (2016). A review of communication, driver characteristics, and controls aspects of cooperative adaptive cruise control (CACC). *IEEE Transactions on Intelligent Transportation Systems* 17(2), 491–509.

Duranton, G., and Puga, D. (2003). Micro-foundations of urban agglomeration economics. In J. Henderson, V., and Thisse, J. F. (Eds.), *Handbook of Regional and Urban Economics* (Vol. 4), pp. 2063–2117. Elsevier, Amsterdam.

European Road Safety Observatory (2016). Advanced Driver Assistance Systems. https://ec.europa.eu/transport/road_safety/sites/roadsafety/files/ersosynthesis2016-adas15_en.pdf (accessed April 5, 2019).

Fagnant, D. J., and Kockelman, K. M. (2015). Preparing a nation for autonomous vehicles: opportunities, barriers and policy recommendations. *Transportation Research Part A: Policy and Practice* 77, 167–181.

Federal Highway Administration (FHWA) (2005). Highway Economics Requirements System, State Version, Technical Report. US Department of Transportation, Washington, DC.

Federal Highway Administration (FHWA) (2008). Freight Benefit/Cost Study: Phase III - Analysis of Regional Benefits of Highway-Freight Improvements, Final Report, FHWA-HOP-08-019. US Department of Transportation, Washington, D.C.

Fernandes, P., and Nunes, U. (2010). Platooning of autonomous vehicles with intervehicle communications in SUMO traffic simulator. *13th International IEEE Conference on Intelligent Transportation Systems*, Funchal, pp. 1313–1318.

Fildes, B., Keall, M., Bos, N., Lie, A. Page, Y., Pastor, C., Pennisi, L., Rizzi, M., Thomas, P., and Tingvall, C. (2015). Effectiveness of low speed autonomous emergency braking in real-world rear-end crashes. *Accident Analysis and Prevention* 81, 24–29.

Fraedrich, E., Heinrichs, D., Bahamonde-Birke, F. J., and Cyganski, R. (2018). Autonomous driving, the built environment and policy implications. *Transportation Research Part A: Policy and Practice* 122, 162–172.

Fung, B. (2015), Driverless Cars are Getting into Accidents, But the Police Reports are Not Being Made Public. *Washington Post*, https://www.washingtonpost.com/news/the-switch/wp/2015/05/11/driverless-cars-are-getting-into-accidents-but-the-police-reports-are-not-being-made-public/?noredirect=onandutm_term=.d07210c506f6 (accessed April 5, 2019).

Glaeser, E. L., and Kahn, M. E. (2003). Sprawl and urban growth. In Henderson, J., and Thisse, J. (Eds.), *Handbook of Regional and Urban Economics,* Vol. 4, pp. 2482–2527. Elsevier, Amsterdam.

Government Accounting Office (2016). Vehicle Cybersecurity: DOT and Industry Have Efforts Under Way, but DOT Needs to Define Its Role in Responding to a Real-world Attack. www.gao.gov/products/GAO-16-350 (accessed April 5, 2019).

Graham, D., and Glaister, S. 2004. Road traffic demand elasticity estimates: a review. *Transport Reviews* 24 (3), 261–274.

Hsu,J.(2017).WhenItComestoSafety,AutonomousCarsAreStillTeenDrivers.*ScientificAmerican*, https://www.scientificamerican.com/article/when-it-comes-to-safety-autonomous-cars-are-still-teen-drivers1/ (accessed April 5, 2019).

Kalra, N., Anderson, J. M. and Wachs, M. (2009). Liability and Regulation of Autonomous Vehicle Technologies. California PATH Research Report, UCB-ITS-PRR-2009-28. California PATH Program, Institute of Transportation Studies, University of California at Berkeley, Berkeley, CA.

Kalra, N. and Paddock, S. M. (2016). *Driving to Safety: How Many Miles of Driving Would It Take to Demonstrate Autonomous Vehicle Reliability?* RAND Corporation, Santa Monica, CA.

Kockelman, K. M., Loftus-Otway, L., Stewart, D., Nichols, A., Wagner, W., Li, J., Boyles, S., Levin, M., and Liu J. (2016). *Best Practices Guidebook for Preparing Texas for Connected and Automated Vehicles.* Center for Transportation Research, University of Texas at Austin, Austin.

Koopman, P., and Wagner, M. (2017). Autonomous vehicle safety: an interdisciplinary challenge. *IEEE Intelligent Transportation Systems Magazine* 9(1), 90–96.

Krugman, P. (1991). *Geography and Trade.* Leuven University Press and MIT Press, Leuven and Boston.

Lee, J., and Park, B. (2012). Development and evaluation of a cooperative vehicle intersection control algorithm under the connected vehicles environment. *IEEE Transactions on Intelligent Transportation Systems* 13(1), 81–90.

McKinsey and Company (2015). Competing for the connected customer: perspectives on the opportunities created by car connectivity and automation. https://www.mckinsey.com/~/

media/mckinsey/industries/automotive%20and%20assembly/our%20insights/how%20 carmakers%20can%20compete%20for%20the%20connected%20consumer/competing_ for_the_connected_customer.ashx (accessed April 5, 2019).

Mervis, J. (2017). Are we going too fast on driverless cars? *Science Magazine*, www.sciencemag. org/news/2017/12/are-we-going-too-fast-driverless-cars (accessed April 5, 2019).

Meyer, J., Becker, H., Bösch, P. M., and Axhausen, K. W. (2017). Autonomous vehicles: the next jump in accessibilities? *Research in Transportation Economics* 62, 80–91.

Milakis, D., Kroesen, M. and van Wee, B. (2018). Implications of automated vehicles for accessibility and location choices: Evidence from an expert-based experiment. *Journal of Transport Geography* 68, 142–148.

Millard-Ball, A. (2018). Pedestrians, autonomous vehicles, and cities. *Journal of Planning Education and Research* 38(1), 6–12.

Mohring, H. (1972). Optimization and scale economies in urban bus transportation. *American Economic Review* 62(4), 591–604.

National Highway Traffic Safety Administration (NHTSA) (2013). Preliminary Statement of Policy Concerning Automated Vehicles, https://www.nhtsa.gov/staticfiles/rulemaking/pdf/ Automated_Vehicles_Policy.pdf (accessed April 5, 2019).

National Highway Traffic Safety Administration (NHTSA) (2015). Critical reasons for crashes investigated in the national motor vehicle crash causation survey, https://crashstats.nhtsa. dot.gov/Api/Public/ViewPublication/812115 (accessed April 5, 2019).

Nechyba, T. J., and Walsh, R. P. (2004). Urban sprawl. *Journal of Economic Perspectives* 18(4), 177e200.

Nodine, E. E., Lam, A. H., Najm, W. G., and Ference, J. J. (2011). Safety impact of an integrated crash warning system based on field test data. *22nd International Technical Conference on the Enhanced Safety of Vehicles (ESV)*. Washington, D.C., pp. 1–9.

Ohnsman, A. (2014), Automated cars may boost fuel use, Toyota scientist says. *Bloomberg Press*, July 16, 2014, https://www.bloomberg.com/news/articles/2014-07-16/automated- cars-may-boost-fuel-use-toyota-scientist-says (accessed April 5, 2019).

Prud'homme, R., and Bocarejo, J.P. (2005). The London congestion charge: a tentative economic appraisal. *Transport Policy* 12(3), 279–287.

Science and Technology Select Committee (2017). Connected and autonomous vehicles: the future? https://publications.parliament.uk/pa/ld201617/ldselect/ldsctech/115/115.pdf (accessed April 5, 2019).

Shay, E., Khattak, A. J., and Wali, B. (2018). Walkability in the connected and automated vehicle era: a US perspective on research needs. *Transportation Research Record* 2672(35), 118–128.

Shladover, S. E., Nowakowski, C., Lu, X. -Y. and Ferlis, R. (2015). Cooperative adaptive cruise control: definitions and operating concepts. *Transportation Research Record* 2489(1), 145–152.

Shladover, S. E., Su, D., and Lu, X. -Y. (2012). Impacts of cooperative adaptive cruise control on freeway traffic flow. *Transportation Research Record* 2324(1), 63–70.

Society of Automotive Engineers (SAE). (2014), *Levels of Driving Automation Are Defined in New SAE International Standard J3016*. https://cdn.oemoffhighway.com/files/base/acbm/ ooh/document/2016/03/automated_driving.pdf (accessed April 5, 2019).

Stephens, T. S., Gonder, J., Chen, Y., Lin, Z., Liu, C. and Gohlke, D. (2016). Estimated Bounds and Important Factors for Fuel Use and Consumer Costs of Connected and Automated Vehicles, Technical Report, National Renewable Energy Laboratory, https://www.nrel.gov/ docs/fy17osti/67216.pdf (accessed April 5, 2019).

Sternlund, S., Strandroth, J., Rizzi, M., Lie, A. and Tingvall, C. (2017). The effectiveness of lane departure warning systems: a reduction in real-world passenger car injury crashes. *Traffic Injury Prevention* 18(2), 225–229.

Taeihagh, A., and Lim, H. S. M. (2019). Governing autonomous vehicles: emerging responses for safety, liability, privacy, cybersecurity, and industry risks. *Transport Reviews* 39(1), 103–128.

United States Department of Transportation (US DOT) (2015). Benefits Estimation Framework for Automated Vehicle Operations, FHWA-JPO-16-229, https://rosap.ntl.bts. gov/view/dot/4298/dot_4298_DS1.pdf (accessed April 5, 2019).

Volpe Center (2014). *Development of a Multimodal Program Plan for Automated Vehicles*. Volpe Center, Cambridge, MA.

Wadud, Z., MacKenzie, D. and Leiby, P. (2016). Help or hindrance? the travel, energy and carbon impacts of highly automated vehicles. *Transportation Research A: Policy and Practice* 86, 1–18.

Walker, J. (2018). The self-driving car timeline: predictions from the top 10 global automakers. https://www.techemergence.com/self-driving-car-timeline-themselves-top-11-automakers (accessed April 5, 2019).

Wardrop, M. (2009). Driverless vehicles could be on Britain's roads in 10 years. http://www.telegraph.co.uk/technology/news/6058498/Driverless-vehicles-could-be-on-Britains-roads-within-10-years.html (accessed April 5, 2019).

10

The Role of Technology and Rebound Effects in the Success of the Sustainable Development Goals Framework

David Font Vivanco and Tamar Makov

1. Introduction

The sustainable development goals (SDGs) were created to guide governments facing multiple pressing challenges, such as climate change, hunger, and poverty. Each of the 16 sustainable development goals (excluding SDG17, global partnership for sustainable development) focuses on a topic area, and collectively the SDGs aim to support a balanced development agenda which incorporates economic, social, and environmental considerations. However, several have noted that there might be hidden barriers to achieving SDGs (Allen et al. 2018; Le Blanc 2015; ICSU-ISSC 2015; Nilsson et al. 2016). Since financing SDGs will require an estimated investment of US$90 trillion over 15 years (SDG-FL 2017), such hidden barriers could come at a high cost.

While SDGs are generally conceptualized as separate, isolated elements within a broader framework, many of the goals are inherently linked, such that progress toward one goal could affect progress toward other goals (ICSU-ISSC 2015). For instance, reducing the reliance on fossil fuels by increasing large-scale bioenergy production (SDG7) could counteract progress toward food security (SDG2) via competition over fertile land (Murphy et al. 2011). As such, conflicting interlinkages could constrain or even eliminate the ability to progress toward multiple goals simultaneously (Nilsson et al. 2016). Moreover, there is concern that "since the SDG framework does not reflect interlinkages, it is possible that the framework as a whole might not be internally consistent—and as a result not be sustainable" (ICSU-ISSC 2015 p. 9). Given the potential meaningful impact of SDG interlinkages, a growing number of scholars have recently begun to map and investigate them. Yet a broader understanding of the mechanisms driving conflicting interlinkages is still lacking (Nilsson et al. 2016). In particular, so-called rebound effects have been shown to reduce the overall benefits delivered from technological change (Sorrell 2007). Since technology and technological change are central components to many SDG strategies, it is possible that rebound effects could meaningfully impact the ability to achieve desired goals either directly or via interlinkages.

Rebound effects can be defined as behavioral and systemic responses to technological change that counteract, to some extent, the expected environmental benefits delivered by the new technology (Sorrell 2007). Well-documented rebound examples include people driving more in response to improved fuel efficiency in passenger vehicles, or setting thermostats to higher indoor temperatures during winter in response to heating efficiency improvements (Greening et al. 2000). Beyond direct impacts, responses to technological change in one area could also spill over to other areas. For example, greater energy efficiency could reduce the costs of desalination and lower water prices, which in turn could lead to an increase in water demand. Thus, rebound effects can also mediate interlinkages between different SDGs. The rich array of theories, quantitative methods, and evidence of rebound research (Jenkins et al. 2011; Sorrell 2007), provide a useful framework for examining hidden barriers for SDG progress.

In this chapter, we explore the role of rebound effects as mediators of interlinkages between SDGs, and their potential implications on the SDG framework as a whole. Our aim is to illustrate the potential usefulness of utilizing the rebound paradigm to shed light on hidden barriers which might impede progress towards SDGs and specifically, the mechanisms driving environmental-economic-social trade-offs. Identifying mechanisms is key for formulating informed management strategies to tackle barriers and improve the overall cohesiveness of the SDG framework. This chapter is structured as follows: section 2 outlines the main reinforcing and conflicting linkages between goals and targets. Section 3 addresses the reliance of SDGs on technology and technological change. Section 4 offers a general introduction to rebound effects, including disciplinary theory, typology, and the state-of-the-art in rebound research. Section 5 focuses on a subset of SDGs and explores the role of rebound effects in mediating both conflicting and reinforcing linkages. Section 6 presents a summary and conclusions.

2. Reinforcing and Conflicting Linkages Between SDGs

SDGs were designed to provide governments with an overarching plan to promote sustainable development. However, while each SDG centers on a specific issue, many of the goals are linked, and progress toward one could impede or enhance progress toward others (Allen et al. 2018; Le Blanc 2015; ICSU-ISSC 2015; McCollum et al. 2018). As a result, to effectively achieve the desired goals and targets policymakers tasked with implementing SDGs must navigate a complex array of positive and negative interdependencies.

For example, SDG7 aims to ensure access to affordable, reliable, sustainable and modern energy for all. This can be done, in part, by replacing fossil fuels with renewable energy sources, such as large-scale biofuel production. Biofuels, however, can compete with food production (Murphy et al. 2011) and interfere with attempts to end hunger (SDG2). Additionally, large-scale introduction of biofuels could lower oil prices and subsequently increase oil demand and carbon emissions (Hochman et al. 2010). At the same time, attempting to end hunger (SDG2) as well as increasing protection for land (SDG15) or marine ecosystems (SDG14) could impose limits on renewable energy production and pose barriers to reducing fossil fuels dependency (SDG7) (Gill 2005; Nonhebel 2005).

Given the potentially far-reaching implications that interlinkages might have on reaching SDG targets, various methods have been adopted to map and investigate them. In particular, network analysis has emerged as a useful tool for teasing out interlinkages and identifying which SDGs are most likely to affect and be affected by other goals (Le Blanc 2015; Zhou and Moinuddin 2017).

Yet not all linkages are negative in nature. In fact, there seem to be more cases where SDGs complement each other such that progress toward one reinforces progress toward others. For example, replacing fossil fuels with renewable energy sources (SDG7) by promoting electric vehicles would lead to reductions in both carbon emissions and urban air pollution (Brady and O'Mahony 2011). Subsequently, such a shift would help improve health and well-being (SDG3), climate action (SDG13), and sustainable cities and communities (SDG11). Moreover, an overarching shift toward energy efficiency and cleaner energy production will likely make manufacturing in general more sustainable (Allan et al. 2007), facilitating progress toward responsible consumption and production (SDG12). Indeed, several have noted that beyond the existence of interlinkages, it is important to consider the nature of the linkages (i.e., be they positive or negative), and the magnitude of their impact. McCollum et al. (2018) for example, focus specifically on linkages between energy transformations (SDG7) and nonenergy SDGs. Ranking each linkage on a scale of -3 (goals fully cancel each other out) to +3 (goals reinforce one another till they cannot be separated), they show that overall, positive linkages between energy and nonenergy SDGs outweigh negative ones in both number and magnitude.

Furthermore, Allen et. al. (2018) use a case study of 22 nations in the Arab region to illustrate how a multicriteria analysis (MCA) decision framework can be used to help policymakers prioritize SDG targets based on their urgency, system wide impact, and alignment with current policies. Lastly, a recent report by the Institute for Global Environmental Strategies (Zhou and Moinuddin 2017) provides a comprehensive review of SDG interlinkages studies and a novel approach to measure the extent to which SDGs are interconnected. Specifically, the report maps networks of positive and negative interlinkages between 108 targets in nine Asian countries according to four key properties Figure 10.1: degree centrality (number of interactions a target has with other targets), eigenvector centrality (number of interactions a target has with other influential targets), closeness centrality (how close a target is to other targets in the network), and betweeness centrality (extent to which a target bridges other unconnected targets).

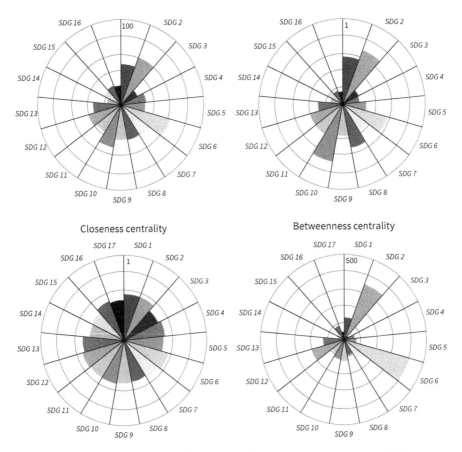

Figure 10.1 Degree, eigenvector, closeness, and betweenness centrality of SDG interlinkages. Degree centrality indicates the number of interactions of a target with other targets, eigenvector centrality indicates the number of interactions of a target with other influential targets, closeness centrality indicates how close a target is to other targets in the network, and betweenness centrality indicates the extent to which a target bridges other unconnected targets. Values correspond to average values for all individual targets within each SDG, for nine Asian countries: Bangladesh, Cambodia, China, India, Indonesia, Japan, the Philippines, Republic of Korea, and Vietnam.
Source: Zhou and Moinudden (2017).

Based on the combined scores of these four properties, the report suggests that six of the SDGs are central in terms of their overall influence within the network, namely SDG1 (end poverty), SDG2 (end hunger), SDG6 (water and sanitation for all), SDG7 (energy for all), SDG10 (reduce inequalities), and SDG12 (sustainable consumption and production). Furthermore, the report identifies key specific targets as the most influential overall: target 2.3 (double agriculture productivity), target 2.4 (build sustainable food production systems), target 6.1 (universal access to safe drinking water), target 6.2 (universal access to sanitation and hygiene), target 7.1 (universal access to energy), and target 9.1 (develop resilient infrastructure). Some authors refer to highly interconnected targets as "core targets," which would facilitate integrated thinking and policymaking (Le Blanc 2015).

A common theme among these core targets is that they rely, to a great extent, on technological change. Since technological change often involves rebound effects, it is possible that rebound effects play an important role in mediating linkages between SDGs. To explore the scope of potential rebound implications, in the next chapter we assess the extent to which each SDG relies on technological change.

3. The Reliance of SDGs on Technology

According to the United Nations, "technology and innovation are central to the implementation of the 2030 Agenda and the Sustainable Development Goals" (ESC 2016). Nonetheless, not all challenges currently facing humanity have technological solutions. Therefore, to ascertain how much SDGs rely on technology, we systematically assessed the degree to which specific goals and targets require technological change. The details of our assessment are presented in Table 10.1 and summarized in Figure 10.2.

Our review indicates that technological change is a fundamental component of many targets across most SDGs. For example, all targets under energy (SDG7) relate in some way or another to technological change. Similarly, other environmental SDGs, such as those pertaining to water and sanitation (SDG6), industrialization (SDG9), sustainable cities (SDG11), and responsible consumption and production (SDG12), also rely largely on technology. Interestingly, "softer," socially oriented goals also include technological components. For example, gender equality (SDG5) specifically calls for "the use of enabling technology, in particular information and communications technology target to promote the empowerment of women" (IAEG-SDGs 2016). In fact, only 4 out of the 17 SDGs do not seem to necessitate wide scale technological change or use of specific technologies to achieve progress. Furthermore, as mentioned previously, targets involving technology tend to be more interconnected.

Past work demonstrates that, even when the adoption of new, more efficient technologies is successful, behavioral and systemic responses can make actual benefits fall short of expectations. For example, a 50% increase in energy efficiency does not always lead to a 50% reduction in energy consumption. This is because increased efficiency tends to stimulate increased demand. Such increases in demand can subsequently reduce or even negate the saving potential of more sustainable technologies or products, a phenomenon generally known as rebound effect (Greening et al. 2000). Since many SDG targets rely on technology, it is important to consider the implications of rebound effects and if and how they might spill over from one goal to another and mediate interlinkages.

Table 10.1 Description of specific goals and targets for each of the sustainable development goals that directly rely on technological change for their achievement.

Sustainable development goal (SDG). In parentheses, the number of specific goals and targets.	Specific goals and targets explicitly requiring technological change for their achievement (adapted names).
SDG1 - No poverty (7)	None.
SDG2 - Zero hunger (8)	2.3. Increase agricultural productivity. 2.4. Ensure sustainable food production systems and implement resilient agricultural practices. 2.a. Increase investment, including technology development, to enhance agricultural productive capacity in developing countries.
SDG3 - Good health and well-being (13)	3.9. Reduce hazardous chemicals and air, water, and soil pollution and contamination.
SDG4 - Quality education (10)	None.
SDG5 - Gender equality (9)	5.b. Enhance the use of enabling technology, in particular information and communications technology, to promote the empowerment of women.
SDG6 - Clean water and sanitation (8)	6.1. Achieve universal and equitable access to safe and affordable drinking water. 6.3. Improve water quality by reducing pollution, eliminating dumping, and minimizing release of hazardous chemicals and materials, reducing the proportion of untreated wastewater and substantially increasing recycling and safe reuse. 6.4. Increase water-use efficiency across all sectors and ensure sustainable withdrawals and supply of freshwater. 6.a. Expand capacity-building support to developing countries in water- and sanitation-related activities and programs, including water harvesting, desalination, water efficiency, wastewater treatment, recycling, and reuse technologies.
SDG7 - Affordable and clean energy (5)	7.1. Ensure universal access to affordable, reliable, and modern energy services. 7.2. Increase substantially the share of renewable energy in the mix 7.3. Improve energy efficiency. 7. Facilitate access to clean energy research and technology, including renewable energy, energy efficiency, and advanced and cleaner fossil-fuel technology, and promote investment in energy infrastructure and clean energy technology. 7.b. Expand infrastructure and upgrade technology for supplying modern and sustainable energy services.
SDG8 - Decent work and economic growth (12)	8.2. Increase economic productivity through diversification, technological upgrading, and innovation. 8.4. Improve resource efficiency in consumption and production.

Continued

Table 10.1 *Continued*

Sustainable development goal (SDG). In parentheses, the number of specific goals and targets.	Specific goals and targets explicitly requiring technological change for their achievement (adapted names).
SDG9 - Industry, innovation, and infrastructure (8)	9.2. Promote inclusive and sustainable industrialization in developing countries 9.4. Upgrade infrastructure and retrofit industries to make them sustainable, with increased resource-use efficiency and greater adoption of clean and environmentally sound technologies and industrial processes. 9.5. Enhance scientific research; upgrade the technological capabilities of industrial sectors. 9.a. Facilitate sustainable and resilient infrastructure development in developing countries through enhanced technological and technological support. 9.b. Support technology development, research, and innovation in developing countries. 9.c. Increase access to information and communications technology and strive to provide universal and affordable access to the Internet in least developed countries.
SDG10 - Reduced inequalities (10)	None.
SDG11 - Sustainable cities and communities (10)	11.1. Ensure access for all to adequate, safe, and affordable housing and basic services and upgrade slums. 11.2. Provide access to safe, affordable, accessible, and sustainable transport systems. 11.6, Reduce the environmental impact of cities, including air quality and municipal and other waste management. 11.c. Support least developed countries, including technological assistance, in building sustainable and resilient buildings.
SDG12 - Responsible consumption and production (11)	12.2. Achieve the sustainable management and efficient use of natural resources. 12.3. Reduce food waste at the retail and consumer levels and reduce food losses along production and supply chains. 12.4. Achieve the environmentally sound management of chemicals and all wastes throughout their life cycle. 12.5. Reduce waste generation through prevention, reduction, recycling, and reuse. 12.a. Support developing countries to strengthen their technological capacity to move toward more sustainable patterns of consumption and production.
SDG13 - Climate action (5)	13.1. Strengthen resilience and adaptive capacity to climate-related hazards and natural disasters. 13.b. Raising capacity for effective climate change-related planning and management in least developed countries and small island developing states.

Table 10.1 *Continued*

Sustainable development goal (SDG). In parentheses, the number of specific goals and targets.	Specific goals and targets explicitly requiring technological change for their achievement (adapted names).
SDG14 - Life below water (10)	14.1. Prevent and significantly reduce marine pollution. 14.7. Foster the sustainable use of marine resources, including through sustainable management of fisheries and aquaculture. 14.a. Increase scientific knowledge, develop research capacity, and transfer marine technology.
SDG15 - Life on land (12)	None.
SDG16 - Peace, justice and strong institutions (12)	None.
SDG17 - Partnerships for the goals (19)	17.6. Enhance access to science, technology, and innovation and enhance knowledge-sharing, including through a global technology facilitation mechanism. 17.7. Promote the development, transfer, dissemination, and diffusion of environmentally sound technologies to developing countries on favorable terms. 17.8. Fully operationalize the technology bank and science, technology, and innovation capacity-building mechanism for least developed countries and enhance the use of enabling technology, in particular information and communications technology.

Source: Goals and targets are taken from the Inter-Agency and Expert Group on Sustainable Development Goal Indicators (2016).

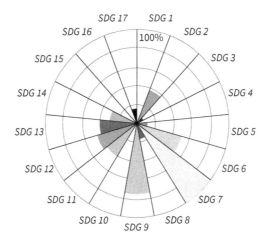

Figure 10.2 Extent to which each of the SDGs relies on technological change, based on the share of goals and targets within each SDG that explicitly require technological change according to the authors' criteria
Source: own elaboration

4. General Introduction to Rebound Effects

As illustrated in the previous section, interlinkages between SDGs are often mediated by technological change. For instance, increases in household appliances' energy efficiency, can create reinforcing effects for mitigating climate change (SDG13) through carbon emission reductions, while simultaneously facilitating access to energy services through lower operating costs (SDG7). The same technological change can however, mediate conflicting effects as well. For example, when more efficient appliances lower operating costs households' disposable income increases. Households can then use this access income to consumer more goods and services (SDG12) adding to overall environmental impacts, such as water and land use (SDG6 and SDG15). This paradox, by which energy efficiency induces additional demand for energy and other resources, has been studied in detail through the concept of rebound effects. While rebound effects are typically moderate, under extreme cases, rebound effects can also be known as "backfire effects" (Saunders 2000), that completely negate resource savings from efficiency improvements and jeopardise the achievement of sustainability targets. Therefore, a better understanding of how rebound effects mediate reinforcing and conflicting effects is critical to assess the overall effectiveness of technological change in meeting SDGs.

4.1 General Concept, Definitions, and Disciplinary Perspectives

The so-called rebound effect has its origins in the seminal works of the English economist William Stanley Jevons, particularly his much-cited book *The Coal Question* (Jevons 1865). Jevons argued that efficiency gains related to the use of coal by steam engines actually led to increased overall coal consumption in Britain, rather than a decrease as conventional wisdom would suggest. Such a seemingly counterintuitive argument was later on branded as the "Jevons' Paradox" (Giampietro and Mayumi 2008; Wirl 1997). In short, Jevons postulated that energy efficiency gains in steam engines would decrease the cost of producing steam by requiring less coal, leading to decreased price of all goods that required steam, such as steel, and rail transport. The decreased price of these goods would induce additional consumption, given the unsaturation in demand for such goods, the overall effect being the increased consumption of coal (see Figure 10.3).

Building on Jevons work, Brookes (1979) argued that energy savings calculations did not account for a number of economic aspects, such as shifts in production factors and energy-activity dependencies via prices. Similarly, Khazzoom (1980) postulated that efficiency improvements in household appliances led to de-facto reductions in the unit price of energy, which then spurred behavioral changes in the form of increased consumer demand for energy services, or a "rebound effect." Today, a widely accepted definition of the rebound effect is the reduction in the expected energy savings when

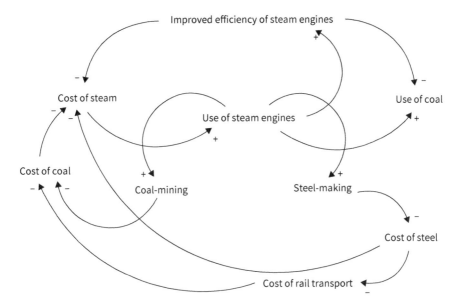

Figure 10.3 Feedbacks from the improved energy efficiency of steam engines
Source: Based on Sorrell (2009a).

the introduction of a technology that increases the energy efficiency of providing an energy service is followed by behavioral and systemic responses to changes in prices, income, and factors of production (Greening et al. 2000).

While energy economics is the cradle of the rebound effect, during the 1990s growing concerns over climate change led other disciplines to adopt the idea and expand on it. For example, in sustainability science, and especially industrial ecology, scholars coined the term "environmental rebound effect" (ERE) to refer to broader changes in environmental efficiency and the use of broader environmental metrics (Font Vivanco et al. 2016b). In ecological economics, some argue in favor of using metrics of energy quality such as exergy, on the basis that the increased availability of high quality energy has been an important driver behind economic growth in the past (Ayres and Warr 2005; Cleveland et al. 1986; 2000). Ecological economists have also stressed the importance of time use as a consumption factor, arguing that lower time costs (e.g., faster vehicles) can also lead to increased demand through a time rebound effect. For example, faster vehicles may increase the overall distance commuters are willing to travel (Jalas 2002). From a sociotechnological perspective, transition economics scholars poise that changes in technology could potentially introduce transformative changes to society. For example, a transformative technological change such as the internet could alter consumer preferences or social institutions, and rearrange the organization of production, leading to what they call, transformational or frontier rebound effects (Greening et al. 2000; Jenkins et al. 2011).

More recently, research in consumer behavior and behavioral economics has expanded the concept of rebound beyond economics to encompass so-called sociopsychological or mental rebound effects (Girod and de Haan 2009; de Haan et al. 2005; Santarius and Soland 2018). The main reasoning behind mental rebound is that consumption patterns have a normative basis, which defines their acceptable financial, social, and environmental costs. As such, when the environmental cost of a product or service dropped, consumers might feel that they have a "moral license" to consume more of it (Tiefenbeck et al. 2013). For example, Sun and Trudel (2017) show that when consumers can recycle paper, they tend to use more of it. Moreover, since consumers do not always know (let alone fully understand) the true costs of their consumption choices (Shepon et al. 2013), they might assume that purchasing sustainable products or services compensates for the purchase of less environmentally-friendly ones (Holmgren et al. 2018).

Given that the definition of rebound effect has been re-interpreted by many disciplines, a more comprehensive definition of the concept would be the environmental consequences of additional demand caused by behavioral and systemic responses to improvements in environmental efficiency mediated by technological change. The trade-offs of such a broader definition in terms of clarity, applications, and so on, are further discussed in Font Vivanco et al. (2016b).

4.2 Rebound Triggers, Drivers, Mechanisms, and Indicators

Rather than something one can measure empirically, the rebound effect is a construct which helps scholars, policymakers, and others gain a better understanding of the actual environmental consequences of technological change. Thus, to quantify rebound effects one must first isolate the impact technological change has on a given economic/environmental metric, from a myriad of other confounding variables, such as economic growth, policies, and social aspects. In addition, it is useful to consider different stages and the specific mechanisms that could lead to rebound effects (see Figure 10.4).

For example, the purchase of an electric vehicle (EV) can be motivated by the desire of drivers to lower their carbon emissions per km driven (efficiency change in Figure 10.4). When overall EV ownership costs are lower than those of conventional cars due to purchase subsidies and lower operational costs (Font Vivanco et al. 2016c), switching to EV could increase drivers' effective income (change in consumption factor in Figure 10.4). As drivers' effective income grows, they might increase their demand for car travel as well as other products (economic mechanisms in Figure 10.4), which will have consequences in terms of carbon emissions (indicators in Figure 10.4). Relatedly, switching to an EV can also be seen as a way to reduce air pollutants in cities, and thus rebound effects can be measured across multiple indicators (Font Vivanco et al. 2014). In the context of SDGs, the rebound effect framework

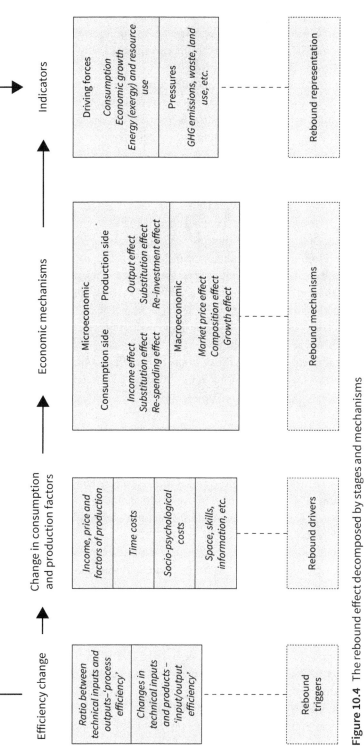

Figure 10.4 The rebound effect decomposed by stages and mechanisms
Source: own elaboration

can be used to analyze responses to various technical developments which go beyond energy efficiency. For example, SDG5 fosters the use of information and communications technology (ICT) to promote the empowerment of women, and ICT has been widely linked to rebound effects (Gossart 2015).

Rebound mechanisms have received particular attention, given that these are at the core of rebound modeling and allow to unravel the drivers behind additional demand. A generally accepted classification is provided by Greening et al. (2000), who divides rebound mechanisms according to two main axes. The first axis relates to the level of analysis, whether micro or macroeconomic. Microeconomic effects relate to short-term, isolated responses from economic agents, and are classified into direct and indirect effects. Direct effects relate to the increased/decreased demand for the product or process the efficiency of which has been changed, such as a household leaving the lights on for longer after buying energy-efficient light bulbs (Schleich et al. 2014). Indirect effects relate to other products the consumption of which increase/decrease as a result of the original technical change, such as increased consumption of transport as a result of the economic savings of energy-efficient lighting. Individual responses at the level of agents can induce further macroeconomic effects, namely market price, composition, and growth effects (Jenkins et al. 2011), for example when increased electricity demand causes a rise in the electricity price (Lemoine 2017). The second axis relates to whether the efficiency change takes place at the level of consumers or producers. Following the example of the efficient light bulbs, a producer could re-invest savings from reduced energy bills into additional capital.

4.3   State of the Art in Rebound Research

In recent years, research on rebound effects has been used to address complex sustainability issues involving economic, social, and natural systems (Font Vivanco et al. 2016b; Walnum et al. 2014). Following, we outline frontier research that can guide future analyses on pressing issues, including the role of rebound effects in mediating SDG interlinkages.

Recent applications of the ERE concept have unveiled the complex linkages between technological change and the use of multiple resources. For instance, energy efficiency improvements have been shown to cause backfire for some resources (but not others), such as minerals and water (Freire-González and Font Vivanco 2017). Similarly, new technologies, such as electric cars and high-speed trains, can also cause a wide range of rebound effects, from a limited offsetting to backfire, depending on the indicators measured (Font Vivanco et al. 2014; 2015). Such differences between footprint-type indicators are largely driven by resource use along supply chains, where resource intensity greatly differs between industries and such intensity is reflected in the product's footprint. Such supply-chain effects and resource intensity aspects are thus of great importance for the study of SDGs interlinkages. Moreover, developments in economic-environment models, such as life-cycle impact assessment

(LCA) and environmentally-extended input-output models, allow to gain insight into the impacts of rebound effects on human and ecosystem health (Weidema et al. 2008) and resource-specific rebounds, such as water (Berbel et al. 2015; Steduto et al. 2018) and material (Buhl 2016) rebound.

Another emerging research is the study of macro-economic rebound effects, which incorporate the economy-wide consequences of economic responses to changes in output and factor prices (Lemoine 2017). In particular, computable general equilibrium (CGE) models have been widely used for rebound analysis (Dimitropoulos 2007; Duarte et al. 2018), including the assessment of circular economy strategies (Winning et al. 2017), energy efficiency policies (Barker et al. 2007a; b), and bioenergy policies (Dandres et al. 2012). Recent advances in terms of the role of material stocks (Winning et al. 2017), increased technology detail (Bosello et al. 2011), and impact assessment methods (Igos et al. 2015), offer valuable research venues for rebound analysis. Regarding SDGs, macro-economic rebound effects could mediate interlinkages via prices and/or investment, such as energy efficiency improvements (SDG6) leading to constrained capital investments (Turner 2009) and decreased economic growth (SDG8), or additional energy demand and associated resources (SDG12) from decreased energy prices (Allan et al. 2007).

Lastly, increasing efforts are being dedicated to better understand human behavior beyond simple rational choice models and static assumptions about consumer preferences through the so-called mental rebound concept (Girod and de Haan 2010). This concept describes psychological mechanisms, which lead consumers to reappraise their responsibility, perceived control, and the consequences of their choices in response to technological changes (Santarius and Soland 2018). For example, an owner of an EV might feel entitled to environmentally-unfriendly behavior, such as buying an additional car for longer trips (Santarius and Soland 2018). Such moral licensing could also spill over from one domain to the other. Tiefenbeck et al. (2013) for example, found that while feedback on water consumption reduced household water use (SDG6), it led to an increase in energy consumption (SDG7). In contrast, spill overs could also have positive ramifications. For example, the purchase of sustainable products (SDG13) can increase consumers' environmental awareness and preferences (Suffolk and Poortinga 2016). Therefore, mental rebounds could drive reinforcing and conflicting linkages as well as mediate linkages arising from "pure" economic mechanisms.

5. Rebound Effects—Mediating Reinforcing and Conflicting SDG Linkages

As shown in section 2, SDGs make up dense and complex networks of reinforcing and conflicting linkages of varying degree and nature. Moreover, many SDGs require development and diffusion of both novel and existing technologies (see section 3). In turn, technological change could spur a myriad of rebound effects (see section 4), which in turn can reduce or negate the potential of achieving an SDG via technological change. In this section, we consolidate these three seemingly unrelated aspects to examine whether and to what extent rebound effects mediate conflicting linkages between SDGs and constrain progress toward desired outcomes.

Due to the amount and complexity of SDG and their interlinkages, the wide range of technologies involved, and the various types of rebound effects that can emerge, we limit our current analysis to prime examples where SDGs linkages can be more easily linked to rebound effects. To this end, we focus on SDGs that are both central in terms of interlinkages influence (Figure 10.1) and rely heavily on technological change (Figure 10.2): SDG6, water and sanitation for all (very high degree, eigenvector, closeness, and betweenness centrality and high technology reliance), SDG7, energy for all (high degree, closeness, and betweenness centrality and very high technology reliance), and SDG12, sustainable consumption and production (high degree, closeness, and betweenness centrality and high technology reliance). It is important to note that Figure 10.1 refers to average metrics from nine Asian countries while Figure 10.2 contains subjective information about the reliance of each SDG on technology.

5.1 SDG6: Water and Sanitation for All

Ensuring the availability and sustainable management of water and sanitation also requires technological solutions. As specifically outlined in targets 6.3 and 6.a (see Table 10.1), ensuring access to clean and safe water requires improving water efficiency across all sectors, as well as broader deployment of desalination and secondary (i.e., postconsumer) water treatment technologies. Yet efficiency improvements or the adoption of new technologies that increase overall water supply (e.g., desalination) can lead to rebound effects.

In agriculture, for example, researchers have identified several rebound mechanisms including expansion of irrigated acreage (Scheierling et al. 2006), a shift to more intensive and higher value crops (Ward and Pulido-Velazquez 2008), and an increased number of irrigation events (Contor and Taylor 2013; see Berbel et al. 2015 for a comprehensive review). Thus, due to rebound effects, efficiency improvements in water management do not necessarily lead to an equal decline in absolute water use. However, when such responses are considered in the broader context of SDG, water efficiency savings can be seen as a net addition to overall water supply, with positive implications for food production and security (SDG2) as well as economic growth (SDG8). Conversely, greater water availability in agriculture could also lead to expansion of agricultural land use and negatively impact biodiversity (SDG15, terrestrial ecosystems and biodiversity).

Rebound effects could also limit the potential to reduce domestic water consumption. Yet the extent to which rebound influences water savings behavior in residential settings has not been thoroughly examined. In theory, rebound could emerge in response to cost reductions (e.g., lower water bills) or changes in consumer attitude toward water usage. For example, greywater systems for water reuse or high efficiency washing machines typically lower consumers' water bills (Friedler and Hadari 2006) (Davis 2008). Such reductions in cost could then increase water consumption or lead to re-spending rebound.

Rebound could also emerge in response to wide-scale adoption of technologies that address shortages in water supply (e.g., desalination, water reuse). Specifically, greater availability of water supply could weaken public perceptions of water scarcity and lead to a mental rebound diminishing consumers' motivation to engage in water saving behavior. For example, after years of public media campaigns depicting water frugality as consumers' civic duty, over the past decade, the state of Israel has shifted focus from minimizing demand to increasing supply through the construction of several large sea water desalination plants. Today, sea water desalination is the country's single largest freshwater source. Yet as supply has grown, per capita demand for water has also steadily increased from 234 m³/cap to 257 m³/cap in 2016 (a 9% increase per capita overall) (Martin et al. 2017). According to Katz (2016, p. 1) "with the recent advent of large-scale desalination in Israel, public perception regarding the importance of conservation has diminished and consumption has increased—this, despite periodic drought conditions and critically low levels of water reserves." In other words, it would seem that the rise in water consumption steams, at least in part, from a mental rebound effect where the adoption of desalination reduced the efficacy of water conservation appeals.

5.2 SDG7: Energy for All

SDG7 requires substantial diffusion and development of a wide range of technologies, primarily an increase in renewable energies (goal 7.2) and improved energy efficiency (goal 7.3). Substantial evidence indicates that energy efficiency, in areas such as transport, lighting, heating, and cooling, leads to rebound effects in terms of energy use and carbon emissions (Greening et al. 2000; Jenkins et al. 2011; Sorrell 2007). While most estimates suggest that rebound effects do not completely offset energy savings and are in the range of 10 to 30% (Greening et al. 2000; Sorrell 2007), some studies applying approaches outside traditional energy economics report backfire effects where rebound completely negates all efficiency savings (Font Vivanco 2016; Font Vivanco et al. 2016b; Sorrell 2009).

Typically, energy related rebound stems from mechanisms related to increased effective income and mental rebound (see section 4.2). For example, energy-saving lightbulbs lower the costs of illuminating households. As a result, consumers may become more careless and abandon routine energy saving behaviors such as switching the light off (Frondel and Lohmann 2011). Alternatively, lower costs could reduce consumers electricity bills, allowing them to re-spend what they saved on other energy services (Fouquet and Pearson 2012). While such additional spending increases emissions (goal 13.1) and resource use (goal 12.2), it can also reflect improvements in energy access (goal 7.1), and health and well-being (SDG3) especially in developing countries where demand for illumination is unsaturated (van den Bergh 2011). As such, rebound effects related to energy efficiency can have a positive impact of progress toward other SDGs. Additional examples of mediating rebound effects from energy efficiency improvements are shown in Figure 10.5. Much like energy savings, when renewable energies are cheaper than conventional ones, income-related rebound effects (i.e., re-spending) can emerge. In addition, since renewable energies are seen as "greener," they can also lead to mental rebound effects (Pehnt 2006; Santarius 2012; Santarius and Soland 2018).

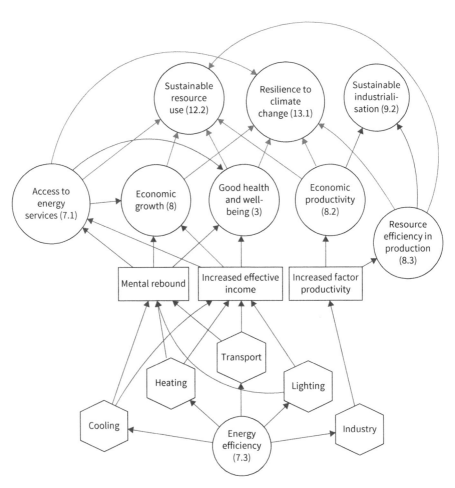

Figure 10.5 Example of reinforcing and conflicting linkages between SDGs (number within brackets) mediated by rebound effects in response to energy efficiency improvements. Blue and red arrows indicate, respectively, reinforcing and conflicting relationships.

Source: own elaboration

5.3 SDG12: Responsible Consumption and Production

SDG12 requires various technological solutions to increase material efficiency, reduce food waste, and increase recycling and reuse. Here, too, rebound effects may limit the ability to reach desired targets. For example, greater utilization of (cheaper) recycled materials in packaging can result in a price rebound effect, where companies are able to increase their output, or re-invest in additional production (Dace et al. 2014). As a result, some expected benefits of greater material efficiency, and specifically recycling, might be negated. Relatedly, the availability of economically-favorable recycled construction materials (e.g., recycled concrete), could encourage construction of larger buildings (Bahn-Walkowiak et al. 2012). While larger, more cost-effective buildings could help progress toward safer and more affordable housing (SDG11, sustainable cities and communities), such progress may come at the expense of an overall increase in resource use and at a considerable carbon cost (Mahasenan et al. 2003).

Technology can also aid in reducing food waste, for example via optimized packaging and refrigeration (Parfitt et al. 2010). Food waste reduction can however lead to economic savings which could then be used toward other purchases (Bernstad Saraiva Schott and Andersson 2015; Papargyropoulou et al. 2014). Estimates suggest that rebound effects from household food waste reduction could be as high as 50–60% for carbon emissions (Druckman et al. 2011; Salemdeeb et al. 2017). Since food waste is more typical in high-income households (Parfitt et al. 2010), however, the positive implications for poverty alleviation (SDG1) and improved nutrition (SDG2) are likely limited. Therefore, overlooking potential rebound effects could lead policy makers to grossly overestimate expected progress toward SDG targets.

Finally, in recent years the internet has revolutionized markets for used goods and lowered transaction costs for both sellers and buyers (Makov et al. 2019). Product reuse is considered a key component of sustainable consumption and production and is commonly assumed to lower demand for new products and subsequently their production. Yet there is evidence to suggest that the benefits of reuse could similarly fall short of expectations. First, despite concerns that sales of non-new products (e.g., used, refurbished) can cannibalize sales of new ones, research suggests that such 1:1 substitution is likely limited in scope (Cooper and Gutowski 2017; Frota Neto et al. 2016; Guide Jr. and Li 2010; Thomas 2003). Instead, lower prices in secondary markets might encourage consumers to purchase products they wouldn't have been able to afford otherwise. Relatedly, consumers selling their used possessions can then re-spend the proceeds earned to purchase new products (Chu and Liao 2010). Makov and Font Vivanco (2018), for example, estimate that rebound effects resulting from second hand smartphone markets in the United States amount to 29% on average, and could reach as much as 46% for some smartphone models. In other product categories, reuse may even lead to an overall increase in consumption (Ghose et al. 2006;

Thomas 2003; Zink and Geyer 2017). Altogether, rebound effects related to reuse could curb the potential of advancing resource and energy efficiency goals.

6. Concluding Remarks

In its current setting, the SDG framework lacks a systems perspective which reflects potential interlinkages between goals and targets. This flaw limits the ability to design and implement internally consistent strategies to promote effective sustainable development. A growing body of work indicates that there are multiple conflicting and reinforcing interlinkages between the SDGs. Although several of the most interlinked SDGs rely heavily on technological change, only limited attention has been given to the role technology might play in mediating such linkages. Rebound research offers a rich array of theories, tools, and methods, and provides a useful framework to identify and characterize conflicting and reinforcing SDG linkages, specify their underlining mechanisms, and assess the nature and magnitude of their impacts. As such, applying rebound theories and methods in the context of the SDG framework could improve our understanding of SDG interlinkages, help investigate the consistency of the framework as a whole, and assist in identifying appropriate management strategies to mitigate undesired effects.

For example, to address environmental spill-overs caused by rebound effects, the International Council for Science and the International Social Science Council (2015, p. 10) suggests to associate goals with a resource intensity target, "so that implementing the goal did not undermine targets in the climate or other environmentally-related goals." A better understanding of rebound effects could help quantify such resource intensity targets by estimating the environmental savings that are "taken back." Given the effectiveness of economic instruments, such as taxes and trading schemes, in limiting resource use and mitigating rebound effects (Font Vivanco et al. 2016a), it is recommended that such tools be considered in specific goals and targets to ensure the overall consistency of the SDG framework. For example, SDG8 (economic growth, employment, and decent work) could include additional targets for taxes and other economic instruments oriented to mitigate rebound. Even more, the quantification of economy-wide rebound effects has proven useful to determine industry-wide resource taxes necessary to avoid rebound, such as energy taxes in response to energy productivity increases (Saunders 2013).

The many conflicting linkages caused via rebound effects pose complex challenges to those tasked with implementation of the SDG framework. Particularly, ensuring sustainable levels of resource use will likely require additional economic resources adding to the already significant price tag of US$90 trillion over 15 years (SDG-FL 2017). In cases where conflicting rebound interlinkages are so strong that targets cannot be simultaneously achieved, policymakers may be forced to prioritize between environmental, economic, and social goals. In contrast, in other cases rebound effects driven via reinforcing mechanisms, could create a positive spillover and simultaneously advance progress towards several goals and targets. In such cases, policy makers should focus and promote the activation of rebound mechanisms as valuable tools promoting the overall success of the SDG framework.

References

Allan, G., Hanley, N., McGregor, P., Swales, K., and Turner, K. (2007). The impact of increased efficiency in the industrial use of energy: a computable general equilibrium analysis for the United Kingdom. *Energy Economics* 29(4), 779–798.

Allen, C., Metternicht, G., and Wiedmann, T. (2018). Prioritising SDG targets: assessing baselines, gaps and interlinkages. *Sustainability Science* 1–18.

Ayres, R. U., and Warr, B. (2005). Accounting for growth: the role of physical work. *Structural Change and Economic Dynamics* 16(2), 181–209.

Bahn-Walkowiak, B., Bleischwitz, R., Distelkamp, M., and Meyer, M. (2012). Taxing construction minerals: a contribution to a resource-efficient Europe. *Mineral Economics* 25(1), 29–43.

Barker, T., Ekins, P., and Foxon, T. (2007a). Macroeconomic effects of efficiency policies for energy-intensive industries: the case of the UK climate change agreements, 2000–2010. *Energy Economics* 29(4), 760–778.

Barker, T., Ekins, P., and Foxon, T. (2007b). The macro-economic rebound effect and the UK economy. *Energy Policy* 35(10), 4935–4946.

Berbel, J., Gutiérrez-Martín, C., Rodríguez-Díaz, J. A., Camacho, E., and Montesinos, P. (2015). Literature review on rebound effect of water saving measures and analysis of a spanish case study. *Water Resources Management* 29(3), 663–678.

Bernstad Saraiva Schott, A., and Andersson, T. (2015). Food waste minimization from a life-cycle perspective. *Journal of Environmental Management* 147, 219–226.

Bosello, F., Campagnolo, L., Eboli, F., Parrado, R., and Portale, E. (2011). Extending energy portfolio with clean technologies in the ICES model. *14th Annual Conference on Global Economic Analysis, Venice*, Citeseer, 16–18.

Brady, J., and O'Mahony, M. (2011). Travel to work in Dublin: the potential impacts of electric vehicles on climate change and urban air quality. *Transportation Research Part D: Transport and Environment* 16(2), 188–193.

Brookes, L. (1979). A low energy strategy for the UK. *Atom* 269, 3–8.

Buhl, J. (2016). Indirect Effects from Resource Sufficiency Behaviour in Germany. In: Santarius, T., Walnum, H., and Aall, C. (Eds.), *Rethinking Climate and Energy Policies. New Perspectives on the Rebound Phenomenon*, PP. 37–54. Springer International Publishing, Basel.

Chu, H., and Liao, S. (2010). Buying while expecting to sell: The economic psychology of online resale. *Journal of Business Research* 63(9–10), 1073–1078.

Cleveland, C. J., Hall, C. A. S., and Kaufmann, R. (1986). Energy and the US Economy: A biophysical perspective. *Science* 225(4665), 890–897.

Cleveland, C. J., Kaufmann, R. K., and Stern, D. I. (2000). Aggregation and the role of energy in the economy. *Ecological Economics* 32(2), 301–317.

Contor, B. A., and Taylor, R. G. (2013). Why improving irrigation efficiency increases total volume of consumptive use. *Irrigation and Drainage* 62(3), 273–280.

Cooper, D. R., and Gutowski, T. G. (2017). The environmental impacts of reuse: a review. *Journal of Industrial Ecology* 21(1), 38–56.

Dace, E., Bazbauers, G., Berzina, A., and Davidsen, P. I. (2014). System dynamics model for analyzing effects of eco-design policy on packaging waste management system. *Resources, Conservation and Recycling* 87(0), 175–190.

Dandres, T., Gaudreault, C., Tirado-Seco, P., and Samson, R. (2012). Macroanalysis of the economic and environmental impacts of a 2005–2025 European Union bioenergy policy using the GTAP model and life cycle assessment. *Renewable and Sustainable Energy Reviews* 16(2), 1180–1192.

Davis, L. W. (2008). Durable goods and residential demand for energy and water: evidence from a field trial. *RAND Journal of Economics* 39(2), 530–546.

de Haan, P., Mueller, M. G., and Peters, A. (2005). Does the hybrid Toyota Prius lead to rebound effects? analysis of size and number of cars previously owned by Swiss Prius buyers. *Ecological Economics* 58(3), 592–605.

Dimitropoulos, J. (2007). Energy productivity improvements and the rebound effect: an overview of the state of knowledge. *Energy Policy* 35(12), 6354–6363.

Druckman, A., Chitnis, M., Sorrell, S., and Jackson, T. (2011). Missing carbon reductions? exploring rebound and backfire effects in UK households. *Energy Policy* 39(6), 3572–3581.

Duarte, R., Sánchez-Chóliz, J., and Sarasa, C. (2018). Consumer-side actions in a low-carbon economy: a dynamic CGE analysis for Spain. *Energy Policy* 118, 199–210.

Font Vivanco, D. (2016). The rebound effect through industrial ecology's eyes: the case of transport eco-innovation. PhD diss. Institute of Environmental Sciences (CML), Faculty of Science, Leiden University.

Font Vivanco, D., Freire-González, J., Kemp, R., and Van Der Voet, E. (2014). The remarkable environmental rebound effect of electric cars: a microeconomic approach. *Environmental Science and Technology* 48(20), 12063–12072.

Font Vivanco, D., Kemp, R., and Van Der Voet, E. (2015). The relativity of eco-innovation: environmental rebound effects from past transport innovations in Europe. *Journal of Cleaner Production* 101, 71–85.

Font Vivanco, D., Kemp, R., and van der Voet, E. (2016a). How to deal with the rebound effect? a policy-oriented approach. *Energy Policy* 94, 114–125.

Font Vivanco, D., McDowall, W., Freire-González, J., Kemp, R., and van der Voet, E. (2016b). The foundations of the environmental rebound effect and its contribution towards a general framework. *Ecological Economics* 125, 60–69.

Font Vivanco, D., Tukker, A., and Kemp, R. (2016c). Do methodological choices in environmental modeling bias rebound effects? a case study on electric cars. *Environmental Science and Technology* 50 (20), 11366–11376.

Fouquet, R., and Pearson, P. J. G. (2012). The long run demand for lighting: elasticities and rebound effects in different phases of economic development. *Economics of Energy and Environmental Policy* 1(1), 83–100.

Freire-González, J., and Font Vivanco, D. (2017). The influence of energy efficiency on other natural resources use: an input-output perspective. *Journal of Cleaner Production* 162, 336–345.

Friedler, E., and Hadari, M. (2006). Economic feasibility of on-site greywater reuse in multi-storey buildings. *Desalination* 190(1–3), 221–234.

Frondel, M., and Lohmann, S. (2011). The European Commission's light bulb decree: another costly regulation? *Energy Policy* 39(6), 3177–3181.

Frota Neto, J. Q., Bloemhof, J., and Corbett, C. (2016). Market prices of remanufactured, used and new items: evidence from eBay. *International Journal of Production Economics* 171, 371–380.

Ghose, A., Smith, M. D., and Telang, R. (2006). Internet exchanges for used books: an empirical analysis of product cannibalization and welfare impact. *Information Systems Research* 17(1), 3–19.

Giampietro, M., and Mayumi, K. (2008). The Jevons paradox: the evolution of complex adaptive systems and the challenge for scientific analysis. In: Polimeni, J. M., Mayumi, K., Giampietro, M., and Alcott, B. (Eds.), *The Jevons Paradox and the Myth of Resource Efficiency Improvements*, pp. 79–140. Earthscan, London.

Gill, A. B. (2005). Offshore renewable energy: ecological implications of generating electricity in the coastal zone. *Journal of Applied Ecology* 42(4), 605–615.

Girod, B., and de Haan, P. (2009). *Mental Rebound* (Rebound research report no. 3). Eidgenössische Technische Hochschule (ETH) Zürich.

Girod, B., and de Haan, P. (2010). More or better? a model for changes in household greenhouse gas emissions due to higher income. *Journal of Industrial Ecology* 14(1), 31–49.

Gossart, C. (2015). Rebound effects and ICT: a review of the literature. *ICT Innovations for Sustainability* 310, 435–448.

Greening, A., Greene, D. L., and Difiglio, C. (2000). Energy efficiency and consumption: the rebound effect- a survey. *Energy Policy* 28(6–7), 389–401.

Guide Jr., V. D. R., and Li, J. (2010). The potential for cannibalization of new products sales by remanufactured products. *Decision Sciences* 41(3), 547–572.

Hochman, G., Rajagopal, D., and Zilberman, D. (2010). The effect of biofuels on crude oil markets. *AgBioForum* 13(2), 112–118.

Holmgren, M., Andersson, H., and Sörqvist, P. (2018). Averaging bias in environmental impact estimates: evidence from the negative footprint illusion. *Journal of Environmental Psychology* 55, 48–52.

Igos, E., Rugani, B., Rege, S., Benetto, E., Drouet, L., and Zachary, D. S. (2015). Combination of equilibrium models and hybrid life cycle-input–output analysis to predict the environmental impacts of energy policy scenarios. *Applied Energy* 145, 234–245.

Inter-Agency and Expert Group on Sustainable Development Goals Indicators (IAEG-SDGs) (2016). *Report of the Inter-Agency and Expert Group on Sustainable Development Goal Indicators*. United Nations, New York.

International Council for Science and the International Social Science Council (ICSU-ISSC) (2015). *Review of the Sustainable Development goals: The Science Perspective*. International Council for Science (ICSU), Paris.

Jalas, M. (2002). A time use perspective on the materials intensity of consumption. *Ecological Economics* 41(1), 109–123.

Jenkins, J., Nordhaus, T., and Shellenberger, M. (2011). *Energy Emergence: Rebound and Backfire as Emergent Phenomena*. Breakthrough Institute, Oakland, CA.

Jevons, W. S. (1865). *The Coal Question. An Inquiry Concerning the Progress of the Nation and the Probable Exhaustion of Our Coal-Mines*. Macmillan, Cambridge.

Katz, D. (2016). Undermining demand management with supply management: moral hazard in israeli water policies. *Water* 8(4), 159.

Khazzoom, J. D. (1980). Economic implications of mandated efficiency in standards for household appliances. *Energy Journal* 1(4), 21–40.

Le Blanc, D. (2015). Towards integration at last? the sustainable development goals as a network of targets. *Sustainable Development* 23(3), 176–187.

Lemoine, D. (2017). General equilibrium rebound from improved energy efficiency. University of Arizona Working Paper, Tucson.

Mahasenan, N., Smith, S., and Humphreys, K. (2003). The Cement Industry and Global Climate Change: Current and Potential Future Cement Industry CO2 Emissions. In: *Greenhouse Gas Control Technologies - 6th International Conference*, pp. 995–1000.

Makov, T., and Font Vivanco, D. (2018). Does the circular economy grow the pie? the case of rebound effects from smartphone reuse. *Frontiers in Energy Research* Frontiers 6: 39, 1–11.

Makov, T., Fishman, T., Chertow, M., and Blass. V. (2018). What affects the second-hand value of smartphones: evidence from eBay. *Journal of Industrial Ecology* 23, 176–187.

Martin, P., Tal, S., Yeres, J., and Ringskog, K. (2017). *Water Management in Israel*. International Bank for Reconstruction and Development / The World Bank, Washington, DC.

McCollum, D. L., Echeverri, L. G., Busch, S., Pachauri, S., Parkinson, S., Rogelj, J., Krey, V., Minx, J. C., Nilsson, M., Stevance, A.-S., and Riahi, K. (2018). Connecting the sustainable development goals by their energy inter-linkages. *Environmental Research Letters* 13(3), 033006.

Murphy, R., Woods, J., Black, M., and McManus, M. (2011). Global developments in the competition for land from biofuels. *Food Policy* 36, S52–S61.

Nilsson, M., Griggs, D., and Visbeck, M. (2016). Policy: map the interactions between sustainable development goals. *Nature* 534(7607), 320–322.

Nonhebel, S. (2005). Renewable energy and food supply: will there be enough land? *Renewable and Sustainable Energy Reviews* 9(2), 191–201.

Papargyropoulou, E., Lozano, R., K. Steinberger, J., Wright, N., and Ujang, Z.B. (2014). The food waste hierarchy as a framework for the management of food surplus and food waste. *Journal of Cleaner Production* 76, 106–115.

Parfitt, J., Barthel, M., and Macnaughton, S. (2010). Food waste within food supply chains: quantification and potential for change to 2050. *Philosophical transactions of the Royal Society of London. Series B, Biological Sciences. The Royal Society* 365(1554), 3065–3081.

Pehnt, M. (2006). Dynamic life cycle assessment (LCA) of renewable energy technologies. *Renewable energy* 31(1), 55–71.

Salemdeeb, R., Font Vivanco, D., Al-Tabbaa, A., and zu Ermgassen, E. K. H. J. (2017). A holistic approach to the environmental evaluation of food waste prevention. *Waste Management* 59, 442–450.

Santarius, T. (2012). Green Growth Unravelled. How Rebound Effects Baffle Sustainability Targets When the Economy Keeps Growing. Heinrich Boell Foundation and Wuppertal Institute for Climate, Environment and Energy, Berlin.

Santarius, T., and Soland, M. (2018). How technological efficiency improvements change consumer preferences: towards a psychological theory of rebound effects. *Ecological Economics* 146, 414–424.

Saunders, H. D. (2000). A view from the macro side: rebound, backfire, and Khazzoom-Brookes. *Energy Policy* 28(6–7), 439–449.

Saunders, H. D. (2013). Historical evidence for energy efficiency rebound in 30 US sectors and a toolkit for rebound analysts. *Technological Forecasting and Social Change* 80(7), 1317–1330.

Scheierling, S. M., Young, R. A., and Cardon, G. E. (2006). Public subsidies for water-conserving irrigation investments: hydrologic, agronomic, and economic assessment. *Water Resources Research* 42(3). doi.org/10.1029/2004WR003809

Schleich, J., Mills, B., and Dütschke, E. (2014). A brighter future? quantifying the rebound effect in energy efficient lighting. *Energy Policy* 72, 35–42.

SDG-FL. (2017). Opening remarks of H.E. Peter Thomson, President of the UN General Assembly at High Level SDG Action Event "SDG Financing Lab." SDG Financing Lab. United Nations, New York.

Shepon, A., Israeli, T., Eshel, G., and Milo, R. (2013). EcoTime: an intuitive quantitative sustainability indicator utilizing a time metric. *Ecological Indicators* 24, 240–245.

Sorrell, S. (2007). The Rebound Effect: An Assessment of the Evidence for Economy-Wide Energy Savings from Improved Energy Efficiency. Project Report. UK Energy Research Centre, London.

Sorrell, S. (2009). Jevons paradox revisited: the evidence for backfire from improved energy efficiency. *Energy Policy* 37(4), 1456–1469.

Steduto, P., Wang, Y., Williams, J., Molle, F., Allen, R. G., Garrick, D., Wheeler, S. A., Ringler, C., Udall, B., Perry, C. J., and Grafton, R. Q. (2018). The paradox of irrigation efficiency. *Science* 361(6404), 748–750.

Suffolk, C., and Poortinga, W. (2016). Behavioural changes after energy efficiency improvements in residential properties. In: Santarius, T., Walnum, H., and Aall, C. (Eds.), *Rethinking Climate and Energy Policies*, pp. 121–142. Springer International Publishing, Cham.

Sun, M., and Trudel, R. (2017). The effect of recycling versus trashing on consumption: theory and experimental evidence. *Journal of Marketing Research* 54(2), 293–305.

Thomas, V. M. (2003). Demand and dematerialization impacts of second-hand markets. *Journal of Industrial Ecology* 7(2), 65–78.

Tiefenbeck, V., Staake, T., Roth, K., and Sachs, O. (2013). For better or for worse? empirical evidence of moral licensing in a behavioral energy conservation campaign. *Energy Policy* 57, 160–171.

Turner, K. (2009). Negative rebound and disinvestment effects in response to an improvement in energy efficiency in the UK economy. *Energy Economics* 31(5), 648–666.

van den Bergh, J. J. M. (2011). Energy conservation more effective with rebound policy. *Environmental and Resource Economics* 48(1), 43–58.

Walnum, H., Aall, C., and Løkke, S. (2014). Can rebound effects explain why sustainable mobility has not been achieved? *Sustainability* 6(12), 9510–9537.

Ward, F. A., and Pulido-Velazquez, M. (2008). Water conservation in irrigation can increase water use. *Proceedings of the National Academy of Sciences of the United States of America* 105(47), 18215–18220.

Weidema, B. P., Wesnaes, J., Hermansen, J., Kristensen, T., and Halberg, N. (2008). *Environmental Improvement Potentials of Meat And Dairy Products.* JRC Scientific and Technical Reports, (P. Eder and L. Delgado, eds.). European Commission, Luxembourg.

Winning, M., Calzadilla, A., Bleischwitz, R., and Nechifor, V. (2017). Towards a circular economy: insights based on the development of the global ENGAGE-materials model and evidence for the iron and steel industry. *International Economics and Economic Policy* 14(3), 383–407.

Wirl, F. (1997). *The Economics of Conservation Programs.* Kluwer Academics Publishers, Dordrecht, The Netherlands.

Zhou, X., and Moinuddin, M. (2017). *Sustainable Development Goals Interlinkages and Network Analysis: A Practical Tool for SDG Integration and Policy Coherence.* Institute for Global Environmental Strategies, Hayama, Kanagawa, Japan.

Zink, T., and Geyer, R. (2017). Circular economy rebound. *Journal of Industrial Ecology* 21(3), 593–602.

THEME II

HEALTH

11

Vaccine Innovation and Global Sustainability

Governance Challenges for Sustainable Development Goals

Cristina Possas, Reinaldo M. Martins, and Akira Homma

1. Introduction

Vaccines are recognized in international scientific publications as the preventive medical measures of greater impact on health, being more cost-effective and with very low risk (Bregu et al. 2011; Greenwood et al. 2011). This impact has been confirmed by data provided by National Immunization Programs worldwide and by international specialized organizations indicating sharp reduction of mortality, particularly in the poorest developing countries. In consequence, human vaccines have already been included as a crucial component in sustainable development goals (SDGs) (United Nations Development Programme 2016), and there is evidence indicating that they play a key role in meeting several of these goals, as we will study in this chapter. A report from the Centers for Disease Control and Prevention (CDC) described the benefits of vaccination of a 2009 birth cohort in the United States through 18 years of age and estimated that 20 million cases of vaccine-preventable disease will not occur, 42 000 early deaths related to these diseases will be avoided, and $76 billion in direct and indirect costs will be averted (Centers for Disease Control and Prevention 2009; Meissner 2016; Zhou et al. 2014). These findings in the United States, a country with adequate sanitation infrastructure and an organized health system, indicate that this impact could be significantly greater in the developing countries deprived of these basic resources. The economic benefit identified by CDC contrasts to the comparatively small costs for vaccine purchases: pricing is $7.8 billion, based on CDC costs, and $11.6 billion at private sector pricing. Considering the high effectiveness and high cost-benefit of immunizations and aiming to expand and accelerate access to immunization strategies to the world population, new global immunization programs have emerged, supported by the World Health Organization (WHO), Pan-American Health Organization (PAHO), United Nations Children's Fund (UNICEF), Global Alliance for Vaccines and Immunization (GAVI), national governments, and international nongovernmental organizations.

The millennium development goals (MDGs) for 2000–2015 (United Nations 2015) were incorporated by governments worldwide and had a strong global mobilization power on promoting development and social initiatives, engaging national leaders in elaborating and monitoring these goals. This mobilization was facilitated since the targets were quantifiable and could potentially be attained. Although the two health related goals, MDG4 (reduce under-five mortality from 1990 to 2015 by two-thirds) and MDG5 (reduce maternal mortality from 1990 to 2015 by three-quarters) had not been met by 2015, significant progress has been made and child and maternal mortality approximately halved. In spite of these achievements, 19,4 million infants worldwide are still missing out on basic vaccines, Following MDGs, the United Nations formulated in 2015 a new global strategy, SDGs for 2016–2030 (United Nations 2015), with 17 goals, of which only one (SDG3) is directly related to health.

The target of SDG3 is to "ensure healthy lives and promote well-being for all at all ages" (panel 1). Its 13 subtargets include two ones that could be met by national governments: two-thirds less maternal mortality and a third less noncommunicable disease (NCD) mortality, all of them strongly related to vaccines, as indicated in this chapter. These subtargets also include ending preventable newborn and under-five deaths; ending HIV/AIDS, tuberculosis, malaria, and neglected tropical diseases, besides other nonvaccine related subtargets. In this chapter, we argue that a major component can be crucial for attaining SDG3 and other 13 vaccine-related SDGs and should not be minimized; that is, innovation, technological development, and production of vaccines. We discuss how this component should be incorporated into monitoring the subtargets of these goals.

Scientific advances have emerged in diverse biotechnology-related fields, such as nanotechnology and gene editing, which could contribute to the development of new vaccines, new diagnostic tools, and biopharmaceuticals designed to meet global public health challenges: Zika, dengue, influenza, HIV/AIDS, and other neglected diseases affecting particularly the developing countries. In this scenario, breakthroughs in fields such as immunology, virology, and genetics have contributed to the development of new vaccine technologies against several infectious diseases with enormous impact on life expectancy and quality of life of the populations at a global scale, particularly in the developing countries.

The growing importance of immunotherapy and in particular a therapeutic vaccine technology are crucial in the search for a cure for HIV/AIDS, cancer, and other infectious and chronic diseases, such as Parkinson and Alzheimer's disease.

These breakthroughs are increasingly blurring the boundaries between preventive and therapeutic strategies, with a potential for a broad range of technological applications in new innovative vaccines (Possas et al. 2016). These advances have been accompanied by increasing global demands posed by demographic growth and an aging population in emerging economies.

We provide in this chapter an overview of the main research challenges in the direction of more effective vaccines with reduced adverse events, stressing the great potential impacts of innovative vaccines on global sustainability, benefiting particularly the poorest countries in a global context, permeated by sharp social inequalities. In this chapter we examine the progress of the MDGs and SDGs and the gaps in the context of global response to emerging and neglected infectious diseases. These diseases affect particularly the poorest countries, emphasizing governance issues related to innovation, technological development, and production of vaccines.

This chapter presents and discusses these research challenges in four related topics: 1) global demand for vaccines against emerging and neglected diseases, stressing the main challenges posed by the new global epidemiological scenario; 2) vaccine innovation and global sustainability, examining crucial issues in technological development and production of vaccines, particularly in developing countries; 3) MDGs and the progress and gaps in improving global access to vaccines, discussing the major constraints in access to immunization that have persisted in the MDGs strategy and the crucial role that SDGs can play in integrating STI into global immunization strategies; and 4) results of vaccine challenges for achieving SDG3, presenting and discussing vaccine performance indicators and the main global governance and regulatory challenges to meet them.

2. Materials and Methods

We provide an analytical evaluation of global vaccine performance indicators and science, technology, and innovation (STI) policy issues, based on updated information from international organizations' reports, published papers, and international databases. We also discuss current vaccine policy gaps and recommend STI governance strategies to overcome these gaps.

2.1 Global Demand for Vaccines Against Emerging and Neglected Diseases

Despite the social gains resulting from development and urbanization in many countries, including developing countries, and from the success of vaccines against infectious diseases and National Immunization Programs, many complex health problems persist and are challenging current SDGs. Changing conditions have favored the global emergence and resurgence of several infectious diseases with complex dynamic cycles, such as Zika, HIV/AIDS, influenza, tuberculosis, dengue, yellow fever, chikungunya, malaria, leishmaniases, leptospirosis, hantavirus pulmonary syndrome, Ebola, and many others could still show up (Possas et al. 2016; Røttingen et al. 2017).

The rapid spread of these diseases worldwide challenges the national health systems, particularly affecting developing countries like Brazil, a nation plagued by social exclusion and environmental degradation and with a rapidly aging population.

New complex and detrimental eco-social conditions have emerged, related to environmental change and to international population mobility favoring the dissemination of new pathogenic agents. These conditions have been aggravated by the intensification of international travel and by the increasing number of refugees from countries with endemic diseases and not covered by vaccination and have led in the last five decades to the unexpected phenomena of emergence and re-emergence of infectious diseases such as Ebola, Zika, dengue, and chikungunya, requiring innovative vaccines and vaccination strategies, and new therapeutic strategies for drug-resistant diseases (Possas et al. 2016).

Moreover, it has been estimated that around 60% of infectious and parasitic diseases are zoonose-related (WHO 2015), a scenario aggravated in the last decades by environmental change and mobility of vectors and wild animals to urban areas, favoring emergence and resurgence of infectious diseases.

From this perspective, new global scientific and technological initiatives supporting the development of new vaccines will be key to deal with emerging and re-emerging infectious diseases and to support global surveillance.

2.2 Vaccine Innovation and Global Sustainability: STI Issues

The global vaccine innovation landscape has undergone drastic changes in the past three decades. New innovative vaccines have emerged in the market, such as dengue, Zika, and Ebola vaccines, which has resulted in an increasing number of multipatented vaccines (Possas et al. 2015), with important implications for global sustainability. The global vaccine market is estimated at over US$30 billion (Possas et al, 2016). Although this segment represents a small part of the overall sales of the pharmaceutical market (about 3%), the growth rate of the vaccine market has been extraordinary, around 15% per year, while the global pharmaceutical industry is growing at rates around 7% per year. It is estimated that the global vaccine market is expected to reach about US$100 billion in 2025 (Possas et al. 2016). An important role is played by the start-ups, as these small biotech companies are expanding their participation in vaccine development and technology transfer (Homma and Possas 2014).

The human vaccine market is organized in an oligopolistic structure, with nearly 80% of global human vaccine sales coming from five large multinational corporations, represented by the International Federation of Pharmaceutical Manufacturers and Associations (IFPMA). These corporations have also increased their participation in developing countries' markets, playing an important role in Global Health Initiatives (Homma and Possas 2014), and focusing mainly on joint-development and technological transfer projects.

On the other hand, emerging countries' manufacturers, represented by the Developing Countries Vaccine Manufacturers Network (DCVMN), have increased their participation in the global market (with 80% of the world population they account for 20% of the market). The BRICS' countries manufacturers (Brazil, Russia, India, China, and South Africa) now play a crucial role in the supply of vaccines to developing countries, particularly for some basic and combination vaccines, and China is expected to play an increasingly important role in the global market (Kaddar et al. 2014). Now they supply about 70% of UNICEF's vaccine procurement in volume of doses and their increased participation in the developing countries' vaccine market has contributed to lower these vaccines' prices, due to increased competition.

The limited number of vaccine suppliers leads to a critical situation in the human vaccine markets and requires constant negotiations and management between the main players—manufacturers, governments and international organizations—in order to assure sufficient global supply of vaccines, particularly in the developing world. GAVI has promoted, through "market-shaping" initiatives, global mobilization and a great effort in reducing vaccine prices.

The global pharmaceutical market for vaccines is spiraling. The pharmaceutical multinationals are increasingly investing in new biotechnology: besides high investments in innovation and technological development of new drugs and vaccines, these enterprises are also rapidly increasing their market share, supporting new start-up companies. The international scenario in science and technology has therefore been marked by accelerated advancement in diverse aspects of biotechnology, and in different fields of human activity.

On the other hand, the increasing international demand for food and meat-derived products and the economic relevance of the global agribusiness have led to the rapid evolution of the veterinary vaccine market, with increasing cross-fertilization with the human vaccine market (Possas et al. 2016). Global surveillance on animal diseases is key to support decisions on meat imports worldwide. The World Organization for Animal Health (OIE), an intergovernmental organization responsible for improving global animal health, elaborates since 1990 a list of countries free of certain animal diseases, such as foot-and-mouth disease (or aphthous fever), and this list has important implications for market decisions related to meat import.

The veterinary vaccine market is a highly competitive oligopolistic market, integrated by few multinational enterprises, related in most cases to the human vaccine market. The capacity of innovation and market differentiation by veterinary vaccines' producers are key to assure entrepreneurial competitiveness and access to new quality products. The creation of healthy animals is critical for human public health, since there are several zoonosis related to pathogenic agents that can be transmitted to humans.

Biotechnology is defined as the application of the life sciences to chemical synthesis. Drugs produced by organic synthesis have been the main biotechnology products of the pharmaceutical industry in recent decades. Nevertheless, since the early 1980s, small start-up biotechnology enterprises began to conduct research on large molecules, derived from or modified from biological systems. New biotechnology-based products are currently the focus of many pharmaceutical companies, both in terms of research and development (R&D) and product sales. The importance of these biotechnology products can also be illustrated by the fact that 190 biotech companies were acquired by large pharmaceutical companies in the last three decades, at a cost of $393.3 billion (Homma and Possas 2014).

In developed countries, there is a clear division of labor related to position of the diverse actors in the flow from basic research to product development and production. The public sector is the main investor in basic science, generating new knowledge and discoveries, while the private sector concentrates in the activities related to product development. The big pharmaceutical companies invest 20% of net sales in product development ($2 billion annually), and 50% of the drugs and vaccines developed in the world are now biotechnology-based. They are also expanding their market, investing in the acquisition of consolidated companies (Kaddar et al. 2014), in an oligopolistic trend: Pfizer bought Wyeth; GSK bought Novartis/vaccines and sold its Oncology component for Novartis, in a strategy to strengthen and consolidate new areas of expertise; Bayer bought from Merck & Co. the division consumer goods and health for $14.2 billion (Homma and Possas 2014). This oligopolistic trend has important implications for global sustainability impacting on vaccine STI activities and access to vaccines by developing countries, as we will discuss later.

3. MDGs: Progress and Gaps in Improving Access to Vaccines

Successful international immunization strategies, such as the eradication of smallpox, the significant reduction of polio's incidence, and the rapid decrease in measles deaths illustrate well the global impact of vaccines (Homma et al. 2013). Immunization against vaccine-preventable diseases currently averts nearly 3 million deaths per year worldwide and additional 1.5 million deaths per year could be prevented by improving access to vaccines (WHO 2013), which might lead to a great impact on global mortality in the next two decades.

The 2015 MDGs Report (United Nations 2015) indicated this progress in attaining goals despite significant remaining challenges. MDG goal 4, "reduce child mortality," has been partially attained with the global number of deaths of children under five nearly halved, reduced from 12.7 million in 1990 to 6 million in 2015, dropping from 90 to 43 deaths -per 1,000 live births. Since the early 1990s, the reduction rate of under-five mortality has more than tripled globally. In sub-Saharan Africa, the annual rate of reduction of under-five mortality was over five times faster during 2005–2013 than it was during 1990–1995. The global measles vaccine coverage has increased from 73% in 2000 to 84% in 2013. Measles vaccination helped prevent nearly 15.6 million deaths between 2000 and 2013. The number of globally reported measles cases declined by 67% for the same period. About 84% of children worldwide received at least one dose of measles vaccine in 2013, up from 73% in 2000 (United Nations 2015).

MDG goal 5 has been also partially attained. Since 1990, the maternal mortality ratio has declined by 45% worldwide, and most of the reduction has occurred since 2000. In Southern Asia, the maternal mortality ratio declined by 64% between 1990 and 2013, and in sub-Saharan Africa it fell by 49%. Finally, MDG6 has also been partially attained. New HIV infections fell by nearly 40% between 2000 and 2013, from an estimated 3.5 million cases to 2.1 million. By June 2014, 13.6 million people living with HIV were receiving antiretroviral therapy (ART) globally, an exponential increase from just 800,000 in 2003. ART averted 7.6 million deaths from AIDS between 1995 and 2013. Over 6.2 million malaria deaths were averted between 2000 and 2015, primarily of children under five years of age in sub-Saharan Africa. The global malaria incidence rate has fallen by an estimated 37% and the mortality rate by 58%. More than 900 million insecticide-treated mosquito nets were delivered to malaria-endemic countries in sub-Saharan Africa between 2004 and 2014. Tuberculosis prevention, diagnosis and treatment interventions saved nearly 37 million lives between 2000 and 2013. The tuberculosis mortality rate fell by 45% and the prevalence rate has been reduced by 41% between 1990 and 2013 (WHO 2015; United Nations 2015).

Despite remarkable progress in partially meeting MDGs, many gaps persist according to WHO estimates (WHO 2012; 2015) and remain as a challenge for meeting the SDGs: respiratory infections, diarrheal diseases, HIV/AIDS, and tuberculosis remain among the major causes of world deaths from infectious and parasitic diseases (Figure 11.1), indicating that a great demand persists for developing new innovative vaccines against most of these diseases and also novel strategies for therapeutic vaccines and immunotherapy, considering the leading participation of chronic and degenerative diseases, such as cardiovascular diseases and cancer, in global mortality.

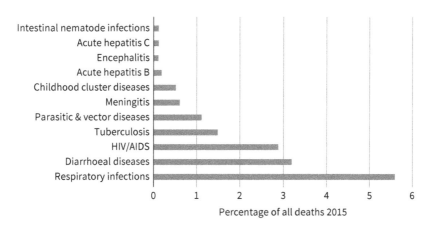

Figure 11.1 Estimated causes of world deaths related to infectious and parasitic diseases, 2015
Source: World Health Organization (2015).

Moreover, some of the MDG outcomes seem to have resulted from increased international aid and humanitarian efforts providing vaccines and pharmaceuticals to the poorest countries, rather than sustained local government policies (Buse and Hawkes 2014; 2015; Copestake and Williams 2014). There is also evidence indicating that humanitarian care, despite its relevance, has been insufficient. Increased global support and incentives to STI should have been provided to developing countries' vaccine manufacturers in the areas of innovation, technological development, and technology transfer. These missing STI strategies were certainly a major gap and could have had a significant impact in attaining MDG. The need to integrate STI into global immunization programs and the recognition of the crucial role that developing countries' manufacturers can play in global access to vaccines are certainly important lessons to be learned for the successful implementation of SDG strategies.

Recent breakthroughs in biotechnology have contributed to the development of new and innovative human and veterinary vaccines, a field with increasing cross-fertilization between both areas (human health and animal health), with wide application in diverse economic and social sectors such as health, agriculture, and livestock. In this scenario, recombinant vaccines, biopharmaceuticals, and drugs have radically changed the strategies for prevention and treatment of a broad range of infectious, parasitic, and chronic-degenerative diseases. They have emerged in a market where competition, global cooperation, and technology transfer have sharply increased access to vaccines, incorporated into National Immunization Programs in the developing countries. Nevertheless, contrasting with these advances, two major constraints to global access vaccines remain and will be discussed in this chapter: gaps in local technological development capacity of vaccines by emerging countries' manufacturers (Homma et al. 2013), and persisting regulatory barriers such as intellectual property constraints to access to multipatented vaccines, impacting on their social sustainability (Possas et al. 2015).

4. Results: Vaccine Challenges for Achieving SDGs

Immunization is one of the most cost-effective ways to ensure long and healthy lives. Every year, vaccines save 2 to 3 million lives, and millions more are protected from disease and disability. Since 2000, GAVI partners have helped countries to immunize almost 640 million children, saving over 9 million lives in the long term. In spite of these advances, it should be noted that only one of the UN SDGs for 2016–2030, SDG3, "ensure healthy lives and promote well-being for all at all ages," is directly related to vaccines, including both drugs and vaccines (3.b.1).

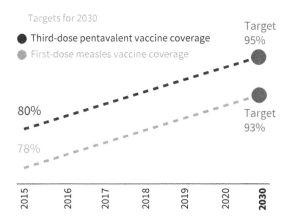

Figure 11.2 Target 2030 for global routine immunization coverage: indicator of children third-dose pentavalent vaccine coverage/first-dose measles coverage

Sources: For 2015–2017 indicators and target for 2020, data from WHO/UNICEF (2017a, b) in GAVI (2018). For target for 2030, elaborated by the authors.

4.1 SGG3: Vaccine Performance Indicators

We have identified in our search three major performance indicators directly related to SDG3 proposed by GAVI (GAVI 2018), as described here.

Measles vaccination helped prevent nearly 15.6 million deaths between 2000 and 2013. The number of globally reported measles cases in children declined by 67% for the same period. In spite of significant progress in this area, third-dose pentavalent vaccine coverage remains, as indicated in Figure 11.2, one of the major challenges.

This indicator, proposed by GAVI, measures the percentage of children reached with the third dose of a vaccine containing antigens against diphtheria, tetanus, and pertussis (DTP), such as pentavalent, and the first dose of measles vaccine in GAVI-supported countries.

Universally present in the routine schedules of GAVI-supported countries, coverage estimates for these two vaccines provide a reliable indicator of the proportion of children with access to basic immunization services.

4.2 Indicator of Average Global Coverage Across Basic Children Vaccines

This indicator measures the breadth of protection, providing the percentage of children reached by last dose of seven vaccines recommended by GAVI (Figure 11.3) and of three vaccines specific to certain regions.

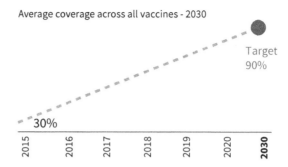

Figure 11.3 Target 2030: indicator for average global coverage across basic children vaccines

Sources: For 2015–2017 and target for 2020, data from WHO/UNICEF (2017a, b) in GAVI (2018). For 2030, target set by the authors.

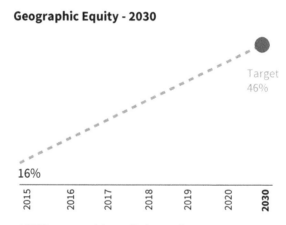

Figure 11.4 Target 2030: geographic equity in vaccine coverage across countries' districts (percentage of countries meeting equity benchmark)

Sources: For 2015–2017 and 2020 target, data from WHO/UNICEF (2017a, b) in GAVI (2018). For 2030, target set by the authors.

4.3 Indicator of Geographic Equity Across Countries' Districts

This indicator provides the percentage of GAVI-supported countries in which coverage with a third dose of pentavalent vaccine is equal to or greater than 80% across all districts (Figure 11.4). As part of an increased global effort to ensure accurate subnational data for measuring equity, WHO and UNICEF have started to report geographically disaggregated coverage data on an annual basis.

4.4 Performance Indicator for SDG Target 3.8

GAVI has proposed a universally applicable vaccine performance indicator to be one of the measures of SDG target 3.8: "By 2030, achieve universal health coverage, including financial risk protection, access to quality essential health-care services and access to safe, effective, quality and affordable essential medicines and vaccines for all."

The proposed performance indicator to achieve SDG3.8 target is to "reach and sustain 90% national coverage and 80% in every district with all vaccines in national programmes." This indicator is universal in its application, since every country measures immunization coverage. As it focuses on the scale up of access to vaccines in the national schedule, the indicator reinforces country-led development.

The indicator is already agreed by all UN member states through the Global Vaccine Action Plan (GAVI 2012), which was endorsed at the 2012 World Health Assembly. No additional monitoring would be required at country level (GAVI 2018).

4.5 Vaccine Contributions to SDGs: Performance Indicators

It should also be noted that vaccines are related to 14 of the 17 SDGs. In addition to the SDG3 goal, vaccines play a crucial role in achieving seven other SDGs, besides SDG3 (Table 11.1) and are also related to six other goals (Table 11.2).

Table 11.2 focuses on six SDGs considered less strongly related to vaccines and vaccine STI performance indicators, consisting mainly to infra-structure and socioeconomic targets. Notwithstanding, as indicated in the following section, they are important components of vaccine strategies.

5. Discussion

Our results indicate that 14 of the 17 SDGs are vaccine-related. Accelerating access to innovative vaccines with reduced adverse effects (Seib et al. 2017; Yeaman et al. 2017; Iaqub 2018; Rottingen 2017) can play a crucial role in global sustainability, reducing morbidity and mortality at very low costs and benefiting particularly the poorest countries. Nevertheless, although there are now several promising innovative vaccines in pipeline (Table 11.3), many challenges persist to global access to these vaccines, particularly affecting developing countries, such as regulatory barriers and governance constraints. We discuss here these obstacles and strategies needed to achieve these goals.

We describe in Table 11.3 selected promising projects of vaccine development against the main emerging and neglected infectious diseases. This table lists

Table 11.1 8 SDGs strongly related to vaccines and vaccine STI performance indicators

SDG1. No poverty–Vaccine-related goal

Rationale: Immunization has a direct impact on reducing poverty. Vaccinated, healthy children can go to school and grow up to become productive adults, and parents can work instead of caring for sick children. In GAVI-supported countries, for every US$1 spent on immunization, US$18 are saved in healthcare costs, lost wages, and lost productivity due to illness. When considering the broader value of people living longer, healthier lives, the return on investment rises to US$48 per US$1 spent (GAVI 2018).
Recommended vaccine performance indicator: Percentage of children enrolled in poverty-reduction programs immunized (indicator of global routine immunization coverage in children: third-dose pentavalent vaccine coverage/first-dose measles coverage).

SDG2. Zero hunger–Vaccine-related goal

Rationale: Immunization is an important component of zero hunger strategy. Malnourished children are more vulnerable to die from infectious diseases. Many can be prevented by immunization, such as pneumonia, diarrhea, and measles. Immunization and good nutrition result in healthy families.
Recommended vaccine performance indicator: Percentage of children enrolled in zero hunger programs immunized (indicator of global routine immunization coverage in children: third-dose pentavalent vaccine coverage/first-dose measles coverage).

SDG3. Good health and well-being–Vaccine-related goal

Rationale: Immunization is one of the most cost-effective ways to protect the health of children. It is an essential element of sound primary healthcare. Today, vaccines save 2 to 3 million lives every year (WHO 2019). Millions of children are alive and healthy thanks to vaccines.
Recommended vaccine performance indicators: 1) Indicator of global routine immunization coverage in children: third-dose pentavalent vaccine coverage/first-dose measles coverage; 2) indicator of average global immunization coverage across basic children vaccines; 3) indicator of geographic equity of children immunization across countries' districts.

SDG4. Quality education–Vaccine-related goal

Rationale: Vaccines protect child health and support cognitive development, enabling children to learn more and have more opportunities. Research shows that vaccinated children get better marks and test scores at school. The benefits flow both ways: educated parents are more likely to have healthy vaccinated children.
Recommended vaccine performance indicator: Percentage of school enrolled children immunized (indicator of global routine immunization coverage in children: third-dose pentavalent vaccine coverage/first-dose measles coverage).

SDG5. Gender equality–Vaccine-related goal

Rationale: Human papillomavirus (HPV) vaccines prevent most cervical cancers (70%) among adolescent girls. Better women's health in the next generation contributes to reduce gender inequality, particularly in developing countries. Every 2 minutes a woman dies from cervical cancer. More than 85% of deaths from cervical cancer occur in low- and middle-income countries (WHO 2018a; LaVigna et al. 2017).
Recommended vaccine performance indicator: percentage of adolescent girls immunized against HPV.

Table 11.1 *Continued*

SGD 9. Industry, innovation and infrastructure–Vaccine-related goal

Rationale: New vaccines require fast-tracked innovation and can prevent pandemics before they start and better prepare the world for disease outbreaks. The fast-track development of the Ebola vaccine in 2015–2016 contributed to a global strategy to develop effective tests, vaccines, and medicines on a much faster pace during epidemics. Innovation for new vaccines can reduce adverse effects and inform testing and regulatory systems that better prepare the world against emerging health threats. Vaccine manufacturers in the developing countries play a key role in global vaccine production and availability.

Recommended vaccine performance indicators: 1) Percentage of innovative vaccines against emerging and neglected infectious diseases available in developing countries; 2) percentage of developing countries' manufacturers (public and private) involved in STI and production of new vaccines against emerging and neglected infectious diseases; 3) percentage of funded projects for new vaccines against emerging and neglected diseases in developing countries.

SDG10. Reduced inequalities–Vaccine-related goals

Rationale: High childhood immunization coverage reduces diseases that often keep people and families in poverty, and gives children of all backgrounds the chance of a healthier and more productive future. By prioritizing areas with low immunization coverage, GAVI support brings basic healthcare to underserved communities.

Recommended vaccine performance indicators: Percentage of children enrolled in inequality reduction programs immunized (indicator of global routine immunization coverage in children: third-dose pentavalent vaccine coverage/first-dose measles coverage).

SDG17. Partnership for the goals–Vaccine-related goals

Rationale: Over the last two decades, the Vaccine Alliance's innovative public–private partnership has transformed immunization progress. Immunization rates today are higher than ever. New vaccines are reaching developing countries at almost the same time as rich countries, often at a fraction of the price. None of this would have been possible without collaboration and innovation across the private and public sectors.

Recommended vaccine performance indicators: Percentage of public–private partnerships in vaccine development projects in developing countries.

Sources: GAVI (2018); WHO/UNICEF (2017a, b).

promising vaccines in development for emerging and neglected diseases that could impact on vaccine-related SDGs. In the current global scenario of rapid aggravation of the COVID-19 pandemic, international demands are spiraling for acceleration in the timelines of vaccine development for the disease. As of May 2020, there are 47 projects for COVID-19 vaccines in initial stages of development and it is expected that in 12 to 18 months a new vaccine could be available in the market. Due to the complexity of the multiple steps required to develop a vaccine, even with the adoption of "fast track" approaches to accelerate regulatory procedures, these vaccines will not be developed in time to control the 2020 pandemic. However, COVID-19 is likely to become an endemic disease and it will be of great importance to have a safe and effective vaccine for its prevention.

Table 11.2 6 SDGs related to vaccines and vaccine STI performance indicators

SDG6. Clean water and sanitation

Rationale: Immunization and hygiene are important components of clean water and sanitation strategy and can prevent some of the leading causes of child death in developing countries, such as diarrhea, cholera, and other infectious diseases. Collectively, these three interventions play a fundamental role in ensuring children can thrive and survive, no matter where they live.

Recommended vaccine performance indicators: percentage of countries with immunization and hygiene programs included in clean water and sanitation strategies.

SDG7. Affordable and clean energy

Rationale: Vaccines should be included as a priority in affordable and clean energy strategy. Vaccines need to be properly stored and transported to stay safe and effective, but up to 90% of health facilities in developing countries lack proper equipment. GAVI's cold chain equipment optimization platform gives countries access to solar direct drive and energy efficient freezers. These are not only more reliable and cost-effective but also more environmentally friendly. GAVI's support includes monitoring devices that ensure optimum energy usage and reduce wastage.

Recommended vaccine performance indicators: Percentage of health facilities in developing countries with vaccines properly stored and transported.

SDG8. Decent work and economic growth

Rationale: Vaccinated, healthy children free up parents' time so they are able to work. In the long term, healthy children grow into a productive workforce and become strong contributors to the economy. In GAVI-supported countries, every US$1 spent on immunization generates US$18 in savings on healthcare costs, lost wages and lost productivity. When considering the broader benefits of people living longer, healthier lives, the return rises to US$48 (GAVI 2018).

Recommended vaccine performance indicators: 1) Percentage of children of workers in developing countries immunized (indicator of global routine immunization coverage in children: third-dose pentavalent vaccine coverage/first-dose measles coverage); 2) percentage of adults in the labor force with completed routine immunization.

SDG11. Sustainable cities and communities

Rationale: By 2030, nearly 60% of the global population will be living in cities. Immunization is a cost-effective way to protect people living in densely populated urban areas against disease outbreaks. Improved vaccination coverage can reduce or stop the rising risk of epidemics and foster the growth of sustainable cities.

Recommended vaccine performance indicators: 1) Percentage of children living in cities immunized (indicator of global routine immunization coverage in children- third-dose pentavalent vaccine coverage/first-dose measles coverage); 2) percentage of adolescents girls living in cities with completed HPV immunization; 3) percentage of adults living in cities with completed routine immunization.

SDG13. Climate action

Rationale: The impact of climate change permeates diverse areas related to well-being, livelihood, health, and security of people and societies, particularly for vulnerable and marginalized communities. Vaccines are critical to building people's resilience to and mitigating the risk of outbreaks of diseases tied to climate change, such as yellow fever, particularly in urban or postdisaster settings.

Recommended vaccine performance indicators: Percentage of population in developing countries immunize against yellow fever.

Table 11.2 *Continued*

SDG16. Peace justice and strong institutions

Rationale: Strong health systems and institutions, with immunization as a core component, can help communities cope with emergencies and keep vulnerable populations healthy. GAVI's health system strengthening support to fragile states is directly contributing to stronger and more efficient systems and institutions, contributing to social cohesion and advancing equity. GAVI's new policy on fragility, emergencies, and refugees help the organization to swiftly respond to countries facing these challenges. Recommended vaccine performance indicators: Percentage of countries with National Immunization Programs in full operation and adequate infrastructure.

Source: GAVI (2018); WHO/UNICEF (2017a, b).

Table 11.3 Novel innovative vaccines for emerging and neglected diseases that could impact 14 vaccine-related SDGs: selected promising projects

COVID-19

Moderna - mRNA technology.
CanSino/ Beijing Institute of Biotechnology - Adenovirus as an expression vector.

Dengue

Sanofi CYD-TDV vaccine registration and pricing, entering now the market after 20 years of development.

Pneumococcal vaccine

Merck: V114 is being evaluated in two phase 3 clinical trials.

Pfizer: phase 3 clinical trial testing its own next-generation pneumococcal vaccine.

RSV vaccine

Novavax: ResVax RSV, a vaccine for protecting infants from RSV via maternal immunization. Phase 3 study.

Human papillomavirus (HPV) vaccine

VGX-3100 vaccine for treating cervical dysplasia caused by HPV. Phase 3 clinical study.

Malaria

GSK/Path: RTS, S malaria vaccine. Registration after 28 years of development.

Diarrhea

Takeda Pharmaceuticals: phase 2 trial, bivalent norovirus vaccine candidate.

Vaccine was well-tolerated and induced immune responses that persisted for one year after vaccination. Following these promising results, one of the vaccine formulations has been selected to move forward to phase 3 study.

Influenza

Sanofi Fluzone–Marketed.

University of Washington School of Medicine: breakthrough research for development of novel universal DNA influenza vaccine.

HIV

National Institute of Allergy and Infectious Diseases (NIAID): VCR01 Phase IIb and III.

Target: Overcome barriers and develop a clinically effective vaccine with more than 50% efficacy, improved safety, and good tolerability profile, with reduced adverse effects. This result would be a breakthrough when compared with the previous efficacy of 31% of the HIV vaccine in the former Thailand trial.

Sources: WHO (2018b); Thanh Le et al. (2020).

These projects are crucial to meet three specific targets related to SDG3:

Target 3.1. By 2030, reduce the global maternal mortality ratio to less than 70 per 100,000 live births.

Target 3.2. By 2030, end preventable deaths of newborns and children under five years of age, with all countries aiming to reduce neonatal mortality to at least as low as 12 per 1,000 live births and under-five mortality to at least as low as 25 per 1,000 live births.

Target 3.3. By 2030, end the epidemics of AIDS, tuberculosis, malaria and neglected tropical diseases and combat hepatitis, water-borne diseases and other communicable diseases.

Innovative next-generation vaccines are designed to overcome several safety and efficacy issues involved in vaccine development. These new vaccines in the product pipeline include DNA vaccines, recombinant viral vector vaccines, and individualized vaccines based on the phenotype and genotype immune response of an individual.

It is therefore of utmost importance to support these innovative projects against emerging and neglected diseases in the context of global STI policy, both at the national and international levels. It will be necessary 1) to accelerate vaccine development, to provide adequate conditions and incentives for their introduction into the market; and 2) to create adequate conditions for global accessibility by 2030 to 90% of these new innovative vaccines, particularly by populations in the poorest countries.

5.1 Regulatory Barriers to Vaccine STI

Many emerging developing countries have qualified National Regulatory Authorities for evaluation and approval of pharmaceutical products, such as drugs, kits for diagnosis, and monitoring of diseases and vaccines. In Brazil, the National Regulatory Authority (ANVISA) is recognized for the high quality of product evaluations and updated regulations and procedures. Nevertheless, in spite of significant advances, regulatory barriers remain a major constraint for vaccine development and production, with several steps in bureaucratic procedures and delays in evaluation and approval.

These barriers can be particularly detrimental for project approval and registration of vaccines and other products for new and resurgent infectious diseases. More flexible and agile mechanisms for speeding up clinical trials evaluation, such as "fast track," "expedited review," "priority review," "accelerated approval," and "rolling review" (USFDA 2018; US Dep HHS et al. 2013) are often missing in these countries, so the progress for the development of a new emergency vaccine could be slower than needed.

Another important legal barrier is intellectual property, favoring unbalance in the international trade, supported by an international system for patenting life-saving products, which has emerged after the Uruguay Round in 1984 and the creation of the World Trade Organization (WTO). This trend toward increasingly strong intellectual property rights regimes related to international trade has been recently confirmed and strengthened by the WTO Agreement on Trade Related Aspects of Intellectual Property Rights (TRIPS), particularly by the TRIPS Plus bilateral commercial agreements between the United States and other trading partners, or between Europe and trading partners. These bilateral agreements have imposed on the partner countries even stronger IPR constraints related to trade. They have impacts on the production of vaccines by manufacturers in developing nations and have affected their availability (Possas et al. 2015).

In this context, intellectual property rights and patents emerge as a crucial issue for vaccine development in these countries (Kaddar et al. 2013; Milstien et al. 2007; Padmanabhan et al. 2010; Possas 2013). A global strategy to support a sustainable progress in innovation and access to new human and veterinary vaccines will require overcoming these legal and regulatory challenges. Intellectual property provides incentives to innovators but may contribute to rising costs and delays in vaccine development and production.

An important area affected by patent protection are adjuvants. Contrasting with the accelerated incorporation of new technologies in the development of new vaccines, allowing better performance and purity, they are all multipatented and their utilization by vaccine manufacturers in emerging countries such as Brazil has been constrained and the case of patented adjuvants illustrates well these regulatory barriers. Vaccine adjuvants are an important component in the development of new vaccines and play a crucial role in boosting the immune response necessary to vaccine protection (Mbow et al. 2010; Sette and Rappuoli 2010).

Figure 11.5 indicates the concentration of vaccine patent deposits in few countries: China (36%), the United States (33%), and the European Patent Office (EPO) (8%), accounting for 77% of vaccine patent deposits.

There seems to be a consensus in the international scientific and technological communities that access to patented vaccine adjuvants is a critical issue and a major constraint, limiting the possibility of development of promising vaccines by developing countries' researchers and manufacturers (Mbow et al. 2010; Possas et al. 2015; Sette and Rappuoli 2010). There is enough evidence indicating that some multinational companies have been unwilling to share information on these adjuvants, protected by patents with external investigators in their international collaborations and even after technology transfer agreement (Bregu et al. 2011).

This constraint contributes to a global impasse in vaccine development, since access to adjuvant information protected by patent could contribute to significantly accelerate vaccine development. Patent barriers and lack of access to new vaccine technologies have also had adverse effects on vaccine development for the so-called diseases of the poor (neglected diseases), for which there is limited commercial interest.

In attempting to overcome these constraints, some nonprofit organizations, with support from international funding agencies, have established a new Global Adjuvant Development Initiative (GADI) laboratory, located at the University of Lausanne,

Top 10 Priority Countries
Vaccine Patent Deposits

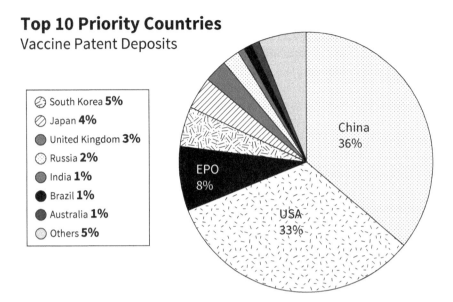

South Korea **5%**
Japan **4%**
United Kingdom **3%**
Russia **2%**
India **1%**
Brazil **1%**
Australia **1%**
Others **5%**

China 36%

EPO 8%

USA 33%

Figure 11.5   The top 10 priority countries for vaccine patent deposits

Source: Federal University of Rio de Janeiro School of Chemistry Information System on the Chemical Industry (SIQUIM); Derwent Innovations Index, in Possas et al. (2016).

under the guidance of WHO, GADI has been conceived to support new strategies and mechanisms to make adjuvants available to developing countries´ vaccine manufacturers.

Adjuvants have emerged as an alternative route for vaccine development, as new antigens with purer and smaller molecules may not elicit the immune response necessary for long-term vaccine protection (Possas et al. 2015). The malaria vaccine RTS provides a good example of the crucial part new adjuvants can play, as it proved successful in providing protection against clinical malaria only when combined with a powerful adjuvant, AS02 or AS01 (Sette and Rappuoli 2010).

"Patent pools" and incentives facilitating access to patent information on new vaccines and new adjuvants by developing countries vaccine manufacturers can therefore be crucial strategies to meet vaccine-related SDGs.

5.2 Global Governance of Vaccine Development and SDGs: Future Challenges

Conceiving and negotiating with public and private stakeholders long-term global governance strategies to support vaccine development is key to meet SDGs. WHO high vaccine coverage requirements to control or eradicate diseases, such as polio or measles, or the need to control epidemics, such as yellow fever or Meningococcus, creates a strain on production capacity. The development of conjugate vaccines against *Pneumococcus*, and combination meningococcal A/C/W/Y, hepatitis B virus, *Haemophilus influenzae* type b, human papillomaviruses, and pentavalent vaccine, DPT/HB/Hib, as well as new vaccines, such as HIV, dengue, and Zika, are requiring from developing countries more effective governance strategies to accelerate vaccine availability for their populations, through public–private partnerships and technology transfer agreements (Homma et al. 2013; Sette and Rappuoli 2010).

International awareness of vaccines' impact and increased global demand for vaccines have highlighted the need for global strategies to strengthen local capacity for vaccine R&D in emerging economies, and to increase access to vaccines through policies supporting free and universal access to vaccines.

In this context, the Decade of Vaccines (DoV) initiative (Possas et al. 2016) was launched at the World Economic Forum in Davos in 2010, signed by international agencies such as WHO, UNICEF, the US National Institute of Allergy and Infectious Diseases (NIAID), and the Bill and Melinda Gates Foundation. The Decade of Vaccine—Global Action Vaccine Plan 2011–2020 was conceived with the mission: "to extend, by 2020 and beyond, the full benefits of immunization to all people, regardless of where they are born, who they are, or where they live" (GVAP 2012, p. 1). This declaration was supported by a commitment by the Gates Foundation to donate $10 billion to R&D and to delivering vaccines for the poorest countries.

The DoV initiative gained significant international support and visibility. Two years later, after consultations with DoV stakeholders, including industry groups, a Global Vaccine Action Plan (GVAP) was launched by the 194 Member States of the 65th World Health Assembly in May 2012, aiming to deliver universal access to immunization (WHO 2013).

Following the collaborative DoV strategies, the GVAP brought together multiple stakeholders to achieve the ambitious goals of the plan: the leadership of the Gates Foundation, GAVI Alliance, UNICEF, US National Institute of Allergies and Infectious Diseases (NIAID), and WHO, mobilizing many partners: governments, health professionals, academia, manufacturers, funding agencies, development partners, civil society, media, and the private sector.

Significant progress in global access to healthcare and vaccines has been made with MDGs, and current trends allow us to anticipate extraordinary advances for SDG health goals, particularly in the burden of diseases and mortality (Horton 2013; Jamison et al. 2013; United Nations 2014; Yamey et al. 2014). Nevertheless, it should be noted that actions and resources will not be sufficient for the success of GVAP and SDGs if these strategies do not conceive an effective global strategy to support manufacturers in the developing world to overcome the main innovation and technological gaps and also to deal with the challenges posed by the Intellectual Property Rights (IPR) and regulatory barriers that delay and hinder vaccine development and production. These new global governance strategies, knowledge-based and "mission-oriented" (Mazzucato 2015; Mazzucato and Penna 2015; Mazzucato 2018) to innovation, should be conceived to promote new incentives, such as prize awards and patent pools (Homma et al. 2013; Padmanabhan et al. 2010; Possas et al. 2015; 2016; Mazzucato 2018), to accelerate technological transfer and to strengthen local technological capacity for development and production of new and complex multipatented vaccines. This should not be a top-down global governance strategy, but must be flexible and consider the complexity of the long-term investments in new platforms and the need for negotiation with the multiple actors involved in global vaccine policy.

6. Conclusion: STI Policy Recommendations for Vaccine-Related SDGs

The main issue discussed in this chapter, in a global situation marked by emergence of new diseases, including COVID-19, is how vaccine STI should be better designed to address the vaccine coverage targets of the SDGs. Several countries, even the developing ones, have included in their public budgets specific resources for immunization programs, searching to assure their long-term sustainability. Support from the Gates Foundation and other international organizations, such as WHO, UNICEF, and governments from developed and developing countries, has been key. These global strategies have contributed to significant advances in immunization and eradication of infectious diseases. Many countries have already attained high immunization coverage, saving the lives of millions of children, with significant progress in attaining partially MDGs and SDGs, which have proved to be strong driving forces for global mobilization.

Nevertheless, despite significant progress in this area many challenges persist in coverage and related STI issues, requiring more coordinated efforts. The poorest countries are now increasingly incorporating new vaccines, with financial support from the GAVI Alliance, but to achieve SDGs it is necessary to obtain more autonomy and to strengthen the local capacity of developing countries' vaccine R&D institutes and manufacturers to incorporate new technologies for production of innovative products. The multinational companies have the intellectual property of these new technologies, but they do not have sufficient production capacity to meet the global demand for these products. Patents and intellectual property issues have been an important barrier to the incorporation of vaccine production technologies by the laboratories and manufacturers in developing countries. Another important issue are the increasing international regulatory requirements which are pressuring developing countries' manufacturers for exponential investments in compliance, modernization, and capacity-building to meet these requirements and resources are often insufficient.

These constraints to vaccine innovation, development, and production in the developing countries are certainly contributing to stagnation of vaccine coverage in several countries and sharp decline of vaccine coverage in some countries, like Brazil. It is urgent to analyze the causes of this phenomenon and find ways to avoid recrudescence of several immune-preventable diseases.

Many challenges therefore remain and highlight the crucial role of the state and international organizations in formulating new vaccine innovation policies, sustainable long-term financing, and adequate regulation, and in promoting scientific collaboration, public–private partnerships, and social welfare.

The main challenge in an increasingly complex global epidemiological scenario is to develop safe vaccines, assuring their availability and adequate coverage, so STI is urgently needed to develop novel vaccines with minimal adverse effects.

More effective public and private financing mechanisms, supported by partnerships and incentives such as patent pools and innovation awards, are urgently needed in order to accelerate the development of vaccines with lower adverse effects.

The reduction of adverse effects would contribute to minimize the recent social phenomenon of vaccine refusal for fear of these effects. This refusal which has been aggravated by exacerbated perception of adverse effects of vaccines in social networks in the internet, has emerged in the last decade as a major issue in developed and also developing countries.

Reducing regulatory barriers and intellectual property constraints are also a major issue to accelerate novel vaccine development and reduce obstacles to access to vaccines by populations in poorest countries.

Nevertheless, STI alone is not sufficient to provide safer vaccines and promote increased coverage. It is necessary to go far beyond STI to achieve these targets: barriers should be urgently overcome in areas such as vaccine STI financing, regulation production, and quality of immunization services, particularly in middle-income countries with significant public and private production capacity.

Vaccines developed by novel STI strategies will require more adequate immunization services reaching the poorest areas with mobile units and home visits, and support by adequate capacity-building and infrastructure.

We advocate for a paradigm shift in global governance of vaccine development production and provision, which we recommend should be more knowledge-based, supported by long-term, mission-oriented innovation strategies focused on specific long-term projects and platforms. More effective innovation policies and on-the-ground initiatives, promoting public–private partnerships and supporting new start-ups, should be conceived. In order to provide follow-up for vaccine development in SDG3, more specific indicators should be conceived such as the ones recently proposed by GAVI and in elaboration.

A substantial increase in global vaccine research, development, and innovation funding and incentives for innovation, such as awards and patent pools, should be promoted, supporting public–private partnerships and technological transfer to manufacturers in emerging countries. The new SDG targets might be not be achievable if attention is not focused on vaccines as a powerful strategy for prevention, avoiding the burden of expensive treatments. The very high costs and social burden of new treatments for diseases, such as HIV/AIDS and hepatitis C treatment, illustrate this argument well.

Considering the complexity of the issues involved in vaccine innovation, and the development, regulation, and patent protection to achieve SDGs, it is urgent to intensify the existing collaborations between academy, industry, research agencies, regulators, and government representatives and developing countries' manufacturers.

References

Bregu, M., Draper, S. J., Hill, A. V., and Greenwood B. M. (2011). Accelerating vaccine development and deployment. *Philosophical Transactions of the Royal Society B* 366(1579), 2841–2849.

Buse, K., and Hawkes, S. (2014). Health post-2015: evidence and power. *Lancet* 383, 678–679.

Buse, K., and Hawkes, S. (2015). Health in the sustainable development goals: ready for a paradigm shift? *Globalization and Health* 11(13), 1–8.

Centers for Disease Control and Prevention (2009). *Epidemiology and Prevention For Vaccine-Preventable Disease*. 11th ed. Public Health Foundation, Washington, DC.

Copestake, J., and Williams, R. (2014). Political-economy analysis, aid effectiveness and the art of development management. *Development Policy Review* 32, 133–153.

Global Alliance for Vaccines and Immunizations (GAVI) (2012). Global Vaccine Action Plan (GVAP). http://www.who.int/immunization/global_vaccine_action_plan/GVAP_doc_2011_2020/en/ (accessed April 5, 2019).

Global Alliance for Vaccines and Immunizations (GAVI) (2018). Sustainable Development Goals. https://www.gavi.org/about/ghd/sdg/ (accessed April 5, 2019).

Greenwood, B., Salisbury, D., and Hill, A. V. S. (2011). Vaccines and global health. *Philosophical Transactions of the Royal Society B* 366(1579), 2733–2742.

Homma, A., and Possas, C. (2014). Panorama da biotecnologia em saúde. *Facto* 40 (abr.-jun.).

Homma, A., Tanuri, A., Duarte, A.J., Marques, E., de Almeida, A., Martins, R., and Possas, C. (2013). Vaccine research, development, and innovation in Brazil: a translational science perspective. *Vaccine* 31S, B54–B60.

Horton, R. (2013). Offline: ensuring healthy lives after 2015. *Lancet* 381, 1972.

Jamison, D. T., Summers, L. H., Alleyne, G., Arrow, K. J., Berkley, S., Binagwaho, A., Bustreo, F., Evans, D., Feachem, R. G. A., Frenk, J., Ghosh, G., Goldie, S. J., Guo, Y., Gupta, S., Horton, R., Kruk, M. E., Mahmoud, A., Mohohlo, L. K., Ncube, M., Pablos-Mendez, A., Reddy, K. S., Saxenian, H., Soucat, A., Ulltveit-Moe, K. H., and Yamey, G. (2013). Global health 2035: a world converging within a generation. *Lancet* 382 (9908), 1898–1955.

Kaddar, M., Milstien, J., and Schmitt, S. (2014). Impact of BRICS? investment in vaccine development on the global vaccine market. *Bulletin of the World Health Organization* 92 (6), 436–446.

Kaddar, M., Schmitt, S., Makinen, M., and Milstien, J. (2013). Global support for new vaccine implementation in middle-income countries. *Vaccine* 31 (Suppl. 2), B81–B96.

LaVigna, A. W., Triedman, S. A., Randall, T. C., Trimble, E. L., and Viswanathan, A. N. (2017). Cervical cancer in low and middle income countries: Addressing barriers to radiotherapy delivery. *Gynecologic Oncology Reports* 22, 16–20.

Mazzucato, M. (2015). *The Entrepreneurial State: Debunking Public vs. Private Sector Myths*. Anthem Press, London.

Mazzucato, M. (2018). *The People's Prescription: Re-Imagining Health Innovation to Deliver Public Value*. UCL Institute for Innovation and Public Purpose, London.

Mazzucato, M., and Penna C. (2015). *Mission-Oriented Finance For Innovation: New Ideas For Investment-Led Growth*. Rowman and Littlefield International, London.

Mbow, M. L., De Gregorio, E., Valiante, N. M., and Rappuoli, R. (2010). New adjuvants for human vaccines. *Current Opinion in Immunology* 22(3), 411–416.

Meissner, H. C. (2016). Immunization policy and the importance of sustainable vaccine pricing. *Journal of the American Medical Association* 315(10), 981–982.

Milstien, J., Gaule, G., and Kaddar, M. (2007). Access to vaccine technologies in developing countries: Brazil and India. *Vaccine* 25 (44), 7610–7619.

Padmanabhan, S., Amin, T., Sampat, B., Cook-Deegan, R., and Chandrasekharan, S. S. (2010). Intellectuasssl property, technology transfer and manufacture of low-cost HPV vaccines in India. *Nature Biotechnology* 28, 671–678.

Possas, C. A. (2013). Propriété intellectuelle et le SIDA dans les pays em développement: inno-vation et accès aux produits pharmaceutiques. In: Possas, C. and Larouzé, B. (Eds.), *Propriété intellectuelle et politiques publiques pour l'accès aux antirétroviraux dans les pays du Sud*, pp. 37–50. Agence Nationale de Recherche sur le SIDA et les Hépatites Virales (ANRS), Paris.

Possas, C., Antunes, A., Mendes F. M. L., Menezes Martins, R., and Homma, A. (2016). Innovation and intellectual property issues in the "Decade of Vaccines." In Singh, H. B., Alok, J., and Keswani, C. (Eds.), *Intellectual Property Issues in Biotechnology*, pp. 181–192. CABI International, Wallingford/Boston, MA.

Possas, C., Antunes, A. M. S., Mendes, F. M. L., Schumacher, S. O. R., Martins, R. M., and Homma, A. (2015). Access to new technologies in multi-patented vaccines: challenges for Brazil. *Nature Biotechnology* 33, 599–603.

Røttingen, J. A., Gouglas, D., Feinberg, M., Plotkin, S., Raghavan, K. V., Witty, A., Draghia-Akli, R., Stoffels, P., and Piot, P. (2017). New vaccines against epidemic infectious diseases. *New England Journal of Medicine* 376, 610–613.

Seib, K., Pollard, A. J., Wals, P., Andrews, R. M., Zho, F., Hatchett, R. J., Picketing, L. K., and Orestein, W. A. (2017). Policy making for vaccine use as a driver of vaccine innovation and development in the developed world. *Vaccine* 35(10), 380–1389.

Sette, A., and Rappuoli, R. (2010). Reverse vaccinology: developing vaccines in the era of geno-mics. *Immunity* 33, 530–541.

Thanh Le, T., Andreadakis, Z., Kumar, A., Gómez Román, R., Tollefsen, S., Saville, M., and Mayhew, S. (2020). The COVID-19 vaccine development landscape. *Nature Reviews Drug Discovery* 19, 305–306. doi: 10.1038/d41573-020-00073-5

United Nations (2014). Secretary General: The Road to Dignity by 2030: Ending Poverty, Transforming All Lives and Protecting the Planet. Synthesis Report of Secretary-General on the Post-2015 Sustainable Development Agenda. https://digitallibrary.un.org/record/785641 (accessed April 5, 2019).

United Nations (2015). The Millennium Development Goals Report 2015. http://www.un.org/millenniumgoals/2015_MDG_Report/pdf/MDG%202015%20rev%20(July%201).pdf (accessed April 5, 2019).

United States Department of Health and Human Services (HHS), Food and Drug Administration (FDA), Center for Drug Evaluation and Research (CDER), and Center for Biologics Evaluation and Research (CBER) (2013). Guidance for Industry: Expedited Programs for Serious Conditions: Drugs and Biologics, http://www.fda.gov/downloads/Drugs/GuidanceComplianceRegulatoryInformation/Guidances/UCM358301.pdf (accessed April 5, 2019).

United Nations Development Programme (2016). 2030 Sustainable Development Goals. United Nations, Geneva.

United States Food and Drug Administration (USFDA) (2018) Fast Track, USFDA https://www.fda.gov/forpatients/approvals/fast/ucm405399.htm (accessed April 5, 2019).

World Health Organization (WHO) (2012). *Projections of Mortality and Causes of Death, 2015 and 2030*. WHO, Geneva.

World Health Organization (WHO) (2013). *Global Vaccine Action Plan, 2011–2020*. WHO, Geneva.

World Health Organization (WHO) (2015). *WHO's Vision and Mission in Immunization and Vaccines 2015–2030*. WHO, Geneva.

World Health Organization (WHO) (2018a). Accelerating cervical cancer elimination. Report by the Director-General, EB144/28. Cervical Cancer Initiative. WHO, Geneva.

World Health Organization (WHO) (2018b). *Vaccines in Pipeline*. WHO, Geneva.

World Health Organization (WHO) (2019). Immunization. WHO, Geneva. https://www.who.int/news-room/facts-in-pictures/detail/immunization (accessed May 5, 2020).

World Health Organization (WHO)/United Nations Children's Fund (UNICEF) (2017a). WHO/UNICEF Estimates of National Immunization Coverage. http://www.who.int/immunization/monitoring_surveillance/routine/coverage/en/index4.html (accessed May 5, 2020).

World Health Organization (WHO)/United Nations Children's Fund (UNICEF) (2017b). WHO/UNICEF Joint Reporting Process. http://www.who.int/immunization/monitoring_surveillance/routine/reporting/en/ (accessed May 5, 2020).

Yamey, G., Shretta, R., and Binka, N. (2014). The 2030 sustainable development goal for health. *British Medical Journal* 349, g5295.

Yaqub, O. (2018) Variation in the dynamics and performance of industrial innovation: what can we learn from vaccines and HIV vaccines? *Industrial and Corporate Change* 27(1), 173–187,

Yeaman, M. R., and Henessey, Jr. J. P. (2017) Innovative approaches to improve anti-infective vaccine efficacy. *Annual Review of Pharmacology and Toxicology* 57, 189–222.

Zhou, F., Shefer, A., Wenger, J., Messonnier, M., Wang, L. Y., Lopez, A., Moore, M., Murphy, T. V., Cortese, M., and Rodewald, L. (2014). Economic evaluation of the routine childhood immunization program in the United States, 2009. *Pediatrics* 133(4), 577–585.

12

The Role of Development-Focused Health Technology Assessment in Optimizing Science, Technology, and Innovation to Achieve Sustainable Development Goal 3

Janet Bouttell, Eleanor Grieve, and Neil Hawkins

1. Introduction

Sustainable development goal 3 (SDG3) aims to ensure healthy lives and promote well-being for all at all ages (United Nations 2018a). The twelve targets within SDG3 include three health targets carried forward from the millenium development goals (MDGs (United Nations 2018b) and nine new targets (United Nations 2018a). The three targets brought forward from the MDGs are to reduce maternal mortality, to reduce mortality in neo-nates and children under five years of age, and to end epidemics of HIV/ADIS, tuberculosis, malaria, and neglected tropical diseases, and combat water-borne diseases and other communicable diseases. The additional nine targets concern premature mortality from noncommunicable diseases and poor mental health; substance abuse; road traffic accidents; family planning; universal health coverage; death from hazardous chemicals, pollution, and contamination; tobacco control; research and development (R&D) into vaccines and medicines that primarily affect developing countries; substantially increase health financing and workforce; and strengthen early warning and risk management.

The likely contribution of science, technology, and innovation (STI) to the achievement of the SDG targets varies according to the particular target. Some targets require scientific endeavor to discover new diagnostics or treatments; for example, target 3, insofar as it concerns tuberculosis (World Health Organization 2018a). Other targets primarily require an investment in workforce; for example, investment in skilled personnel to address the target on maternal mortality, or political will, for example, to enact regulations or legislation required by the World Health Organization Framework Convention on Tobacco Control (World Health Organization 2018b). Not all STI needs be what the Economic and Social Council of the United Nations (ESC) refer to as "highly advanced and efficient 'hard' technologies" (United Nations ESC 2016). "Soft" or "social" technologies, which concern "mindsets, attitudes and behaviour" (United Nations ESC 2016), will also be important. In many cases, technology exists which would help achieve the relevant SDG target, but what is required is an innovative approach to improve the reach of the technology, geographically or into sections of the population where take-up has historically been low (Hay Burgess et al. 2006).

Health Technology Assessment (HTA) is a multidisciplinary approach which seeks to evaluate a health technology across a range of criteria of including incremental cost and impact on health outcomes. Technology is broadly defined (encompassing both hard and soft technology) and includes pharmaceuticals, vaccines, screening programs, and devices (including diagnostic tests), as well as health service delivery and processes. As HTA aims to assess the likely improvement in health outcomes a technology may deliver and the incremental cost of that technology, it is an ideal tool to inform decision-makers which STI projects are best able to help achieve SDG3 goals in a budget constrained environment. HTA activities can be divided into "use-focused HTA" concerned with whether and how an extant technology should be used in a given context, and "development-focused HTA," concerned with whether and how a candidate technology should be developed. Use-focused HTA compares the clinical benefit that an available technology is likely to deliver (generally, according to evidence from clinical trials) to the cost of the technology and makes a recommendation to a decision-maker based on an assessment of opportunity cost and other local criteria. Development-focused HTA differs in that the technology is under development, perhaps still at concept stage. Development-focused HTA aims to inform the developers of the technology about a wider range of questions including how the technology should be designed, used, and/or priced. Development-focused HTA is an expanding field, and we believe that the tools of development-focused HTA could be usefully employed to prioritize the development of technologies which impact most upon the SDG3 targets given budgetary constraints. This chapter was based on an extensive literature review designed to identify features, activities, and methods of development-focused HTA and examples of their application. This was supplemented by a targeted literature review of the application of HTA in low- and middle income country (LMIC) settings.

In this chapter we first set out how STI can contribute to the achievement of SDG3. Then we explain use-focused HTA in more detail and show some examples of how it has already contributed to the achievement of the health-related MDGs. Next, we contrast development-focused health technology assessment (HTA), giving some examples of methods used in the small but growing body of literature in this area. Finally, we show, using a case study, how the methods of development-focused HTA could be used to assess a diagnostic technology for a neglected tropical disease. This aim fits within SDG3.3, which calls for the end of the epidemics of AIDS, tuberculosis, malaria, and neglected tropical diseases as well as action to combat hepatitis, water-borne diseases, and other communicable disease by 2030. The relevant indicator for our case study is indicator 3.3.5 which measures the number of people requiring interventions against neglected tropical diseases.

2. Science, Technology, and Innovation and Its Contribution to Meeting SDG3

According to the Economic and Social Council of the United Nations, technology and innovation are central to the implementation of the sustainable development goals (SDGs) (United Nations ESC 2016). The remainder of this section discusses how STI can directly impact upon the achievement of SDG3. It does not address the indirect impact of STI through linkages between the SDGs. For example, STI that improves sanitation and water quality (SDG6) will lead to improvements in health as it will reduce water-borne disease, and STI that leads to improved nutrition (SDG2) will make populations more resilient against disease. STI that improves health will enable the achievement of other SDGs. For example, healthy children are able to attend school (SDG4) and a healthy workforce promotes economic growth (SDG8).

It is also important to remember the role of both hard and soft technologies. Hard or cutting-edge development may be required where there is no existing treatment, vaccine, or diagnostic for a particular disease. For example, the World Health Organization has stated that the SDG for tuberculosis (TB) (SDG indicator 3.3.2) "cannot be achieved" without "a major technological breakthrough" by 2025 (World Health Organization 2018a). Other examples of STI that fall under the hard technology category would be advances in communications technology (such as the use of drones, tablets, and smart phones in facilitating disease surveillance, digital tele-health, or improving data collection) or biotechnology (such as technologies to improve specimen processing, enrich diagnostic targets, and amplify signals for rapid detection (Peeling and Mabey 2010) and the use of "omic"-based technologies to treat neglected tropical diseases (University of Glasgow 2018). Sometimes, the introduction of one technology leads to the identification of a previously hidden unmet clinical need. For example, Peeling and Mabey (2010) describe the introduction of a point-of-care test for malaria which revealed that most children presenting with fever in malaria-endemic areas did not have malaria but other infections, and that mortality was higher among the children without malaria. Further, hard technology development was then required to develop point-of-care (POC) tests for the diagnosis of multiple diseases (Peeling and Mabey 2010).

In addition to the development of new technologies, useful STI may include what the Economic and Social Committee of the United Nations describe as "frugal" innovation—taking an existing technology and making it less expensive and more accessible—or "hybrid" innovation—repurposing an existing technology (United Nations ESC 2016). Accessibility is a key issue with diagnostic tests. There may be no laboratory at the place where diagnosis is required or the laboratory may be of poor quality (Peeling and Mabey 2010). Patients may have travelled a long distance to seek medical attention and may not be able to wait or return for results (Peeling and Mabey 2010). In 2003, a WHO special program for research and training in tropical disease (Kettler et al. 2004) developed the ASSURED criteria for diagnostic tests specifying the characteristics of tests which would maximize their accessability in LMIC. ASSURED stands for Affordable, Sensitive and Specific, User-friendly (simple, few steps, minimal training), Robust and rapid (storage at room temperature, result available in under 30 minutes), Equipment-free or minimal (e.g., solar powered), and Deliverable to those who need it. Thus, the repurposing of an existing laboratory-based diagnostic test to meet the ASSURED criteria would be an example of hybrid STI.

Soft technology tends to receive less attention and funding than hard (World Health Organization 2013). Yet there is much potential for the achievement of SDG3 through better targeting of existing resources. For example, a diagnostics forum convened by the Bill and Melinda Gates Foundation concluded that there was more impact to be gained from increasing access to existing diagnostics rather than from improvements in test performance (Hay Burgess et al. 2006). STI is still relevant here, but the innovation is the process or system, perhaps improvement in methods used to deliver the existing technology, developing innovative ways of overcoming barriers and challenges in adoption behavior or infrastructure (United Nations ESC 2016). An example would be using an innovative approach like a "vaccination day" to expand uptake of a vaccine (United Nations ESC 2016). For diagnostics and treatments, reach, and therefore impact, are often reduced due to logistics challenges leading to frequent stock-outs (Peeling and Mabey 2010). A relevant soft innovation here would be to develop innovative logistics solutions to improve the supply chain.

Investing in a larger and more highly skilled workforce forms the basis of target 3c of SDG3. Specifically, the target calls for a substantial increase in health financing and the recruitment, development, training, and retention of the health workforce in developing countries (United Nations 2018a). Even this target could involve soft STI in the sense of innovative approaches to recruitment, training, deploying, or supporting the workforce. For example, an innovation in India was the development of a mobile diabetes unit so that the same workforce could be deployed more widely (World Health Organization 2013). This approach may have wider applicability both within diabetes and in other disease areas, and the report recommended HTA be used to determine whether extension would be advisable (World Health Organization 2013).

The role of both hard and soft technology development is recognized in the Astana Declaration adopted in October 2018 (World Health Organization 2018c). This declaration recognized primary health care (PHC) as the cornerstone of a sustainable health system for health-related SDGs which, among others, will be driven by technology: "We support broadening and extending access to a range of health care services through the use of high-quality, safe, effective and affordable medicines, including, as appropriate, traditional medicines, vaccines, diagnostics and other technologies ... We will use a variety of technologies to improve access to health care, enrich health service delivery, improve the quality of service and patient safety, and increase the efficiency and coordination of care. Through digital and other technologies, we will enable individuals and communities to identify their health needs, participate in the planning and delivery of services and play an active role in maintaining their own health and well-being."

This section has shown how many different kinds of STI could impact upon the targets contained in SDG3. In particular, it has highlighted that there is potentially more scope for impact from better use of existing technology so that the innovation required is in the process or the system. The next section will explain how HTA uses a range of tools to assess technology of any kind in terms of its impact on health and its cost. The assessment provides information to decision-makers to allow them to compare technologies with a view to selecting the most appropriate for their purpose.

3. What Is Health Technology Assessment?

Healthcare resources are finite in every setting and, irrespective of the financing and organization of a country's healthcare system, policymakers are faced with the challenge of deciding what to cover and under what circumstances. HTA is defined by World Health Organization as "a multidisciplinary process to evaluate the social, economic, organizational and ethical issues of a health intervention or health technology" (World Health Organization, n.d.). Cost-effectiveness analysis is an important component of HTA.

4. How Has HTA Been Used in LMIC?

LMICs are increasingly beginning to develop HTA processes to assist in their healthcare decision-making. Rather than focusing on the evaluation of new technologies, as in high income countries (HICs), HTA has been used in LMICs to inform decisions about what universal health coverage (SDG indicator 3.8) should comprise. The decision-maker (typically the department of health in a given country) has a restricted budget and wishes to deliver the most health impact within that budget (subject to other constraints such as equity and legal obligations). Given the limited resources in most LMICs and HTA's ability to inform decision-makers about clinical and cost-effectiveness, the value of HTA to help make better resource allocation decisions is widely acknowledged (World Health Organization 2013; Pan American Health Organization 2012; World Health Organization Regional Office for South East Asia 2013). For example, in India, HTA has been used to inform national clinical guidelines and quality standards to improve the quality of care delivery; and in sub-Saharan Africa, countries such as Ghana, Tanzania, and Zambia, which are transitioning toward universal health coverage (UHC), are seeking to use HTA for health benefit package design and other purchasing decisions. In Thailand, a country at the forefront of UHC, HTA has been used to inform decision-making for over a decade to define the benefits package and the National List of Essential Medicines.

5. What Is Development-Focused HTA?

In recent years, it has been recognized that the techniques of HTA could also be used to assist the developers of health technologies during the development process (Annemans et al. 2000; Ijzerman and Steuten 2011) as well as to prioritize R&D expenditure (Bennette et al. 2016; Andronis 2015; De Graaf et al. 2015). The form of HTA used to inform developers of health technologies has been termed "early HTA" in the academic literature (Ijzerman and Steuten 2011). We prefer to use the label "development-focused HTA" as it is the audience, rather than the timing of the HTA, which drives many of the differences. Table 12.1 lists key differences between use-focused and development-focused HTA, and these are explained in the paragraphs that follow using an example of a vaccination program.

5.1 Illustration of Features of Use-Focused and Development-Focused HTA

Vaccination programs can potentially reduce the impact of infectious disease in LMICs. The following paragraphs contrast a use-focused HTA exercise and a development-focused exercise concerning a hypothetical vaccination program.

5.1.1 Use-Focused HTA

The *audience* for a use-focused HTA of a vaccination program would generally be a national decision-maker (perhaps a ministry of health or a national reimbursement agency) whose *underlying objective* would be to maximize health given the budget at their disposal. The *specific decisions to be informed* would be whether to implement the program and in what populations. Due to the *timing* of the assessment, the vaccine itself would be an established technology and its effectiveness would be known from the *available evidence*. The price of the vaccine would also be known (although it may potentially be open to negotiation). The *decision space* in this exercise would be relatively narrow. The jurisdiction and disease are both fixed, although it will probably be necessary to consider different subgroups of the population where prevalence or effectiveness of the vaccine may vary. The use-focused HTA would use the existing evidence as well as information relevant to the local context such as the prevalence of the disease and the likely uptake of the vaccine. The HTA exercise would involve modeling the impact of the vaccine on disease rates to determine how health outcomes would differ with the vaccine program in place compared to the current position. The resource-use with the vaccine in place would be modeled and compared

Table 12.1 Features of use-focused and development-focused health technology assessment (HTA)

Feature	Use-focused (traditional) HTA	Development-focused HTA
Target audience	Reimbursement agencies, insurers, national governments	Commercial and academic technology developers, commercial investors, and public sector research funders such as nongovernmental organizations
Underlying objective	Maximize health	Maximize financial or societal return on investment
Core decision rule	Reimburse when value meets established criteria	Fund programs with greatest return
Specific decisions HTA designed to inform	Reimbursement decisions Optimizing guidelines Price revisions	Technology design Reimbursement strategy Trial design Go/no go decisions Research prioritization
Available evidence	Evidence base fixed Specific to indication Specific to context	Evidence base fluid Evidence may be limited Needs to be highly context specific
Timing	Around the time of approval and postapproval Technology is in usable form	Pre and during development Technology may be concept or prototype
Decision space	Targeted at specific decision-makers Indication/s fixed Comparator and patient groups defined by local practice and licensing	Potentially multiple indications and comparators Multiple funders/users to be considered Diffusion of technology relevant May be narrowed through specific project criteria
Resources for HTA	Committed Limited number of technologies reviewed	May be committed if evaluation required within project May be limited

Source: authors.

to current resource-use. This analysis would produce an estimate of the clinical and cost-effectiveness of the vaccination program, which could then be used to inform the one-off decision about whether or not to implement the program. This decision would be informed by the decision-makers *underlying decision rule* in order to decide whether or not to implement the program. As the national decision-making body is likely to have commissioned the HTA exercise, *resources available for analysis* are likely to have been adequate to undertake a comprehensive analysis.

5.1.2 Development-Focused HTA

By way of contrast, the *timing* of a development-focused HTA may precede the discovery research for a vaccine or may be undertaken when there is a prototype vaccine, but there are *decisions to be made* about whether it is worthwhile to continue the development or where to prioritize its implementation. There may be several prototype vaccines in different disease areas and the question is which one could have the most impact. The *target audience* for the HTA analysis may be a public or charitable research funder allocating funds across a portfolio of projects or a commercial developer/investor. The *underlying objective* of these two groups would potentially differ with public or charitable research funders looking to maximize health given the budget at their disposal and commercial developers/investors looking to maximize financial return on investment. The timing of the assessment would determine the *available evidence* but this is unlikely to be large-scale technology-specific evidence even at prototype stage. Evidence is more likely to come from bench studies, similar technologies or assumptions informed by input from experts. In contrast to use-focused HTA the *decision space* in this exercise may be very wide. There may be scope to use the vaccine in many different geographical areas, populations and dosages. As in use-focused HTA, the analysis may involve modeling the health and cost impact of the vaccine but a number of plausible scenarios may be modeled incorporating evidence and assumptions as described earlier. The modeling process would be an iterative process, revised a number of times reflecting evidence generated and with increasing sophistication as the decision-space becomes narrower through the development process. Rather than informing a single *decision*, the analysis informs ongoing discussions. Even if the analysis showed that the technology in the current form in the selected scenarios did not look promising, this may not mean that it should be abandoned—it may instead indicate that other settings are preferable or an improved design is required.

5.2 Activities of Development-Focused HTA

The features identified in the preceding section drive important differences between development and use-focused HTA in the analytic tools used. Synthesis of clinical evidence and economic evaluation are the mainstay of use-focused HTA, as the assessment tends to focus on comparative effectiveness, cost-effectiveness, and budget impact associated with well-defined interventions. In contrast, development-focused HTA draws on a broad range of multidisciplinary methods due to the wide range of the decisions that development-focused HTA is intended to inform. Development-focused HTA can be considered as contributing to three iterative and interlinked assessments of clinical value, economic value, and the business case (see Figure 12.1). Each assessment is informed by evidence

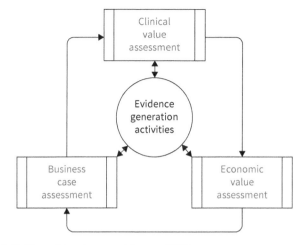

Figure 12.1 Activities of development-focused HTA
Source: authors

generation activities and drives evidence generation activities as gaps in evidence become apparent.

Arrows between assessments indicate an iterative process. Two-headed arrows between evidence generation and the assessment activities indicate that in addition to the evidence informing the assessment, the assessments identify gaps which require evidence generation activities.

The clinical value assessment considers what impact the technology might have on clinical practice and health (and wider social) outcomes. The economic value assessment builds on the clinical value assessment to consider the economic impact of changes in healthcare resource use and other economic value drivers such as productivity effects. It may also include consideration of the potential pricing of the new technology, volume of sales, fixed and variable costs of production, and distribution in order to produce estimates of net margin. Finally, the business case assessment considers whether the continuing development is likely to deliver a positive financial or societal return on investment. This assessment will consider the costs associated with continuing development, the value of the technology, if development is successful, and the estimated probability of success at potential development break points. It may also usefully involve the joint assessment of portfolios of candidate technologies rather than a set of individual assessments. The business case assessment may go beyond what is generally regarded as HTA; however, we argue that it is useful to see it as an extension of HTA rather than a independent activity. These assessments will typically be iterative, being updated as further information becomes available. They are also likely to become more intensive as the costs of ongoing development increase. Finally, individual assessments may require a range of analytic techniques and it is important not to view a particular assessment as being synonymous with a

Table 12.2 Activites and methods of development-focused HTA

Activity	Relevant Analytic Methods
Clinical Value assessment	Epidemiological analysis Health Impact Assessment
Economic Value assessment	Cost-effectiveness analysis (e.g., cost-utility analysis, cost-benefit analysis, cost-consequence analysis, cost-minimization analysis) Value of information analysis
Business Case assessment	Return on investment analysis Societal return on investment analysis Strategic planning methods such as political, economic, social, technological (PEST) and strengths, weaknesses, opportunities, threats (SWOT) Horizon scanning Scenario analysis
Evidence generation	Qualitative methods (e.g., focus groups, workshops, questionnaires, surveys, interviews) Literature reviews and synthesis (including clinical and nonclinical studies) Expert panels/elicitation (e.g., Delphi method) Multicriteria decision analysis (including analytic hierarchy process)

Source: authors.

particular analysis; for example, economic value assessment does not always involve some form of cost-effectiveness analysis, as instead it is driven by what is appropriate for the particular assessment.

5.3 Analytic Methods of Development-Focused HTA

Table 12.2 presents analytic methods of development-focused HTA which have been employed in undertaking the three assessments, as well as methods of evidence generating activities. We illustrate a selection of appropriate methods through the continuation of the vaccination example and through the case study, which will be discussed later in the chapter.

5.4 Clinical Value Assessment

The clinical value assessment essentially compares clinical practice and outcomes in two or more future worlds: one without the new technology and one or more (if there are multiple options for employing the new technology) with. The assessment of clinical value requires an understanding of the epidemiology of the disease, its current treatment, and expected costs and outcomes in the local context. This may be based on published epidemiological studies and published treatment guidelines, opinions from local experts, and, if feasible, local primary research. In all cases, it is crucial that the evidence is relevant to the LMIC setting(s) that we are interested in. In practice, this epidemiological analysis in relation to a potential vaccination program would involve understanding the prevalence of the disease, relevant subtypes, and populations affected in relevant geographical locations. Information required would be the numbers of patients affected, the impact on their health, and the current diagnostic and treatment options. These may be many and various depending upon the context. Practical considerations may require a narrowing of the settings investigated in depth, although it is useful to have an overview of all potential relevant areas.

Next, a health impact assessment of the potential impact of the technology on the disease and treatment pathway will be required. In development-focused HTA, where there is likely to be a paucity of trial or other empirical data relevant to the new technology, alternative methods of evidence generation are often required. These are described in the evidence generation section. As well as an estimate of the clinical effectiveness of the technology itself (e.g., the success of the vaccine in preventing the disease), issues around uptake and diffusion are relevant to the ability of the technology to prevent disease and help countries achieve the relevant SDG target. This illustrates the difference between hard and soft technology—hard technology in developing an effective vaccine, soft technology in designing the logistics and education around the distribution of the vaccine and in monitoring and improving uptake. An initial health impact assessment may make a range of estimates of effectiveness and uptake, including a perfect vaccine with 100% uptake. This assists developers in setting the benchmark performance required for the vaccination program to be clinically and/or cost-effective.

It is important also to note that a new technology may offer incremental value in many ways each of which may be valued in a different way by different users or other stakeholders. A technology may directly improve health outcomes, may facilitate service and process improvements, and/or reduce resource use. Technology development in a LMIC context may involve developing a variant of an existing technology at a price where it becomes cost-effective in LMIC countries (frugal STI). For example, OneBreath was developed initially as a ventilator that could be stockpiled by the US government in case of an influenza type epidemic, but because of its characteristics (cheap, easy to operate, low power requirement), it was suited to resource constrained environments in LMIC and its first target market is India (Yock et al. 2015). It may also involve redesigning technologies so that they are more appropriate for LMIC settings. An example would be portable diagnostics which are designed to be taken to more remote settings such as the Daktari CD4 antibody count for monitoring viral load in HIV positive individuals in sub-Saharan Africa (Yock et al. 2015). This technology has since been superseded and the company is no longer in business, but the value of portable diagnostics remains. The case study discussed later in the chapter has a similar value proposition. It concerns a paper technology currently in development at the University of Glasgow. The value of the potential technology comes from its ability to function accurately despite high levels of humidity. Poor literacy and/or eyesight is another difficulty in some LMIC environments, and technologies which take this into account can improve health outcomes. For example, Doseright is a low-cost clip which can be inserted into a syringe which allows safe administration of drugs by lay people or in areas where literacy is low (Yock et al. 2015). A new technology can sometimes bring benefits which indirectly improve health outcomes such as a new diagnostic test which accurately distinguishes viral and bacterial infections thus allowing overtreatment with antibiotics to reduce, contributing to a reduction in antimicrobial resistance (Peeling and Mabey 2010).

5.5 Economic Value Assessment

Economic value assessment extends the clinical value assessment by accounting for healthcare and other resource usage. It may also include productivity and other wider impacts. In use-focused HTA, economic evaluation is typically undertaken to determine whether the technology, at a given price and given current evidence, represents a cost-effective use of healthcare resources. In development-focused HTA, economic evaluation plays a different role. It is used to support a range of decisions including whether further development should be funded and how studies should be designed.

Cost-benefit, cost-utility, cost-effectiveness, cost-consequence, and cost-minimization analysis (Drummond et al. 2015) are all forms of economic value assessment in that all compare the difference in outcomes brought by a new technology to the difference in costs. All methods calculate costs in the same way, but they measure outcomes differently. In a cost-benefit analysis, outcomes are measured in monetary terms. The principle challenge is determining who should value the outcomes. In a cost utility analysis, outcomes are expressed in quality adjusted life years (QALYs) or disability adjusted life years (DALYs). This form of economic evaluation is preferred in the Gates Reference Case for LMIC (Claxton et al. 2014). This is because the outcome measures, QALYs, and DALYs used in cost-utility analysis are generic measures which facilitate comparisons across technologies and disease areas. In cost-effectiveness analysis, outcomes are expressed as a single disease-specific measure, for example, cost per infection avoided. In cost-consequence analysis outcomes are expressed across multiple measures. A particular challenge with both these forms of analysis is the difficulty in making comparisons between indications and determining an acceptable threshold for willingness to pay. Finally, in cost-minimization analysis it is assumed that outcomes are either equal or superior with the new technology and hence it is sufficient to simply compare costs. The challenge with this analysis is that it is only appropriate when it is safe to assume that outcomes are either equal or superior with the new technology.

Cost-effectiveness analysis may be used in development-focused HTA in two main ways. It can be used to indicate whether a technology is likely to be regarded as cost-effective, and hence used at a given price for the new technology. Alternatively, it can be used to estimate the maximum price at which the new technology is likely to be deemed cost-effective. This has been referred to as "headroom" analysis (Chapman 2013). Where these analyses are based on cost-effectiveness or cost-utility analysis, an estimate of an acceptable threshold willingness to pay will be required. A threshold of three times a country's per capita gross domestic product (GDP) has historically been used in LMICs. However, research is ongoing to determine values for the acceptable threshold that better reflects the opportunity cost or "shadow price" of investment in other healthcare technologies. Threshold analysis is a similar formulaic approach which assumes a price for the technology then investigates what values the other parameters need to take in order for the technology to remain cost-effective. These approaches are very simple and quick to perform and can be based on expert opinions or assumptions so they are ideal for undertaking extensive scenario or sensitivity analysis. Any factors which are not modeled can be assessed qualitatively alongside the simple modeling (Chapman 2013). For example, in the assessment of a diagnostic test for hepatitis C in development, Chapman (2013) discussed the potential impact of a move away from routine testing by biopsy, which could limit the market for the new test. Scenario analysis examining different settings (for example, high/low prevalence and urban/rural setting) and varying sensitivity and specificity in local settings can be undertaken.

In addition to providing information about the potential economic value of a new technology, cost-effectiveness analyses can be used to provide estimates of the potential value of additional information from further research. This is known as value of information (VOI) analysis. It can be conducted from a commercial or a societal perspective. When conducted from a societal perspective, the analysis can be used to estimate the net impact on health benefit arising from an increase in the probability of selecting the optimal treatment resulting from further research reducing uncertainty. VOI can aid study design and investment decisions (Vallejo-Torres et al. 2008).

In our example of a development-focused HTA of a vaccination program, a number of scenarios would be worth considering. If commercial developers were developing the vaccine, they may use a form of headroom analysis based on cost-utility analysis to estimate the price they may be able to charge for the vaccine in order for it to remain below an acceptable cost-effectiveness threshold in a number of target jurisdictions. They could use threshold analysis to determine what the take-up would need to be in order to justify a given price. They could use value of information analysis to design clinical trials which address the areas of most uncertainty (Vallejo-Torres et al. 2008; Wong et al. 2012; Meltzer et al. 2011). A cost-utility analysis undertaken to inform a charitable or public sector funder may assume a price sufficient to cover ongoing unit cost of the vaccine and then calculate the incremental cost effectiveness ratio of different vaccine strategies. If the development of the vaccine seemed likely to have an acceptable incremental cost-effectiveness ratio (i.e., to deliver sufficient health impact for its cost), then the project should be continued. If it is too expensive, developers may need to try to alter the design to improve effectiveness or acceptability to the target population. VOI analysis has been used in a number of research prioritization pilots to help funders choose which clinical trials to fund (Bennette et al. 2016; Carlson et al. 2013; Meltzer et al. 2011). Alternatively, the results of the cost-utility analysis may form part of a multicriteria analysis for project prioritization or the basis for a calculation of societal return on investment.

5.6 Business Case Assessment

The economic analysis can be broadened to consider financial return on investment (ROI) and/or social return on investment (SROI). ROI is used by commercial entities to estimate the likely return on a project. The basic calculation is expected revenues less ongoing costs of production/distribution less development costs still to be incurred to determine whether to continue the development. The commercial viability of a candidate technology will depend on combinations of price and volume that are achievable, fixed, and variable costs of production and development costs (Meltzer et al. 2011). The price or cost estimates from the headroom calculation can be multiplied by the expected volume of sales (from epidemiological analysis) to calculate revenue. SROI is a framework for measuring and accounting for a broader concept of value; it considers aspects valued by different stakeholders such as reductions in inequality, environmental degradation and improvements in well-being by incorporating social, environmental and economic costs and benefits (SROI network 2012). Gains are typically measured in monetary terms (financial ROI), or can also be expressed in terms of social values. Costs remain the same in both cases. A societal return on investment analysis can be undertaken from multiple stakeholder perspectives (patient, society, manufacturer). This analysis assesses whether the potential value demonstrated in the clinical and economic value assessments is likely to translate into a viable business opportunity. Both ROI and SROI can be used as means to prioritize projects within portfolios, generally as input to a multicriteria process. For example, a research team looking at investments in the development of biomarkers in the prevention of type 2 diabetes considered attributes including reduction in downstream costs, added quality-adjusted survival, cost of test, feasibility of treat-all option, competition, and ease of implementation in an MCDA exercise (De Graaf et al. 2015).

Both commercial developers and public or charitable funders need to be aware of the current competitive environment, as this will impact upon the potential viability of the technology in development. Commercial developers may not be able to launch a product if the market opportunity is taken by an equivalent or better technology. A public or charitable funder may find that the competitor technology is a more cost-effective solution for their need. Methods for the assessment of the market opportunity include strategic planning methods such as political, economic, social, technological (PEST) and strengths, weaknesses, opportunities, threats (SWOT) (Cosh et al. 2007). These methods are essentially means of structuring qualitative discussions among the development team to ensure that all relevant macro and micro economic considerations are addressed. Relevant information may also be gained from literature review and analysis (Chapman 2013), formal horizon scanning systems (Ijzerman and Steuten 2011), and/or on-the-ground research (Yock et al. 2015).

As with assessment of the clinical value, decision space is a key factor and the business case will vary with each disease area and local context. Moreover, the strategic planning methods are likely to identify conditions in the wider environment which could favor or impede the success of the development. Scenario analysis involves the development and consideration of multiple scenarios reflecting uncertainties about the context and future events. These scenarios could inform stakeholder analysis and be built in to future economic evaluation in order to provide a range of options for consideration throughout the development process (Joosten et al. 2016). This is a good illustration of development-focused HTA's role in informing discussions and decisions throughout the development process.

5.7 Evidence Generation

Each of the assessments (clinical value, economic value, and business case) involved in development-focused HTA generally involve consultation with a wide range of stakeholders including patients, physicians, and policymakers. In the very early stages evidence may be primarily based on literature review and consultation with the technology development team. Consultations with stakeholders may involve informal or formal qualitative methods such as focus groups or interviews or may take the form of structured expert elicitation. Informal methods are appropriate (and may be the only option) when resources are limited. However, a particular concern in expert elicitation is that estimates are often too optimistic (particularly if the expert is involved with the development team) and underrepresent uncertainty. Thus, formal methods have developed which attempt to address this concern. Formal methods can be divided into two main groups: structured expert elicitation (SEE), which is used to obtain estimates of relevant population parameters, such as what is the expected treatment effect for a new technology; and multiple criteria decision analysis (MCDA), which provides a solution to a decision-problem, such as which is the optimum choice among a set of treatment options with different characteristics.

Delphi methods are commonly used to elicit expert opinions and typically involve two rounds of questions to individual experts (Gosling 2014). In the first round, experts respond without knowledge of other responses. Responses are then shared and a second round of estimates is sought from the entire group. Estimates may include probability distributions or ranges. It is possible to compensate for optimism-bias and underestimation of uncertainty through sensitivity analysis in economic evaluation or qualitative consideration of a broader range of options in scenario analysis. Responses may also be weighted according to their experience in the field. Best practices for the conduct of expert elicitation are discussed by Iglesias et al. (2016).

Several online tools are available including SHELF (Oakley and O'Hagan 2010) and MATCH uncertainty elicitation (Morris et al. 2014). The availability of these tools will aid the conduct of expert elicitation in resource limited LMIC settings. Structured methods of expert elicitation have been used to identify current treatment pathways and standard of care (Davey et al. 2011); to specify the technical features of current technology (Terjesen et al. 2017); to estimate likely levels of test performance (Haakma 2011); to estimate treatment effect (Girling et al. 2007; Østergaard and Møldrup 2010) to estimate the effect of a new test on discharge rates (Kip et al. 2016); and to consider the likely position of a test in a diagnostic pathway (Breteler 2012).

Multicriteria decision analysis (MCDA) describes approaches which seek to take explicit account of multiple criteria in a decision-making process (Wahlster et al. 2015). In the context of development-focused HTA, experts will typically be asked to indicate their preferences among current and one or more versions of the new technology. A set of criteria relevant to the decision problem will be identified. Each option will be scored on each criteria. Then, in consultation with the experts, weights will be assigned to each criteria and some form of aggregation conducted to identify the optimum treatment. MCDA techniques have been used to support development-focused HTA in various ways. Several studies have used the techniques to support design decisions through the assessment of the relative importance of features concerning safety, ease of use, and aspects of performance (Hilgerink et al. 2011; Hummel et al. 2012). Analytic hierarchy process (a form of MCDA) has also been used to predict the health economic performance of a new surgical technique and populate performance criteria for a diagnostic technology in development (Koning 2012). The MCDA process described by Koning established that the technology under development, an electronic nose for the diagnosis of infection disease, was not sufficiently accurate or rapid to offer advantages in well-resourced settings, but may be viable in lower resource settings. Another use of MCDA has been to generate the criteria and key issues for two key stakeholder groups in the development of computer chips simulating organ functions for use in biotechnology development (Middelkamp et al. 2016).

The preceding sections have summarized the activities and methods of development-focused HTA. The following case study provides an example of how those activities and methods can be put into practice to inform the developers of a point-of-care diagnostic test.

6. Case Study: Point-of-Care Diagnostic Device

This technology is a multiplexed DNA-based assay using paper which can enable the rapid detection of infectious agents to guide treatment. It has been developed by a multidisciplinary team at the University of Glasgow, funded by the Engineering and Physical Sciences Research Council, the National Institute for Health Research, and the Global Challenges Research Fund, public funding bodies in the United Kingdom.

It could be useful in multiple disease areas and across multiple locations and would be of particular use in areas where coinfections are endemic. With limited ability to recall patients in rural areas, POC tests allow decisions to be made at the point of contact so treatments can be administered there and then. It is aimed at hard-to-reach settings, targeted where there are coinfections. This device is initially being designed to diagnose malaria and schistosomiasis (a neglected tropical disease) in Uganda, an area the development team know well. If successful, this project would aid Uganda in achieving SDG3.3 by reducing the incidence of malaria and schistosomiasis. Research has demonstrated the analytical and clinical validity of the diagnostics tests, which now need engineering work to improve speed and ease of use. In the longer term, the project team (including researchers from Rwanda and Uganda and a representative of the Ugandan Ministry of Health) aim to create a sustainable, not-for-profit entrepreneurial activity around the manufacture of a pipeline of diagnostic products in Africa.

Use of development-focused HTA will include:

- epidemiological and health impact assessment to inform the development team about other infections and settings where the technology could be used effectively;
- qualitative methods to understand the context , to identify human resource pathways (personal costs, time costs, productivity impacts, spillover costs, and outcomes) and to develop a context specific economic logic model;
- modeling of the costs and benefits of the tests targeted at malaria and schistosomiasis to estimate the thresholds of prevalence and test performance required to make reimbursement and clinician/patient acceptance more likely and estimate societal return on investment for a number of plausible scenarios;
- value of information analysis to estimate the value of further research and the value of the information from the test; and
- choice experiments (similar to multicriteria decision analysis) to identify relevant behavioral aspects and inform the development team about likely product uptake.

7. Discussion

It is clear that STI has a role to play in the achievement of many of the targets of SDG3. Our case study and much of our discussion has been firmly anchored in the achievement of target 3.3, where the diagnostic test aims to improve diagnosis and treatment of malaria and a neglected tropical disease, first in Uganda, then elsewhere. The diagnostics and treatments in this case exist and the new diagnostic is innovative in that it aims to put the test into a point-of-care form to increase the reach of the technology. STI in the form of discovery science is required in order to achieve some of the targets of SDG3 (targets for the reduction of TB); however, in many cases it is not this form of STI which will have most impact. Technology development in many areas has to be focused on innovations which improve the impact of what we already have. We may need to be innovative in the design of distribution channels, optimizing the effectiveness of the workforce or in our use of communication technology. This is exemplified by the Covid-19 pandemic where developers needed to rapidly evaluate and optimize or adapt existing technologies.

STI is undertaken by an increasingly diverse group of stakeholders and innovative partnerships. International funders and prominent NGOs such as the Bill and Melinda Gates Foundation, US National Institutes of Health, Wellcome Trust, UK Department of International Development, and the European Commission (Peeling and Mabey 2010) continue to play a crucial part in funding STI for the achievement of SDG3. The business case for STI aimed at LMIC is complex as financial returns comparable to those in HIC are not generally achievable in countries with lower levels of health spending (The Lewin Group 2005). However, there is now a greater willingness on the part of major international companies to work with partners to address pressing global health issues (Droppert and Bennett 2015). For example, much of the burden of tuberculosis (TB) is concentrated in a handful of middle-income countries—Brazil, Russia, India, China, and South Africa (the BRICS), which "places TB awkwardly in the interstitial space between donor-driven innovation and market-led R&D" (Silverman et al. 2018). The Center for Global Development, a US-based nonprofit think-tank and the Office of Health Economics, a UK-based consultancy, are working on a new approach for BRICS cooperation against tuberculosis with the aim of driving investment for better TB treatment (Silverman et al. 2018). They are endeavoring to develop a market-focused approach by helping BRICS governments to inform private-sector pharmaceutical companies about "the kind of innovation they would value and at the kind of price they would be willing to consider." There is also increasing focus on involving local entrepreneurs from LMIC in the innovation process (United Nations ESC 2016). Networking bodies are of importance in bringing together innovators and providers of finance (United Nations ESC 2016). Product development partnerships, such as the Foundation for Innovative New Diagnostics (Peeling and Mabey 2010), play a similar role to foster collaboration between disease experts, scientists, and engineers.

Development-focused HTA can contribute in three main ways toward optimizing STI to achieve SDG3: prioritization within portfolios, optimizing individual projects, and improving structures and communication between multiple stakeholders. Funders have a fixed amount of resources available, and development-focused methods of portfolio evaluation (such as multicriteria decision analysis, financial or social return on investment) can help target these funds to ensure they are focused on the projects which deliver the most health impact for an acceptable cost.

The second role development-focused HTA could play is illustrated by the case study and involves the optimization of an individual technology during development rather than a choice between projects. A key aspect of this optimization is evidence gathering in the specific context where the technology will be used.

Development-focused HTA evidence gathering methods include tools (such as epidemiological analysis, qualitative methods, and MCDA) to gain a thorough understanding of the human factors, infrastructure, and healthcare organization of the context where the technology is to be deployed. It is therefore critical for grants supporting the development of STI in LMICs to include funds to allow the development team to spend time in developing country settings (Peeling and Mabey 2010). This should occur early enough in the development process to allow the product profile to be tailored appropriately (Peeling and Mabey 2010). This process allows indigenous knowledge to be incorporated in STI and allows for local innovators to be fully integrated into STI development where appropriate (United Nations ESC 2016). Also critical is an understanding of the value placed on different aspects of the STI by different stakeholders. Detailed user interaction can ensure underrepresented and/or vulnerable groups have their values and priorities reflected in the STI (United Nations ESC 2016). This type of approach will address the behavioral and systematic factors sometimes neglected in STI development (United Nations ESC 2016).

The achievement of SDG3 will require efforts in STI from multiple stakeholders including local populations, funders, commercial organizations, academic and scientific research institutes, engineers, communications specialists, epidemiologists, healthcare workers, and politicians. Development-focused HTA can play a part in facilitating cooperation and communication between these stakeholders, as it offers transparent and systematic methods. Evidence-generating methods gather many perspectives, and these can be synthesized using methods such as logic models and societal return on investment calculations. To date, we are aware that the Bill and Melinda Gates Foundation have expressed interest in the exploration of these methods. In the future, we would hope that this interest can develop more widely and that development-focused HTA can optimize STI in the achievement of SDG3.

References

Andronis, L. (2015). Analytic approaches for research priority-setting: issues, challenges and the way forward. *Expert Review of Pharmacoeconomics and Outcomes Research* 15, 745–754.

Annemans, L., Genesté, B. and Jolain, B. (2000). Early modelling for assessing health and economic outcomes of drug therapy. *Value in Health* 3, 427–434.

Bennette, C. S., Veenstra, D. L., Basu, A., Baker, L. H., Ramsey, S. D. and Carlson, J. J. (2016). Development and evaluation of an approach to using value of information analyses for real-time prioritization decisions within SWOG, a large cancer clinical trials cooperative group. *Medical Decision Making* 36, 641–651.

Breteler, M. (2012). Scenario analysis and real options modeling of home brain monitoring in epilepsy patients. MA thesis, University of Twente, Netherlands.

Hay Burgess, D. C., Wasserman, J. and Dahl, C.A., (2006). Global health diagnostics. *Nature* 444(1s), 1–2.

Carlson, J. J., Thariani, R., Roth, J., Gralow, J., Henry, N. L., Esmail, L., Deverka, P., Ramsey, S. D., Baker, L. and Veenstra, D. L. (2013). Value-of-information analysis within a stakeholder-driven research prioritization process in a US setting: an application in cancer genomics. *Medical Decision Making* 33, 463–471.

Chapman, A. M. (2013). The use of early economic evaluation to inform medical device decisions: an evaluation of the Headroom method. PhD diss., University of Birmingham, UK.

Claxton, K. P., Revill, P., Sculpher, M., Wilkinson, T., Cairns, J. and Briggs, A. (2014). *The Gates Reference Case for Economic Evaluation.* The Bill and Melinda Gates Foundation, Seattle.

Cosh, E., Girling, A., Lilford, R., Mcateer, H. and Young, T. (2007). Investing in new medical technologies: a decision framework. *Journal of Commercial Biotechnology* 13, 263–271.

Davey, S. M., Brennan, M., Meenan, B. J., McAdam, R., Girling, A., Chapman, A. and Lilford, R. (2011). A framework to manage the early value proposition of emerging healthcare technologies. *Irish Journal of Management* 31(1), 59–75.

De Graaf, G., Postmus, D. and Buskens, E. (2015). Using multi-criteria decision analysis to support research priority setting in biomedical translational research projects. *BioMed Research International*, article 191809.

Droppert, H., and Bennett, S., (2015). Corporate social responsibility in global health: an exploratory study of multinational pharmaceutical firms. *Globalization and health* 11(1), 15.

Drummond, M. F., Sculpher, M. J., Claxton, K., Stoddart, G. L., and Torrance, G. W. (2015). *Methods for the Economic Evaluation of Health Care Programmes.* Oxford University Press, Oxford.

Girling, A. J., Freeman, G., Gordon, J. P., Poole-Wilson, P., Scott, D. A. and Lilford, R. J. (2007). Modeling payback from research into the efficacy of left-ventricular assist devices as destination therapy. *International Journal of Technology Assessment in Health Care* 23(2), 269–277.

Gosling, J. P. (2014). *Methods for Eliciting Expert Opinion to Inform Health Technology Assessment.* Vignette Commissioned by the MRC Methodology Advisory Group. Medical Research Council (MRC) and National Institute for Health Research (NIHR), Leeds, UK.

Haakma, W. (2011). Expert elicitation to populate early health economic models of medical diagnostic devices in development. MA Thesis, University of Twente, Netherlands.

Hilgerink, M. P., Hummel, M. J., Manohar, S., Vaartjes, S. R. and Ijzerman, M. J. (2011). Assessment of the added value of the Twente Photoacoustic Mammoscope in breast cancer diagnosis. *Medical Devices* 4, 107–115.

Hummel, J. M., Boomkamp, I. S., Steuten, L. M., Verkerke, B. G. and Ijzerman, M. J. (2012). Predicting the health economic performance of new non-fusion surgery in adolescent idiopathic scoliosis. *Journal of Orthopaedic Research* 30(9), 1453–1458.

Kettler, H., White, K. and Hawkes, S. (2004). *Mapping the Landscape of Diagnostics for Sexually Transmitted Infections.* http://www.who.int/tdr/publications/documents/mapping-landscape-sti.pdf (accessed April 9, 2020).

Kip, M., Steuten, L. M., Koffijberg, H., Ijzerman, M. J. and Kusters, R. (2016). Using expert elicitation to estimate the potential impact of improved diagnostic performance of laboratory tests: a case study on rapid discharge of suspected non–ST elevation myocardial infarction patients. *Journal of Evaluation in Clinical Practice* 24(1), 31–41.

Koning, R. (2012). Bacterial pathogen identification in blood samples with electronic nose technology in a clinical setting: an early medical technology assessment of the µDtect. MA Thesis, University of Twente, Netherlands.

Iglesias, C. P., Thompson, A., Rogowski, W. H. and Payne, K. (2016). Reporting guidelines for the use of expert judgement in model-based economic evaluations. *PharmacoEconomics* 34, 1161–1172.

Ijzerman, M. J., and Steuten, L. M. (2011). Early assessment of medical technologies to inform product development and market access. *Applied Health Economics and Health Policy* 9, 331–347.

Joosten, S., Retèl, V., Coupé, V., van den Heuvel, M., and Van Harten, W. (2016). Scenario drafting for early technology assessment of next generation sequencing in clinical oncology. *BMC Cancer* 16(1), 66.

Meltzer, D. O., Hoomans, T., Chung, J. W. and Basu, A. (2011). Minimal modeling approaches to value of information analysis for health research. *Medical Decision Making* 31, E1–E22.

Middelkamp, H. H., van der Meer, A. D., Hummel, J. M., Stamatialis, D. F., Mummery, C. L., Passier, R. and Ijzerman, M. J. (2016). Organs-on-chips in drug development: the importance of involving stakeholders in early health technology assessment. *Applied In Vitro Toxicology* 2(2), 74–81.

Morris, D. E., Oakley, J. E. and Crowe, J. A. (2014). A web-based tool for eliciting probability distributions from experts. *Environmental Modelling and Software* 52, 1–4.

O'Hagan, A. (2019). SHELF: the sheffield elicitation framework (version 4.0). http://www.tonyohagan.co.uk/shelf/ (accessed April 9, 2020).

Østergaard, S., and Møldrup, C. (2010). Anticipated outcomes from introduction of 5-HTTLPR genotyping for depressed patients: an expert Delphi analysis. *Public health genomics* 13(7–8), 406–414.

Pan American Health Organization (2012). Resolution CSP28.R9: Health Technology Assessment and Incorporation into Health Systems. https://www.who.int/medical_devices/assessment/resolution_amro_csp28.r9.pdf (accessed April 5, 2019).

Peeling, R., and Mabey, D. (2010). Point-of-care tests for diagnosing infections in the developing world. *Clinical Microbiology and Infection* 16(8), 1062–1069.

Silverman, R., Chalkidou, K. and Towse, A. (2018). *As UN General Assembly Highlights Tuberculosis Fight, Will BRICS Lead on the R&D Agenda?* Center for Global Development. https://www.cgdev.org/blog/un-general-assembly-highlights-tuberculosis-fight-will-brics-lead-rd-agenda (accessed April 5, 2019).

Social Return on Investment (SROI) Network (2012). *A Guide to Social Return on Investment.* http://www.socialvalueuk.org (accessed April 5, 2019).

Terjesen, C. L., Kovaleva, J. and Ehlers, L. (2017). Early assessment of the likely cost effectiveness of single-use flexible video bronchoscopes. *PharmacoEconomics-Open* 1(2),133–141.

The Lewin Group (2005). *The Value of Diagnostics: Innovation, Adoption and Diffusion into Health Care.* The Lewin Group, Inc., Falls Church, VA, USA.

United Nations (2018a). *Sustainable Development Goal 3: Ensure Healthy Lives and Promote Well-Being for All at All Ages.* https://sustainabledevelopment.un.org/sdg3 (accessed April 5, 2019).

United Nations (2018b). United Nations Millennium Development Goals. http://www.un.org/millenniumgoals/ (accessed April 5, 2019).

United Nations, Economic and Social Council (2016). *Multi-Stakeholder Forum on Science, Technology and Innovation for the Sustainable Development Goals: Summary by the Co-Chairs.* United Nations Official Document. United Nations, New York. http://www.un.org/ga/search/view_doc.asp?symbol=E/HLPF/2016/6&Lang=E (accessed April 5, 2019).

University of Glasgow (2018). Using technology to treat tropical diseases, *MyGlasgow News* (October 15). https://www.gla.ac.uk/myglasgow/news/peopleprojects/headline_617368_en.html (accessed April 5, 2019).

Vallejo-Torres, L., Steuten, L. M., Buxton, M. J., Girling, A. J., Lilford, R. J. and Young, T. (2008). Integrating health economics modeling in the product development cycle of medical devices: a Bayesian approach. *International Journal of Technology Assessment in Health Care* 24, 459–464.

Wahlster, P., Goetghebeur, M., Kriza, C., Niederländer, C. and Kolominsky-Rabas, P. (2015). Balancing costs and benefits at different stages of medical innovation: a systematic review of Multi-criteria decision analysis (MCDA). *BMC Health Services Research* 15(1), 1–12.

Wong, W. B., Ramsey, S. D., Barlow, W. E., Garrison, L. P. and Veenstra, D. L. (2012). The value of comparative effectiveness research: projected return on investment of the RxPONDER trial (SWOG S1007). *Contemporary Clinical Trials* 33, 1117–1123.

World Health Organization (2013). *Main Messages from World Health Report 2013: Research for Universal Health Coverage.* http://www.who.int/whr/2013/main_messages/en/ (accessed April 5, 2019).

World Health Organization. (2018a). *Global Tuberculosis Report.* http://www.who.int/tb/publications/global_report/en/ (accessed April 5, 2019).

World Health Organization (2018b). *Framework on Tobacco Control.* http://www.who.int/fctc/en/ (accessed April 5, 2019).

World Health Organization (2018c). *Declaration on Primary Health Care.* http://www.who.int/primary-health/conference-phc/declaration (accessed April 5, 2019).

World Health Organization (n.d.). *Health Technology Assessment.* http://www.who.int/medical_devices/assessment/en/ (accessed April 5, 2019).

World Health Organization Regional Office for South East Asia (2013) WRCfS-E. Resolution SEA/RC66/R4: Health Intervention and Technology Assessment in Support of Universal Health Coverage. WHO, New Delhi.

Yock, P. G., Watkins, F. J., Denend, L. and Kurihara, C. Q. (2015). *Biodesign.* Cambridge University Press, Cambridge.

13

Anti-Malarial Drug Development and Diffusion in an Era of Multidrug Resistance

How Can an Integrated Health Framework Contribute to Sustainable Development Goal 3?

Freek de Haan and Ellen H.M. Moors

1. Introduction

Sustainability has grown into a paradigm of grand societal and environmental challenges that cannot be ignored by governments, firms, and nongovernmental organizations (NGOs). At the same time, sustainability challenges these groups by requiring short-term investments while the benefits are for the long term. To reach a better environmental, economic, and social future for the world, the United Nations set targets for 2030 by introducing the sustainable development goals (SDGs). The close and mutual relation between sustainability and health is a central focus of SDG3, which aims to "ensure healthy lives and promote well-being for all at all ages" (UN 2015). The pharmaceutical industry is a major element of global healthcare and has the potential to play a key role toward the achievement of SDG3.

Sustainable healthcare, however, is challenged by flaws in the current pharmaceutical system. On the one hand, there is pressure from society to develop innovative medicines that address unmet medical needs. On the other hand, the development of these medicines is expensive while health budgets are inherently limited. Excessive health expenses will inevitably endanger access and affordability of treatment, and so there is also pressure to reduce costs and mitigate health expenses. As a result, pharmaceutical innovation systems are increasingly struggling to deliver new medicines for unmet medical needs at affordable prices (Moors et al. 2014).

Although these are challenges with a global reach, they are even more prominent for countries with limited access to resources, the so-called low and middle-income countries (LMICs).[1] A substantial number of the SDG3 targets aim to improve

---

[1] Low-income economies are defined as those with a gross national income (GNI) per capita, calculated using the World Bank Atlas method, of $3,895 or less; middle-income economies are those with a GNI per capita between $3,896 and $12,055; high-income economies are those with a GNI per capita of $12,056 or more (IAMCR 2019).

healthcare in LMICs. We highlight two of those targets. First, SDG target 3.3 proposes "to end the epidemics of AIDS, tuberculosis, malaria and neglected tropical diseases and to combat hepatitis, water-borne diseases and other communicable diseases by 2030." While the burden of these diseases are low in most affluent countries, they remain prevalent in LMICs (Bhutta et al. 2014). Second, SDG target 3.8 emphasizes the "achievement of universal health coverage, including financial risk protection, access to quality, essential health-care services and access to safe, effective, quality and affordable essential medicines for all" (UN 2015). Here again, people living in LMICs would benefit most from fulfillment of this target, as they are generally lagging behind when it comes to health coverage.

Unfortunately, there are important barriers to establishing a sustainable pharmaceutical system for LMICs. The development of new medicines is resource intensive but commercially unattractive when aiming for poor populations (Arrow et al. 2004; Ubben and Poll 2013). Another complicating factor is the number of stakeholder groups along the pharmaceutical value chain in LMICs. The interests of these stakeholders are often poorly aligned or even conflicting (Yadav et al. 2007). Finally, the healthcare systems in LMICs are heterogeneous in nature and often function suboptimally as they are fragmented and characterized by a shortage of resources (Mills 2014). As a result, medicines often fail to flow to the ones in need. Added together, these conditions complicate the achievement of global access to affordable medicine and the realization of SDG targets 3.3 and 3.8.

Although various solutions have been suggested in the literature to the challenges noted here, in practice a combination of reforms will be necessary to make the development and diffusion of medicines for LMICs sustainable (Moors et al. 2014). Such reforms include a larger, open-access role for public research institutes in the discovery, clinical testing, and safety evaluations of new drugs. Furthermore, innovative business models and collaborations are essential to address healthcare challenges that affect people in LMICs, and a broader view that encompasses policy solutions to tackle global health challenges is needed. This chapter applies an innovation systems perspective to analyze a worrying development that requires coordinated action at drug development, diffusion, and subsequent implementation stages. The challenge is that malaria parasites have started to develop multidrug resistance to the global first-line drug combinations (Imwong et al. 2017). As a result, malaria is becoming increasingly difficult to treat and innovative solutions toward a sustainable anti-malarial drug development and diffusion system are urgently required.

Malaria is a tropical infectious disease caused by one-celled plasmodium parasites, which are transmitted through the bites of mosquitos. Of the four types of plasmodium parasites, p. falciparum is the most prevalent and by far the most deadly. In past decades, impressive progress was made in malaria control and the global burden has been reduced over 50% since the year 2000. Still, nearly half a million people die each year because of the disease, mainly children under five in sub-Saharan Africa (WHO 2017). This is particularly tragic because malaria is well treatable and effective drugs do exist. Unfortunately, the most effective anti-malarial medicines often fail to reach the infected patient, especially in remote areas with limited access to quality healthcare.

Historically, the African continent has been the center of attention and funding for malaria control. However, this focus is shifting—at least partially—now that resistance to the global first-line anti-malarial drugs has emerged and is spreading through the Greater Mekong Subregion (GMS) in Southeast Asia. Although malaria incidence and mortality in the GMS is at historically low levels, these recent developments imply a serious threat to all gains against the disease (Phyo and von Seidlein 2017). Moreover, there is risk that resistance will also spread to South Asia and Africa, where the majority of malaria burden is situated. This could create a public health crisis because no alternative treatment with comparable efficacy against *p. falciparum* parasites is expected in the coming years (Wells et al. 2015). According to conservative estimations, the scenario of resistance spreading to Africa could cause over 100.000 excess deaths annually and the overall extra costs could be more than USD 400 million per year (Lubell et al. 2014).

Hence, new anti-malarial medicines and diffusion strategies are urgently required to restore anti-malarial efficacy in the GMS and to prevent the spread of resistance to other continents. These goals are particularly critical for achieving the SDG3 targets of ending the epidemics of malaria and achieving universal health coverage. In the remainder of the chapter, we focus on anti-malarial drug development and diffusion in the context of emerging drug resistance. By taking a systems perspective, we answer the research question: Which reforms in the anti-malarial drug development and diffusion system are required to address the emergence of multidrug resistance?

The next section presents a sustainable drug development and diffusion framework for analysis of innovative medicines in LMICs. Then we examine malaria in relation to the SDGs, explain the state of the art of malaria treatment, and provide the methodology for this chapter. The following section applies the integrated framework to evaluate the established anti-malarial drug development and diffusion system and discusses implications of emerging multidrug resistance. Finally, the findings are discussed and further policy implications are provided.

2. Sustainable Drug Development and Diffusion Framework for LMICs

While healthcare in most LMICs is still insufficient, the sustainability of healthcare systems in affluent parts of the world is also under pressure, owing to increasing health costs and an aging population. Although most inventions of new treatment modalities are made by academic institutions, drug development has almost exclusively become the activity of large pharmaceutical companies (Moors et al. 2014). The output of new drugs and vaccines, however, has been decreasing dramatically for the last decades. At the same time the prices of new drugs and vaccines have risen steadily, leading to access problems for many patients, especially in LMICs.

In addition, the development and use of medicines is extensively institutionalized through regulations, guidelines, and standards. For example, randomized controlled trials to assess the safety and efficacy of medicines are costly but often mandatory for regulatory approval and market access (Howick 2011; Rafols et al. 2014; Timmermans and Berg 2003). Moreover, active ingredients can be expensive, investments are necessary to produce in line with international standards, and pharmaceutical companies require profit margins to invest in future R&D and to remain commercially viable. The high development and production costs for innovative medicines provoke high end-user prices, which has been a prominent topic in societal debates. In resource restricted settings, the prices of innovative medicines are often too high for governments and patients. As a result, governments and patients in LMICs remain dependent on donor subsidies or lack the access to essential medicines. Moreover, counterfeits (fake medicines) have been entering the medicine market in many settings (Dondorp et al. 2004; Newton et al. 2001). Furthermore, cultural habits of citizen communities, perceived side effects, and inadequate information provision negatively influence the willingness of patients to take drugs (O'Connell et al. 2012). This is all leading to *availability, affordability, accessibility,* and *acceptability* problems for many patients in LMICs. Table 13.1 provides an overview of the central concepts for a sustainable pharmaceutical system in LMICs.

Accordingly, system reforms are necessary to sustainably address medical needs of people living in LMICs. Disruptive reform of the current pharmaceutical innovation system would offer alternative ways of developing drugs. While there is still a strong focus on fixing the flaws in the current system, in practice, a combination of reforms is necessary to make drug development and distribution to patients in LMICs sustainable. Sustainable drug development is described as a process in which medicines are accepted, safe, affordable, and accessible by the ones in need (Nwaka and Ridley 2003). Additionally, sustainability requires responsible deployment of scarce resources to reduce negative future effects such as accelerated development of drug resistance. The patent system, short-term incentives and reward systems, and the lack of capacity of academic institutions to develop innovative drugs are associated with an unsustainable system (Moors et al. 2014; Munos 2009). At the patient levels unsustainability is provoked by availability of low-quality drugs and inadequate utilization of drugs.

Sustainable access to medicine requires a multisectoral approach involving governments, the pharmaceutical industry, NGOs, and multilateral organizations (Meijer et al. 2013). Drug development, then, asks for open transfer and new forms of cooperation through, for example, public private partnerships (PPPs). Innovative and market focused policy approaches are required to improve access to medicines in LMICs (Waning 2011). These should involve sustainable funding, efficient drug regulation and registration practices, reimbursement programs, and effective supply of medicines to both public and private outlets (Arrow et al. 2004; Yadav et al. 2007).

Table 13.1 Overview concepts of sustainable drug development and diffusion

Concept	Description	Anti-malarial S,T,I implications
Availability	Initiating and performing R&D to address unmet medical needs. Sustainable drug development should aim to create open access to share knowledge and to stimulate open innovation for drugs, for example through public–private consortia. Availability also involves capacity issues within regulatory health authorities and drug registration processes to enable market access.	Pragmatic short term approaches are required to tackle the spread of multidrug resistant malaria. Product development consortia can prelude the availability of more sustainable anti-malarial drugs. A promising example is the recent establishment of a consortium which aims to develop and test triple artemisinin combination therapies (TACT). Moreover, accelerated regulatory pathways may be required to ensure rapid availability of innovative medicines.
Affordability	Innovative medicines should be affordable to governments and to patients in resource restricted settings. Pricing strategies (both in public and private sectors) should therefore take the economic possibilities of the patient population in consideration. External funding with subsidies and reimbursement can improve affordability of drugs. In order to become sustainable, this requires ongoing funding rather than incidental programs. Patents and product competition can also affect affordability.	Subsidies such as the Global Fund ensure affordability of medicines in the public sectors of most malaria endemic countries. Eligibility to these subsidy schemes is also required for rapid and sustainable uptake of new anti-malarial medicines, such as TACT. However, the most significant challenges relate to private sectors affordability. In most settings, this requires significant system reformations.
Accessibility	Accessibility deals with the distribution of interventions throughout supply chains: from the central levels to health practitioners and patient populations. Usually access starts with inclusion of effective and safe medicines in both normative global guidelines and in national treatment guidelines. Sustainable access also requires functioning supply chains and sufficient health system capacity. Both public and private sector channels should be taken in consideration to become sustainable. Moreover, access to low-quality alternatives should be avoided on the medicine markets.	Healthcare systems and drug supply chains in malaria endemic settings are heterogeneous in nature. Therefore, no single solutions toward sustainable access to anti-malarials exists. However, inclusion of more sustainable drugs, such as TACT, in global and national guidelines is a crucial first step. Moreover, investments in structural improvements of (private sector) drug distribution networks is required to ensure accessibility of sustainable anti-malarials. Active market regulation may be required to avoid access to outdated, substandard, or counterfeit alternatives.

Continued

Table 13.1 *Continued*

Concept	Description	Anti-malarial S,T,I implications
Acceptability	Drugs need to be accepted by prescribers and patients. This can be achieved by offering drugs without too many side effects and risks. Moreover, the drugs should be adapted to end-user preferences. This requires user involvement and strategic consideration about demand issues such as branding and packaging in an early stage. Acceptability also deals with information provision and promotional activities to inform prescribers and patient populations.	During development of new anti-malarials, including TACT end-users preferences such as taste, tablet size, and perceived side effects should be considered at an early stage to facilitate rapid uptake. Also appropriate branding and packaging of the new medicines is required. After market introduction, training and promotional activities can be useful tools in informing target populations about the risks of malaria resistance and the importance of switching to new anti-malarials.

Source: Based on Moors et al. (2014); Munos (2009); Warren (2018).

3. Malaria: A Multifaceted Global Health Issue

Malaria is a mosquito-borne tropical disease caused by plasmodium parasites. Of the four types of plasmodium parasites, *plasmodium falciparum* is by far the most common and the most deadly type (WHO 2017). In the remainder of the chapter, *plasmodium falciparum* malaria will simply be referred to as malaria.

3.1 Malaria and the SDGs

Major improvements have been made in global malaria control since the year 2000. The burden has been reduced by more than half, and malaria has been eliminated in 19 countries. Moreover, global investments in malaria management and control have more than doubled between 2005 and 2014 (Cibulskis et al. 2016).

Nonetheless, malaria is still a significant health issue with 3.4 billion people in 91 countries at risk of the disease. Each year, more than 200 million people are infected with malaria and the disease takes an estimated 445,000 deaths per year (WHO 2017). This is unacceptably high when considering that malaria is well preventable and treatable. Malaria mortality is mainly prevalent in children under five years of age in sub-Saharan Africa, the region with highest malaria transmission. These young African children are most vulnerable to malaria because they have not yet acquired any form of immunity toward the disease, in contrast with adolescents and adults. In settings with lower transmission rates, including the GMS, malaria is less of a pediatric disease but also affects adults, often forest workers and migrant populations in border areas (WHO 2016).

The clinical representation of a malaria infection is usually classified into two levels of intensity: uncomplicated malaria and severe malaria. Uncomplicated malaria refers to a common malaria infection where the patient presents febrile symptoms such as headache, vomiting, chills, and diarrhea. These symptoms overlap with many other tropical diseases (Luxemburger et al. 1998). Hence, malaria is often under- or overdiagnosed when reliable diagnostics are not available. Symptoms of severe malaria include coma, organ failure, and eventually death (Bartoloni and Zammarchi 2012). A case of severe malaria requires intravenous medication for direct relief of symptoms, while global guidelines recommend a three day oral treatment for uncomplicated malaria (WHO 2017).

The ongoing malaria challenges have explicitly been addressed in the SDGs. The SDG target 3.3 aims to end the epidemics of malaria and other (infectious) diseases by 2030. However, malaria is a poverty-related disease and the disease also affects other areas of sustainable development. Reducing the malaria burden could also benefit other sustainability goals including poverty (SDG1) and inequality reduction (SDG10) and improving education through reducing disease related absence (SDG4). An increasing malaria burden because of multidrug resistance would simultaneously affect these goals in a negative manner. Aligned with the SDGs, WHO formulated a global technical strategy (GTS) 2016–2030 (see Table 13.2). This GTS report suggests ways to reduce global malaria mortality and case incidence by 90% in 2030, with intermediate objectives for the years 2020 and 2025 (WHO 2016). Moreover, malaria should be eliminated in at least 35 countries and, importantly, re-establishment of the disease needs to be prevented. These milestones seem ambitious and we are not yet on track to achieving them (WHO 2018).

A number of strategies and interventions are essential in malaria control and in proceeding toward the SDG and GTS targets. First, preventive measures such as insecticide treated bed nets (ITN) and indoor residual spraying (IRS) have proven extremely effective in improving malaria control (Bhatt et al. 2016). Second, in response to the overlap in symptoms between malaria and other diseases, WHO now recommends diagnostic confirmation of all suspected cases of malaria before administering treatment. Malaria diagnosis is typically conducted by means of microscopy or rapid diagnostic tests (RDT). Global availability of effective medicines is another strategic cornerstone for effective malaria control. These days, the treatment of uncomplicated falciparum malaria globally relies on a three-day oral course of artemisinin combination therapies (WHO 2001).

Malaria surveillance is another fundamental cornerstone for management of the disease. Surveillance refers to the generation of data in order to identify outbreaks and populations at risk, but also to target interventions and to evaluate control programs. However, such data are often considered limited or unreliable owing to poor information technology structures, lack of administrative traditions in LMICs, and

Table 13.2 Global technical strategy for malaria 2016–2030

	Milestones		Targets
Goals	2020	2025	2030
Reduce malaria mortality rates globally compared with 2015	≥ 40%	≥ 75%	≥ 90%
Reduce malaria case incidence globally compared with 2015	≥ 40%	≥ 75%	≥ 90%
Eliminate malaria from countries in which malaria was transmitted in 2015	≥ 10 countries	≥ 20 countries	≥ 35 countries
Prevent re-establishment of malaria in all countries that are malaria-free	Re-establishment prevented	Re-establishment prevented	Re-establishment prevented

Source: WHO (2016).

the absence of central coordination. Therefore, malaria surveillance is put on the agenda as one of the cornerstones for achieving the GTS goals (WHO 2016).

3.2 Malaria Treatment: The State of the Art

Artemisinin combination therapy (ACT) constitutes the regime of first-line treatment for uncomplicated malaria throughout the world. ACT combines an artemisinin derivative, the most potent class of anti-malarial drug, and a less aggressive but longer acting partner drug. The rationale behind deployment of combination therapies is to synergize efficacy of both individual compounds while, at the same time, risk of parasites developing resistance to the individual compounds is being reduced (White 1999).

ACT has only been available since the beginning of this millennium. For decades, the anti-malarial market had been dominated by chloroquine, a cheap and widely available drug costing approximately USD $0.10 per dose (Arrow et al. 2004). Unfortunately, chloroquine resistance emerged in Asia and spread throughout the world. In response, malaria endemic countries switched to alternative first-line treatments such as sulfadoxine-pyrimethamine and mefloquine, but soon these replacement drugs started to fail as well (Faurant 2011). In the late 1990s there was only one anti-malarial compound left without declining efficacy: artemisinin, an extract of the Chinese sweet wormwood plant. It was clear that the future of malaria treatment had to rely on this ingredient and its derivatives (artesunate, artemether, dihydro-artemisinin).

The artemisinin derivatives are extremely potent in eliminating malaria parasites and in rapidly reducing the symptoms of a malaria infection. However, the compounds also have a short plasma half-life, which means that it is rapidly removed from the body. As a result, chances of parasites surviving the treatment are high, which may subsequently lead to recrudescence of the infection. Moreover—and this is considered an even larger threat—the short plasma half-life significantly reinforces the risk of drug resistance emerging (White 1999). These resistance mechanisms are a consequence of genetics and natural selection when the treatment regime fails to eliminate all parasites, for example because of inappropriate drug use (Phyo et al. 2016).

Given the fact that artemisinin is the last available anti-malarial and is crucially important in the battle against malaria, there has been a general consensus that artemisinin resistance would have to be avoided at all costs. This led to the guiding principle that artemisinin and its derivatives ought to be exclusively deployed in combination with a partner drug (usually an "older" anti-malarial with a history of resistance) (WHO 2001). The rationale is that the artemisinin compound of the ACT would reduce the majority of the parasite burden within the first three days of treatment. The partner drug component eliminates the remaining parasites in the following days and prevents recrudescence of the infection. WHO recommends a three-day oral course of ACT in all areas of resistance to traditional anti-malarials since the year 2001. In 2006, this was further extended to all cases of diagnostically confirmed uncomplicated malaria (WHO 2006).

Upon this policy change and when the transition to ACT was started, the global anti-malarial market was rapidly polluted by dangerous compounds that did not meet quality requirements. Artemisinin was still often being prescribed without partner drug, despite the risk of resistance, but also substandard and even counterfeit ACTs became commonly available on the market (Bosman and Mendis 2007; Newton et al. 2001). In response, WHO stressed that the prescription of artemisinin monotherapies (without a partner drug) should be avoided to reduce risk of artemisinin resistance. Also, unfinished courses of ACT and the use of drugs that did not meet quality standards were strongly discouraged in global treatment guidelines (WHO 2006). Moreover, the development of quality assured, fixed-dose ACT (the individual components combined in one pill) was initiated. The idea was that fixed-dose ACT would improve adherence to the full treatment regime and avoid the selective intake of compounds, as was common with coblistered drug combinations.

Despite all these preventive measurements, artemisinin resistance has emerged in Cambodia, and is spreading throughout the GMS (Imwong et al. 2017; Noedl et al. 2008; Phyo et al. 2016). Along the way, artemisinin resistant parasite lineages also managed to acquire resistance to partner drugs. As a result, malaria has become difficult to treat in the region. Switching to other ACT combinations has proven to be only a short-term solution before resistance to the new combination is acquired (Dondorp et al. 2017). Unfortunately, new anti-malarial compounds are still years away in the drug pipeline (Wells et al. 2015). Therefore, pragmatic, short term solutions are rapidly required to tackle the direct, urgent threat.

Potential short term solutions are to extend ACT intake from three to seven days, or to deploy multiple first-line ACTs in a region at the same time (Boni et al. 2016; Dondorp et al. 2017; Phyo and von Seidlein 2017). However, malaria experts consider a transition to triple combinations of currently prescribed compounds to be the most feasible short- to middle-term solution. Deployment of these triple artemisinin combination therapies (TACT) could provide direct relieve in the GMS, where double combinations are increasingly failing. Moreover, TACT could prevent spread and the de novo emergence of artemisinin and partner drug resistance in South Asia and Africa. In the long term, new active compounds will be required in the battle against malaria and to achieve the SDG3 and WHO's GTS goals.

4. Methods

This chapter takes a systems perspective and applies the sustainable framework for drug development and diffusion to the problem of emerging anti-malarial drug resistance. The qualitative study is based on desk research, including peer reviewed and grey literature review, and on expert consultations. Peer-reviewed literature was accessed via the Web of Science search engine. A broad search terminology was used to understand the anti-malarial drug development and diffusion systems and the implications of drug resistance. An example of such a search query is ("development" or "implementation" or "diffusion" or "transition") and ("artemisinin" or "artemether" or "artesunate" or "dihydroartemisinin") or ("ACT" and "malaria"). For each publication, the titles and abstracts were examined. The full content of papers that were subjected to relevant topics for anti-malarial drug development and diffusion was scrutinized. Papers that did not meet these criteria (e.g., in-vitro/vivo clinical studies, gene detection studies) were excluded. Moreover, references of all papers were checked for additional publications of interest. Typical journals of relevant studies included *Malaria Journal*, the *New England Journal of Medicine, Drug Discovery Today*, and *Nature*. Grey literature reports and websites were also identified through reference checks and by using similar search terminology in the search engine Google. Examples of selected grey literature are the yearly WHO malaria reports and the websites of the Worldwide Antimalarial Resistance Network (WWARN) and of the Medicines for Malaria Venture (MMV).

Additional insights were generated through expert consultations. Those consultations were conducted with individuals who have been involved professionally in the development, diffusion, and/or supply of (anti-malarial) medicines in LMICs. The experts were consulted in the period August–November 2018 and included representatives from academia and research based consultancy who are or have been active in projects related to anti-malarial drug development and diffusion. They were asked to share their experiences and opinions regarding the availability, affordability, accessibility, and acceptability of anti-malarial medicines. Where relevant to the background of the expert, specific consultation about these themes in light of drug resistance was requested.

The selected literature was screened for relevant topics regarding the established anti-malarial drug development and diffusion systems or to implications of emerging multidrug resistance to this system. Topics were then categorized according to the components of the sustainable drug development and diffusion framework. Moreover, when a topic was relevant to more than one category, this connection was further explored and explicated. The same process of data analysis was also carried out for data generated through the expert consultations. Finally, the results of the desk research were structured and will be presented in the next section. Each subsection first provides a general overview and relates the established subsystems for ACT to one of the framework components. The analysis then evaluates the situation of emerging of multidrug resistance. Finally, the implications of these developments on sustainable drug development are discussed. Interdependence between the framework components is specified throughout each subsection.

The GMS is a region in Southeast Asia and consists of six countries: Thailand, Cambodia, Myanmar, Vietnam, Laos, and China (Yunnan province). The malaria burden in the region has been strongly reduced in the last years: the number of malaria deaths declined 84% between 2012 and 2015 (WHO 2015). These gains are mainly attributed to improved access to effective interventions for malaria prevention, diagnosis, and treatment. Despite these recent improvements, the GMS is also the region where resistance to artemisinin and partner drugs has developed and is spreading. Innovative responses are required to restore anti-malarial efficacy in the region. At the same time, the spread of resistance to South Asia and Africa, where the majority of the malaria burden is situated, could have a large public health impact. This situation touches upon many challenges for sustainable drug development for LMICs.

5. Toward a Sustainable System of Anti-Malarial Drug Development and Diffusion

Earlier in this chapter, a framework for sustainable drug development and diffusion in LMICs was developed. The present section applies this framework to provide an evaluation of the anti-malarial drug system following the framework components availability, affordability, accessibility, and acceptability. For each component, a general overview of the established system for ACT, the current first-line anti-malarial drugs, is provided. Then the implications of emerging multidrug resistance are evaluated. Finally, lessons from the integrated framework are translated into policy implications toward a sustainable system of anti-malarial drug development and diffusion, helping to meet the achievement of the SDG3 targets.

5.1 Availability

Traditionally, research based pharmaceutical companies are the institutions that develop innovative medicines and bring them to the market. However, the development of new medicines is resource intensive while the prospective returns when developing drugs for poor populations is low. Hence, for a long period of time, pharmaceutical companies have failed to address many neglected and poverty related diseases. This is well reflected by the history of anti-malarial medicines (Ubben and Poll 2013). Nearly all medicines that have become first-line treatments in recent history have been the product of governmental research projects (MMV 2016). Remarkably often they have been developed by military institutions during wartime. Examples have been chloroquine and sulfadoxine-pyrimethamine during the Second World War, and mefloquine during the Vietnam War. All these "traditional" anti-malarial medicines appeared relatively cheap to produce and were soon adopted in portfolios of pharmaceutical companies, including Indian generic manufacturers.

The anti-malarial drug development system appeared unsustainable, however, when resistance had emerged to all these traditional drugs at the end of the last millennium and a rapid transition to artemisinin-based combination therapies was required. At the time, artemisinin was only cultivated on a small scale and ACTs were not yet commercially produced. Because of the lengthy cultivation and extraction process of artemisinin, the estimated production price of ACT was twenty- to fifty-fold that of chloroquine (Arrow et al. 2004). As a result, ACT was considered unaffordable for patients and governments in most malaria endemic settings.

Despite the urgent need, most research based pharmaceutical companies were initially reluctant to invest in ACT, although some coblistered combinations were introduced (Ubben and Poll 2013; Wells et al. 2013). The only exemption at the time was Novartis, a large Swiss pharmaceutical company, which had a fixed-dose combination of Artemether and Lumefantrine (AL) in their pipeline. This first fixed-dose ACT was introduced in 2001 under the brand name Coartem® (Spar 2008). In the following years Coartem® was added to the WHO list of essential medicines and received regulatory prequalification approval, which enabled its entry on the global anti-malarial market.

A few years earlier, two organizations were established to extend the anti-malarial pipeline with fixed-dose ACTs. The Medicines for Malaria Venture (MMV) initiative started as a collaboration between multiple European governments and philanthropic organizations. The mission of MMV was to reduce the burden of malaria by discovering, developing, and delivering new effective and affordable antimalarial drugs (MMV 2016; Ubben and Poll 2013). Moreover, Drugs for Neglected Diseases initiative (DNDi) was founded by the MSF Access Campaign. Similar to MMV, it established a collaborative, needs-driven, not-for-profit drug research and development (R&D) organization. Its aim was to develop new or improved treatments for neglected diseases, including malaria (DNDi 2015). In 2015, DNDi transferred all their malaria related activities to MMV.

These nonprofit organizations aimed to developed new medicines for resource restricted populations by combining expertise and capabilities from industry, academia, and NGOs. Funding came from governmental and philanthropic resources, and these partnerships led to the introduction of several fixed-dosed ACTs between

2008 and 2012 (Bompart et al. 2011; Spar 2008; Wells et al. 2013). Different regulatory pathways were pursued, and most of these newly introduced ACTs received WHO prequalification status. Moreover, patent rights were waived for all the fixed-dose ACTs that were launched through these partnerships. This enabled generic pharmaceutical companies to enter the market, eventually leading to price reductions (Orsi et al. 2018). For example, a fixed-dosed combination of artesunate and mefloquine (AS-MQ) has been produced by Indian Cipla since 2008 after a technology transfer by the Brazilian pharma company Farmanguinhos (Wells et al. 2013).

Now that ACT efficacy has started to decline because of resistance, new medicines are again urgently required. Product development partnerships such as those managed by DNDi and MMV have proven to be very useful in managing the development of need-based medicine. They have successfully introduced a number of effective ACTs and it is likely that the near future pipeline of anti-malarials will also have to rely on partnerships. Therefore, they should be considered as important instruments in addressing unmet medical needs in LMICs and achieving the SDG3 targets of ending the epidemics of malaria (target 3.3) and of ensuring universal health coverage by providing access to safe, effective, quality, and affordable malaria products (target 3.8). An encouraging development is therefore the recent establishment of a consortium to start the development of triple ACT or TACT. By combining compounds with reverse resistance mechanisms, these TACT combinations are considered the most feasible short-term solution to address artemisinin and partner drug resistance. Such innovative development models reduce development costs and thereby contribute to increased affordability of new drugs. Moreover, lessons should be learned from past experiences in acquiring regulatory approval and in strategically positioning new anti-malarials on the market. Standard regulatory procedures can, for example, be too lengthy to respond to the development of resistance and so accelerated regulatory pathways may be required. Considering distribution and market related issues in an early stage allows for strategically anticipating on challenges and may therefore improve acceptability and affordability of new medicines. Hence, we stress the importance of complementary, bottom-up studies throughout the development of need-driven medicines to improve prospects of rapid uptake.

5.2 Affordability

The development of medicines is inherently a lengthy and resource intensive process. To remain profitable, pharmaceutical companies put innovative medicines on the markets for prices that are often too high for poor populations. Reducing the costs of development through innovative business models can therefore be an important step toward cheaper medicines. Moreover, sustainable funding programs or procurement subsidies are required for innovative medicines to become affordable for populations in LMICs.

In the early days of ACT, a first solution toward affordability came in the form of a memorandum between WHO and Novartis. Directly after the introduction of Coartem®, an agreement was signed that Novartis would waive profit and deliver Coartem® at cost to WHO. WHO would then distribute the medicines to the governments of malaria endemic countries. In return, WHO would provide Novartis with a quarterly forecast of expected orders (Bosman and Mendis 2007; Spar 2008). This way, Novartis was guaranteed that they would not risk an unsold overproduction, and they could start scaling up production. Economies of scale further reduced the production costs when global demand increased and Novartis managed to exponentially scale up the production of Coartem® (Spar 2008).

The funding structure between Novartis and WHO became the standard for other ACTs that were introduced, including those launched by the product development partnerships. As a result, access to ACT within the governmental controlled public sectors significantly improved. However, ACT was still significantly more expensive than all previous generations of anti-malarials and external subsidies were required to ensure affordability (Arrow et al. 2004). Initially, the funding of public sector procurement was accounted for by multiple donors. These days, most public sector procurement procedures have been institutionalized by a few larger programs such as the Global Fund to fight Aids, Tuberculosis, and Malaria. The Global Fund guarantees sustainable funding to ensure ongoing access to affordable, quality-ensured medicines.

Although many patients do benefit from these subsidies, most patients in developing countries seek care in the private sectors. For many of them, ACT remained unaffordable even after the establishment of the Global Fund, and these problems are unfortunately still ongoing (Littrell et al. 2011). As a result, many patients who purchase their medicines in the private sector still receive outdated, low-,quality or even fake medicines.

Moreover, in many settings even the subsidized ACT remain unaffordable for patient populations. In LMICs, public insurance schemes are absent and healthcare is usually paid out of pocket by patients (Hopkins 2010). Moreover, medicine supply chains in LMICs are complex and consist of many different levels of procurement agents, national and regional wholesalers, traders, and dispensing outlets (Yadav 2009). Price markups throughout the supply chain are common, especially in the private sectors. Within such an environment, heavily subsidized medicines can still be unaffordable for populations who often have an income below the poverty line. This requires increased control over supply chains and touches upon effective accessibility and acceptability issues. In this way, these subsidized medicines are undermining the overall achievement of SDG3 regarding increased affordability of medicines for all in need.

Innovative anti-malarial medicines that are suitable to tackle resistance, such as TACTs, need to become available at affordable prices. Eligibility to subsidies to stimulate affordability of anti-malarial medicines, including the Global Fund, is closely related to the regulatory status of the product (availability) and should therefore be addressed in an early developmental stage. Moreover, and this remains a serious challenge, affordability at the level of end-users in the private sectors should be ensured. These systems challenges not only touch upon the anti-malarial drugs but also affect other diseases such as tuberculosis and HIV/Aids. Hence, achievement of SDG3 by ending (poverty-related) epidemics and realizing universal health challenge requires ongoing attention to affordability of essential medicines.

5.3 Accessibility

When analysing drug distribution in LMICs, a distinction is usually made between private versus public sector accessibility. Private sector supply chains in LMICs often consist of unregulated networks of intermediary traders before medicines reach the retailing outlets (Yadav 2009). As explained earlier, these intermediaries usually require a mark-up, and so the prices of ACTs are incremented throughout the supply chain. Private sector retailers include formal facilities such as pharmacies and private clinics, but also informal facilities such as general stores and drug traders that sell anti-malarials (Yadav et al. 2007). Within the government regulated public sectors, drugs rather tend to flow horizontally from national governments to the level of health districts and eventually to the retailing facilities. Public sector retailers tend to be more compliant to guidelines than their private counterparts (O'Connell et al. 2011). The public sector includes governmental hospitals and health clinics, but also community or village health worker programs (Ajayi et al. 2008). Important to note here is that health systems, drug supply chains, and medicine markets are heterogeneous in nature and strongly context- and country-dependent. Therefore, our analysis will mainly touch upon some general themes instead of going in-depth into a specific local situation.

Once fixed-dose ACTs were produced and registered, they had to be delivered to both public and private sector retailers and prescribers in malaria endemic settings. The important first step was that malaria endemic countries had to include ACT as first-line anti-malarial drug in their national treatment guidelines. This pace was catalyzed after WHO recommended ACT as global first-line treatment for uncomplicated malaria in 2001. Between 2001 and 2007, nearly all malaria endemic countries did so in response to WHO recommendations (Bosman and Mendis 2007). However, time delays from months to years were common between inclusion in treatment guidelines and the start of implementation programs to actively ensure access and promote the uptake of ACTs in local settings.

From the onset, countries experienced many challenges with the distribution of quality assured ACT to end-users. Weak coordination of private sector supply chains and dispensing outlets often prevented quality assured ACT from reaching the patient, while outdated, low-quality, or even counterfeit medicines remained present on the anti-malarial markets (O'Connell et al. 2011; Palafox et al. 2014). Unfortunately, these problems are still ongoing in many endemic settings, and this challenges the achievement of universal health coverage (SDG3.8) by preventing patients from receiving adequate treatment. Moreover, deployment of inadequate medicines, in particular artemisinin monotherapies, contribute to increased drug pressure which puts the ability to end the endemics of malaria (SDG3.3) at risk.

The access problems are especially worrying in the light of emerging artemisinin and partner drug resistance. Lack of accessibility of appropriate, sustainable drugs are, equally to a lack of affordability, likely to stimulate the use of low quality alternatives. This, in turn, increases the risk of resistance. Hence, investments in overall structures of drug distribution in LMIC, with a strong emphasis on private sectors, is required for a more sustainable system. To tackle anti-malarial drug resistance, an important first step is inclusion of sustainable, evidence-based and quality assured drugs in both global and national guidelines. This is directly related to the availability of such medicines and requires a viable drug pipeline in times of multidrug resistance. Beyond that, increased coordination and proper supply chain management, including procurement, stocking, and distribution, is essential to ensure access to sustainable medicines (Palafox et al. 2014). This also includes market regulation to prevent inadequate medicines from being available. Access to appropriate drugs is, logically, dependent on access to those drugs.

5.4 Acceptability

Finally, for drug development and diffusion to become sustainable, evidence-based medicines should be accepted and used rationally. However, many factors complicate acceptability and again these factors may be heterogeneous according to the setting. In general, issues such as perceived safety and side effects are known to affect the acceptability of medicines (Moors et al. 2014). However, in LMICs the willingness and ability to pay for treatment modalities are other important factors that affect acceptability of medicines (Mills 2014). As elaborated in previous subsections, this requires not only alternative business models for production but also well-functioning supply chains.

Many elements have been proposed to affect the acceptance of ACT at the population level including the level of education and knowledge, community perceptions, and cultural and socioeconomic factors (Littrell et al. 2011). However, there is little consensus on effective behavior change strategies to increase uptake of anti-malarial medicines (Smith et al. 2009). This may well relate to the diversity of settings, which all deal with locally different socioeconomic, cultural, and epidemiological situation, but also to local market conditions. A complicating factor toward a sustainable pharmaceutical system is that the interest of the individual patient in selecting appropriate medicines does not necessarily match the interest of the larger population (Arrow et al. 2004). A sick patient may want a cheap and rapid working drug (e.g., artemisinin monotherapies), while the interest of the larger population is that the patient uses the most sustainable drugs with the lowest risk of developing resistance. Such issues, which can even be reinforced by, for example, side effects of the sustainable medicine, are also subject to ethical debates.

Despite the diversity in the acceptability literature, there is some consensus that community engagement programs can be effective in increasing the acceptance of ACT, and therefore in achieving the SDG3 targets (Ajayi et al. 2008; Yeung et al. 2008). The same applies for promotion campaigns that include end-user perceptions, including branding and packaging (Novotny et al. 2016). For sustainable new medicines, such as TACT, these issues should be considered in an early stage to allow for strategic consideration and to anticipate on appropriate market positioning. Finally, information provision and educating providers and populations in malaria endemic settings on disease trends such as drug resistance can be useful to improve acceptability of new medicines. However, the availability of effective and sustainable drugs are a prerequisite to acceptability. Moreover, the accessibility and affordability compared to alternative treatment options are also likely to affect end-user perceptions and the acceptability of new drugs. The acceptance and appropriate use of sustainable medicines is essential in the battle against resistance and therefore deserves strategic consideration by conducting patient acceptance studies to match innovative products with end-user preferences.

6. Discussion

In this chapter, a framework for sustainable drug development and diffusion in LMICs was developed. The framework was then applied to evaluate a global health challenge: the emergence of resistance to the global first-line anti-malarial medicines. The chapter started with the question which system reforms are required to tackle resistance. What becomes clear is that the threat of resistance is multifaceted and that availability, affordability, accessibility, and acceptability of essential medicines is interrelated. We have shown that a system approach is required to tackle this global health challenge and to work toward a sustainable system of anti-malarial drug development and diffusion. This involves collective actions of stakeholder groups along the medicine value chain and should be aligned with the prevalent institutional landscapes (Yadav 2009; Moors et al. 2014).

The assessment elaborately addressed relevant topics for achieving sustainable healthcare (SDG3) and other interlinked SDGs. Particular focus areas were SDG target 3.3, which aims to end the epidemics of a number of diseases, including malaria, by 2030, and SDG target 3.8, which explicates goals for achieving universal health coverage. However, the pathway toward a sustainable anti-malarial development and diffusion system also affects other SDGs. Simultaneously, emerging multidrug resistance does not only threat the achievement of SDG3, but also of these other goals. We mentioned that a reduced malaria burden can be associated with higher productivity and therefore reduces poverty (SDG1), and at the same time reduces inequality between countries (SDG10). Moreover, in many settings malaria is a pediatric disease which mostly affects children. Being sick because of malaria will prevent them from going to school, and therefore reducing the malaria burden through STI approaches will improve education rates (SDG4).

To become sustainable and to tackle the threat of resistance, a number of adaptations in the pharmaceutical system are required. First, sustainable and innovative anti-malarials should be developed and rapidly made available to patients through regulatory pathways (Waning 2011). To realize this, the role of the pharmaceutical industry needs to be reconsidered. Poverty-related diseases are often considered commercially unattractive by pharmaceutical companies, although corporate responsibility can be an incentive. At the same time, these companies are extremely powerful in developing drugs because they are the ones with expertise and resources to do so (Access to Medicine Index 2018). To become sustainable and safeguard future R&D efforts for resource restricted populations, alternative business models such as product development partnerships are required. By combining expertise and resources of both industry and academia, such partnerships can be established and prices of drug development can be reduced. Promising results have been achieved through innovative approach such as MMV and DNDi in managing the development of a number of fixed-dose anti-malarials (Bompart et al. 2011; Ubben and Poll 2013; Wells et al. 2013). In line with this, the recent establishment of the DeTACT project is an encouraging development in the battle against multidrug resistant malaria. This partnership, funded by the UK Department for International Development, aims to pragmatically combine current anti-malarial compounds into TACT and test them in multiple malaria endemic settings. The idea is that selecting compounds with reverse resistance mechanisms can extend the effective lifetime of current compounds. This way, TACT can benefit populations in malaria endemic regions until effective new medicines become available.

To achieve a sustainable system, affordability of next-generation anti-malarials needs to be ensured. Therefore, drugs should be designed and positioned to fit established market structures and preconditions of subsidy schemes. Establishment of subsidy programs such as the Global Fund have been an important step forward to public sector affordability. Particular attention should now be given to affordability in private sector outlets, which generally lag behind its public counterparts. This could involve increased control on private sector supply chains and legal action against disproportional markups and retail prices (Novotny et al. 2016).

Sustainable accessibility at patient levels starts with inclusion of appropriate anti-malarials in global and national guidelines but is also related to efficient supply chains and market conditions (O'Connell et al. 2011). Defining a market positioning strategy to integrate sustainable medicines into established market structures and stimulate procurement of appropriate medicines should therefore be a central focus in the trajectory toward a sustainable system. This is even more important in an era of multidrug resistance, where utilization of inadequate drugs can increase drug pressure and lead to accelerated resistance. Finally, acceptability of appropriate drugs is essential in achieving a sustainable pharmaceutical system. This can be achieved by offering drugs without too many side effects and risks and by adapting drugs to end-user preferences. Promotional activities to inform populations about appropriate and rational drug use can also be effective tools (Novotny et al. 2016). We stress that bottom-up studies to investigate end-user perceptions should be an integrated element throughout the development and implementation of need-driven medicines.

Our analysis shows that the four framework components are interrelated. Affordability and accessibility of innovative medicines are affected by similar distribution and market dynamics. At the same time, ensuring availability in line with institutional requirements is required for affordability and accessibility. Finally, acceptability of medicines will only become an issue if availability, affordability, and accessibility are ensured. The interrelatedness of issues could also be reflected in the SDGs. Ending epidemics that are mostly prevalent in LMICs (SDG3.3) cannot be seen separately from universal access to healthcare (SDG3.8). The same applies to the malaria burden versus poverty (SDG1), education (SDG4), and equality between countries (SDG10). Moreover, water and sanitation conditions (SGD6) and global warming (SDG13) may affect the living conditions of mosquitos and therefore also affect malaria endemicity. Given the interrelations and multifaceted nature, we argue that systemic approaches should be at the root of tackling global health challenges.

7. Conclusion

An integrated health framework toward sustainable drug development and diffusion in LMICs was developed. The framework was applied to evaluate the situation of emerging anti-malarial drug resistance. Some policy implications follow from the analysis. First, business models such as product development partnerships can be powerful instrument and innovative approach in addressing medical needs that are neglected by industry. Therefore, targeted investments in such collaborations can be a useful tool in achieving the SDG3 targets. The second policy implication is that the effective transition to new medicines in LMICs involves many stakeholders and takes place in complex institutional landscapes. Considering demand, distribution, and market-related issues in an early stage of drug development allows for adequate positioning of new medicines to tackle global health challenges. Finally, availability, affordability, accessibility, and acceptability of innovative medicines are important concepts in improving global health and can contribute significantly to the implementation of the SDG3. They are interrelated and should be considered as integrated components of the drug development and diffusion system rather than as isolated themes.

Acknowledgments

The authors want to thank the participants of the interviews and the science and innovation management bachelor students in the course Sustainability, Health and

Medical Technology of Utrecht University for their valuable input. This chapter was made possible by funding from the Department for International Development, UK.

References

Access to Medicine Index. (2018). https://www.accesstomedicinefoundation.org/access-to-medicine-index (accessed April 4, 2019)

Ajayi, I. O., Browne, E. N., Garshong, B., Bateganya, F., Yusuf, B., Agyei-Baffour, P., Doamekpor, L., Balyeku, A., Munguti, K., Cousens, S., and Pagnoni. F. (2008). Feasibility and acceptability of artemisinin-based combination therapy for the home management of malaria in four African sites. *Malaria Journal* 7, 1–9.

Arrow, K. J., Panosian, C. B., Gelband, H. (2004). *Saving Lives, Buying Time: Economics of Malaria Drugs in an Age of Resistance*. National Academies Press, Washington, DC.

Bartoloni, A., and Zammarchi, L. (2012). Clinical aspects of uncomplicated and severe malaria. *Mediterranean Journal of Hematology and Infectious Diseases* 4(1), e2012026.

Bhatt, S., Weiss, D. J., Cameron, E., Bisanzio, D., Mappin, B. and Dalrymple. K. (2016). The effect of malaria control on plasmodium falciparum in Africa between 2000 and 2015. *Nature* 526(7572), 207–211.

Bhutta, Z. A., Sommerfeld, J., Lassi, Z. S., Salam, R. A., and Das. J. K. (2014). Global burden, distribution, and interventions for infectious diseases of poverty. *Infectious Diseases of Poverty* 3(1), 1–7.

Bompart, F., Kiechel, J. R., Sebbag, R., and Pecoul, B. (2011). Innovative public-private partnerships to maximize the delivery of anti-malarial medicines: lessons learned from the ASAQ Winthrop experience. *Malaria Journal* 10(1), 143.

Boni, M. F., White, N. J. and Baird, J. K. (2016). The community as the patient in malaria-endemic areas: preempting drug resistance with multiple first-line therapies. *PLoS Medicine* 13(3), 1–7.

Bosman, A., and Mendis, K. N. (2007). A major transition in malaria treatment: the adoption and deployment of artemisinin-based combination therapies. *American Journal of Tropical Medicine and Hygiene* 77(6), 193–197.

Cibulskis, R. E., Alonso, P., Aponte, J., Aregawi, M., Barrette, A., Bergeron, L., Fergus, C. A., Knox, T., Lynch, M., Patouillard, E., Schwarte, S., Stewart, S., and Williams. R. (2016). Malaria: global progress 2000 - 2015 and future challenges. *Infectious Diseases of Poverty* 5(1), 1–8.

Drugs for Neglected Diseases Initiative (DNDI). (2015). *The Successful Development of a Fixed Dose Combination of Artesunate Plus Amodiaquine Antimalarial*. Drugs for Neglected Diseases Initiative, Geneva.

Dondorp, A. M., Newton P. N., Mayxay M., Van Damme W., Smithuis F. M., Yeung S., Petit A., Lynam A. J., Johnson A., Hien T. T., McGready R., Farrar J. J., Looareesuwan S., Day N. P. J., Green M. D., and White N. J. (2004). Fake antimalarials in Southeast Asia are a major impediment to malaria control: multinational cross-sectional survey on the prevalence of fake antimalarials. *Tropical Medicine and International Health* 9(12), 1241–1246.

Dondorp, A. M., Smithuis. F. M., Woodrow, C., and von Seidlein, L. (2017). How to contain artemisinin- and multidrug-resistant falciparum malaria. *Trends in Parasitology* 33(5), 353–363.

Faurant, C. (2011). From bark to weed: the history of artemisinin. *Parasite* 18(3), 215–218.

Hopkins, S. (2010). Health expenditure comparisons: low, middle and high income countries. *Open Health Services and Policy Journal* 3(1), 111–117.

Howick, J. H. (2011). *The Philosophy of Evidence-Based Medicine*. John Wiley and Sons, New York.

Imwong, M., Suwannasin, K., Kunasol, C., Sutawong, K., Mayxay, M., Rekol, H., Smithuis, F. M., Hlaing, T. M., Tun, K. M., van der Pluijm, R. W., Tripura, R., Miotto, O., Menard, D., Dhorda, M., Day, N. P. J., White, N. J. and Dondorp, A. M. (2017). The spread of artemisinin-resistant plasmodium falciparum in the Greater Mekong subregion: a molecular epidemiology observational study. *The Lancet Infectious Diseases* 17(5), 491–497.

International Association for Media and Communication Research (IAMCR) (2019). Country-Income Classification for IAMCR Membership. https://iamcr.org/income (accessed April 8, 2019).

Littrell, M., Gatakaa H., Evance I., Poyer S., Njogu J., Shewchuk S. T., Munroe E., Chapman S., Goodman C., Hanson K., Zinsou C., Akulayi L., Raharinjatovo J., Arogundade E., Buyungo P., Mpasela F., Adjibabi C., Agbango J., Ramarosandratana B., Coker B., Rubahika D., Hamainza B., Shewchuk, T., Chavasse, D., and O'Connell, K. A. (2011). Monitoring fever treatment behaviour and equitable access to effective medicines in the context of initiatives to improve act access: baseline results and implications for programming in six African countries. *Malaria Journal* 10, 327.

Lubell Y., Dondorp A. M., Guerin P. J., Drake T., Meek S., Ashley E., Day N. P. J., White N. J., and White L. J. (2014). Artemisinin resistance-modelling the potential human and economic costs. *Malaria Journal* 13, 452.

Luxemburger, C., Nosten, F., Kyle D. E, Kiricharoen, L., Chongsuphajaisiddhi, T., and White, N. J. (1998). Clinical features cannot predict a diagnosis of malaria or differentiate the infecting species in children living in an area of low transmission. *Transactions of the Royal Society of Tropical Medicine and Hygiene* 92(1), 45–49.

Medicines for Malaria Venture (MMV). (2016). *Developing New Medicines for Case Management and Vulnerable Populations*. MMV, Geneva.

Meijer, A., Boon, W. P. C., and Moors, E. H. M. (2013). Stakeholder engagement in pharmaceutical regulation: connecting technical expertise and lay knowledge in risk monitoring. *Public Administration*, 9(3), 696–711.

Mills, A. (2014). Health care systems in low- and middle-income countries. *New England Journal of Medicine* 370(6), 552–557.

Moors, E. H. M., Cohen, A. F., and Schellekens, H. (2014). Towards a sustainable system of drug development. *Drug Discovery Today* 19(11), 1711–1720.

Munos, B. (2009). Lessons from 60 years of pharmaceutical innovation. *Nature Reviews in Drug Discovery* 8, 959–968.

Newton P., Proux S., Green M., Smithuis F. M., Rozendaal J., Proux Prakongpan S., Chotivanich K., Mayxay M., Looareesuwan S., Farrar J., Nosten F., White, N. J. (2001). Fake artesunate in Southeast Asia. *Lancet* 357(9272), 1948–1950.

Noedl, H., Se, Y. Schaecher, K., Smith, B. L., Socheat, D., and Fukuda. M. M. (2008). Evidence of artemisinin-resistant malaria in Western Cambodia. *New England Journal of Medicine* 359(24), 2619–2620.

Novotny, J., Singh, A., Dysoley, L., Sovannaroth, S., and Rekol, H. (2016). Evidence of successful malaria case management policy implementation in Cambodia: Results from national ACTwatch outlet surveys. *Malaria Journal* 15, 194.

Nwaka, S., and Ridley, R. G., (2003). Virtual drug discovery and development for neglected diseases through public-private partnerships. *Nature Reviews Drug Discovery* 2(11), 919–928.

O'Connell, Gatakaa H., Poyer S., Njogu J., Evance I., Munroe E., Solomon T., Goodman C., Hanson K., Zinsou C., Akulayi L., Raharinjatovo J., Arogundade E., Buyungo P., Mpasela F., Adjibabi C. B., Agbango J. A., Ramarosandratansa B. F., Coker B., Rubahika D., Hamainza B., Chapman S., Shewchuk T., Chavasse D. (2011). Got ACTs? availability, price, market share and provider knowledge of anti-malarial medicines in public and private sector outlets in six malaria-endemic countries. *Malaria Journal* 10, 326.

O'Connell, Kathryn A., Samandari, G., Phok, S., Phou, M., Dysoley, L., Yeung, S., Allen, H., and Littrell. M. (2012). Souls of the ancestor that knock us out and other tales. a qualitative study to identify demand-side factors influencing malaria case management in Cambodia. *Malaria Journal* 11, 1–13.

Orsi, F., Singh, S., and Sagaon-Teyssier. L. (2018). The creation and evolution of the donor funded market for antimalarials and the growing role of southern firms. *Science, Technology and Society* 23(3), 349–370.

Palafox, B., Patouillard, E., Tougher, S., Goodman, C., Hanson, K., Kleinschmidt, I., Rueda, S. T., Kiefer, S., O'Connell, K.A., Zinsou, C., Phok, S., Akulayi, L., Arogundade, E., Buyungo, P., Mpasela, F., and Chavasse. D. (2014). Understanding private sector antimalarial distribution chains: a cross-sectional mixed methods study in six malaria-endemic countries. *PLoS ONE* 9(4), e93763.

Phyo, A. P., Ashley, E. A., Anderson, T. J. C., Bozdech, Z., Carrara, V. I., Sriprawat, K., Nair, S., McDew White, M., Dziekan, J., Ling, C., Proux, S., Konghahong, K., Jeeyapant, A., Woodrow, C.J., Imwong, M., McGready, R., Lwin, K. M., Day, N. P. J., White, N. J., and Nosten, F. (2016). Declining efficacy of artemisinin combination therapy against p. falciparum malaria on the Thai-Myanmar border (2003–2013): the role of parasite genetic factors. *Clinical Infectious Diseases* 63(6), 784–791.

Phyo, A. P., and von Seidlein, L. (2017). Challenges to replace ACT as first-line drug. *Malaria Journal* 16(1), 1–5.

Rafols, I., Hopkins, M., Hoekman, J., Siepel, J., O'Hare, A., Perianes-Rodrigez, A. and Nightingale, P. (2014). Big Pharma, little science? a bibliometric analysis on Big Pharma's RandD decline. *Technological Forecasting and Social Change* 81, 22–31.

Smith, L. A., Jones, C., Meek, S., and Webster. J. (2009). Review: provider practice and user behavior interventions to improve prompt and effective treatment of malaria: do we know what works? *American Journal of Tropical Medicine and Hygiene* 80(3), 236–335.

Spar, D., and Delacey, B. J. (2008). The coartem challenge. *Harvard Business School Case* 706–037.

Timmermans, S., and Berg, M. (2003). *The Gold Standard: The Challenge Of Evidence-Based Medicine*. Temple University Press, Philadelphia.

Ubben, D., and Poll, E. M. (2013). MMV in partnership: ehe Eurartesim experience. *Malaria Journal* 12(1), 1–10.

United Nations. (2015) The Millennium Development Goal Report. https://sustainabledevelopment.un.org (accessed April 4, 2019).

Waning, B. (2011). Innovative approaches for pharmaceutical policy research in developing countries: the view through a market lens. PhD diss., Utrecht University.

Warren, M. (2018). Access to Medicine Foundation. Guest lecture, Utrecht University, bachelor course Sustainable Health and Medical Technology, September 27, 2018.

Wells, S., Graciela D., and Kiechel, J. R. (2013). The story of artesunate-mefloquine (ASMQ), innovative partnerships in drug development: case study. *Malaria Journal* 12(1), 1–10.

Wells, T. N. C., Hooft Van Huijsduijnen, R., and Van Voorhuis, W. C. (2015). Malaria medicines: a glass half full? *Nature Reviews Drug Discovery* 14(6), 424–442.

White, N. J. (1999). Antimalarial drug resistance and combination chemotherapy. *Philosophical Transactions of the Royal Society B: Biological Sciences* 354(1384), 739–749.

World Health Organization (WHO) (2001). *Antimalarial Drug Combination Therapy*. Report of a WHO Technical Consultation. World Health Organization, Geneva.

World Health Organization (WHO) (2006). *WHO Briefing on Malaria Treatment Guidelines and Artemisinin Monotherapies*. World Health Organization, Geneva.

World Health Organization (WHO) (2015). *Strategy for Malaria Elimination in the Greater Mekong Subregion (2015–2030)*. World Health Organization, Geneva.

World Health Organization (WHO) (2016). *Global Technical Strategy for Malaria 2016–2030*. World Health Organization, Geneva.

World Health Organization (WHO) (2017). *World Malaria Report 2017*. World Health Organization, Geneva.

World Health Organization (WHO) (2018). *World Malaria Report 2018*. World Health Organization, Geneva.

Yadav, P. (2009). Countering Drug Resistance in the Developing World: An Assessment of Incentives across the Value Chain and recommentations for Policy Interventions. Center for Global Development, Washington, DC.

Yadav, P., Sekhri, N., and Curtis. K. (2007). Barriers to access: an assessment of stakeholder risks and incentives in the value chain for artemisinin combination therapy (ACT) treatments. SSRN. doi.org/10.2139/ssrn.1008307

Yeung, S., van Damme, W., Socheat, D., White, N. J., and Mills. A. (2008). Access to artemisinin combination therapy for malaria in remote areas of Cambodia. *Malaria Journal* 7, 1–14.

14

Digital Health

How Can It Facilitate Progress on Meeting Sustainable Development Goals in China?

Simon K. Poon, Yiren Liu, Ruihua Guo, and Mu Li

1. Introduction

Since the start of the internet era in the mid-1990s, digital technologies have been recognized as one of the most important drivers behind the economic growth in the past two decades (Muhleisen 2018). Digital technologies have not only led to efficiency gains by producing more with lesser human resource, they also have shown the potential to transform societies through their impact on organizational, social, and economic structures. The applications of digital technologies in healthcare (also known as digital health) have made significant progress in transforming healthcare in recent years, including enabling healthcare providers to deliver better care, empowering patients to manage their own health, and strengthening governments' abilities to transform their health systems.

In this chapter, our research goal is to provide an assessment of current digital health situation in China and how digital health may facilitate China to meeting a number of sustainable development goals (SDGs). To achieve this goal, we have developed a new conceptual measurement model based on our earlier research as the methodological basis to guide our review of the recent effort of the China's National Health Information Platform and our assessment from each construct based on the measurement model to evaluate potential connections between digital technology and health (SDG3), inclusive and equitable education (SDG4), and reduction of inequality (SDG10).

Digital health refers to IT adopted within clinical and healthcare workflows and processes (Lupton 2013). As defined by the US Food and Drug Administration (2018), digital health technology has a broad scope, including:

- Mobile health, which refers to mobile devices or apps that support medical and health practices (Ryu 2012).
- Health information technology (HIT), which focuses on data storage, interaction, and analysis. Electronic health records (EHRs), personal health records (PHRs), and electronic prescribing (E-prescribing) are examples of HIT (Lester et al. 2016).

- Wearable health devices, which include a range of technologies or sensors placed on or inside the body to enable long-term prevention and surveillance of disease and health conditions (Chan et al. 2012).
- eHealth, which refers to information communication technology (ICT) or Internet of Thing (IoT) to deliver internet-based healthcare services, with emphasis on industry and e-commerce (Gu et al. 2018).
- Telemedicine/telehealth, used interchangeably with eHealth, to provide face-to-face contact, bypassing the limitations of time and space (Della Mea 2001).
- Precision medicine to identify effective approaches for the patient or target groups based on analyzing gene sequencing, environmental factors, and living behaviors (Jameson and Longo 2015).
- Personalized medicine, often referred to as precision medicine, in which treatments and preventions developed uniquely based individuals' clinical, behavioral, and environmental information (Collins and Varmus 2015).

All forms of these digital health technologies have evolved and been widely applied in different contexts, in the attempt to improve health services and patient care globally. In China, the most significant progress in digital health has been establishment of the National Health Information Platform, known as the "4631-2 project," first announced in 2013 by the Chinese National Health and Family Planning Commission (National Health Commission of the People's Republic of China 2018a), as part of building a government administration and information management system. The earlier version of a similar project, "3521," for electronic health (e-health), was subsequently revised to the 4631-2 project (P. Li et al. 2017). The aim of the 4631-2 project is to create a comprehensive and three-dimensional national health and family planning resource system based on the existing integrated Chinese and Western public health information systems, the primary healthcare management information system, and the medical health public service system (State Council, The People's Republic of China 2013).

In this national project (Figure 14.1), the "4" represents four levels of health information platforms: 1) national population health management, 2) provincial population health information, 3) prefecture-level population health regional information, and 4) district and county population health regional information. The "6" represents six application services: 1) public health, 2) medical services, 3) medical insurance, 4) drug management, 5) family planning, and 6) integrated health management. The "3" represents three fundamental databases systems: 1) electronic health records, 2) electronic medical records, and 3) a national database containing the healthcare-related cases of the entire population. The "1" represents the congregated population health digital network, and finally, the "2" refers to the population health information standard and information safety protection systems. This plan aligns with the challenges as well as opportunities identified toward achieving the 2030 SDGs (World Bank and the Development Research Center of the State Council 2013, p. 202).

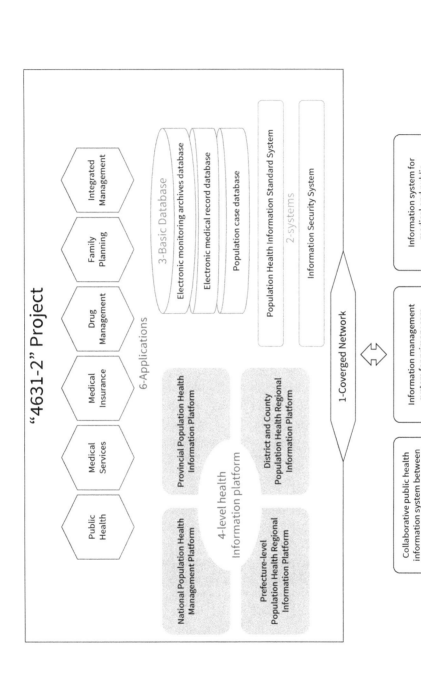

Figure 14.1 Framework for the National Health Information Project

Source: The State Council, The People's Republic of China (2013).

In 2015, the United Nation General Assembly adopted the SDGs as a development agenda to 2030. The 17 goals cover almost every aspect of sustainable development for a country (Ministry of Foreign Affairs of the People's Republic of China 2017).

By taking a systemic perspective, we aim to address the following question: How well does the digital health in China further facilitate the achievement of SDG targets 3.4, 3.7, 4.2, and 10.2? Section 2 establishes the assessment framework for the analysis of digital health maturity. Section 3 applies the assessment framework to evaluate the digital health progress in China in respect to SDG3, SDG4, and SDG10. Section 4 attempts to make connections to specific targets, namely 3.4, 3.7, 4.2, and 10.2. Finally, section 5 summarizes implications and discusses opportunities and challenges.

2. Method: Development of the Assessment Framework

The benefits of digital health for the achievement of SDGs can be articulated and evaluated via an appropriate assessment framework. The development of the assessment framework described in this chapter underwent a two-phase development process.

In phase 1, a preliminary assessment was established to provide a landscape of digital health and well-being in China based on the literature published between 2005 and 2017 (Poon and Li 2017, unpublished data). The review was organized into several broad dimensions of digital health: developmental progress, health application, and medical discipline. The landscape review was presented to a multidisciplinary research cluster who are interested in well-being and health in China via a research workshop in February 2017, followed by online discussions among participants of the workshop. Four future research focuses were established based on consensus of all participants:

1. Regulation: relating to legal framework to support digital health, promotion of national adopted standards, and strategies to address inequity caused by digital divide.
2. Knowledge and skills development: relating to effort to develop educational programs to manage the lack of teaching faculty and research in medical informatics.
3. Application: relating to efforts into implementation and translational research (social and cultural factors in digital health adoption).
4. Technology: relating to efforts for fostering medication technology innovation, including assessment and evaluation.

In phase 2, we articulate the four research focuses into an assessment framework on the basis of the eHealth Readiness Conceptual Measurement Model (Phillips et al. 2017). Using the same resource-based view (Newbert 2008), we converted the four research focuses from phase 1 into four dimensions of capabilities. In this study the benefits of digital health for facilitating the achievement of SDGs can be articulated and evaluated in four dimensions: 1) translation, 2) education, 3) transformation, and 4) technology, which would bridge the gap between digital health and SDGs (Figure 14.2).

2.1 Translation

Translation refers to assessing the ability of digital health to affect clinical or healthcare outcomes, such as reducing mortality and improving communicable or noncommunicable disease management. The indicators directly facilitate healthy lives and promote well-being at all ages. In particular, improvement of health indicators for the elderly and disabled could decrease inequality. Assessment is organized as follows:

- From the population perspective: to improve clinical indicators and health services though digital health interventions, such as mHealth, for diabetes management.
- From the healthcare provider perspective: to improve efficiency and work flow, decrease medication error, and enable precision medicine, less invasive procedures, and more.
- From the management and governance perspective: to organize resource allocation for disease prevention and control, particularly for higher disease burdens and needs, and support information sharing, such as across health services and medical insurance.

2.2 Education

Education refers to evaluating digital health in increasing health literacy for providers and patients. Digital health can overcome geographic and time constraints and therefore has the potential to ensure inclusive and equitable quality education and promote gender equality. Effective and equal health information delivery could enhance better health management and accelerate the achievement of universal health coverage.

- From a healthcare providers' perspective: to provide training to improve the ability of the health workforce to apply digital health.
- From a population perspective: to add value through improved health literacy.

2.3 Transformation

Transformation refers to exploring changes in management characteristics to facilitate digital healthcare delivery and behavior change, which will reflect in equity of

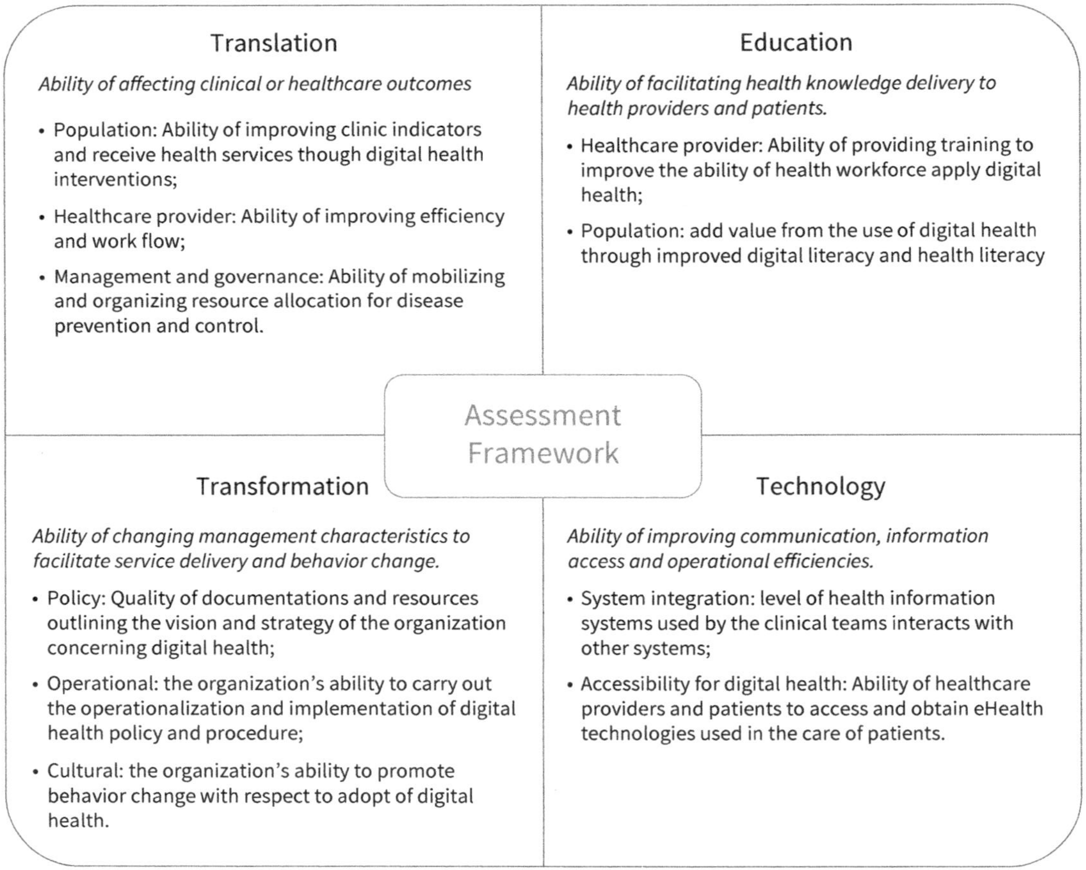

Translation

Ability of affecting clinical or healthcare outcomes

- Population: Ability of improving clinic indicators and receive health services though digital health interventions;
- Healthcare provider: Ability of improving efficiency and work flow;
- Management and governance: Ability of mobilizing and organizing resource allocation for disease prevention and control.

Education

Ability of facilitating health knowledge delivery to health providers and patients.

- Healthcare provider: Ability of providing training to improve the ability of health workforce apply digital health;
- Population: add value from the use of digital health through improved digital literacy and health literacy

Assessment
Framework

Transformation

Ability of changing management characteristics to facilitate service delivery and behavior change.

- Policy: Quality of documentations and resources outlining the vision and strategy of the organization concerning digital health;
- Operational: the organization's ability to carry out the operationalization and implementation of digital health policy and procedure;
- Cultural: the organization's ability to promote behavior change with respect to adopt of digital health.

Technology

Ability of improving communication, information access and operational efficiencies.

- System integration: level of health information systems used by the clinical teams interacts with other systems;
- Accessibility for digital health: Ability of healthcare providers and patients to access and obtain eHealth technologies used in the care of patients.

Figure 14.2 The four-dimensional assessment framework for the digital health
Source: authors.

educational distribution and improvement of health conditions. Evaluation can be approached in the following ways:

- From a policy perspective: quality of documentation and resources outlining the vision and strategy of the organization concerning digital health.
- From an operational perspective: the organization's ability to carry out the operationalization and implementation of digital health policy and procedures.
- From a cultural perspective: the organization's ability to promote behavioral change with respect to adoption of digital health.

2.4 Technology

Technology refers to the improvement of system capacity, and integration and accessibility of systems could amplify and facilitate communication between providers, users, and policymakers. Improvement of information access and operational efficiencies would ensure sustainable education coverage, reduce national inequity, and catalyze universal health coverage. Progress in this dimension is anticipated to correlate with increased efficiency in administrative tasks, availability of patient information, timely information technology support, or provision of appropriate devices and internet infrastructure to allow healthcare providers to access patient and other health information. Assessment will be as follows:

- System integration: level of health information systems such as electronic medical records (EMRs), electronic health records (EHRs), and hospital information system (HIS) used by clinical teams interacting with data transmission and data integration, for reporting, data sharing across different services, and more.
- Accessibility for digital health: ability of healthcare providers and patients to access and obtain eHealth technologies used in patient care.

3. Application Assessment to the Three SDGs

3.1 SDG 3: Ensure Healthy Lives and Well-Being for All

Health is a fundamental human right and a key indicator of sustainable development. Poor health threatens the rights of children to education, limits people's economic opportunities. To accelerate progress and address health challenges, digital health solutions need to be adopted that work for people, families, communities and nations.

3.1.1 Translation Dimension

Digital health could translate to healthy lives by improving noncommunicable disease management, smoking cessation, surgery accuracy, inpatient and outpatient service delivery, and disease prevention and regulation (targets 3.4., 3.8, and 3a). Systems have been developed such as WE-CARE, an intelligent mobile tele-cardiology system for anomalies detection and cardiovascular disease (CVD) diagnosis (Huang et al. 2014), and mHealth apps that focus on diabetes, hypertension, and hepatitis (Hsu et al. 2016). These target the high disease burden, such as the 109.6 million patients (Hu and Jia 2018), 244.5 million (Wang, Z. et al. 2018), and more than 90 million (WHO 2018) in mainland China. Mobile phone text messaging interventions also have the capability for smoking cessation (Augustson et al. 2017). The first robotic surgery performed on an orthopedic operation, a minimally invasive digital technique, was successfully conducted in China and could improve functional and social outcomes despite the high cost of robotic devices (15 million yuan or US$2.31 million) (China Daily 2018).

Clinical decision systems (Milcent 2016) and bar-code screening (CHIMA 2018) can minimize human errors in inpatient service delivery. Internet hospitals and online hospital platforms have the potential to provide widely accessible outpatient services, such as online appointments and consulting, hospital navigation, and so on, with doctors from an internet hospital spending 10 minutes or more to communicate with each patient (Tu et al. 2015). However, development is not yet established and distribution is uneven, with 65% of internet hospitals in major cities (Xie et al. 2017). In 2015, there were more than 1,000 internet medical consulting facilities covering 21 municipalities in Guangdong (Tu et al. 2015). For the prevention and control of infectious disease, the Infectious Disease Reporting Information Management System (IDRIMS) was established after the SARS outbreak in 2003 (Salemink et al. 2017). Another example is the use social media data collected from Sina Weibo, the Chinese version of Facebook (Yang et al. 2014) to predict the outbreak of flu five days earlier than the Chinese National Influenza Center. In 2016, the cases of tuberculosis in China were decreased by 3.2% compared to the cases in 2015. There are 836,000 cases of tuberculosis that have been reported, among which 93% were successfully cured by Chinese Center for Disease Control and Prevention system (Ministry of Foreign Affairs of the People's Republic of China 2017).

3.1.2 Education Dimension

From 273 WeChat questionnaires, eHealth literacy played a critical role in users' health outcomes (X. Zhang et al. 2018). Based on data from 1396 questionnaires, the health literacy of Chinese was found to be associated with self-management, awareness, and sufficient access to health information from multiple sources (Liu et al. 2015). In addition, 44 of 234 mHealth apps had a potential for better self-management by providing complementary education to improve outcomes (Hsu et al. 2016). However, awareness for hypertension in China was 42.1% of 785 participants, and only 58.7% of 461 participants had regularly self-monitored blood pressure (Cai et al. 2017), which was low compared to Mongolia (Li et al. 2016).

3.1.3 Transformation Dimension

Some progress has been made for transforming digital into universal health. From the policy perspective, an increased awareness of health information systems by government has been achieved, such as the Outline of National Health Information Development Plan 2003–2010 and Healthy China 2030 (World Bank and the Development Research Center of the State Council 2013, p. 202). These focus on the health needs and equality of the population as well as achieving sustainable development, directly aligned with the implementation of the 2030 Agenda for Sustainable Development. Also, with the establishment of the 4631-2 project (National Health Commission of the People's Republic of China 2018a), use of integrated digital health systems is increasing.

However, lack of financial incentive and policy could lead to a lack of motivation for physicians to adopt digital health (Li et al. 2017). Also, there are unequal benefits and financial protection between medical insurance systems, which impede implementation of digital health (Meng et al. 2015) and affect healthcare delivery (Yip et al. 2012). Among over three thousand policies and documents identified in 2016, Hsu et al. (2016) found only 11 official releases that could directly or indirectly influence the operation of medical apps. However, the application of digital health in the realization of universal health is supported and the overall trend is toward improvement. Nevertheless, a lack of workforce and cross-disciplinary collaboration in hospitals could delay development of digital health. For example, there is a shortage of experienced IT personnel to operate and maintain electronic systems such as EHRs (Li et al. 2017) and telemedicine (Zhao et al. 2018), and a dominated technology-driven approach in many healthcare organizations (L. Li et al. 2017; P. Li et al. 2017) has led to user needs being disregarded in the design process (Liang et al. 2018). This example highlights the importance of complementary investments in both digitization and digitalization are needed, hence complementary investment in training and organizational is a critical enabler for any health technology investment.

Culture influences digital health transition. On the one hand, Chinese patients with mental illness have seldom sought professional healthcare due to stigma, negative cultural beliefs, and lack of access to care (Li et al. 2014). Thus, mobile mental health apps have the potential to encourage patients to seek professional help (Li et al. 2014), even though most current health-related apps focus on physical health. On the other hand, face-to-face encounters with patients, which involve looking, smelling, asking, and feeling the pulse that are particularly valued in traditional Chinese medicine (TCM), are problematic for the development of telemedicine. In addition, philosophical differences in diagnosis, assessment, and treatment between Western medicine and TCM also challenge the adoption of digital health (Hsu et al. 2016; Poon and Poon 2014; Waldram 2000). However, efforts are underway to integrate digital health into the culture of TCM with the introduction of a computerized remote traditional Chinese pulse diagnosis system to identify disease based on pulse tracing the language of life (Dong et al. 2010).

3.1.4 Technology Dimension

Technology-enabled data integration and process integration are essential to incorporate digital health into public health and hospital working flow management (Li et al. 2017). Without standards for system design, low-quality data (incomplete and/or duplicated) and poor interfaces lead to "information isolated islands" with a lack of compatibility and data integration between systems (Wang et al. 2018). The implementation of EMR systems has increased in recent years. Of 848 tertiary hospitals in China, 65% remain in adoption levels 1, 2, and 3 (the highest is level 7), which reflects a low level of clinical information sharing (Shu 2014). Similarly, fragmented applications of hospital information system (HIS) may not be integrated with clinical practice (Hsu et al. 2016; Zeng et al. 2017), with clinical care systems and basic public health services often separated (X. Li et al. 2017). Deficiencies in integrated digital data impede the implementation of daily primary healthcare practice and clinical decision-making (Li et al. 2017), especially SDG3 targets such as 3.2, aiming to end preventable deaths of newborns and children under five years of age. Relying on paper-based or isolated systems of vaccination information system was found to impede the uptake surveillance of vaccination (Li et al. 2017). Also, lack of technical mechanisms focusing on surveillance of internet medical services (Zeng 2017) and system quality control (Shi et al. 2014) can impede the capacity and accessibility of the EMR and HIS systems. Patient information privacy is of concern. In a study including 598 hospitals, 67.6% of 34,155 items from EMR were compliant, yet only 30.9% provided individual information protection (Zeng 2017).

3.2 SDG 4: Ensure Quality Education, for All

SDG4 addresses inclusive and equitable quality education and promotes lifelong learning opportunities for all. This involves improving both the health service providers' and the users' capabilities to utilize digital health for quality services.

3.2.1 Translation Dimension

Digital health technologies, which inherently have the capacity for information sharing by improving accessibility, could translate into equitable education and benefit learning. Yet there remains a significant gap in continuing education between China and the United States from the perspective of the way medical informatics conferences are conducted (Liang et al. 2017). However, China has made progress in patient education delivery, which is reflected in improved clinical outcomes. Through knowledge delivery, an eight-week home-based telehealth exercise program has significantly reduced heart failure and improved quality of life of the patients, measured by six-minute walking distance and resting heart rate (Zhao 2010; Peng et al. 2018).

3.2.2 Education Dimension

Telemedicine could provide long-distance training for healthcare providers, such as village doctors (Zhao 2010). There are 1,220 Chinese hospitals registered in the directories of the National Health Committee telemedicine center network (Lu 2016). However, acceptance for telemedicine is associated with age and gender, and healthcare providers and medical students are usually better prepared for digital health informatics (Chen et al. 2017). In a study of adults over the age of 55, there was a low acceptance of mobile phone use and inadequate health literacy (Ma et al. 2016). Variations in digital literacy could be information recall (password), functional familiarity (computer skills), and technology accessibility (downloading apps) (Lau 2017). The rate of use of the WeChat (similar to Facebook) was 61.7% of 1,097 respondents in an mHealth survey for patients with glaucoma, with higher acceptance rates in younger participants (Lin et al. 2014; Dai et al. 2017).

Similarly, in dental practices, 67% of 354 dentists reported having their own websites or bulletin board services online for advertisements and patient education purposes, but only 26% of 285 computerized respondents reported internet access in their office (Hu et al. 2009). In a study of examining the predictors of users' mhealth adoption intention and the gender differences in the intention, it was found that males always had a higher level of mHealth adoption intention than females (Zhang et al. 2013). Providing technical training to clinician to upskill their technical knowledge could lead to better health information dissemination among clinicians and also for patient adoption.

3.2.3 Transformation Dimension

Progress has been made to provide governmental support for more than 1.6 million people who have accessed tele-education (Zhao 2010). Nevertheless, village doctors, the primary healthcare providers in rural areas, are often an older group with lower digital literacy than other primary healthcare doctors (Li et al. 2017). Pathways to transform digital health technology into inclusive and equitable quality education are lacking (Lin et al. 2014), and these are coupled with unequally distributed health resources and infrastructure. Further, there are insufficient financial incentives and learning opportunities to improve digital literacy (Lu 2016).

Currently, there is no marketing environment in China for health education delivery through telemedicine. Most telemedicine companies prefer to invest in teletherapy rather than education delivery. The emerging telemedicine market has been dominated by local Chinese companies such as SUNPA, Huawei, Newsoft, and Kingway, whose products include interactive teleconsultation, medical image reading, medical image acquisition, surgical guidance, and distant education (Zheng and Rodriguez-Monroy 2015). Although low bandwidth transmission causes time lags and limited image sharing by telemedicine, recent networks such as 4G bandwidth have been developed (CHIMA 2018; Gao 2011). However, the deficiencies of both health technicians and necessary infrastructures in rural hospitals, which are usually the receivers of health education, limit the application of telemedicine on a large scale (Li et al. 2017).

3.3 SDG 10: Reduce Inequalities

Reducing inequality requires transformative change. Greater investments are needed in health, education and social protection for vulnerable people in the society. Digital health could help to close the gap in access to health services.

3.3.1 Translation Dimension

To reduce age inequalities, telemedicine has been developed for senior patients with diabetic retinopathy and ophthalmic screening, such as the Beijing Eye Public Health Care Project, which is a telemedicine-based eye care system for the elderly (Xu et al. 2012). A study found that of 11,987 teleconsultations improved the accessibility of diagnosis for patients with neoplasms (19%), injuries (13.9%), and circulatory diseases (10.3%) in western China (Wang et al. 2016). Treatment plans changed significantly in 6,591 (55.0%) patients as a result of the use of telemedicine. In addition, online consulting and home-based healthcare are also believed to improve healthcare among disabilities, yet few studies have been conducted in this field.

3.3.2 Education Dimension

There is rapid progress in the implementation of telemedicine in hospitals, but the digital divide, which describes the uneven diffusion of internet services within countries, persists in China (Hong et al. 2018), although telemedicine could improve health literacy in remote areas and narrow geographic inequality (Wang et al. 2016). By 2017, 13,000 health facilities had implemented telemedicine (Central People's Government of People's Republic of China 2018). However, the distribution of these health facilities is clearly uneven (Figure 14.3). On the other hand, there has been progress in digital literacy, especially in mobile phone use. In 2014, 96.7% of cataract patients had inadequate computer skills to seek online health information as measured by routine use of computer and search engines (X. Lin et al. 2014). To date, of 18,215 participants, 6.5% had used the internet in the past month, and 83% owned a mobile phone (Hong et al. 2017).

3.3.3 Transformation Dimension

In 2016, to enhance equality between people with different incomes, the Chinese government released legislation banning any third-party apps from providing appointments for patients to avoid profiting. All appointments were to be made directly through official hospital websites or apps instead of through third-party providers (Hsu et al. 2016). Increasing the adoption of digital health technologies across the country might rely on awareness of the value of home-based healthcare, but remote and rural areas are known to have low rates of health literacy and low self-management

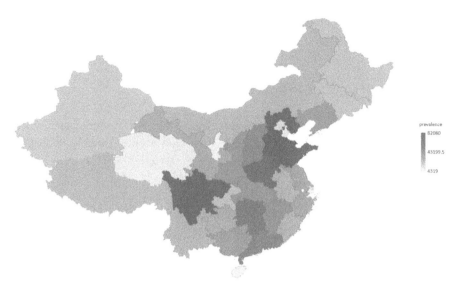

Figure 14.3 Distribution of health institutions in China 2018
Source: Extracted from the National Health Commission of the People's Republic of China (2018).

awareness in health (Drury et al. 2018). Therefore, the adoption rate in remote and rural areas is expected to be lower than in cosmopolitan areas.

3.3.4 Technology Dimension

To minimize the digital divide, frugal innovations addressing healthcare are needed and many digital health technologies, such as information communication technologies (ICTs), can meet this requirement (Zhang 2018). For example, cloud-based EHR was developed in Inner Mongolia to help village doctors deliver cost-effective services by improving record-keeping, to better manage chronic diseases (Liu et al. 2014). Digital health technology also contributes to significantly reduced laboratory costs and hospital stays time (Liu et al. 2014). A study in western China found that on average annual telemedicine costs are $74,156, which is less than the traditional visit cost of $271,200 for patients to travel to the major hospitals or $387,407 for specialists to travel to rural hospitals. The cost savings were achieved using telemedicine (Wang et al. 2016).

4. Discussion: Making Connections Between Digital Health and Specific SDG Targets

As illustrated in the preceding section, digital health has played a growing role across China in many forms despite the technical, legal, ethical, and financial uncertainties. Digital technologies are gradually changing the way healthcare professionals, patients, and the public communicate and interact with each other. For example, mHealth, enabled through smartphones, tablets, and computers, provides health-care professionals, patients, and the general public with means that can be useful to support everyday health and disease management and professional practices. In addition, innovative and cost-effective digital solutions are being embraced by health systems that struggle to cope with growing demand for services due to demographic changes and the rise of chronic diseases and multiple morbidities.

China has attained some of the health-related SDGs ahead of schedule, and is making steady progress in achieved other health related SDGs targets. Here are some examples to demonstrate how digital health can further facilitate the progress.

SDG3 target 3.4 specifies a reduction of one-third premature mortality from non-communicable diseases through prevention and treatment and promote mental health and well-being by 2030 (indicator 3.4.1, mortality rate attributed to cardiovascular disease, cancer, diabetes, or chronic respiratory disease.) Noncommunicable chronic diseases (NCDs) are the leading causes of death in China, accounting for close to 70% of the disease burden and over 80% of the total mortality. The four leading NCDs in China are cardiovascular diseases, diabetes, cancer, and chronic obstructive pulmonary diseases (Ma et al. 2016). How can digital health solutions contribute to achievement of the SDG3.4? Notably, by the reduction of noncommunicable and chronic diseases and the achievement of universal access to good quality healthcare in terms of translation, education, and technology.

Diabetes is one of the most common chronic diseases in China. The 2010 China National Chronic Disease Risk Factor Surveillance reported that the overall prevalence of diabetes was 9.7% in Chinese adults aged 18 years and over. The estimated prevalence of diabetes in the Chinese population is similar to the US population (11.3%) even though overweight and obesity, the main risk factors for diabetes, are much more common in the United States (Ma et al. 2106). Many early eHealth solutions were designed as support tools for diabetes patients to self-manage their condition. The increased use of smartphones and growing number of mobile applications and websites are available to help patients with their daily diabetes management and improve health outcomes. Many digital solutions can also gather data on patients' diets, physical activity, and medication, blood glucose levels, and body weight. Some mobile device apps are designed to provide coaching and patient education. The ability to log and enter relevant data, whether manually or automatically (via smartphones, tablets, or computers), and to update or change information in real time, enables diabetes patients to exert greater control over their condition. The solutions from mHealth have proven to be beneficial and played an increasing role in diabetes management (Hsu et al. 2016).

SDG3 target 3.7 states "by 2030, ensure universal access to sexual and reproductive healthcare services, including for family planning, information and education, and the integration of reproductive health into national strategies and programmes" (indicator 3.7.1, proportion of women of reproductive age, 15–49 years, who have their need for family planning satisfied with modern methods). Data on married rural to urban female migrants, extracted from the Monitor Survey of Migrants (including employment, residence, family planning, reproductive health services, and other public services), revealed that although close to 90% women used some forms of contraceptive methods, and 56% reported that they had to access the contraceptives from their permanent registered rural household residence. Only 40% could access contraceptives from health providers in their current urban residence (Guo et al. 2016), which indicated barrier and unmet contraceptive needs in migrant women of reproductive age.

SDG4 target 4.2 states "by 2030, ensure that all girls and boys have access to quality early childhood development, care and pre-primary education so that they are ready for primary education" (indicator 4.2.1, proportion of children under five years of age who are developmentally on track in health, learning, and psychosocial well-being, by sex). China has made significant progress in child survival and health through its efforts of achieving the millennium development goals (MDGs). Between 1996 and 2015, the under-five mortality rate in China declined from 50.8 per 1000 live births to 10.7 per 1000 live births. However, 181,600 children still died before their fifth birthday. Great inequity exists in child mortality across regions and in urban versus rural areas. The western region had the highest under-five mortality rate and neonatal mortality rate in 2015, with an estimated under-five mortality rate of 18.5 (95% UR 12.6–25.2) deaths per 1,000 live births and neonatal mortality rate of 9.5 (6.8–13.7) per 1,000 live births. In the eastern region, the under-five mortality rate was 5.8 and neonatal mortality rate was 3.1, respectively, similar to the rates in the United States and Canada. (He et al. 2017).

Digital health, within the broader Chinese National Health Information Project, will contribute to improved data availability for monitoring the achievement some of these SDG3 and SDG4 targets, in several ways addressing translation, transformation, and technology, such as:

- linking health facility data and community/population data so people can access health service in where they are reside;
- enabling data disaggregation in order to identify priorities for disease burden, resource allocations, and technical support by population groups, regional, and rural/urban locations, and closing the gaps;
- improving facility based reporting on issues such as service quality, supplies and medicine, human resources, and financial resources, and collating and sharing the information collected in different facilities.

For example, the current child health services, including regular health check-ups for infants and young children up to three years of age, are provided by the local community health centers (CHCs) and the information is kept at the local CHCs. Most times when families move the information is not followed or forwarded to the healthcare providers in their new locations. Digital health can make the information sharing easier to facilitate the tracking of child health status, growth, and development. Further, mHealth has been well accepted in the parenting and infant young child feeding space in China. On the one hand, it fills in the gap of the need for knowledge and support of new parents, at the same time it helps ease the workload of health professionals (Jiang et al. 2018).

SDG10 target 10.2 states that "by 2030, empower and promote the social, economic and political inclusion of all, irrespective of age, sex, disability, race, ethnicity, origin, religion or economic or other status." An effective healthcare system has a critical role in the prevention and control of the chronic disease epidemic that is currently raging out of control in rural China. Historically, the dual urban–rural healthcare system structure in China has led to an underresourced healthcare system in the rural areas of the country and many other significant disadvantages that affect access to health services and quality of services by about 50% of the total Chinese population. Many of the healthcare practitioners working in the rural primary healthcare services, such as village doctors and township hospitals doctors, are not qualified physicians by Western standards; they work with limited medical training, have basic equipment and devices, and a restricted list of medicines. In addition, although the healthcare financing of rural China has been constantly improved through the National Cooperative Medical Scheme since its establishment in 2003, there is a significant disparity in healthcare expenditure between urban hospitals and rural healthcare facilities, particularly in relation to the huge chronic disease burden in the rural population (Wu 2016). While the Chinese government continues to address the health financing issue through health financing reforms, digital health has an important role to play in improving the quality of care and addressing the inequality of health outcomes for the patients in rural setting through translation, education, transformation, and technology. For example, these can be achieved by:

- providing telemedicine/teleconsultation for diagnosis and management plans by doctors from major urban hospitals;
- delivering tailored online medical training and education program for rural primary healthcare providers; and
- setting up electronic record-keeping systems for tracking patients medication, laboratory tests, and other investigation results.

5. Remarks: Opportunities and Challenges

Digital health plays an increasingly important role in China despite its potential uncertainty. Based on the four-dimensional assessment framework, China has made steady progress in achieving SDG target indicators, such as prevention of communicable diseases, increasing accessibility of health education, and reducing unevenly distributed health resources. The implementation of national health information reforms in the recent decade has improved healthcare coverage, but financial incentives for healthcare providers to fully embrace digital health are yet to be realized. The Chinese population health information system needs to be strengthened by systematic integration of many related health systems and has the capability to bridge Western medicine and traditional Chinese medicine, to integrate surveillance of noncommunicable diseases, to monitor medical services at different levels and in different regions, and to address healthcare provision in terms of supply, quality, and uptake.

The government's support for public education and use of information technology in education can be linked to the establishment of the 4631-2 project, the Healthy China 2030 initiative, and the 13th Five-Year Plan. Digital technology is changing how users and providers communicate and interact, and is reshaping health knowledge delivery. Many digital technologies directly and indirectly enable users to monitor and understand their personal health data, thus creating a new relationship between patients and the health system. The focus now needs to be on addressing the readiness and cost-effectiveness of digital health innovation to ensure that this will translate to health outcomes and full educational potential, independent of age or gender. The promise of digital health could change the achievement of universal health, sustainable education, and national equality.

Furthermore, a number of fundamental building blocks would require additional attention.

National strategies should be revised to take account of the health inequity caused by the digital divide in rural and urban regions. With the anticipation of rapid development of digital health, there is a demand for more evidence-based health technology evaluation and assessment methodologies. Hence, efforts are needed for implementation and translational programs, particularly in understanding the social and cultural aspects for digital health to be adopted and used efficiently. To secure adequate knowledge and skills development for the digital health future, a roadmap for increased well-trained medical informatics faculty to facilitate teaching and research in China is necessary.

Finally, mainstreaming digital health in health promotion, disease monitoring, and management means that digital health technology has to be affordable and accessible for everyone. It must not increase the digital divide by exacerbating health inequalities and excluding vulnerable groups. Tailoring digital health to particular population groups, such as adolescents and elderly people, and providing accessible information and education at minimal cost for people in underresourced communities, is important for the sustainability of digital health.

Acknowledgment

We would like to thank the China Studies Centre at the University of Sydney for the support of the initial China digital health landscape review and the assistance in preparing this chapter.

References

Augustson, E., Engelgau, M. M., Zhang, S., Cai, Y., Cher, W., Li, R., and Bromberg, J. E. (2017). Text to quit China: an mHealth smoking cessation trial. *American Journal of Health Promotion* 31(3), 217–225.

Cai, L., Dong, J., Cui, W. L., You, D. Y., and Golden, A. R. (2017). Socioeconomic differences in prevalence, awareness, control and self-management of hypertension among four minority ethnic groups, Na Xi, Li Shu, Dai and Jing Po, in rural southwest China. *Journal of Human Hypertension* 31(6), 388–394.

Central People's Government of People's Republic of China. (2018). Thirteen thousand organizations have implemented telemedicine in China. http://www.gov.cn/xinwen/2018-01/08/content_5254212.htm (accessed April 5, 2019).

Chan, M., Estève, D., Fourniols, J. Y., Escriba, C., and Campo, E. (2012). Smart wearable systems: current status and future challenges. *Artificial Intelligence in Medicine* 56(3), 137–156.

Chen, P., Xiao, L., Gou, Z., Xiang, L., Zhang, X., and Feng, P. (2017). Telehealth attitudes and use among medical professionals, medical students and patients in China: a cross-sectional survey. *International Journal of Medical Informatics* 108, 13–21.

China Daily. (2018). The 19th CPC National Congress. http://www.chinadaily.com.cn/china/19thcpcnationalcongress/index.html (accessed April 5, 2019).

China Hospital Information Management Association (CHIMA). (2018). CHIMA 2018 Hospital Internet Application. http://www.chima.org.cn (accessed April 5, 2019).

Collins, F. S., and Varmus, H. (2015). A new initiative on precision medicine. *New England Journal of Medicine* 372(9), 793–795.

Dai, M., Xu, J., Lin, J., Wang, Z., Huang, W., and Huang, J. (2017). Willingness to use mobile health in glaucoma patients. *Telemedicine and e-Health* 23(10), 822–827.

Della Mea, V. (2001). What is e-Health (2): the death of telemedicine? *Journal of Medical Internet Research* 3(2), e22.

Dong, Z., Xiang, H., and He, W. (2010). Remote diagnosis in traditional Chinese medicine using wireless sensor networks. In: Qingling, L., Fei, Y., and Yun, L. (Eds.), *Proceedings of the 2010 Third International Symposium on Information Processing*, pp. 255–257. IEEE Computer Society, Washington, D.C.

Drury, P., Roth, S., Jones, T., Stahl M., Medeiros D. (2018). *Guidance for Investing in Digital Health*. Report No. 52, Asian Development Bank, pp. 45–49. https://www.adb.org/sites/default/files/publication/424311/sdwp-052-guidance-investing-digital-health.pdf (accessed April 5, 2019).

Gao, X. (2011). The anatomy of teleneurosurgery in China. *International Journal of Telemedicine and Applications* 2011, 353405.

Gu, D., Yang, X., Li, X., Jain, H., and Liang, C. (2018). Understanding the role of mobile internet-based health services on patient satisfaction and word-of-mouth. *International Journal of Environmental Research and Public Health* 15(9), e1972.

Guo C., Pang L. H., Zhang G., and Zheng X. Y. (2016). Migrant health: reproductive health awareness and access. In: Li, M., and Wu, Y. F. (Eds.), *Urbanisation and Public Health in China*, pp. 213–232. Imperial College Press, London.

He, C., Liu, L., Chu, Y., Perin, J., Dai, L., Li, X., Lei, M., Kang, L., Li, Q., Scherpbier, R., Guo, S., Rudan, I., Song, P., Chan, K.Y., Guo, Y., Black, R. E., Wang, Y., and Zhu, J. (2017). National and subnational all-cause and cause-specific child mortality in China, 1996–2015: a systematic analysis with implications for the Sustainable Development Goals. *The Lancet Global Health* 5(2), e186–e197.

Hong, Y. A., and Zhou, Z. (2018). A profile of eHealth behaviors in China: Results from a national survey show a low of usage and significant digital divide. *Frontiers in Public Health* 6, 274–274.

Hong, Y. A., Zhou, Z., Fang, Y., and Shi, L. (2017). The digital divide and health disparities in China: evidence from a national survey and policy implications. *Journal of Medical Internet Research* 19(9), e317.

Hsu, J., Liu, D., Yu, Y. M., Zhao, H. T., Chen, Z. R., Li, J., and Chen, W. (2016). The top Chinese mobile health apps: a systematic investigation. *Journal of Medical Internet Research* 18(8), e222.

Hu, C., and Jia, W. (2018). Diabetes in China: epidemiology and genetic risk factors and their clinical utility in personalized medication. *Diabetes* 67(1), 3–11.

Hu, J., Yu, H., Luo, E., Song, E., Xu, X., Tan, H., and Wang, Y. (2009). Are Chinese dentists ready for the computerization of dentistry? a population investigation of China's metropolises. *Journal of the American Medical Informatics Association* 16(3), 409 LP–4.

Huang, A., Chen, C., Bian, K., Duan, X., Chen, M., Gao, H. and Xie, L. (2014). WE-CARE: an intelligent mobile telecardiology system to enable mHealth applications. *IEEE Journal of Biomedical and Health Informatics* 18(2), 693–702.

Jameson, J. L., and Longo, D. L. (2015). Precision medicine-personalized, problematic, and promising. *Obstetrical and Gynecological Survey* 70(10), 612–614.

Jiang, H., Li, M., Wen, L. M., Baur, L. A., He, G., Ma, X., and Qian, X. (2018). A short message service intervention for improving infant feeding practices in Shanghai, China: planning, implementation, and process evaluation. *The Journal of Medical Internet Research mHealth and uHealth* 6(10), e11039.

Lau, A. Y., Piper, K., Bokor, D., Martin, P., Lau, V. S., and Coiera, E. (2017). Challenges during implementation of a patient-facing mobile app for surgical rehabilitation: feasibility study. *The Journal of Medical Internet Research Human Factors* 4(4), e31.

Lester, M., Boateng, S., Studeny, J., and Coustasse, A. (2016). Personal health records: beneficial or burdensome for patients and healthcare providers? *Perspectives in Health Information Management* 13(Spring), 1–12.

Li, G., Wang, H., Wang, K., Wang, W., Dong, F., Qian, Y. and Wang, B. (2016). Prevalence, awareness, treatment, control and risk factors related to hypertension among urban adults in Inner Mongolia 2014: differences between Mongolian and Han populations. *BMC Public Health* 16(1), 294–304.

Li, H., Zhang, T., Chi, H., Chen, Y., Li, Y., and Wang, J. (2014). Mobile health in China: current status and future development. *Asian Journal of Psychiatry* 10, 101–104.

Li, L., Du, K., Xin, S., and Zhang, W. (2017). Creating value through IT-enabled integration in public organizations: a case study of a prefectural Chinese Center for Disease Control and Prevention. *International Journal of Information Management* 37(1), 1575–1580.

Li, P., Xie, C., Pollard, T., Johnson, A. E. W., Cao, D., Kang, H. and Zhang, Y. (2017). Promoting secondary analysis of electronic medical records in China: summary of the PLAGH-MIT Critical Data Conference and Health Datathon. *The Journal of Medical Internet Research Medical Informatics* 5(4), e43.

Li, X., Lu, J., Hu, S., Hu, S., Cheng, K., De Maeseneer, J., and Jiang, L. (2017). The primary health-care system in china. *The Lancet* 390(10112), 2584–2594.

Liang, J., He, X., Jia, Y., Zhu, W., and Lei, J. (2018). Chinese mobile health APPs for hypertension management: a systematic evaluation of usefulness. *Journal of Healthcare Engineering* 2018, 7328274, 1–14.

Liang, J., Wei, K., Meng, Q., Chen, Z., Zhang, J., and Lei, J. (2017). The gap in medical informatics and continuing education between the United States and China: a comparison of conferences in 2016. *Journal of Medical Internet Research* 19(6), e224.

Lin, C. W., Abdul, S. S., Clinciu, D. L., Scholl, J., Jin, X., Lu, H. and Li, Y. C. (2014). Empowering village doctors and enhancing rural healthcare using cloud computing in a rural area of mainland China. *Computer Methods and Programs in Biomedicine* 113(2), 585–592.

Lin, X., Wang, M., Zuo, Y., Li, M., Lin, X., Zhu, S., and Lamoureux, E. L. (2014). Health literacy, computer skills and quality of patient-physician communication in Chinese patients with cataract. *PLoS ONE* 9(9), 2–6.

Liu, G. G., Chen, Y., and Qin, X. (2014). Transforming rural health care through information technology: an interventional study in China. *Health Policy and Planning* 29(8), 975–985.

Liu, Y. B., Liu, L., Li, Y. F., and Chen, Y. L. (2015). Relationship between health literacy, health-related behaviors and health status: a survey of elderly Chinese. *International Journal of Environmental Research and Public Health* 12(8), 9714–9725.

Lu, B. (2016). Analysis on the regional distribution of telemedicine in china. *Value in Health* 19(7), A836–A836.

Lupton, D. (2013). The digitally engaged patient: Self-monitoring and self-care in the digital health era. *Social Theory and Health* 11(3), 256–270.

Ma, J. X., Li, J. H., Ge, Z., and Wu, Y. F. (2016). Surveillance of non-communicable disease risk factors. In: Li, M., and Wu, Y. F. (Eds.), *Urbanisation and Public Health in China*, pp. 69–88. Imperial College Press, London.

Ma, Q., Chan, A. H. S., and Chen, K. (2016). Personal and other factors affecting acceptance of smartphone technology by older Chinese adults. *Applied Ergonomics* 54, 62–71.

Meng, Q., Fang, H., Liu, X., Yuan, B., and Xu, J. (2015). Consolidating the social health insurance schemes in China: towards an equitable and efficient health system. *The Lancet* 386(10002), 1484–1492.

Milcent, C. (2016). Evolution of the health system: inefficiency, violence, and digital healthcare. *China Perspectives* (4), 39–50.

Ministry of Foreign Affairs of the People's Republic of China. (2017). *China's Progress Report-on Implementation of the 2030 Agenda for Sustainable Development.* http://www.chinadaily.com.cn/specials/China'sProgressReport2(CN).pdf (accessed April 5, 2019).

Mühleisen, M. (2018). The long and short of the digital revolution. *Finance and Development,* 55(2), 4–8.

National Health Commission of the People's Republic of China. The People's Republic of China. (2018a). *Guide on Accelerating the Construction of Population Health Informationization.* (Report No. 38[2012]). http://www.nhc.gov.cn/zwgk/wtwj/201304/e1b9fd5596ce-4a5e8123337552358b38 (accessed April 5, 2019).

National Health Commission of the People's Republic of China. The People's Republic of China. (2018b). The number of medical and health institutions in China at the end of June 2018. http://www.nhfpc.gov.cn/mohwsbwstjxxzx/s7967/201808/b9745940d99a-4645b63afb6f5f8e1369.shtml (accessed April 5, 2019).

Newbert, S. L. (2008). Value, rareness, competitive advantage, and performance: a conceptual-level empirical investigation of the resource-based view of the firm. *Strategic Management Journal* 29(7), 745–768.

Peng, X., Su, Y., Hu, Z., Sun, X., Li, X., Dolansky, M. A., and Hu, X. (2018). Home-based tele-health exercise training program in Chinese patients with heart failure: a randomized controlled trial. *Medicine* 97(35), e12069.

Poon, J., and Poon S. K. (2014). Data Analytics for Traditional Chinese Medicine Research. Springer International, Switzerland.

Phillips, J., Poon, S. K., Yu, D., Lam, M., Hines, M., Brunner, M., Power, E., Keep, M., Shaw, T., and Togher, L. (2017). A conceptual measurement model for eHealth readiness: a team based perspective. In: Sarkar, N. (Ed), *AMIA Annual Symposium Proceedings Vol. 2017*, pp. 1382–1391. American Medical Informatics Association, Washington DC.

Ryu, S. (2012). Book review: mHealth: new horizons for health through mobile technologies: based on the findings of the second global survey on eHealth. *Healthcare Informatics Research* 18(3), 231–233.

Salemink, K., Strijker, D., and Bosworth, G. (2017). Rural development in the digital age: a systematic literature review on unequal ICT availability, adoption, and use in rural areas. *Journal of Rural Studies* 54, 360–371.

Shi, C., Hui-Ting, L., Hui, P., Yang, J. J., Li, J., and Wang, Q. (2014). Current electronic medical record in China. In: Carl, K. C., Yan, G., Ali, H., Mihhail, M., Bruce, M., Yasuo, O., Cristina, S., and Kenichi, Y. (Eds.), *Computer Software and Applications Conference Workshops (COMPSACW), 2014 IEEE 38th International Computer Software and Applications Conference Workshops*, pp. 668–672. IEEE, Vasteras, Sweden.

Shu, T., Liu, H., Goss, F. R., Yang, W., Zhou, L., Bates, D. W., and Liang, M. (2014). EHR adoption across China's tertiary hospitals: a cross-sectional observational study. *International Journal of Medical Informatics* 83(2), 113–121.

State Council, The People's Republic of China. (2013). *Guide on Accelerating the Construction of Population Health Informationization*. http://www.gov.cn/zhuanti/2013-12/09/content_2593451.htm (accessed April 5, 2019).

Tu, J., Wang, C., and Wu, S. (2015). The internet hospital: an emerging innovation in China. *The Lancet Global Health* 3(8), e445–e446.

United States Food and Drug Administration (2018). Digital Health. https://www.fda.gov/medicaldevices/digitalhealth (accessed April 5, 2019).

Waldram, J. B. (2000). The efficacy of traditional medicine: current theoretical and methodological issues. *Medical Anthropology Quarterly* 14(4), 603–625.

Wang, T. T., Li, J. M., Zhu, C. R., Hong, Z., An, D. M., Yang, H. Y., and Zhou, D. (2016). Assessment of utilization and cost-effectiveness of telemedicine program in western regions of China: a 12-year study of 249 hospitals across 112 cities. *Telemedicine Journal and E-Health* 22(11), 909–920.

Wang, Z., Chen, Z., Zhang, L., Wang, X., Hao, G., Zhang, Z and Wang, J. (2018). Status of hypertension in China: results from the China Hypertension Survey, 2012–2015. *Circulation* 137(22), 2344–2356.

World Bank and the Development Research Center of the State Council, People's Republic of China. (2013). *China 2030: Building a Modern, Harmonious, and Creative Society*. http://www.worldbank.org/content/dam/Worldbank/document/China-2030-complete.pdf (accessed April 5, 2019).

World Health Organization (WHO). (2018). Hepatitis. http://www.wpro. who.int/china/mediacentre/factsheets/hepatitis/en/ (accessed April 5, 2019).

Wu, Y. F. (2016). Major challenges the rural China healthcare system is facing in the epidemic of chronic diseases. In: Li, M., and Wu, Y. F. (Eds.), *Urbanisation and Public Health in China*, pp. 181–196. Imperial College Press, London.

Xie, X., Zhou, W., Lin, L., Fan, S., Lin, F., Wang, L., and Chen, Y. (2017). Internet hospitals in China: cross-sectional survey. *Journal of Medical Internet Research* 19(7), e239, 1–6.

Xu, L., Jonas, J. B., Cui, T. T., You, Q. S., Wang, Y. X., Yang, H., Li, J. J, Wei, W. B., Liang, Q. F., Wang, S., Yang, X. H., and Zhang, L. (2012). Beijing eye public health care project. *Ophthalmology* 119(6), 1167–1174.

Yang, N., Cui, X., Hu, C., Zhu, W., and Yang, C. (2014). Chinese social media analysis for disease surveillance. In: Bilof, R. (Ed.), *Identification, Information and Knowledge in the Internet of Things (IIKI), 2014 International Conference on*, pp. 17–21. IEEE. Beijing, China.

Yip, W. C. M., Hsiao, W. C., Chen, W., Hu, S., Ma, J., and Maynard, A. (2012). Early appraisal of China's huge and complex health-care reforms. *The Lancet* 379(9818), 833–842.

Zeng, D-H, Fan, G-H, and Xiao, F. (2017). Analysis of status quo of construction of medical information integration platform in Chinese tertiary hospitals. *Chinese Journal of Health Policy* 10(7): 75–78.

Zhang, X. (2018). Frugal innovation and the digital divide: developing an extended model of the diffusion of innovations. *International Journal of Innovation Studies* 2(2), 53–64.

Zhang, X., Guo, X., Lai, K., Guo, F., and Li, C. (2013). Understanding gender differences in mhealth adoption: a modified theory of reasoned action model. *Telemedicine and eHealth* 20(1), 39-46.

Zhang, X., Yan, X., Cao, X., Sun, Y., Chen, H., and She, J. (2018). The role of perceived e-health literacy in users' continuance intention to use mobile healthcare applications: an exploratory empirical study in China. *Information Technology for Development* 24(2), 198–223.

Zhao, H., Li, G., and Feng, W. (2018). Application of "internet+" in healthcare. In: Yihu, W., Xun, W., and Wancheng, Y. (Eds.), *2018 International Conference on Engineering Simulation and Intelligent Control* (ESAIC), pp. 368–371. IEEE. Changsha, China.

Zhao, J., Zhang, Z., Guo, H., Ren, L., and Chen, S. (2010). Development and recent achievements of telemedicine in China. *Telemedicine and e-Health* 16(5), 634–638.

Zheng, X., and Rodriguez-Monroy, C. (2015). The development of intelligent healthcare in China. *Telemedicine Journal and E-Health: The Official Journal of the American Telemedicine Association* 21(5), 443–448.

15

Responsive and Responsible Science, Technology, and Innovation for Global Health

Nora Engel, Agnes Meershoek, and Anja Krumeich

1. Introduction

Toward the end of the era of the millennium development goals (MDGs), it became clear that these goals, especially health-related MDGs, were not always achieved. There were many explanations for these disappointing outcomes. Some of them related to the way in which the goals were set without sufficient involvement from low resource settings (Fehling et al. 2013). Some associated the lack of success with the reductionist way in which the goals framed health problems, too limiting in scope and in the health issues that they addressed, while ignoring the importance of strengthening health systems (Magrath et al. 2009; Haines and Cassels, 2004; Miranda and Patel, 2005). Others argued that the MDGs paid inadequate attention to the promises of technology. Where technology did come into focus, it was approached as an unproblematic solution for a narrowly defined health problem—a silver bullet or quick win (Richard et al. 2011; Fehling et al. 2013). In this chapter, we discuss how science, technology, and innovation (STI) are conceptualized for the sustainable development goals (SDGs) and explore the challenges of this approach.

The problem with viewing technologies as magic silver bullets—that is, straightforward solutions with extreme effectiveness—as was done in the MDGs is that it disregards all the work required to make any technology function. Social scientists have criticized imaginaries of silver bullets for their lack of problematizing technology and technology determinist view on technology. Such a perspective assumes that technological development determines social change following an inevitable course, whereas the making of facts, science, and technology is always a collective process. Similarly problematic from this viewpoint is the idea that technology drives human development and pushes it forward, and thereby allows leapfrogging or skipping other development steps (such as building road or sanitation infrastructure or strengthening health or laboratory systems). Many social scientists have also critiqued the related ideas of technology transfer and diffusion underlying much technology design for global health (Hadolt et al. 2011; Pfotenhauer and Jasanoff 2017;

Prasad 2014; Engel 2012); that technologies can be developed in setting A and transferred to setting B and made to work there as long as implementation problems can be overcome. Commonly, technologies that are seen as silver bullets or as those that allow leapfrogging are also believed to be more readily transferable to other settings, because they do not rely on (or leapfrog) infrastructure and contextual systems. The linear understanding of technology development that underlies these approaches accepts technology as something given, to which systems need to adapt. This assumes that we can find universal solutions that can be brought to scale if only they overcome implementation challenges of different contexts (e.g., function in weak healthcare systems where infrastructure, staff and capacities might be absent). Yet this idea of technology transfer disregards some of the fundamental arguments by science and technology studies scholars, namely that technology does not exist separately from its context of use and thus will always be shaped by and in return shape the setting in which it is being used. STI and society are nonseparable; they coproduce each other.

In contrast to the MDGs, the SDGs entail a universal and equitable approach to global health, defining health problems as multidimensional issues and explicitly calling for technologies that are made to work in particular cultures, communities, and health systems around the world. Sustainable development goal 3 (SDG3), "ensure healthy lives and promote well-being for all at all ages," continues a focus on maternal and child health as well as infectious diseases that is broader than the MDGs, by aiming to curb "the epidemics of AIDS, tuberculosis, malaria and neglected tropical diseases and combat hepatitis, water-borne diseases and other communicable diseases" (subtarget SDG3.3). In addition, it also targets noncommunicable diseases, mental health, substance abuse, traffic accidents, and health threats from hazardous environmental pollution. To that end, SDG3 suggests supporting research and development (R&D) of appropriate technology, explicitly mentioning vaccines and medicines for diseases affecting mainly developing countries.

This chapter asks how STI are conceptualized in SDG3 in particular and the SDGs more generally, and how they are assumed to tackle health inequalities. Our aim is to discuss, inspired by insights from science and technology studies (STS), the challenges with the way in which STI are conceptualized in SDG policies and how to strengthen existing efforts toward galvanizing STIs for reaching SDGs. In a first step, we discuss UN, policy, and media documents related to the SDGs. In this narrative literature review we draw particularly on the UN documentation of SDG3 and its thematic review, as well as the policy documents, conference reports, and online media related to the technology facilitation mechanism (TFM) and an annual, multistakeholder STI forum.

In a second step, we discuss examples from the authors' ethnographic studies on development and implementation of point-of-care (POC) diagnostics and cookstoves. Specifically, these include 1) an ethnographic study of diagnostic processes at point of care across major diseases in South Africa between September 2012 and June 2013, consisting of 101 semistructured interviews and seven focus group discussions with doctors, nurses, community health workers, patients, laboratory technicians, policymakers, hospital managers, and diagnostic manufacturers (Engel, Davids et al.

2015); 2) an ethnographic study of diagnostic processes at point of care in India's public and private sector, peripheral laboratories, communities, and homes, consisting of 78 semistructured interviews and 13 focus group discussions with healthcare providers (doctors, nurses, specialists, traditional healers, and informal providers), patients, community health workers, test manufacturers, laboratory technicians, program managers, and policymakers between January and June 2013 (Engel, Ganesh, Patil, Yellappa, Pant Pai, Vadnais and Pai 2015); and 3) an interdisciplinary project, consisting of an ethnography on codesign of cookstoves in non-notified urban slums in India. This 12-month project (February 2015–February 2016) employed a participatory and cyclic design, using the observation-reflect-plan-act cycles (ORPA) as discussed by Kemmis, McTaggart, and Nixon (2014). During the 40 intense field visits, data were collected using a variety of methods, such as observations, informal interviews, semistructured interviews, prioritization workshops, community forums, and photo voice. Collected data and data collections methods were collaboratively discussed with and reflected upon by the different stakeholders involved, including designers, engineers, slum inhabitants, and researchers. This resulted in the planning for the next phase of the project and started a new cycle with the implementation of these activities (Ghergu 2018).

Based on these analyses we propose, in a concluding step, a responsive and responsible approach to STI. Such an approach is based on a thick description of the different settings and of stakeholder perceptions and needs, and takes the coproduction of knowledge, innovation, and society, as well as the undesired effects, into consideration.

2. The Promises Attached to STI for Development

The TFM and the annual, multistakeholder STI forum were initiated by the United Nations and its member states in 2015 as part of the efforts of reaching all SDGs. These two activities have an explicit focus on STI without being focused on one specific SDG alone. Our analysis reveals that the imaginaries—the visions, symbols, and feelings people associate with a phenomenon—attached to STI for development in the SDGs as conceptualized by the United Nations have moved from silver bullets to needs-driven and inclusive STI for unlocking the revolutionary potential of technology.

In SDG3, "ensure healthy lives and promote well-being for all at all ages," STI is implied in words such as "interventions," "prevention," and "treatment." Specific technologies that are mentioned include medicines, vaccines, health coverage, and essential healthcare services, but not, for instance, diagnostic or monitoring technologies. In the thematic review of SDG3, target 3.b, it is highlighted that global health-related R&D is not aligned to global health needs and that there is an insufficient focus on health needs of developing countries:

> The current landscape of health research and development (RandD) is insufficiently aligned with global health demands and needs. As little as 1% of all funding for health RandD is allocated to diseases that are predominantly incident in developing countries. (World Health Organization and UNFPA 2017, p. 7)

Other than that, there is no explicit mentioning of the role of STI for meeting SDG3.

Instead, the role of STI for reaching all the SDGs is explicitly mentioned in the TFM and an annual, multistakeholder STI forum. A background paper on the UN technology initiatives by the UN Inter-Agency Working Group on a TFM attested great gaps and fragmentation in the various UN initiatives related to technology. It argues for the establishment of a platform for content coordination and a multistakeholder forum (Liu et al. 2015). The STI forums are convened by the president of the Economic and Social Council (ECOSOC), cochaired by two member states and prepared by the UN Inter-Agency Working Group on a TFM in collaboration with ten representatives from civil society, the private sector, and the scientific community.

In the notes of the cochairs of the first STI forum taking place in New York on June 6–7, 2016, the transformative and revolutionary nature and potential of technology for all economic sectors and societies is mentioned and the need to take advantage of this potential for social and economic inclusion, sustainability, and peace (United Nations 2016). In this line, it is also mentioned how new technologies, such as mobile phones, allow leapfrogging, and how a lack of infrastructure can foster leapfrogging because infrastructure development is path-dependent. (United Nations 2016). In the second STI Forum, Bill Gates (New York, May 15, 2017) voiced his optimism in how innovation can help meet the SDGs in a video address to the attendees:

> Science and technology aren't silver bullets, but they can unlock miracles. The sustainable development goals provide a framework to focus those miracles where they're needed most: on the needs of the poorest and the most vulnerable. (United Nations Department of Economic and Social Affairs 2017)

While the understanding of technology seems to have moved on from a traditional view of technology as silver bullets, technology is still unanimously agreed to bear revolutionary potential. Yet there is also recognition that technology without provision of incentives, community empowerment, and based on a thorough understanding of the setting in which it has to function has only limited transformative potential. Alternative imaginaries that are being highlighted are those of needs-driven and inclusive STIs with a deep understanding of the conditions that they are meant to improve.

Instead of silver bullets being applied to social problems, the report of the first STI forum differentiates needs-driven STI:

> The focus should be on how social needs can drive and transform science, technology and innovation, a shift from the dominant model of science, technology and

innovation being "applied to" social problems. That entails new ways of looking at
the interface between society and science, technology and innovation, new kinds
of social expertise to be associated with science, technology and innovation activities, institutionalized community and civil society participation and new kinds
of science, technology and innovation policy and practice overall. (United Nations
2016, p. 3).

This will require new approaches to the policy–science interface and redesign of science systems to address problems relevant to reach the SDGs and coherence between
STI policies. This new approach involves recognizing different sources of knowledge
(for example, indigenous knowledge) to mobilize technology in such a way that it
helps with achieving the SDGs.

Related to this understanding, the forum also emphasized inclusive STI: STI promises advancing human progress and leaving no one behind in reaching the SDGs, but
it can also create more divisions and further marginalize vulnerable groups. STI,
therefore, can cause gaps and divisions but also entails the promise to overcome the
very divisions it created. For instance:

ICT tools can help communities develop inclusive technologies that reflect their
own priorities and needs. New technologies have also enabled the development of
collaborative models and learning platforms based on open-source applications
and the sharing of data. (United Nations 2016, p. 8).

Ensuring access to STIs, participatory technology assessment, and inclusive participation of multiple stakeholders in innovation and entrepreneurship is therefore
viewed as important to counteract these divisions.

To conclude, the UN documentation, policy documents, and conferences around
STIs for reaching the SDGs underline a general recognition that technologies are no
silver bullets and that their revolutionary potential needs to be unlocked for reaching
the SDGs. But how precisely this should and can be done is not so clear. Among the
ideas mentioned are multistakeholder participation, innovation that focuses on the
needs of the poor and is inclusive, social technologies, strengthening of the science–
policy interface, redesign of science systems to address problems relevant to reach
the SDGs, and coherence between STI policies. There is also recognition among the
speakers of the first STI forum that needs-driven and inclusive STI will require a
systems approach, a deep understanding of the environments in which STI will be
integrated, and strengthening of STI capabilities—all of which will require a lot of
investments and resources. Technology is seen as an opportunity, an enabler, and a
revolutionizer. But technology is also problematized—at least to a certain extent—as
a potential problem in itself, which in isolation will not only not have transformative impact, but can even be something that might hinder progress by creating more
divisions if not done in an inclusive manner. The lack of adequate technological infrastructure, such as internet access, is mentioned as an example of creating a digital

divide that has held back entire countries and is particularly impactful on vulnerable groups, including women, indigenous peoples, and persons with disabilities.

While this different conceptualization of STI is a promising starting point, it is not so clear what it requires to do needs-driven and inclusive STI. Overall, there is little attention to the work it takes to make technologies function in such a way that they are effective for the local settings and practices in which they are applied. In the next section we use two empirical examples to discuss some of the important characteristics and challenges when attempting to do needs-driven or inclusive STI, and argue for the importance of making an approach to innovation processes explicit and negotiable.

3. Challenges with the Promise of Needs-Driven and Inclusive STI

In this section, we draw from results of the author's ethnographic research on two examples of STI, POC testing and participatory design of cookstoves. We discuss some of the challenges with the promises of and proposed solutions for needs-driven and inclusive STI as conceptualized in the SDGs by the United Nations and their implications for innovations. These two empirical examples are cases wherein actors make conscious efforts toward needs-driven and inclusive STI: POC diagnostics are needs driven in the sense of being designed for the local point of care and participatory design of cookstoves represents an attempt of inclusive STI. If successful, both STIs would support reaching SDG3 as they address diagnostic delays resulting in increased morbidity, the emergence of drug resistance and mortality, and indoor air pollution resulting in respiratory diseases. Yet our point is not to discuss in what way POC diagnostics or cookstoves can help reaching the SDGs, but to use these examples to reflect on the challenges of this particular approach (i.e., needs-driven and inclusive) to STI and to contrast these experiences with the conceptualizations and proposed solutions of the United Nations. In both cases, the challenges we encountered were not necessarily anticipated by those designing these STIs. The discussion of these challenges underlines that there is no easy fix of unlocking the imagined potential of STIs for development, but that it requires continuous work, negotiation, and reflection.

3.1 Point-of-Care Testing

POC testing has caused a lot of hope and enthusiasm among global health experts in recent years (Peeling and Mabey 2010; Tucker et al. 2013). Largely for its potential to bring more diagnostic precision to settings without easy access to laboratory-based testing and to allow healthcare workers to make better informed treatment decisions on the spot or while patients wait. It is hoped that this would prevent delayed diagnoses, mistreatment and misdiagnosis, and resulting mortality and drug resistance, and with it directly support achieving SDG3.3 ("by 2030, end the epidemics of AIDS, tuberculosis, malaria and neglected tropical diseases and combat hepatitis, water-borne diseases and other communicable diseases"). The technologies that enable successful POC testing in clinics, hospital wards, communities, and homes are usually envisioned as small devices or use-and-throw/disposable testing kits. It is believed that they should be simple to operate and maintain, and rapid and cheap, to fit resource-constraint settings and busy clinics (Peeling et al. 2006; Pant Pai et al. 2012). As such, they are thought to be innovations that focus on the needs of the poor and are inclusive.

Yet being simple and rapid does not necessarily mean these tests are used as envisioned at POC and are inclusive; that is, accessible to those who need them. The Xpert MTB/RIF, a molecular test that allows the diagnosis of tuberculosis (TB) in 90minutes, a TB test that was thought to revolutionize diagnosing TB at the point of care (WHO 2010), did not necessarily translate into rapid treatment decisions and cuts in diagnostic delay (Albert et al. 2016). In an ethnographic study of diagnostic processes at point of care in South Africa, Engel and colleagues found that the time from providing a sample to acting on the result of a Xpert MTB/RIF was in fact often several days. This happened because a backlog was created, since only a limited number of samples can be run at the same time (4, 16, or 48 samples depending on the machine—the latter being only available in very few, high through-put laboratories). As a result, patients were told to come back in one to three days instead of waiting for 90 minutes, which is the time required for the machine to analyze a sample (Engel, Davids et al. 2015). What is more, the GeneXpert MTB/Rif is widely underutilized and due to cost and operational constraints only available to limited patient groups and in more centralized locations. This creates differential access and therefore challenges reaching SDG3.3. But the GeneXpert also offers a better diagnosis for those who can reach the diagnostic and generated funding and interest for development of other TB diagnostics, contributing more indirectly to SDG3. In an ethnographic study of diagnostic processes at point of care in India, Engel and colleagues found that most of the rapid tests for common infectious diseases are used in laboratories and not in hospital wards or doctor's offices. In the laboratory, either the test is not made to work within a single patient encounter or the rapidity is compromised. These tests included, for instance, rapid tests for HIV, dengue, malaria, and typhoid. High workloads and equipment shortages are leading to delays and in settings with shorter turnaround times, these rapid tests were too costly or unavailable (Engel, Ganesh, Patil, Yellappa, Vadnais, Pant Pai and Pai 2015). The detailed ethnographic analyses of diagnostic processes across different healthcare settings revealed that diagnosing at the POC in India is challenged by available healthcare infrastructure (including material and financial scarcity and human resource constraints); relationships between providers, and between providers and patients; and modified behaviors and practices that often favor empirical treatment guided by clinical experience over treatment guided by testing (Engel, Ganesh, Patil, Yellappa, Pant Pai, Vadnais and Pai

2015). This means that even if rapid diagnostics are made available, access to these diagnostic services in a way that they ensure timely treatment and thereby support reducing the epidemics of these communicable diseases and access to quality essential healthcare services (SDG3.3 and 3.8) is not guaranteed. It also shows the importance of health system strengthening and funding (SDG3.C and 3.D). Yet brushing this off as mere implementation challenges that do not allow realization of the revolutionary potential of these diagnostic technologies for reaching SDG3.3 misses the point: namely, that making a diagnostic technology work takes continuous work in any environment.

This work shapes how the technology is used, and its added value, and therefore society shapes technology. At the same time, the decisions concerning design of new diagnostics have consequences for patients, laboratory workers, physicians, program officers, community workers, and policymakers. How the new technology will look, what samples it requires, what it can detect, who is eligible to test, the costs involved, time required, and user efforts shape the experience of these actors and the type of work they need to put in. If STI is to be made useful for reaching SDG3, technology development needs to conceptualize the infrastructure and capacity of the setting and try to align the design to these specificities of the setting, but also to the social relations among the many different actors (for instance in the case of POC diagnostics: clinicians, lab technicians, nurses, community health workers, patients, test manufacturer, suppliers, policymakers, bureaucrats, donors, and regulators) and things (such as testing equipment, databases, patient files, internet connections, courier systems, alternative testing options, and cost) that are involved and their agendas and interests. In other words, it needs to be *responsive* to the needs of that context. Those needs might be multiple, divergent, shifting, and hard to identify.

3.2 Cookstoves

Participatory approaches focusing on the coproduction of STIs are increasingly seen as a way to design and implement technologies that are responsive to the specific needs of a particular context. Designing STI in dialogue with the intended users (professionals, community, policymakers) is believed to ensure that the innovation is inclusive, in the sense that involvement of users in the design and implementation process guarantees that the STI will be taken up by those who need it most and will be put to work as intended. In the field of indoor air pollution, for instance, many participatory projects have been initiated over the last two decades. These projects were a response to the numerous earlier projects in which improved cookstoves were distributed among beneficiaries, but of which the long-term impact was nil, as people did not use them or altered the stove design after installation undoing its positive effects on indoor air pollution (Barnes et al. 2012). In the participatory projects, different stakeholders, such as community members and designers, jointly work on development of cookstoves that are adjusted to the needs, preferences, and interests of community members in a specific setting (Honkalaskar et al. 2013; Makonese and Bradnum 2017). These newly designed cookstoves are supposed to reduce indoor air pollution, a major cause of chronic obstructive pulmonary disease (COPD) in many low-income countries (LICs). As such this will directly support achieving SGD 3, SDG3.9.1 in particular ("by 2030, substantially reduce the number of deaths and illnesses from hazardous chemicals and air, water and soil pollution and contamination"; "mortality rate attributed to household and ambient air pollution").

These participatory experiences show that local cooking practices, family size, cooking utensils, type of available cooking fuel, housing structures, and specific spatial layout of neighborhoods matter for what an appropriate design for a specific setting is. Addressing these issues does improve uptake of technology to certain extent (Ghergu et al. 2018; Urmee and Gyamfi 2014) and thus contributed to reaching SDG3.9. At the same time, these projects demonstrate some serious limitations of this form of participatory approach in achieving responsiveness to a particular context and thus not realizing the promises of improved cookstove use, and therefore limiting the potential impact these technology could have on reaching SDG3.9.

In an interdisciplinarly project, consisting of an ethnography on codesign of cookstoves in non-notified urban slums in India, Ghergu et al. showed that involving community members in the entire process of development and implementation of the stoves and actually addressing their needs appeared to be very challenging, if not impossible (Ghergu et al. 2018; Sushama et al. 2018). Pointing out the risk of chronic and other lethal lung diseases, epidemiologists and health professionals identified the design of a better cooking device as a major need of these slum dwellers (Sushama et al. 2018). Community members did agree that indoor air pollution was a nuisance and that they encountered irritated airways and blackened walls in their homes. Yet they were not always keen to participate in activities aimed at the cocreation of a new type of stove (Sushama et al. 2018). On some occasions, they cancelled planned cocreation meetings to attend to other, more pressing matters, including uncertain financial situations and unstable job opportunities resulting in continuous relocation. In addition, the lack of legal protection in non-notified slums created a constant threat of slum demolition and eviction that could not be disregarded. Moreover, in monsoon time, many community members are confronted with acute health issues, such as dengue fever, chikungunya, and malaria, while dealing with flooded homes (Ghergu et al. 2018). Clearly, situations that required immediate action were prioritized over coproduction of cooking stoves.

These distractions from the cookstove project are serious concerns that are all related to other more urgent or pressing needs of this specific local context. This illustrates that needs in a needs-driven approach are not simply given, but that identifying and framing needs can be done from many different perspectives. It raises the question how needs can or should be identified and who is qualified to do so. In the field of indoor air pollution and related health effects, the issue is often framed from an epidemiological perspective (World Health Organization 2016; Barnes et al. 2012; Sinha 2002). Areas in which people cook indoors with stoves that produce a high level of air pollution have a much higher incidence of COPD and thus the epidemiological/public health need for combatting COPD. But as discussed in the previous paragraph, community members themselves define their needs differently. They encounter many existential uncertainties that they consider more important. Interestingly, when asked for their most pressing need, community members in this project mentioned secure access to electricity. Electricity is seen as a source that can bring some light into their daily struggle to survive. The outsiders' focus on investing community members time in the cocreation of a cooking device clearly lacked responsiveness, because needs were driven from outside. A more responsive approach would therefore include reflection on which needs are being identified and by whom, and make those negotiable.

Aside from the challenges in ensuring needs-driven STI, inclusive approaches to STI are also not without challenges and can have unintended consequences. In the cooking stove project, slum dwellers, professionals, and other users were included in the process of cocreation. During the cocreation process a prototype was developed that was acceptable for the users in terms of taste, cooking traditions, size, combustion, and so on (Ghergu et al. 2018). However, it was quite heavy and could not easily be transported by the slum dwellers, who, forced by labor insecurities, frequently moved from slum to slum to follow jobs. One solution that emerged was to install the cookstoves permanently in the houses along with a chimney that would decrease indoor air pollution even further. This required investment by the landlords who were at that point invited to participate in the cocreation process as well. As the owners of the houses rented by the slum dwellers, they were a more stable stakeholder in the development of new cookstoves. Yet the landlords doubted this permanent solution, because the land on which their houses stood was owned by external landowners who were usually not involved in what was going on in the slums. Approaching these landowners led in one case to the eviction of the slum dwellers families, as the landowner found the attention that was being paid to problems with houses on his land annoying and potentially risky. The well-intended organization of the cocreation process thus ultimately resulted in eviction and homelessness. Responsibility can be difficult to achieve, even when input from all stakeholders is ensured. Continuously monitoring intended as well as unintended consequences is essential.

To conclude this section on the challenges with the promise of needs-driven and inclusive STI, these examples show the challenges of developing and implementing POC diagnostics and cookstoves in a needs-driven and inclusive, *responsive* manner. Making STI such as POC diagnostics and cookstoves function and therefore supportive in achieving SDG3 requires continuous hard work. But being responsive to various contexts also requires ongoing reflections about *responsibility*, whose needs are being addressed, who is being included, and monitoring of (un)intended consequences and thus unforeseen results.

4. Proposed Approach to STI: Responsive and Responsible

To support realizing the potential that is believed to be inherent to STI for reaching SDG3, and based on these analyses and our understanding of the nature of contemporary global health problems, we propose a *responsive* and *responsible* approach to STI that is based on a thick ethnographic description of different settings and stakeholder perceptions and needs, and that takes the coproduction of knowledge, innovation, and society into consideration as well as the unintended effects. "Responsive" means to respond to the needs of the context in design, development, and implementation of STIs. This implies, for instance, that needs are not defined beforehand or externally driven, but reflected upon and negotiated with those affected and involved. Responsive therefore includes both a needs-driven and an inclusive approach to STIs. "Responsible" refers to ongoing reflections about one's own responsibility as a researcher, innovator, or implementer about whose needs are being addressed with STI; who is being included/excluded in design, development, evaluation, and implementation; and what the (un)intended consequences and unforeseen results of these activities are. This might require changes to current funding models and evaluation practices for global health innovations to account for this ongoing reflection and iteration in (re)defining needs.

Results of ethnographic examinations allow making design, evaluation, and implementation decisions that are based on an in-depth understanding of the different settings, actors, things, and practices involved. This approach is of course no guarantee that resulting STIs will be more responsive and responsible and more likely to achieve SDG3, but at least the process will 1) include more attention to different viewpoints, needs, and practices of various stakeholders and the complex dynamics between local and global factors that characterize contemporary global health challenges; and 2) take into account the nonlinear nature of STI design through ongoing reflections about the innovation process, as well as the responsibility that is being assigned to others through the STI. These we consider essential ingredients for a responsive, needs-driven and inclusive, and responsible approach to STIs. Next, we elaborate on these two points.

First, we believe that an ethnographic approach is particularly suited to address contemporary global health challenges, such as indoor air pollution or diagnostic delays, which are characterized by a mix of factors related to biological characteristics (e.g., disease strains, patient bodies, and bacteria), global trends, and local conditions. A detailed, ethnographic analysis of the social determinants of health (including social, political, economic, systemic, structural, and cultural dimensions) that constitute local practice and context is crucial for tackling health risks and health inequalities and therefore for reaching SDG3. Yet these dimensions are entangled, and shape and change each other in ways unique for a particular context and global and historical processes interact with this local interplay of the social determinants of health (Krumeich and Meershoek 2014). Ethnographic methods that allow for a thick description of the different user and developer settings and of stakeholder perceptions and needs are particularly suited for disentangling, and hence understanding the interplay between local and global health determinants. The setting to be researched does not necessarily have clear boundaries. How the ethnographer delineates the setting is part of trying to be responsive, by making sure that interests of the primary problem holders are not overlooked, and that the selection of stakeholders that are included is subject of constant reflection. As these dimension are economic, cultural, political, financial, medical, biological, and social in nature, thorough ethnographic analysis requires the combined effort of scholars from different disciplines. Whereas traditionally ethnography is a method most commonly associated with anthropology, we argue that its ability to provide thick description of multiple settings and of stakeholder perceptions and needs should be used in interdisciplinary ways as well. The in-depth ethnographies that we propose can be time-consuming and require investment and flexibility to incorporate ongoing findings into the innovation processes. What is more, existing power imbalances between stakeholders can make it difficult to implement innovations that are reaching those in need or the most vulnerable. Political factors are hard to change and often require long-term change that is difficult to realize within the scope of research or innovation projects.

Secondly, our responsive and responsible approach to STI also acknowledges the nonlinear nature of STI design and implementation. STS scholarship has pointed out that the development of any STI is rooted in the designer's and users' assumptions of how reality works and that knowledge, innovation, and society coproduce each other (Bijker 2009; Hyysalo et al. 2016; Jasanoff 2004). As such any innovation represents the norms, values, routines, practices, and ways of doing and organizing things of the designers and users. Considering the unique nature of contexts, as discussed in this chapter, STI as such it is not by definition aligned to the societies, communities, or settings into which it is introduced. To allow for the global up-take of an innovation in a variety of unique settings—to make it work in each specific context, in fact—work needs to be done. In the process of making the innovation responsive to the needs and the peculiarities of that context, knowledge and insights change or new knowledge and insights are added. We adopt, in other words, the understanding that knowledge and innovation are not static, but change and are shaped by the context in which they are implemented and ensure this through ongoing reflections about the innovation process and one's own responsibility and those being assigned to others.

This also entails a very different understanding of technology transfer than the linear view associated with silver-bullet thinking. Yet in the United Nations' SDG documentation analyzed in this chapter, there is a tendency to fall back into the language and imaginaries of technology transfer wherever transfer is elaborated on. The definition of technology facilitation entails technology transfer: "facilitation: the concept of 'facilitation' refers to both direct interventions to match supply and demand, transfer specific technologies, and indirect, broader policy interventions aimed at improving enabling environment for science, technology and innovation (STI)" (Liu et al. 2015, p. 2). And the report of the first STI forum concludes that "specific guide-lines on technology transfer to developing countries that address the conditions of transfer, evaluation and replication of technology are needed" (United Nations 2016, p. 8). The risk of emphasizing technology transfer in the SDGs, excluding other, lo-cally available technologies or eco-friendly alternatives, has recently been raised in connection with environmental degradation (Imaz and Sheinbaum 2017). And, we argue, this essentially goes against a needs-driven and inclusive approach to STI as conceptualized in the SDGs.

This shows that while this different conceptualization of STI in the SDGs is a promising starting point, it is not so clear what it requires to do needs-driven and inclusive STI. Among the activities mentioned in the United Nations' SDG doc-umentation are strengthening of the science–policy interface to support needs-driven STI, innovation that focuses on the needs of the poor and is inclusive, and multistakeholder participation. The analysis of the POC diagnostics and cook-stove examples shows that these activities are easier said than done. Instead, they need to be paired with continuous substantial investment, research capacity, and reflection in order to familiarize oneself with the specificities of multiple user settings, different actors, their needs, sociomaterial relations, capacities, and in-frastructure; to negotiate how multiple needs should be identified, framed, and prioritized; and to realize that even when input from all stakeholders is ensured, the consequences can be undesired (such as homelessness and eviction in the case of the participatory design of cookstoves). Overall, more policy attention needs to be given to the work it takes to make technologies responsive and func-tion effectively and the (un)intended consequences. In other words, also in policy the focus should be on technology development in context and how to create technologies that are flexible and adjustable instead of focusing solely on tech-nology transfer.

To ensure that STI, such as POCs and cookstoves, is responsive and responsible to the needs of that context and better realize the potential of contributing to SDG3, we propose to make it a policy requirement to examine ethnographically the mul-tiple settings and stakeholder perceptions and to determine the needs involved, how knowledge and innovation travel to different contexts, and how they change in interaction with that context. We propose to continuously reflect on whose needs are being addressed and who is being included on what terms and with what con-sequences, in participatory approaches to engage users and communities. This also has implications for global evidence-making policies on STI. We need different, more varied notions of what counts as evidence. To ensure that innovations are responsible as well as responsive, it is essential to not only evaluate to what extent the innovation was successful in achieving its intended outcomes (as is commonly done in drug or vaccine trials and diagnostic laboratory studies), but to also study how a technology aligns with the context, what the undesired outcomes are, and what the impact has been in terms of reduction or increase of inequality and injus-tice in health and well-being.

References

Albert, H., Nathavitharana, R. R., Isaacs, C., Pai, M., Denkingerm, C. M. and Boehme, C. C. (2016). Development, roll-out and impact of Xpert MTB/RIF for tuberculosis: what lessons have we learnt and how can we do better? *European Respiratory Journal* 48, 516–525.

Barnes D. F., Kumar, P., and Openshaw, K. (2012). *Energy Sector Management Assistant Programme, and Worldbank. Cleaner Hearths, Better Homes: New Stoves For India and the Developing World*. Oxford University Press, New Delhi.

Bijker, W. E. (2009). How is technology made? that is the question. *Cambridge Journal of Economics* 34(1), 63–76.

Engel, N. (2012). New diagnostics for multi-drug resistant Tuberculosis in India: innovating control and controlling innovation. *BioSocieties* 7(1), 50–71.

Engel, N., Davids, M., Blankvoort, N., Pai, N. P., Dheda, K. and Pai, M. (2015). Compounding diagnostic delays: a qualitative study of point-of-care testing in South Africa. *Tropical Medicine and International Health* 20(4), 493–500.

Engel, N., Ganesh, G., Patil, M., Yellappa, V., Pant Pai, N., Vadnais, C. and Pai, M. (2015). Barriers to point-of-care testing in India: results from qualitative research across different settings, users and major diseases. *PLoS ONE* 10(8), e0135112.

Engel, N., Ganesh, G., Patil, M., Yellappa, V., Vadnais, C., Pant Pai, N. and Pai M. (2015). Point-of-care testing in India: missed opportunities to realize the true potential of point-of-care testing programs. *BMC Health Services Research* 15(1), 550.

Fehling, M., Nelson, B. D., and Venkatapuram, S. (2013). Limitations of the millennium development goals: a literature review. *Global Public Health* 8(10), 1109–1122.

Ghergu, C. T., Meershoek, A., Sushama, P., Onno, C. P. V. and Luc, P. D. (2018). Participation in co-design: in search of a recipe for improved cookstoves in urban Indian slums. *Action Research,* 1476750317754125.

Hadolt. B., Hörbst, V. and Müller-Rockstroh, B. (2011). Biomedical techniques in context: on the appropriation of biomedical procedures and artifacts. *Medical Anthropology* 31(3), 179–195.

Haines, A., and Cassels, A. (2004). Can the millennium development goals be attained? *The BMJ* 329(7462), 394–397.

Honkalaskar, V. H., Bhandarkar, U. V. and Sohoni, M. (2013). Development of a fuel efficient cookstove through a participatory bottom-up approach. *Energy, Sustainability and Society* 3(1), 16.

Hyysalo, S., Jensen, T. E., and Oudshoorn, N. (2016). *The New Production of Users: Changing Innovation Collectives and Involvement Strategies*. Routledge, New York.

Jasanoff, S. (2004). States of Knowledge. The co-production of science and social order. In: Urry, J. (ed.) *International Library of Sociology*. Routledge, London.

Krumeich, A. and Meershoek, A. (2014). Health in global context: beyond the social determinants of health? *Global Health Action* 7(23506). doi: 10.3402/gha.v6i0.23506

Liu, W, Kanehira, N., and Alcorta, L. (2015). *An Overview of the UN Technology Initiatives*. United Nations, New York.

Magrath, I., Seffrin, J., Hill, D., Burkart, W., Badwe, R. and Ngoma, T. (2009). Please redress the balance of millennium development goals. *The BMJ* 338. doi.org/10.1136/bmj.b2533

Makonese, T., and Bradnum, C. M. S. (2017). Public participation in technological innovation: the case of the Tshulu stove development programme. *Journal of Energy in Southern Africa* 28, 13–24.

Miranda, J. J., and Patel, V. (2005). Achieving the millennium development goals: does mental health play a role? *PLoS Medicine* 2(10), e291.

Pant Pai, N., Vadnais, C., Denkinger, C., Engel, N. and Pai, M. (2012). Point-of-care testing for infectious diseases: diversity, complexity, and barriers in low- and middle-income countries. *PLoS Medicine* 9(9), e1001306.

Peeling, R. W., Holmes, K. K., Mabey, D., and Ronald, A. (2006). Rapid tests for sexually transmitted infections (STIs): the way forward. *Sexually Transmitted Infections* 82 (suppl 5), v1–v6.

Peeling, R. W., and Mabey, D. (2010). Point-of-care tests for diagnosing infections in the developing world. *Clinical Microbiology and Infection* 16, 1062–1069.

Pfotenhauer, S., and Jasanoff, S. (2017). Panacea or diagnosis? imaginaries of innovation and the "MIT model" in three political cultures. *Social Studies of Science* 47(6), 783–810.

Prasad, A. (2014). *Imperial Technoscience: Transnational Histories of MRI in the United States, Britain and India*: MIT Press, Cambridge, MA.

Richard, F., Hercot, D., Ouédraogo C, Delvaux, T., Samake, S., van Olmen, J, Conombo, G., Hammonds, R., and Vandemoortele, J. (2011). Sub-Saharan Africa and the health MDGs: the need to move beyond the "quick impact" model. *Reproductive Health Matters* 19(38), 42–55.

Sinha, B. (2002). The Indian stove programme: an insider's view—the role of society, politics, economics and education. *Boiling Point* 48, 23–26.

Sushama, P., Ghergu, C., Meershoek, A., de Witte, L. P., van Schayck, O. C. P., and Krumeich, A. (2018). Dark clouds in co-creation, and their silver linings practical challenges we faced in a participatory project in a resource-constrained community in India, and how we overcame (some of) them. *Global Health Action* 11(1), 1421342. doi: 10.1080/16549716.2017.1421342

Tucker, J. D., Bien, C. H. and Peeling, R. W. (2013). Point-of-care testing for sexually transmitted infections: recent advances and implications for disease control. *Current Opinion in Infectious Diseases* 26(1), 73–79.

United Nations (2016). *Multi-Stakeholder Forum on Science, Technology and Innovation for the Sustainable Development Goals: Summary by the Co-Chairs*. United Nations, New York.

United Nations Department of Economic and Social Affairs (2017). STI Forum: Video message from Bill Gates. https://youtu.be/66VZnC4eUak (accessed April 4, 2019).

Urmee, T.and Gyamfi, S. (2014). A review of improved cookstove technologies and programs. *Renewable and Sustainable Energy Reviews* 33, 625–635.

World Health Organization (WHO) (2010). WHO endorses new rapid Tuberculosis test. https://www.who.int/mediacentre/news/releases/2010/tb_test_20101208/en/ (accessed April 4, 2019).

World Health Organization (WHO) (2016). *Burning Opportunity: Clean Household Energy for Health, Sustainable Development, and Wellbeing of Women and Children*. WHO, Geneva.

World Health Organization (WHO) and United Nations Population Fund (UNFPA) (2017). 2017 High Level Political Forum Thematic Review of SDG3: Ensure healthy lives and promote well-being for all at all ages. https://sustainabledevelopment.un.org/content/documents/14367SDG3format-rev_MD_OD.pdf (accessed April 4, 2019).

16

A Systemic Perspective on the Global Sanitation Challenge

Insights from Sociotechnical Dynamics in Nairobi's Informal Settlements

Mara J. van Welie and Bernhard Truffer

1. Introduction

The lack of sanitation supply in cities in low-income countries is one of the most persistent development challenges. The sanitation target of the millennium development goals (MDGs) to half the proportion of the population without sustainable access to basic sanitation was not reached (WHO 2015). Only 95 countries met the MDGs regarding sanitation, mostly in developed regions. None of the least developed countries did, which means that in total 700 million people missed the MDG target. The large majority of these people lived in Southern Asia and sub-Saharan Africa. Sub-Saharan Africa as a region has been lagging mostly behind in terms of the percentage of the population that had access to improved sanitation: less than 20% in 2015 (WHO 2015). Adequate sanitation plays an important role to ensure health: worldwide in 2012, 871.000 deaths were caused by inadequate water, sanitation, and hygiene (WASH) services (WHO 2016, p. 72).

The lack of sanitation is especially persistent in cities in the Global South.[1] The world is rapidly urbanizing and this creates huge challenges for city planners, who are not able to keep pace with the number of people moving into cities in search for work and life opportunities. New city dwellers often end up impoverished, living in informal settlements without access to basic services such as sanitation (UN-Habitat 2004). One of the major challenges that hampered the progress toward the sanitation target of the MDGs in cities in the Global South was the narrow focus on providing sanitation hardware/technologies such as latrines, toilets, and waterborne sewerage. The MDGs reinforced this "hardware focus," as the number of toilets was often used to describe improvements in access to sanitation (Andersson et al. 2016).

---

[1] The term Global South in this chapter is not a direct reference to Southern Hemispheres, but applied to differentiate nations in terms of socioeconomic capabilities and related characteristics. We use the term to refer to low income economies according to the World Bank (2018).

The provision of sanitation is much more complex than installing toilets, especially in urban areas. The challenge goes beyond achieving a defecation free environment (Mara 2018). Conventional waterborne, centralized sewer systems are often not appropriate for most cities in the Global South and especially for informal settlements. It is not affordable, requires too much water, and institutional capacity and technological skills are lacking. As a consequence, these systems even contribute to the pollution of the environment because the majority of the wastewater is not treated (Black and Fawcett 2008; Szanto et al. 2012). Decentralized sanitation solutions might be an option (Larsen et al. 2016), but they also face many challenges, such as the need for proper management of fecal sludge. For example, in the MDG era many latrines were installed in cities that were never emptied, which led to overflowing latrines (Koné 2010).

The failure of the MDGs sanitation target has led to a greater recognition that a broader focus is key for the success sanitation services (Andersson et al. 2016). This is also reflected in the SDGs that focus on safely managed sanitation services (subtargets 6.2, 6.3, 6.A, and 6.B are specifically relevant for the global sanitation challenge). Whereas the MDGs focused on improved sanitation access, the sustainable development goal for water and sanitation (SDG6) therefore aims to improve sanitation services beyond a minimum coverage of toilets. For example, in SDG6 the link to the environmental dimension of sustainability has been included, such as the potential of recycling and reuse of wastewater (subtarget 6.3), and there is an implementation target focused on capacity building (subtarget 6.A) (UN 2016; Bartram et al. 2018).

The wider set of sanitation targets in the SDGs is a step toward recognizing the complex problem of providing sanitation services in cities in the Global South. However, progress toward SDG6 is very slow: worldwide, 4.5 billion people still lack safely managed sanitation (WHO 2017, p. 29). This problem thus demands long-term fundamental transformations of urban sanitation services and infrastructures. Innovations (technological, social, and organizational) have the potential to improve sanitation service supply, management, and access in cities in the Global South toward reaching SDG6. However, the potential of many of these innovations has often not been met. For example, innovative sanitation technologies do not fit with regulatory frameworks, are not properly used and maintained, lack acceptance, face cultural resistance, or are perceived inconvenient by users. In a same vein, social innovations, such as behavior change campaigns, are not continued, scaled up, or monitored (Parkinson and Tayler 2003; Markard and Luthi 2010; Jones et al. 2013).

A science, technology, and innovation (STI) policy perspective might potentially be fruitful in order to identify which factors lead to poor innovation performance, and how various forms of policies could address these shortcomings. We adopt a specific STI perspective in this chapter, namely a *sociotechnical systems* perspective that draws on insights from innovation studies and the sustainability transitions literature. Such a systemic perspective enables the analysis of the interplay between technologies, infrastructures, and their associated actor networks, institutions, and user and provider practices. It leads to the identification of barriers to innovation and suggests potential pointers to overcome them. This system perspective thus broadens the perspective on urban sanitation problems, in order to achieve the sustainable development goals (SDGs). In this chapter, we aim to show the use of a sociotechnical system lens as an analytical tool to identify challenges and opportunities of the associated innovation processes. As an illustration, we apply the sociotechnical system frameworks to the case of sanitation in the city of Nairobi, Kenya.

The chapter is structured as follows: in the next section, we present the analytical framework based on a sociotechnical system perspective, which enables the analysis of innovation and transition dynamics in complex basic service sectors in cities in the Global South. In section 3, we introduce the methodology of the study. Section 4 introduces the case of sanitation in Nairobi, discusses the current structure of the sanitation regimes in Nairobi, and analyzes different innovation activities that aim at changing the existing regimes. Finally, we discuss barriers toward a more sustainable urban sanitation sector and how these can be overcome, in order to improve innovation performance and long-term transformation processes of urban services for reaching the SDGs.

2. Basic Service Sectors and Innovation Dynamics as Sociotechnical Systems

Cities in the Global South are characterized by heterogeneity, spatial unevenness, and complexity of basic services and infrastructures such as housing, solid waste management, electricity, water, and sanitation (Bakker et al. 2008; Kooy and Bakker 2008; Furlong 2014). Large inequalities exist in these cities. The poorest residents rely on informal, inefficient, and/or low-quality services to meet their basic needs in the informal settlements, while residents in high-income areas access domestic toilet services connected to waterborne-piped systems. In the context of these complexities, social and organizational innovations and new technologies often fail to successfully improve service supply, management, and access in cities in the Global South. One of the core reasons for these failures stems from the fact that the successful transformation of existing services depends on the alignment of manifold conditions such as regulations, finance, institutions, social issues, and the environment.

An STI perspective is very appropriate for analyzing these complex dynamics. We built on sociotechnical systems which consist of networks of actors and institutions, as well as material artefacts and knowledge (Markard et al. 2012, p. 956). Table 16.1 shows the basic elements of a sociotechnical system, which can be analyzed in more or less detail using particular sociotechnical system concepts.

A sociotechnical system perspective can both help to understand the elements and their interrelationship that constitute basic service sectors, and to understand the diversity of factors that influence innovation processes changing these sectors. We borrow two sociotechnical system concepts in particular: first, existing service systems in cities are conceptualized as strongly aligned *sociotechnical regimes* to

Table 16.1 Basic elements of a sociotechnical system

Elements of a sociotechnical system	Examples
Actors	Individuals, firms, producers, suppliers, customers, policymakers, organizations, users, universities, etc.
Networks	Relationships and links between the actors: industry associations, conferences, clubs, partnerships, etc.
Institutions	Societal and technical norms, rules and regulations, guidelines, standards of good practice, culture, cognitive frames, etc.
Technology	Material artefacts, infrastructures, etc.

These basic elements are used to identify both the structures of the particular sociotechnical system concepts of sociotechnical regimes and technological innovation systems.
Source: By authors with inputs from Markard et al. (2012); Schot et al. (2016).

understand the persistent current state of basic service sectors in cities in the Global South. Second, we use technological innovation systems (TIS) to analyze success conditions for initiatives that aim for the development, diffusion, and use of specific innovations (e.g., new technologies and service models) toward transforming basic service sectors (Markard and Truffer 2008). TIS emphasizes how actors, their interactions in networks, and the role of institutional arrangements promote or hinder innovations (Hekkert et al. 2007; Truffer 2015). In the following part we will elaborate how sociotechnical regime and TIS can be leveraged to better understand the challenges of introducing innovations in sanitation in cities in the Global South.

The concept of a sociotechnical regime denotes the institutionalized set of rules and patterns in a sector related to the elements of a sociotechnical system, which govern and determine how basic services are provided and used (Geels 2004; Fuenfschilling and Truffer 2014). The regime's degree of institutionalization depends on how widely diffused and taken for granted certain characteristics of the sociotechnical system are, how long it has been in place, and to what degree it is contested by different societal actors (Fuenfschilling and Truffer 2014). The sociotechnical regime thus provides the logic and direction for incremental sociotechnical change along established pathways (Markard et al. 2012). The concept of a sociotechnical regime helps to understand the challenges to, and possibilities for, transition pathways toward improved service delivery systems.

To be able to capture the heterogeneity and complexity of basic services in cities in the Global South, we distinguish between two levels of the sociotechnical regime: *service regimes* (e.g., the automobile regime) and the *sectoral regime* (e.g., the individual transport regime). Service regimes can be identified by five dimensions: artefacts and infrastructures, social interaction, rationale and meaning, organizational mode, and time and space (van Welie, Cherunya et al. 2018). A sectoral regime consists of one or more service regimes (e.g., the individual transport regime consists of automobile, bike, and public transport service regimes). When service regimes fit very well together and complement each other at the sectoral level, we refer to the sectoral regime as a "polycentric regime." In contrast, when the service regimes do not complement each other and physical linkages between the infrastructures are missing, we call the regime "splintered." This two-leveled (service and sectoral) regime analysis enables the systematic identification of the complexities of basic services in cities in the Global South, which we illustrate in the next section with the empirical case of sanitation in Nairobi.

The success conditions for introducing new technologies or service offerings can be analyzed by means of the concept of TIS. TISs are "'socio-technical systems focused on the development, diffusion and use of a particular technology" (Bergek et al. 2008, p. 408). TISs are constituted by actors, networks, and institutions that are supporting the development and maturation of new solutions. Innovation success can be related to the presence and quality of six formation processes: knowledge development and diffusion, entrepreneurial activities, legitimation creation, guidance of the search, market development, and resource mobilization, following (Binz et al. 2016), who built on Hekkert et al. (2007) and Bergek et al. (2008). The performance of these processes indicates how well a new technology or product is developing, diffusing, and being used. A functional analysis leads to the identification of system weaknesses, which hinder innovation development (Klein Woolthuis et al. 2005; Jacobsson and Bergek 2011). These system weaknesses give pointers how to improve the success conditions for innovation development.

Transitions can therefore be understood as fundamental changes in the prevailing sociotechnical regime. These changes include technological, social, and practice components, which actors may address from different angles depending on their specific capabilities. The success of such innovation processes depends often on the mobilization of resources from a broad set of actors and their networks, forming appropriate TIS structures (Markard and Truffer 2008). The sustainability transitions literature has provided ample evidence on how TIS emerge and mature. However, most of the studies have analyzed transformation processes in OECD countries (Markard et al. 2012). We will show in the following case study how these insights can be leveraged for the challenging contexts of informal settlements in the Global South, with the case of sanitation in informal settlements of Nairobi.

3. Methodological Approach

We use a case study design built on expert interviews. A total of 104 semistructured interviews were conducted between February and December 2016 in Nairobi, covering experts in government agencies, nongovernmental organizations (NGOs), international agencies, sanitation enterprises, community based organizations (CBOs), nonformalized sanitation service providers, and individual inhabitants. We selected the interviewees based on their knowledge about the sanitation sector, innovative on-site sanitation activities, and/or the utility's pro-poor activities. We used snowball sampling to identify key experts, until saturation in responses was reached. All interviews were recorded, transcribed, and coded with MAXQDA 12 (VERBI-Software 2015), using qualitative analysis. The results were triangulated with various secondary data sources: reports, websites, journal articles, online newsletters, and online articles.

4. Nairobi's Sanitation Sector

We first introduce the case of sanitation in Nairobi, and then outline how the sanitation regime in Nairobi looks. This is followed by the analysis of the innovation strategies of the public utility, NGOs, and enterprises applying an innovation system lens.

4.1 The Case of Sanitation in Nairobi

Nairobi's sanitation sector is characterized by a high variety of access options and conditions, multiple providers, different institutional arrangements, various spatial structures and user practices, and complex formal and informal governance structures (CCN 2007; Nyanchaga and Ombongi 2007; O'Keefe et al. 2015; Cherunya et al. 2018). Inequalities in the city are large, and adequate provision of sanitation services to all the city's inhabitants is a fundamental challenge, especially in the informal settlements where 36% of Nairobi's population lives (Mansour et al. 2017). Despite a wide range of strategies to increase sanitation coverage in informal settlements since several decades, sanitation improvements did barely keep pace with the city's rapid population growth (Szanto et al. 2012). The city is thus facing significant infrastructure and services challenges. We study the initiatives of different actors that try to innovate toward increased coverage of hygienic and sustainable sanitation services in the city's informal settlements. The utility has recently started to expand its operations to informal settlements, while its core business historically focused on high-income neighborhoods (van Welie et al. 2019). Furthermore, social enterprises and NGOs introduce innovative on-site sanitation services, which are more safe, dignified, clean, and well-organized than the on-site sanitation options that are currently in use in informal settlements (van Welie, Truffer et al. 2018).

4.2 Nairobi's Splintered Sanitation Regime

Through an analysis of user and provider practices in the sanitation sector, we identified five service regimes that operate in Nairobi. The five service regimes vary greatly in one or more of their dimensions: 1) the *domestic sewer regime* encompasses a flushing toilet used by one household, connected to the sewer system which is provided and operated by the utility; 2) the *shared on-site sanitation regime* encompasses a shared on-site toilet located either inside a plot or off-plot. It is shared by multiple households and mostly provided and installed by the landlord of the plot or by an NGO; (3) the *public sanitation regime* consists of toilet services in public places, which provide pay-per-use services. They are mostly operated by CBOs or by private enterprises; (4) the *coping sanitation regime* denotes practices of people to relieve themselves in their homes using improvised domestic items or defecation in the open; finally, (5) the *container based regime* consists of toilets equipped with containers or biodegradable bags to collect the feces and the urine. The containers or bags are regularly collected to a place where the waste is treated and the resulting sludge is re-used. Container-based services function as a public pay-per use or as in-home toilets, and are provided by social enterprises. The core dimensions of the five service regimes are summarized and compared in Table 16.2 (for more elaborate descriptions of the five service regimes see van Welie, Cherunya et al. 2018).

Across the city of Nairobi, these five service regimes constitute the sectoral regime. However, the different service regimes are not complementing each other well and physical infrastructures that could align the service regimes are largely lacking. In particular, regulations do not recognize all these service regimes as legitimate ways of service provision. Planners and policymakers, for instance, often ignore the coping sanitation regime, or emptying services for shared and public toilets by manual pit emptiers are considered illegal. Consequently, the interoperability between the service regimes is low; users can only meet their daily basic needs through their own efforts of combining services in different service regimes. The sectoral regime of sanitation in Nairobi can therefore be characterized as a *splintered regime*.

The splintered regime does not look the same in every part of the city. Some neighborhoods are characterized by a single service regime (e.g., high-end areas are solely characterized by the domestic sewer regime), others by a mixture of two or more (e.g., certain low-income residential areas have shared on-site sanitation services and public sanitation services), and in informal settlements often all service regimes coexist. Figure 16.1 provides a conceptual representation of the splintered regime in Nairobi.

The systemic identification of the heterogeneity of Nairobi's sanitation sector helps to overcome the often simplified interpretation of sanitation sectors focused on the internationally dominant paradigm of water-borne sewerage systems (Fuenfschilling and Binz 2018), which only serves a minority of the population in cities in the Global South.

All five service regimes have existed for a long time, their existence is taken for granted by users and providers, and all the service regimes are widely used in the city. It is thus unlikely that one of the service regimes will suddenly replace another or disappear on a city-scale in the near future. The often envisioned transition process toward a domestic sewer regime that covers the whole city seems thus unrealistic in the near future (if ever). Rather, the sanitation sector in Nairobi could benefit from improved alignments between the existing service regimes at the sectoral level. Improving the complementarities and interoperability between the service regimes

Table 16.2 The detailed dimensions of the sanitation service regimes in Nairobi

	Infrastructure & artefacts	Rationale & meaning	Social interaction	Organizational mode	Time & space
Domestic sewer regime	Central sewer system + (pour) flush toilet	Users: comfortable, good image, costly, consumes too much water Provider: sanitation using high quality modern technologies	None	Daily maintenances by households Waste management by utility	Timing users: anytime location: inside the house or on the plot
Shared on-site regime	Latrine + pit or septic tank	Users: accessible, convenient, low costs, dirty conflicts among households Provider: arranging sanitation for tenants	Coordinating access and cleaning among households	Organized by landlords or NGOs Daily maintenance by households Waste management by manual emptiers or private exhauster trucks	Timing users: anytime when on-plot and only during the day-time when off-plot location: off-plot or on-plot
Public sanitation regime	Latrine + pit or septic tank; bio & compost latrines; hanging toilet; central sewer system + pour flush toilet	Users: convenient, costly, dirty, risk of diseases, insecure during the night Provider: business opportunity	Trust building: being a customer Everyday interaction between operator and user	Daily commercial operations by CBO, NGO or enterprise Waste management: manual emptiers; private exhauster trucks, or utility	Timing users: during the day when user has money Location: commercial areas, public residential, hanging over a river
Coping sanitation regime	Cleaning bucket; plastic bag	Convenient option, no costs, useful in the setting of informal settlements, shameful, indignity, bad smell, done secretly, dirty, risk of diseases, insecure, acceptable for children	Coordination within the family, being accompanied by others	Organized by households and individuals No safe disposal of the waste	Timing users: anytime Location: inside the house, close to the home, around shared toilets, at open defecation hotspots (rivers, bushes)

Container based sanitation regime	Waterless system with urine diversion, biodegradable bags, containers	*As in-house service:* Users: convenient, indignity, not appropriate for adults, culturally unfit & uncomfortable for men, useful for children, useful at night *As public service:* Users: convenient, costly, risk of diseases Provider: environmental friendly sanitation, creating value from recycling waste	*As in-house service:* Coordination within the family *As public service:* Trust building, being a customer, a lot of interaction between operator and user	*As in-house service:* Daily operations by household; enterprise collects the waste and re-uses it as fertilizer, biogas, animal feed *As public service:* Daily commercial operations by enterprise; waste is collected and re-used as fertilizer, biogas, animal feed	*As in-house service:* Timing users: anytime, especially at night Location: in-house toilet *As public service:* Timing users: during the day when user has money Location: public locations: commercial areas, public residential

Source: van Welie et al. (2018).

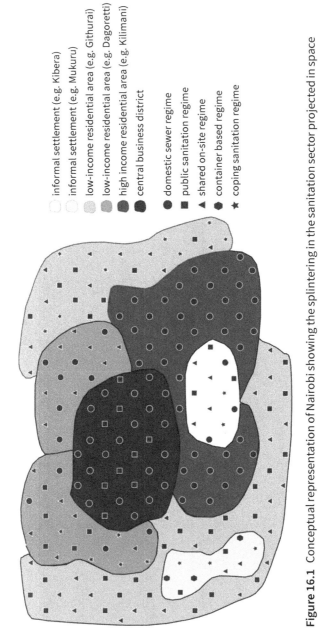

Figure 16.1 Conceptual representation of Nairobi showing the splintering in the sanitation sector projected in space

Source: van Welie et al. (2018a).

Legend:

- informal settlement (e.g. Kibera)
- informal settlement (e.g. Mukuru)
- low-income residential area (e.g. Githurai)
- low-income residential area (e.g. Dagoretti)
- high income residential area (e.g. Kilimani)
- central business district

- domestic sewer regime
- public sanitation regime
- shared on-site regime
- container based regime
- coping sanitation regime

could lead to a transition from a splintered regime toward a well-aligned polycentric regime. This would lead to a sanitation sector that provides higher quality, sustainable, and more justly distributed services. City planning toward SDG target 6.2, which calls for "access to adequate and equitable sanitation and hygiene for all" (UN 2016), should include plans on how to align the various persistent sanitation service regimes in a city. Transition processes to a well-aligned regime require conscious innovation effort at the sectoral level. However, innovations are mostly focused on individual service regimes. Potential synergies and interactions with other regimes are the exception rather than the rule. This leads to incremental improvements of service regimes (e.g., new toilet designs), while misalignments at the sectoral regime are not recognized or improved (e.g., fecal sludge not being collected in an orderly way). As a result, new offerings are too often locally not well-embedded and fail shortly after introduction.

What does it take to be innovative in a splintered regime when considering the lack of alignments? We will elaborate on two initiatives: pro-poor initiatives of the utility and innovative on-site sanitation services developed by NGOs and social enterprises. We gain insights into the barriers that hinder these innovations and identify opportunities toward meeting the SDG sanitation targets in the city.

4.3 Pro-Poor Innovations of the Utility

Nairobi's water and sewerage utility has recently started to expand its operations into the informal settlements, under pressure of the SDGs and Kenya's new constitution in 2010 that gave all citizens the right of access to water and sanitation (Kenya 2010). The informal settlements were a new context for the utility, because historically the utility only operated in high-income areas, where it installed conventional sewer systems and domestic connections. The utility's conventional capabilities and organizational structures were challenged to operate successfully in informal settlements, due to the high complexity associated with multiple informal institutional arrangements, poor infrastructure conditions, inefficient governance structures, very heterogeneous user needs, and widespread poverty.

The utility needed a radical innovation strategy to adapt its organizational mode, and to develop new capabilities to operate successfully in informal settlements. This was challenging, because the utility's conventional capabilities were strongly linked to operate the domestic sewer regime that prevails in most high-income areas of the city, for example the use of centralized infrastructures and domestic connections, monthly payment models, written procedures for applications, and formal interaction with customers. This way of working was highly institutionalized within the utility, as it was done for many years and perceived as normal in all departments of the organization. Different capabilities were needed, however, to successfully operate in the public service regimes, the shared service regimes, and to deal with the impacts of the coping regimes, all three being widely practiced in informal settlements. For instance, capabilities were needed in the form of social skills to interact intensively with customers, the capacity to collaborate with community groups, skills to deal with cartels, flexible (nonwritten) application procedures, and the ability to use a variety of different infrastructures and artefacts, new public service models, and flexible (nonregular) payment systems. These are important capabilities that need to be developed by sanitation providers in the context of the SDGs. Subtarget 6.B emphasizes "support and strengthen the participation of local communities in improving water and sanitation management" as a means of achieving other subtargets of SDG6, such as subtarget 6.2, "access to adequate and equitable sanitation and hygiene for all" (UN 2016).

As a first step, the utility decided to install a separate "pro-poor" unit with a special mandate. This enabled them to hire a number of sociologists, who engaged in collaborations with NGOs in informal settlements besides the otherwise predominant engineers. Based on these new competences, the unit learned to interact with local leaders and residents in informal settlements, introduced flexible payment models, and developed new administrative procedures. The pro-poor unit set up several successful projects in informal settlements, such as public ablution blocks and water kiosks. Despite these successes, these new offerings provoked considerable tensions with the rest of the utility. Interfaces between the new services and the established departments of the utility did not work well. For instance, the operation and management tasks of these projects, such as water meter reading and billing, were handled by the respective business departments, which found it difficult to perform these tasks in the informal settlements. The conventional utility's employees lacked the capacity for intensive customer engagement, did not have the negotiation and social skills, and lacked the time that was needed to successfully perform maintenance tasks in the dense and unplanned settlements.

To solve these problems, the utility decided to re-integrate the pro-poor activities into the conventional organizational structure. The special mandate was replaced by a conventional business mandate, similar to the other departments. An exemplary strategy of the new unit was the set-up of a sewer expansion project in a low-income area. Poor customers could connect to a simplified sewer network for subsidized tariffs. In this project, the utility could leverage its well-established capabilities: installing centralized infrastructure with domestic connections, no collaborations with community groups, and formal interaction with customers. However, this innovative service model did not perform well in terms of serving the urban poor in informal settlements. The project suffered from frequently blocking sewer pipes, because residents dumped solid waste in the toilets as they used to do in pit latrines. The subsidized tariff was furthermore criticized as still too high for the poorest residents. Lastly, the model could not be rolled out in dense, unplanned informal settlements, but only suited relatively well-planned low-income neighborhoods. Therefore, the new strategy did not reach the target of expanding services into the informal settlements and serving the poorest of the poor.

Thus, by solving its internal tensions, the utility did not pay sufficient attention to building up and further developing new capabilities. This led to pro-poor innovation projects that did not fit many of the dimensions of the variety of service regimes in informal settlements. Instead of catering for the development of new service offerings and alignments between different service regimes, they tried to fit the solutions in informal settlements to the context they knew best: the domestic sewer regime. Such narrow focus on domestic connections to centralized sewer systems and ignorance of improving (the linkages between) other sanitation service regimes undermines reaching the SDG6 target of access to adequate and equitable sanitation for all (UN 2016).

4.4 On-Site Sanitation Innovations Along the Sanitation Chain

The second case of innovative activity in Nairobi's sanitation sector relates to initiatives by social enterprises and NGOs that implement innovative on-site sanitation systems. These actors mostly engage with systems that collect, store excreta on the plot where it is generated and treat the waste, as opposed to sewer systems that convey waste away from the plot using pipes (Tilley et al. 2014, p. 173). The innovators in Nairobi engaged in developing integrative offerings that include different segments of the "sanitation chain": user interface, storage, conveyance, treatment, use, or safe disposal of waste (Tilley et al. 2014) (Figure 16.2). It provides a necessary holistic view on sanitation, because focus on only one segment of the chain (e.g., the toilet/latrine) is insufficient to provide safe and effective sanitation services for low-income populations at scale (WSUP 2017). These initiatives thus respond to experiences that were gained in the era of the MDGs, where many programs were implemented that aimed at improving (the number of) latrines, without consideration of how to deal with the waste (Koné 2010; Wald 2017). This led to cities in which overflowing pit latrines became a normality, especially in the informal settlements. The current more holistic

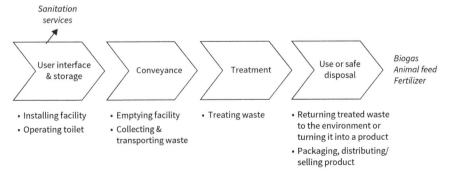

Figure 16.2 Segments, core activities and outputs of the sanitation chain
Source: van Welie et al. (2018b).

view on on-site sanitation among social enterprises and NGOs fits with the SDGs, especially with the focus on reuse and recycling in SDG6 (UN 2016).

However, many of the newer initiatives despite adopting a broader perspective on the sanitation chain, mostly developed in isolation of each other. We therefore propose to adopt a TIS perspective in order to identify capability, coordination, and institutional failures, which might hinder innovation success. The on-site TIS is constituted by all the innovative infrastructures, technologies, actors, and institutions along the sanitation chain present in informal settlements of Nairobi. Not only the core activities of installing and operating sanitation services are included, but also the downstream activities of waste management, treatment, and reuse (Figure 16.2). These activities are essential for the success of the early parts of the sanitation chain.

In Nairobi, three innovative approaches comprise the TIS of on-site sanitation. First, *biocenters*, which are community centers that have several functions, one of which is public sanitation facilities. A biogas reactor is used to treat waste and produce biogas that is used for cooking and to heat showers in the centers. Secondly, personal single-use *biodegradable bags* are used in people's homes or at schools for the storage of excreta. The bag is coated with urea to disinfect the feces directly. The bags are regularly collected and transported to a storage location for the composting process, after which they are reused as a fertilizer by coffee farmers. Lastly, the *container based sanitation* approach is used, which is built around stand-alone waterless toilets that capture waste in (portable) containers. The containers are regularly collected, transported, and the waste is composted and treated. Animal feed and fertilizer are produced and sold to farmers. Each approach's sanitation chain is operated and coordinated by one actor (NGO or social enterprise). Innovative on-site sanitation systems can thus contribute to various types of job creation, and thereby to SDG8, "achieving full and productive employment and decent work for all women and men" (target 8.5) (UN 2016).

A TIS functional analysis for each segment of the sanitation chain revealed that innovation development along the sanitation chain is heterogeneous, as the functions develop differently in each segment of the chain. For example, in the first segment (user interface), many experiments are conducted and knowledge is developed about user's preferences. Also in the third segments (treatment), innovative activity is relatively high, different technologies are being tested, and research is conducted to optimize treatment processes. However, other systemic innovation functions such as resource mobilization, knowledge diffusion, guidance of search, and market development are underdeveloped. When comparing the segments, systemic innovation development in the second (conveyance) segment lags mostly behind.

One important reason for the relatively low innovation success in the on-site sanitation TIS in Nairobi is the lack of interaction between the different actors. Different initiatives are trying to develop their own individual sanitation chain. Consequently, many innovations have been tested in isolation from each other—even though they target the same problems in the same city. This lack of coordination is a barrier for the TIS to scale, which prevents the improvement of public health in Nairobi and hampers progress to reach the SDGs. Working on synergies between the different innovation initiatives would enable to mobilize critical resources that are lacking now for each individual initiative. For instance currently, collection and transportation are taken care of by all sanitation chain operators in different ways and all actors have independently implemented one or more innovative treatment technologies. As a consequence, some initiatives are confronted with insufficient waste streams and are therefore unable to use the full capacity of their treatment facility. To scale the TIS, conveyance and treatment of on-site sanitation waste should therefore increasingly be coordinated. This could also attract actors with an interest in addressing the public health situation in Nairobi, such as the utility, the water board, and the health department of Nairobi County to support the TIS. These actors could, for example, contribute to improving the legitimation of waste collection from on-site sanitation and thereby improve the institutional context for the innovation processes. Or they could help to generate resources for the innovation processes and stimulate market development for reused products.

5. Discussion and Conclusion

This chapter investigated the opportunities and struggles that innovators face in heterogeneous service sectors in cities in the Global South by leveraging core concepts of the STI literature. The conceptualization of Nairobi's sanitation sector as a splintered sectoral regime enabled the identification of five rather incomplete and uncoordinated service regimes. This sectoral regime leads to inefficiencies in service provision and requires users to spend a lot of effort in combining different services on a daily basis. In order to improve the sustainability of this sector, the different service regimes could benefit from improved alignments at the sectoral level. The currently splintered regime could develop toward a better-aligned polycentric regime, and by this providing higher quality, more sustainable, and more justly distributed services. Drawing on two cases of recent innovation attempts, we discussed factors that hindered innovation development at the sectoral level in Nairobi, mostly related to system weaknesses in the form of capability, coordination, and institutional failures.

From the case of the utility, we learn that creating alignments in a splintered sectoral regime might mean that an actor moves from operating successfully in one service regime to learning how to operate in another one. Consequently, an incumbent actor in one service regime can be a niche actor in another service regime, because it suffers from capability constraints and has to learn how to operate in an environment characterized by different types of social interaction, rationales, use of infrastructures at different locations, and organizational modes. This is an important learning tool for various types of actors, such as utilities, local governments, and NGOs that aim at improving basic services in the context of the SDGs. These actors need to understand the interrelated regime dimensions in order to successfully conduct innovation strategies. The established capabilities of a specific actor are often closely aligned with the service regime in which he originally acts. Identifying the interlinkages between capabilities and the (old and new) regime structures can be a useful starting point toward understanding the challenges that individual actors are confronted with in a splintered regime context. The lack of capacity and importance of capacity building in sanitation in cities in the Global South has already been widely recognized, and one of the implementation targets of SDG6 focusses therefore on capacity-building support (Bartram et al. 2018). The sociotechnical perspective enables the systemic identification of capability constraints of different actors that aim to contribute to transformations of heterogeneous basic service sectors in cities. Such an analysis can thus help to specify how capacity building assistance for the SDGs should be effectively used to not only help individual actors, but to contribute to system building.

The case of on-site sanitation innovations showed that various actors develop similar innovations within different service regimes in the city, without deliberately complementing each other. The innovation system analysis led to the identification of coordination failures among the different actors, and the identification of lack of institutional support for on-site sanitation innovations. Such systemic failures are a barrier for SDG6 to be achieved, when assuming that a variety of possible sanitation services and systems are necessary to improve alignments in heterogeneous basic service sectors. However, coordination and institutional issues in cities are often not the main focus of planners, governments, or development organizations that want to support innovation. Coordination and institutional aspects are also not mentioned as a means of implementation target for SDG6 (Bartram et al. 2018). We argue that interventions should stimulate the reaping of synergies between the individual innovation efforts and improve institutional conditions, to support fundamental change of urban basic services. Exemplary interventions could be coordination of innovation activities to increase knowledge diffusion; development of standards to stimulate market formation for reused products; and development of guidelines for citywide sanitation to increase guidance of search for innovators. On-site sanitation innovations challenge several dimensions of the different service regimes, and could thereby contribute to the alignment of the sectoral sanitation regime, for example, by coordinating the collection of waste from various on-site systems. These insights are relevant for other cases of innovation in basic services in cities in the Global South.

We illustrated how factors hindering innovation development toward the SDGs often go beyond technological aspects, but rather represent system weaknesses related to actors, networks, and institutional aspects of sociotechnical systems. A sociotechnical system perspective can help to identify such factors and thereby provide a fresh perspective for planners, service providers, and policymakers toward solving persistent development challenges. The sociotechnical system perspective complements other STI perspectives through the identification of system weaknesses in the form of institutional, coordination, and capability failures. Such a systemic perspective is particularly novel compared to the often used behavior change and engineering lenses to assess the performance of technologies and innovations for development. The illustration of our argument was limited to the sanitation sector. But we maintain that a sociotechnical system analysis can be used to analyze a broad set of development challenges in other basic service sectors where similar complex problems prevail, such as water, electricity, solid waste management, road infrastructure, and housing. This might reveal new insights to support long-term transformations of urban services and infrastructures toward reaching the SDGs.

Acknowledgment

This research was funded by the Swiss National Science Foundation (grant number 10001A_159300).

References

Andersson, K., Dickin, S., and Rosemarin, A. (2016). Towards "sustainable" sanitation: challenges and opportunities in urban areas. *Sustainability* 8(12), 1289.

Bakker, K., Kooy, M., Shofiani, N. E., and Martijn, E. (2008). Governance failure: rethinking the institutional dimensions of urban water supply to poor households. *World Development* 36(10), 1891–1915.

Bartram, J., Brocklehurst, C., Bradley, D., Muller, M., and Evans, B. (2018). Policy review of the means of implementation targets and indicators for the sustainable development goal for water and sanitation. *Nature Partner Journals, Clean Water* 1(1), 1–3.

Bergek, A., Jacobsson, S., Carlsson, B., Lindmark, S., and Rickne, A. (2008). Analyzing the functional dynamics of technological innovation systems: a scheme of analysis. *Research Policy* 37, 407–429.

Binz, C., Truffer, B., and Coenen, L. (2016). Path creation as a process of resource alignment and anchoring: industry formation for on-site water recycling in Beijing. *Economic Geography* 92(2), 172–200.

Black, M., and Fawcett, B. (2008). *The Last Taboo: Opening the Door on the Global Sanitation Crisis.* Earthscan, London.

City Council of Nairobi (CCN) (2007). City of Nairobi Environment Outlook. http://wedocs. unep.org/handle/20.500.11822/8738 (accessed April 4, 2019).

Cherunya, P. C., Truffer, B., and Ahlborg, H. (2018). Anchoring innovations in oscillating domestic spaces: why sanitation service offerings fail in informal settlements. Paper presented at the International Sustainability Transitions (IST) Conference 2018, June 12–14, Manchester, UK.

Fuenfschilling, L., and Binz, C. (2018). Global socio-technical regimes. *Research Policy* 47(4), 735–749.

Fuenfschilling, L., and Truffer, B. (2014). The structuration of socio-technical regimes: conceptual foundations from institutional theory. *Research Policy* 43(4), 772–791.

Furlong, K. (2014). STS beyond the "modern infrastructure ideal": extending theory by engaging with infrastructure challenges in the South. *Technology in Society* 38, 139–147.

Geels, F. W. (2004). From sectoral systems of innovation to socio-technical systems: insights about dynamics and change from sociology and institutional theory. *Research Policy* 33(6–7), 897–920.

Hekkert, M. P., Suurs, R. A. A., Negro, S. O., Kuhlmann, S., and Smits, R. E. H. M. (2007). Functions of innovation systems: a new approach for analysing technological change *Technological Forecasting and Social Change* 74, 413–432.

Jacobsson, S., and Bergek, A. (2011). Innovation system analyses and sustainability transitions: contributions and suggestions for research. *Environmental Innovation and Societal Transitions* 1(1), 41–57.

Jones, S., Greene, N., Hueso, A., Sharp, H., and Kennedy-Walker, R. (2013). *Learning from Failure: Lessons for the Sanitation Sector*. UK Sanitation Community of Practice (SanCoP). http://www.bpdws.org/web/d/DOC_360.pdf%3FstatsHandlerDone%3D1 (accessed April 4, 2019).

Klein Woolthuis, R., Lankhuizen, M., and Gilsing, V. (2005). A system failure framework for innovation policy design. *Technovation* 25, 609–619.

Koné, D. (2010). Making urban excreta and wastewater management contribute to cities' economic development: a paradigm shift. *Water Policy* 12(4), 602–610.

Kooy, M., and Bakker, K. (2008). Splintered networks: the colonial and contemporary waters of Jakarta. *Geoforum* 39(6), 1843–1858.

Larsen, T. A., Hoffmann, S., Lüthi, C., Truffer, B., and Maurer, M. (2016). Emerging solutions to the water challenges of an urbanizing world. *Science* 352(6288), 928–933.

Mansour, G., Oyaya, C., and Owor, M. (2017). Situation analysis of the urban sanitation sector in Kenya. Retrieved from Water and Sanitation for the Urban Poor (WSUP). https://www.wsup.com/insights/situation-analysis-of-the-urban-sanitation-sector-in-kenya/ (accessed April 4, 2019).

Mara, D. (2018). "Top-down" planning for scalable sustainable sanitation in high-density low-income urban areas: is it more appropriate than "bottom-up" planning? *Journal of Water Sanitation and Hygiene for Development* 8(2), 165–175.

Markard, J., and Luthi, C. (2010). Institutional and organizational contexts for sustainable innovations in sanitation. Paper presented at the Water Environment Federation Technical Exhibition and Conference (WEFTEC), Cities of the Future/Urban River Restoration. New Orleans, Louisiana, USA, October 2–6, 2010.

Markard, J., Raven, R., and Truffer, B. (2012). Sustainability transitions: an emerging field of research and its prospects. *Research Policy* 41, 955–967.

Markard, J., and Truffer, B. (2008). Technological innovation systems and the multi-level perspective: towards an integrated framework. *Research Policy* 37, 596–615.

Nyanchaga, E. N., and Ombongi, K. S. (2007). History of water supply and sanitation in Kenya, 1895–2002. In: Juuti, P. S., Katz, J. M., and Vuorinen, H. S. (Eds.), *Environmental History of Water: Global Views on Community Water Supply and Sanitation*, pp. 271–321. IWA Publishing, London.

O'Keefe, M., Lüthi, C., Kamara, T., and Tobias, R. (2015). Opportunities and limits to market-driven sanitation services: evidence from urban informal settlements in East Africa. *Environment and Urbanization* 27(2), 421–440.

Parkinson, J., and Tayler, K. (2003). Decentralized wastewater management in peri-urban areas in low-income countries. *Environment and Urbanization* 15(1), 75–90.

Republic of Kenya (2010). The Constitution of Kenya. http://kenyalaw.org/lex/rest/db/kenyalex/Kenya/The%20Constitution%20of%20Kenya/docs/ConstitutionofKenya%202010.pdf (accessed April 4, 2019).

Schot, J., Kanger, L., and Verbong, G. (2016). The role of users in shaping transitions to new energy systems. *Nature Energy* 1(16054), 1–7.

Szanto, G. L., Letema, S. C., Tukahirwa, J. T., Mgana, S., Oosterveer, P. J. M., and van Buuren, J. C. L. (2012). Analyzing sanitation characteristics in the urban slums of East Africa. *Water Policy* 14, 613–624.

Tilley, E., Ulrich, L., Lüthi, C., Reymond, P. and Zurbrügg, C. (2014). *Compendium of Sanitation Systems and Technologies*. 2nd ed. Swiss Federal Institute of Aquatic Science and Technology (Eawag), Dübendorf, Switzerland.

Truffer, B. (2015). Challenges for technological innovation system research: introduction to a debate. *Environmental Innovation and Societal Transitions* 16, 65–66.

United Nations (UN) (2016). Sustainable Development Goals. https://sustainabledevelopment.un.org/sdgs (accessed April 6, 2019).

United Nations Human Settlements Programme (UN-Habitat) (2004). *The Challenge of Slums. Global Report on Human Settlements 2003.* http://mirror.unhabitat.org/pmss/getElectronicVersion.aspx?alt=1&nr=1156 (accessed April 6, 2019).

van Welie, M. J., Cherunya, P. C., Truffer, B., and Murphy, J. T. (2018). Analyzing transition pathways in developing cities: the case of Nairobi's splintered sanitation regime. *Technological Forecasting and Social Change* 137, 259–271.

van Welie, M. J., Truffer, B., and Gebauer, H. (2019). Innovation challenges of utilities in informal settlements: combining a capabilities and regime perspective. *Environmental Innovation and Societal Transitions* 33, 84–101. doi.org/10.1016/j.eist.2019.03.006

van Welie, M. J., Truffer, B., and Yap, X. S. (2018). Towards sustainable urban basic services in low-income countries: a TIS analysis of sanitation value chains in Nairobi. Paper presented at the International Sustainability Transitions (IST) Conference, Manchester, UK, June 11–14, 2018.

VERBI-Software. (2015). *MAXQDA 12 [computer software].* https://www.maxqda.com (accessed April 4, 2019).

Wald, C. (2017). The economy in the toilet. https://www.nature.com/news/the-new-economy-of-excrement-1.22591 (accessed April 4, 2019).

World Bank (2018). World Bank Country and Lending Groups. https://datahelpdesk.worldbank.org/knowledgebase/articles/906519 (accessed April 4, 2019).

World Health Organization (WHO) (2015). Progress on Sanitation and Drinking Water: 2015 Update and MDG Assessment. https://washdata.org/report/jmp-2015-report (accessed April 4, 2019).

World Health Organization (WHO) (2016). World Health Statistics 2016: Monitoring Health for the SDGs, Sustainable Developmpent Goals. http://www.who.int/gho/publications/world_health_statistics/2016/en/ (accessed April 4, 2019).

World Health Organization (WHO) (2017). Progress on Drinking Water, Sanitation and Hygiene: 2017 Update and SDG Baselines. https://www.unicef.org/publications/files/Progress_on_Drinking_Water_Sanitation_and_Hygiene_2017.pdf (accessed April 4, 2019).

Water and Sanitation for the Urban Poor (WSUP) (2017). *A Guide to Strengthening the Enabling Environment for Feacal Sludge Management.* WSUP, London.

17

The Role of Animal-Source Foods in Sustainable, Ethical, and Optimal Human Diets

Julia de Bruyn, Brigitte Bagnol, Hilary H. Chan, Delia Grace, Marisa E. V. Mitchell, Michael J. Nunn, Kate Wingett, Johanna T. Wong, and Robyn G. Alders

1. Introduction

Achieving sustainable, ethical food systems in support of human and planetary health is among the greatest challenges facing our global community. During the past 50 years, a doubling of the world's population has been met by a three-fold increase in food supply (FAO et al. 2017), largely through improvements in agricultural productivity. Much of this has been due to intensification of land use and increased inputs, such as fertilizers, energy, and irrigation (Fuglie and Wang 2012). Measures of agricultural productivity rarely reflect environmental impacts, despite the fact that current agricultural systems, together with forestry and other land use, account for about one-quarter of global greenhouse gas emissions (Intergovernmental Panel on Climate Change 2014) and close to three-quarters of freshwater use (FAO 2016).

Current food systems face multiple and diverse challenges. Inadequate diets, in terms of quantity and nutritional quality, remain among the leading causes of death and disability globally (Black et al. 2008; Masters et al. 2017). Estimates of the number of chronically undernourished people has risen slowly in recent years, from 777 million in 2015 to 820 million in 2018, marking a reversal of the sustained downward trend reported during the preceding fifteen years (FAO et al. 2019). At the other end of the nutritional spectrum, obesity affected about 108 million children and 604 million adults in 2015 and has doubled in prevalence in 73 countries since 1980 (Global Burden of Disease 2015 Obesity Collaborators 2017). Despite ongoing efforts to prevent and treat micronutrient deficiencies, more than two billion people are estimated to be at risk of insufficient intake of iron, vitamin A, or iodine deficiency (Bhutta and Salam 2012; Péter et al. 2014). Micronutrient deficiencies are particularly prevalent in sub-Saharan Africa and South Asia, but occur in countries of all income levels, and are a major factor in nutrition-related deaths globally (Viteri and Gonzalez 2002).

Determinants of food choices vary widely between individuals and settings, and may include social and cultural influences, resource availability, and geographical context (Drewnowski and Kawachi 2015; Sobal and Bisogni 2009). In high-income countries (HICs), a predominance of urban lifestyles and increasingly complex food systems have led to a marked disconnect between consumers, producers, and the multiple other participants in food value chains. For smallholder farmers, particularly in low- and middle-income countries (LMICs), the duality of being both food producers and consumers often results in a contrasting set of factors influencing diets. As rising incomes and urbanization drive an increase in the demand for "high-value" foods (Drewnowski and Popkin 1997), alongside climate variability, land degradation, and scarce environmental resources, particular attention is focused on consumption of animal-source foods (ASFs).

The inclusion of ASFs in human diets marked a turning point in the history of human development (Gupta 2016). Current research continues to explore the nutritional contributions of ASFs, their interactions with other dietary components, and their role in healthy human growth and development (Grace et al. 2018). This chapter uses the framework of the sustainable development goals (SDGs) to consider the value of ASFs in human diets across a range of settings. It discusses opportunities for multidisciplinary efforts that harness science, technology, and innovation (STI) to promote livestock health and production and to support the role of ASFs in meeting human nutrient requirements while minimizing adverse health implications and negative environmental impacts.

2. Methodology

To gather references relevant to this chapter, co-authors working in research and development (R&D) programs involving ASFs from differing geographies and disciplines were identified. Literature collected by coauthors through current and previous engagement in this area was supplemented by searching scientific databases (Web of Science and Medline) and online document repositories (Food and Agriculture Organization of the United Nations and International Livestock Research Institute) for primary research, review papers and manuals with a focus on nutritional, food safety, gender, and sociocultural dimensions of ASF consumption. The potential relevance of references was examined based on titles and abstracts, and nonrelevant citations excluded. The full text of remaining references was assessed for direct relevance to the theme. Preference was given to literature available via open-access sources.

3. A Framework to Support Sustainable and Nutritious Human Diets

The second SDG highlights the multidimensional nature of food and nutrition security, encompassing not only the quantity of available food but extending to issues of resilience, nutrient content, and food safety. Particular emphasis is given to the food security status of the poor and vulnerable, and to meeting the nutritional needs of children under five years of age, adolescent girls, pregnant and lactating women, and elderly people. Targets in SDG2 (end hunger, achieve food security and improved nutrition, and promote sustainable agriculture) incorporate both agriculture and nutrition, highlighting the importance of food-based approaches to address nutritional challenges—a concept recognized within rural smallholder households for millennia, and an area of growing attention within the international R&D community (Hawkes et al. 2012). In this chapter, we consider the role of ASFs in meeting targets relating to diet quality (2.1 and 2.2), livelihoods of small-scale producers (2.3), and the sustainability of food systems (2.4 and 2.5).

Meeting targets outlined in SDG5 (gender equality), including targets 5.1, 5.4, 5.5, and 5.a, offers opportunities for positive nutritional impact through gender equality and empowerment of women and girls. Across all settings, women of reproductive age face elevated nutrient requirements associated with menstruation, pregnancy, and lactation (Hallberg and Rossander-Hultén 1991; Marangoni et al. 2016). In some settings, women's "nutritionally vulnerable" status may be compounded by their lower social status, which affects food allocation within households and limits access to health services (Girard et al. 2012). Women in LMICs are commonly responsible for food production, harvesting, storage, distribution, and preparation, yet often have limited access to information and a limited ability to make autonomous decisions about these activities. This chapter discusses the gendered nature of livestock-keeping and consumption of ASFs in many resource-poor settings and presents the central role of gender-sensitive communication within development programs.

A several-fold increase in agricultural output and resultant decrease in undernutrition is reported to be possible by granting women access to the same productive resources as their male counterparts (FAO 2011), and women have been shown to be more likely than men to invest in their children's health, nutrition, and education (Quisumbing et al. 1995; Hoddinott and Haddad 1994; World Bank 2007). Targets for SDG5—ensuring women have equal opportunities for leadership, equal rights to economic resources, and full and effective participation in decision-making—offer pathways to improved nutrition and health outcomes not only for women, but for children and households more broadly.

The report outlining the SDGs describes an agenda that balances three dimensions of sustainable development: economic, social, and environmental (UN 2015). Targets in SDG12 (sustainable consumption and production) encompass the efficient management and use of natural resources (12.2), reduction of food waste and losses along supply chains (12.3), and ensuring access to relevant information about sustainable development (12.a). Although substantial current attention is directed toward the sustainability of livestock production and the consumption of ASFs, the predominant focus has been on environmental impacts. Such impacts, and corresponding issues of economic and social sustainability, should be recognized to differ greatly between livestock species, management systems, socioeconomic contexts, and geographical settings. This chapter describes variation in natural resource management in different livestock production systems and proposes opportunities to reduce losses in value chains for ASFs.

Moving from linkages between nutrition security and SDGs to linkages between food safety and SDGs, there are both commonalities and differences. Conceptually, food safety and foodborne disease (FBD) might be assumed to sit within SDG3, relating to good health and well-being; however, the extensive documentation around this goal makes little mention of these areas. This is most likely because the SDGs were being negotiated before the publication of the first landmark study on the health burden of FBD. This study found that the burden was comparable to that of malaria, HIV/AIDs or tuberculosis, that 98% of the burden fell on LMICs, that 97% was due to biological hazards, and that children under five years of age were disproportionately affected (Havelaar et al. 2015). As this evidence becomes available to the development community, it is likely that food safety will assume greater prominence in the SDGs.

Food safety has significant implications for attainment of other SDGs, especially:

- *SDG6: water and sanitation for all.* Many infectious FBDs can be transmitted via water, and people and animals infected with these diseases can contaminate water making it less safe (e.g., cysticercosis, cryptosporidiosis). Lack of clean water for washing food and food equipment, and for hygiene of food-handlers, increases the risk of food being unsafe.
- *SDG1: end poverty.* Ill health is a major factor in causing and maintaining poverty. Foodborne disease is one of the major causes of ill health in LMICs and is associated with a range of costs that fall on poor people and contribute to their remaining in poverty. Beyond the direct costs of illness, disease may also act as a "poverty trap"; that is, a self-reinforcing mechanism causing poor individuals or countries to remain poor (Grace et al. 2017).
- *SDG2: end hunger.* Foodborne disease has multiple complex interactions with nutrition. These include causing illness, which may worsen nutritional status through higher nutrient losses and lower food intake, and direct links to growth impairment, for example through ingestion of aflatoxins (Khlangwiset et al. 2011). Livestock production may result in greater exposure to bacteria in animal faeces that are associated with environmental enteric dysfunction (Zambrano et al. 2014). The most nutritious foods, including ASFs, are also the most implicated in FBD (Hoffman et al. 2017).

4. Animal-Source Foods: Contributions, Constraints, and Concerns

Animal-source foods have nutritional advantages but also some negative implications. Delivering the positives can be challenging in resource-poor settings and the global level of per capita consumption varies widely.

4.1 Nutritional Advantages of ASFs

Animal-source foods are densely packed with macro- and micronutrients, most notably protein, iron, vitamin B12, vitamin A, zinc, and riboflavin (Demment et al. 2003; Murphy and Allen 2003). Animal proteins comprise almost all those classified as having high nutritional quality, meeting the criteria of containing a balanced complement of essential amino acids and having high bioavailability (Wolfe et al. 2018). Animal-source foods contain a variety of micronutrients in forms more readily absorbed and used by the human body than those found in plant-source foods (PSFs), such as iron bound to the *haem* molecule and vitamin A in its active form, retinol, aiding absorption and use of these key nutrients (Neumann et al. 2002). Vitamin B12 is rarely found in adequate quantities in PSFs and its deficiency is a concern when diets do not contain ASFs (Neumann et al. 2002).

Numerous studies have documented positive nutritional and developmental outcomes associated with the inclusion of ASFs in predominantly cereal-based diets. Consumption of ASFs by women has been associated with improvement in iron and vitamins A and B12 deficiencies (Hall et al. 2017), while in children, greater intake or serum levels of calcium, zinc, vitamin A, and vitamin B12 have been documented (Dror and Allen 2011; Rahman et al. 2017). The inclusion of ASFs in the diets of pregnant women has been linked to higher birth weight and length of offspring (Lee et al. 2011), and the addition of ASFs to children's diets has been associated with increased linear growth in both LMICs and HICs (Braun et al. 2016; Krasevec et al. 2017). Improved cognitive function and school performance have also been associated with increased intake of folate, iron, energy, vitamin B12, zinc, and riboflavin from ASFs (Hulett et al. 2014; Neumann et al. 2014).

4.2 Challenges in Achieving Nutritionally Adequate Diets with PSFs Only

In resource-poor settings, it is frequently difficult to achieve a nutritionally adequate diet. Despite high micronutrient levels in many PSFs, bioavailability is commonly limited by the presence of fiber or antinutrients such as phytic acid or oxalate (Gibson 1994; Libert and Franceschi 1987). Lower nutrient density necessitates consumption of larger volumes. To meet recommended protein requirements, a greater proportion of daily caloric intake would be met when proteins are derived from PSFs (25–40%), compared to ASFs (10–20%) (Wolfe et al. 2018). Relying solely on plant-based protein therefore leaves a smaller proportion of the diet for other foods to fulfil multiple nutrient requirements.

Studies in Burkina Faso, Nepal, and Zambia have found that diets are commonly deficient in calcium, iron, zinc, riboflavin, and vitamins A, B6, and B12, but that incorporating milk, fish, meat (including offal such as liver) into diets can make them nutritionally complete (Arimond et al. 2018; Biehl et al. 2016; Zhang et al. 2016). This may be due to an additive effect, but the presence of *haem* iron in ASFs can also have a synergistic effect in enhancing absorption of micronutrients, including non-*haem* iron and zinc (Hallberg and Hulthén 2000). Ultimately, a nourishing diet needs to be composed of a variety of foods and the inclusion of an appropriate amount of ASFs dramatically increases the nutritional efficiency of largely plant-based diets.

4.3   Global Patterns in Consumption of ASFs

Globally, there is great disparity in the consumption of ASFs (Figure 17.1). Average per capita intakes have been consistently low during the past 50 years in most LMICs but are gradually increasing in some LMICs. Barriers to consumption of ASFs include their high cost relative to PSFs, animal disease limiting productivity and thus availability of ASFs to households, the reservation of livestock for other purposes including as savings and for ceremonies, and low levels of nutritional knowledge (Biehl et al. 2016; Wong et al. 2018). A contrasting situation exists in HICs, where intake of ASFs far exceeds nutritional requirements (Walker et al. 2005). Diverse bodies of literature arise from these respective regions, with studies in LMICs pursuing opportunities to increase consumption of ASFs in resource-poor areas, and research from HICs addressing adverse health and environmental effects of excessive consumption of ASFs, especially red and processed meat. Although consumption of ASFs in HICs is projected to remain stable, global consumption is set to continue rising, particularly in areas experiencing strong income growth, including Asia, Latin America, and the Middle East (Henchion et al. 2017).

4.4   Negative Implications of Consumption of ASFs

Despite the nutritional benefits of ASFs, there are many concerns about their production and consumption. Probably the most direct and important negative health effect is through FBDs resulting from consumption of ASFs. Several recent studies implicate ASFs as the single most important source of FBD (Grace 2015b; Hoffman et al. 2017). Most of this is due to biological hazards, with salmonellosis, toxigenic *Escherichia coli*, and the pig tapeworm especially important (Havelaar et al. 2015). Toxins are mainly a problem in plant and marine food, but livestock products may contain aflatoxins and dioxins (Dolan et al. 2010). By contrast, cow's milk and eggs are often associated with allergic responses (Prescott et al. 2013) that can cause severe acute reactions or even clinical signs similar to those of malnutrition (Boye 2012). Lactose intolerance results from deficiency of an enzyme (lactase) and is common in LMICs, but rare before four to five years of age (Vandenplas 2015).

   Animal production can also indirectly affect health through routes not related to dietary intake. Zoonoses are diseases transmissible between animals and

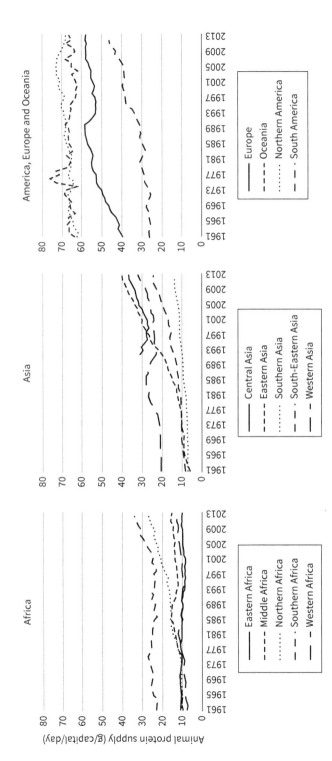

Figure 17.1 Trends in the daily per capita supply of animal protein across different regions of the world, 1961–2013.

Source: FAOSTAT 2017a

people. Participation in livestock farming or value chains brings people in contact with livestock and their secretions and excretions, and these may transmit zoonoses such as brucellosis, tuberculosis, Q fever, or leptospirosis (Grace et al. 2012). In addition, livestock and fish production, especially in intensive systems and if accompanied by land-use change, can lead to the emergence of diseases (Jones et al. 2013). Notable recent emerging diseases include bovine spongiform encephalopathy, highly pathogenic avian influenza, and Middle East respiratory syndrome. Livestock production, especially intensive production in LMICs, is also associated with high use of antimicrobials (Grace 2015a). Antimicrobial resistance is a significant and increasing threat to global health, food security, and development (Robinson et al. 2017).

The effect of production and consumption of ASFs on ecosystems and the environment is also of concern. Up to 18% of anthropogenic greenhouse gas emissions arise from livestock production systems (Herrero and Thornton 2013), including those generated in the production of crops to feed intensively reared animals, that in turn contributes to the ongoing conversion of natural habitats into agricultural lands (Henchion et al. 2017). These land and water requirements put pressure on ecosystems, including wild animal and plant populations, and threaten biodiversity on a global scale (Golden et al. 2011; Jones et al. 2013). In line with SDG12, it is in the interest of current and future human health and well-being to continue examining our patterns of production and consumption of ASFs and look for ways to increase the efficiency and sustainability of use of these valuable resources.

5. Multidisciplinary Approaches Harnessing STI

5.1 Improving Management of Animal Health and Production

The sustainable management of food animals is a complex endeavor that varies according to the livestock species, production systems, and local environmental and socioeconomic circumstances. In a warming world, production of ASFs is challenged by changes to the quality of feed crops and forage materials, availability of water, livestock diseases, animal reproduction, and the need to mitigate greenhouse gas emissions and biodiversity loss (Rojas-Downing et al. 2017). STI have an essential role in climate-smart adaptation and mitigation strategies to optimize livestock production. With water consumption by animals expected to increase by a factor of three by 2050 (Rojas-Downing et al. 2017), there is a need to pursue strategies that reduce water requirements and promote efficient management of available water resources (SDG target 6.4). These include selection of heat-tolerant livestock breeds, and the use of agroforestry to provide shade and to decrease the impact of wind on pasture and soil moisture content.

The rapid increase in intensive livestock production in the twentieth century has been associated with increased use of cereals in livestock feed, with one-third of the global cereal harvest currently consumed by livestock (Rojas-Downing et al. 2017). Scientific risk assessments will play a vital role in determining the best allocation of available feed resources, especially in relation to decisions about the use of foods suitable for consumption by both people and animals (i.e., monogastric omnivores, such as pigs and poultry; SDG target 2.4). Based on quantitative analysis of the extent to which reductions in the amount of human-edible crops fed to animals could increase the supply of nutrients for human consumption, Berners-Lee et al. (2018) found no nutritional case for feeding human-edible crops to animals.

Assessments of the best use of feed resources must also take into account the direct link between the quality of livestock feed and the nutrient content of resulting ASFs. For example, genetic selection and dietary changes have adversely affected the nutritional composition of chicken meat. The modern broiler chicken provides several times more energy from fat than from protein and contains a ratio of omega-6:omega-3 fatty acids of as high as 9:1 (Wang et al. 2009)—shifting away from the lower ratio advised for decreasing the risk of coronary heart disease (Simopolous 2008).

Extensive livestock production is expected to remain important in areas with significant rangeland, where grazing livestock and scavenging poultry and pigs are able to convert fibrous vegetation into high quality nutrients that can be digested by humans. In sub-Saharan Africa, three-quarters of the population is engaged in small-scale farming and 80% of these households keep livestock, which represent a critical asset and provide protection against economic shocks (Marsh et al. 2016). Effective disease surveillance, prevention, and control strategies will be required to deal effectively with the changing distribution of pathogens and associated vectors due to climate change. Mobile phone technology will increasingly play a role in information-sharing and syndromic surveillance of disease (SDG target 17.8).

Developing ASF production systems that use feed and water resources efficiently, while at the same time minimizing the risk of zoonotic and emerging infectious disease (SDG target 3.D), remains a significant challenge. Within the intensive commercial poultry industry, infectious disease emergence and spread is facilitated by (i) genetic homogeneity, which results in reduced genetic potential to resist or tolerate disease at a population level; (ii) high-density housing, which allows close animal-to-animal contact and favors the transmission of virulent disease; and (iii) intensive vaccination programs, which provide selective immune pressures, by limiting the replication of specific organisms and may not be executed properly in resource-limiting settings (Alders et al. 2013; Higgins and Shortridge 1988).

Until in vitro meat equivalents are able to be produced at affordable prices, there is a need for further research into safe, sustainable, and ethical intensive livestock production systems that are appropriate for a range of socioeconomic settings. Genomic research can be used to increase livestock resistance and tolerance against diseases, heat tolerance, and feed conversion efficiency (SDG target 2.5). Ongoing research into the gut microbiome will help to optimize food conversion efficiency without the inclusion of antimicrobial growth promotants in feeds.

5.2 Proposing Alternative Approaches for Assigning Value to ASFs

For the SDGs to be achieved, including strengthening food and nutrition security (SDG2), improving sustainability of production and consumption patterns (SDG12), and combatting climate change (SDG13), there is a need for large-scale change in our political, social and economic systems. Traditionally, the "value" in value chains refers to economic factors (Porter 1985). For true sustainability, the concept of value within agri-food value chains needs to be expanded to reflect contributions to gender-specific human health and nutrition requirements (targets 2.1 and 2.2) and the environmental and animal welfare effects of food production (2.4, 12.1, 12.2, 12.3, 13.1, and 13.2). The success of this approach relies on systems remaining financially profitable (Committee on World Food Security 2016).

Shifting perspectives on how value is assigned within agri-food value chains could lead to improvements in human health, animal health and welfare, and the environment. Industry-led initiatives, with financial recognition of sustainable production, is likely to be the most successful way to promote reduced losses. The Australian sheep meat value chain provides an apt example. Australia is the driest inhabited continent in the world (Australian Bureau of Statistics 2012) with vast areas of rangeland suited to grazing herbivores, such as sheep and kangaroos. This land is of limited value to other forms of agriculture and sheep are raised extensively on rangelands. However, this land is vulnerable to overgrazing (Kimura and Antón 2011) and in an increasingly unpredictable climate, will need to be well managed to ensure farming continues for generations to come.

Australia has the second largest sheep population globally and the industry is dominated by Merinos, a fine-wool breed. As a result, Australia is a leader in the production of muscle meat from older sheep (mutton) and sheep offal (Australian Bureau of Agricultural and Resources Economics and Sciences 2017). From a public health perspective, liver and kidney have the highest concentration of iron of all ASFs, with more than double the levels found in muscle meat. As animals age, the concentration of iron in muscle meat increases—the concentration of iron in mutton is twice that of lamb and, more significantly, nine times higher than an equivalent serve of chicken meat (Food Standards Australia New Zealand 2015). Concurrently, anemia is a worldwide issue and iron deficiency is the most significant contributing factor to this condition. Other micronutrient deficiencies, such as vitamins A and B12, folate, and riboflavin, may also lead to anemia. In Australia, it is estimated one in four pregnant women are anemic (WHO 2015). The most recent Australian National Nutrition and Physical Activity Survey revealed widespread suboptimal intake of micronutrients within the Australian population. The elderly ($\geq$ 71 years old) living in private dwellings have been identified as a cohort with significant inadequate micronutrient intake (Australian Bureau of Statistics and Food Standards Australia New Zealand 2015).

Although mutton and offal are rich sources of protein and micronutrients (Williams et al. 2007; Hutchison et al. 1987), the consumption of these products in Australia has fallen dramatically in recent decades and is currently estimated to be less than 1 kg per person annually. Simultaneously, consumption of pig and poultry meat in Australia has increased markedly. With a per capita consumption of 47 kg in 2016, compared with 4.6 kg in 1965, chicken is currently the most consumed meat (Australian Bureau of Agricultural and Resources Economics and Sciences 2017). This increase in chicken meat consumption relates overwhelmingly to muscle meat, and offal remains rarely eaten. As with mutton and sheep offal, a significant portion of chicken offal is wasted, exported, or diverted from the direct human food chain in Australia (FAOSTAT 2017b), potentially due to negative public perception and limited knowledge of its nutritional benefits (Tucker 2013). Chicken offal currently constitutes 80% of total exports of chicken from Australia due to the low domestic demand compared to Southeast Asia and the Pacific islands (Australian Bureau of Agricultural and Resources Economics and Sciences 2017).

A chicken carcass containing the heart, liver and gizzard contains seven times the amount of vitamin A and folate and triple the amount of vitamin B12 than the muscle meat alone, as well as higher levels of iron, zinc, and protein. From an economic perspective, chicken offal contains more micronutrients than muscle meat relative to the price paid for each, rendering it a more cost-effective source of micronutrients (Chan et al. 2017). This becomes significant within communities where access to ASFs is limited by cost.

Greater use of animal carcasses, including organ meat and older animals, can improve the micronutrient status of people. In turn this leads to reduced preconsumer food loss and allows farmers to manage their land and businesses sustainably. Technology can help to reduce the wastage of offal, through improved feedback systems along the value chain (such as individual electronic identification of animals and voice recognition technology for meat inspectors) and processing techniques to maintain the nutrient content of offal over longer periods. This could include, for example, developing products such as dietary supplements or preprepared meals based on mutton and offal for targeted markets that would increase the consumption of these nutrient-rich ASFs both in Australia and globally.

5.3 Using Effective Gender- and Culturally Sensitive Communication

Differences in culture, gender, and language may obstruct the communication process and misrepresent messages being exchanged, leading to ineffective or a lack of communication (Shrivastava 2012). Gender- and culturally sensitive communication is important because women and men in different contexts and ecological settings have different needs, interests, rights, and responsibilities in relation to natural resources and human health. Effective sharing of knowledge and information underpins progress toward a wide range of SDG targets, including the success of strategies to promote health and well-being for all (SDG3) and to achieve gender equality and female empowerment (SDG5).

There is a gender dimension to the ownership, management and control of resources such as crops and livestock, with women in some settings facing specific constraints related to their educational, economic, social, legal, political, and cultural status. For these reasons, men and women experience different impacts during outbreaks of animal and human disease and in periods of ecological stress, such as floods or droughts. Women commonly have ownership of chicken and ducks, particularly when kept in small numbers, while men are more likely to control pigs, goats, cattle, and larger poultry flocks (Pym and Alders 2012; United Nation Economic and Social Council 2002). Chickens are commonly used by women to meet immediate household expenses, while cattle typically provide a means of storing wealth in the longer term and are often sold only in case of emergency (Doran et al. 1979). High-mortality disease outbreaks, such as due to Newcastle disease virus, are a common seasonal occurrence among free-roaming chicken flocks in LMIC settings, and its impact is recognized to be particularly felt by women.

Gender roles and stereotypes also assign characteristics, values, and status to men and women that influence when, where, how, and what they eat. This is particularly true in relation to ASFs, due to their value and relative scarcity in many settings. ASFs have a high value and are often reserved for visitors, special events and ceremonies, or consumed outside the household. Most food proscriptions and traditional beliefs relate to ASFs and tend to restrict access by children and women (Meyer-Rochow 2009; Trant 1954). Although food may be considered to be evenly distributed within the home, in several societies men have greater freedom of mobility and financial resources that enable them to attend social gatherings outside the household where they may consume ASFs (Wong et al. 2018). Children under one year of age consume a less diverse diet and smaller amounts of ASFs than older children and adults (de Bruyn 2017; Wong et al. 2018).

Using a gender-sensitive approach is essential to support effective participation in nutrition programs (Pandey et al. 2016). Some impact on improved food access and dietary diversity has been demonstrated by agricultural projects aiming to improve the nutritional status of vulnerable or marginalized groups (Herforth and Ballard 2016). Incorporating gender- and culturally-sensitive communication is an important component in facilitating the uptake of both agricultural and nutritional interventions (Kadiyala et al. 2016; Pamphilon and Mikhailovich 2017). The use of information and communication technology (ICT) to promote behavior related health and nutrition has increased globally and has been shown to increase and strengthen extension services in resource-poor settings. Despite the focus on the use of ICT, community radio broadcasts and extension workers remain highly favored as sources of information on farming innovations (Isaya et al. 2016).

It has been observed that development planning has often failed to achieve sustainable development, whereby top-down planning has failed to incorporate indigenous knowledge at the local level (Grenier 1998). Findings from Jamaica suggest that local and traditional knowledge of indigenous populations are often overlooked due to negative views of the legitimacy and reliance of smallholder farmers on these forms of knowledge (Beckford and Barker 2007). Local and traditional knowledge enable adaptation to environmental changes, economic instability and inequities. Many traditional food production systems are complex, developed from having observed and adapted to the agroecological and sociocultural environment over multiple generations, and can contribute significantly to the food and nutrition security of households (Beckford and Barker 2007; Grenier 1998). Despite this, indigenous knowledge is often not incorporated into food security programs (Nigussie 2016). For example, in eastern Tigray in Ethiopia programs are communicated verbally at public meetings by experts via a top-down approach, leading to unequal power relations between community members and policymakers and professionals (Nigussie 2016). Having a culture-centered approach involves communication developed from within the community, instead of originating from outside (Dutta-Bergman 2005). In the case of the design of nutritional interventions, community involvement helps to ensure that the local context, perceptions of food, and determinants of food choices are understood and addressed (Nu and Bersamin 2017).

5.4 Tailoring Sustainable Dietary Choices to Local Conditions

Supporting individuals, communities, and nations to make sustainable dietary choices will require improved collaboration between nutritionists, food scientists, ecologists, agriculturalists, social scientists, farmers, and others within food value chains. Sustainable evidence- and rights-based policymaking on access to food for humans and animals (both domestic and wild) should include tailoring diets by gender, age, and reproductive status, selecting plant and animal species and farming systems according to agro-ecological zones, and building knowledge of nutritional requirements and environmental impacts (Alders et al. 2016). Global trends in production of ASFs will change over time, contributing to reductions in greenhouse gas emissions, declines in ASF consumption per capita in HICs, and more affordable in vitro production costs.

Immediate gains toward SDGs 2, 12 and 13 can be achieved in HICs through more efficient use of ASFs by consuming meat, offal, and bone marrow that would reduce the number of animals required to meet human nutrient requirements (Wingett et al. 2018); and the sustainable harvest of domestic and wild animals that graze rangelands (Alders et al. 2018). A return to these practices, which were commonplace within the past 50 years in many countries, will require strategic nutrition education and awareness-raising activities. Immediate gains can be achieved in LMICs through improved animal health services to enhance production efficiency through reduced losses of ASFs and through targeted improvements to livestock production and nutrition education (to support access to adequate quantities of high quality protein and bioavailable micronutrients from ASFs, especially for mothers and young children (Grace et al. 2018).

6. Conclusions

The diverse range of animal species and production systems across the globe are associated with varying environmental impacts, nutritional contributions, food safety risks, and sociocultural significance. As climate change affects the temporal and geographical distribution of livestock diseases, effective strategies for disease surveillance, prevention, and control are required at national and subnational levels. As population and income growth drives a greater demand for ASFs, genomics and research into the gut microbiome will play a role in optimizing the efficiency of intensive livestock production systems. In many settings, nutrient losses can be reduced and sustainable land management promoted through use of all parts of animal carcasses. This relies on the effective integration of strategies throughout value chains, including financial incentives for sustainable livestock production and education campaigns to generate demand for underutilized food resources such as offal. In resource-poor communities, participatory and gender-sensitive communication strategies are central to responding to constraints faced by male and female livestock-keepers and supporting access to nutritionally-balanced diets. Holistic approaches, which combine the expertise of natural and social scientists with indigenous knowledge, are essential to meet the nutritional needs of current and future populations while supporting biodiversity and conserving the planet's scarce natural resources.

Acknowledgments

We are grateful for funding provided by the Australian Government, especially the Australian Centre for International Agricultural Research and the University of Sydney, in support of research into sustainable and ethical livestock production and food and nutrition security.

References

Alders, R., Awuni, J., Bagnol, B., Farrell, P., and de Haan, N. (2013). Impact of avian influenza on village poultry production globally. *EcoHealth* 11(1), 63–72.

Alders, R., Chan, H., Wong, J., de Bruyn, J., Allman-Farinelli, M., Maulaga, W., Jong, J., and Wingett, K. (2018). "Eating more to consume less: the role of nose to tail consumption of animal-source food in sustainable food systems." Presented at 2nd Planetary Health Alliance Annual Meeting, Edinburgh, May 29–31.

Alders, R., Nunn, M., Bagnol, B., Cribb, J., Kock, R., and Rushton, J. (2016). Chapter 3.1 approaches to fixing broken food systems. In: Eggersdorfer, M., Kraemer, K., Cordaro, J. B., Fanzo, J., Gibney, M., Kennedy, E., Labrique, A., and Steffen, J. (Eds.), *Good Nutrition: Perspectives for the 21st Century*, pp. 132–144. Karger, Basel.

Arimond, M., Vitta, B.S., Martin-Prevel, Y., Moursi, M., and Dewey, K. G. (2018). Local foods can meet micronutrient needs for women in urban Burkina Faso, but only if rarely consumed micronutrient-dense foods are included in daily diets: a linear programming exercise. *Maternal and Child Nutrition* 14, e12461. doi:10.1111/mcn.12461.

Australian Bureau of Agricultural and Resources Economics and Sciences (2017). *Agricultural Commodity Statistics 2017*. Department of Agriculture and Water Resources, Canberra.

Australian Bureau of Statistics (2012). *Year Book Australia 2012*. Australian Bureau of Statistics, Canberra.

Australian Bureau of Statistics and Food Standards Australia New Zealand (2015). *Australian Health Survey: usual nutrient intakes, 2011–2012*. Australian Bureau of Statistics, Canberra.

Beckford, C., and Barker, D. (2007). The role and value of local knowledge in Jamaican agriculture: adaptation and change in small-scale farming. *The Geographical Journal* 173(2), 118–128.

Berners-Lee, M., Kennelly, C., Watson, R., and Hewitt, C. N. (2018). Current global food production is sufficient to meet human nutritional needs in 2050 provided there is radical societal adaptation. *Elementa: Science of the Anthropocene* 6(1), 52.

Bhutta, Z. A., and Salam, R. A. (2012). Global nutrition epidemiology and trends. *Annals of Nutrition and Metabolism* 61(Suppl 1), 1–27.

Biehl, E., Klemm, R. D. W., Manohar, S., Webb, P., Gauchan, D., and West, K. P. (2016). What does it cost to improve household diets in Nepal? using the cost of the diet method to model lowest cost dietary changes. *Food and Nutrition Bulletin* 37, 247–260.

Black, R. E., Allen, L. H., Bhutta, Z. A., Caulfield, L. E., de Onis, M., Ezzati, M., Mathers, C., and Rivera, J. (2008). Maternal and child undernutrition: global and regional exposures and health consequences. *The Lancet* 371(9609), 243–260.

Boye, J. I. (2012). Food allergies in developing and emerging economies: need for comprehensive data on prevalence rates. *Clinical and Translational Allergy* 2, 25.

Braun, K. V. E., Erler, N. S., Kiefte-De Jong, J. C., Jaddoe, V. W. V., van den Hooven, E. H., Franco, O. H., and Voortman, T. (2016). Dietary intake of protein in early childhood is associated with growth trajectories between 1 and 9 years of age. *Journal of Nutrition* 146, 2361–2367.

Chan, H., Wong, J., Thomson, P. C., Darnton-Hill, I., and Alders, R. (2019). "What's in a Chicken? Comparing the nutrient value, potential to meet nutrient requirements and health-cost effectiveness of whole and frozen chickens." Presented at Agriculture, Nutrition and Health Academy Week, Hyderabad, June 24–28.

Committee on World Food Security. (2016). Inclusive Value Chains for Sustainable Agriculture and Scaled Up Food Security and Nutrition Outcomes: Background Paper. FAO, Rome.

de Bruyn, J. (2017). Healthy chickens, healthy children? Exploring nutritional contributions of village poultry-keeping to the diets and growth of young children in rural Tanzania. PhD thesis, University of Sydney.

Demment, M. W., Young, M. M., and Sensenig, R. L. (2003). Providing micronutrients through food-based solutions: a key to human and national development. *Journal of Nutrition* 133, 3879S–3885S.

Dolan, L. C., Matulka, R. A., and Burdock, G. A. (2010). Naturally occurring food toxins. *Toxins* 2, 2289–2332.

Doran, M. H., Low, A. R. C., and Kemp, R. L. (1979). Cattle as a store of wealth in Swaziland: implications for livestock development and overgrazing in Eastern and Southern Africa. *American Journal of Agricultural Economics* 61(1), 41–47.

Drewnowski, A., and Kawachi, I. (2015). Diets and health: how food decisions are shaped by biology, economics, geography, and social interactions. *Big Data* 3(3), 193–197.

Drewnowski, A., and Popkin, B. M. (1997). The nutrition transition: new trends in the global diet. *Nutrition Reviews* 55(2), 31–43.

Dror, D. K., and Allen, L. H. (2011). The importance of milk and other animal-source foods for children in low-income countries. *Food and Nutrition Bulletin* 32, 227–243.

Dutta-Bergman, M. J. (2005). Theory and practice in health communication campaigns: a critical interrogation. *Health Communication* 18(2),103–122.

Food and Agriculture Organization of the United Nations (FAO) (2011). *Women in Agriculture: Closing the Gender Gap for Development*. http://www.fao.org/docrep/013/i2050e/i2050e.pdf (accessed April 4, 2019).

Food and Agriculture Organization of the United Nations (FAO). (2016). *AQUASTAT: Water Uses*. http://www.fao.org/nr/water/aquastat/water_use/index.stm (accessed April 4, 2019).

Food and Agriculture Organization of the United Nations (FAO), International Fund for Agricultural Development (IFAD), United Nations Children's Fund (UNICEF), World Food Programme (WFP) and World Health Organization (WHO). (2019). *The State of Food Security and Nutrition in the World 2019: Safeguarding Against Economic Slowdowns and Downturns*. http://www.fao.org/3/ca5162en/ca5162en.pdf (accessed April 14, 2020).

Food and Agriculture Organization Corporate Statistical Database (FAOSTAT) (2017a). *Food Supply: Livestock and Fish Primary Equivalent*. http://www.fao.org/faostat/en/#data/CL (accessed April 4, 2019).

Food and Agriculture Organization Corporate Statistical Database (FAOSTAT) (2017b). *Commodity Balances: Livestock and Fish Primary Equivalent*. http://www.fao.org/faostat/en/#data/BL (accessed April 4, 2019).

Food Standards Australia and New Zealand. (2015). NUTTAB 2010 Searchable Online Database: Food Standards Australia and New Zealand, Canberra and Wellington.

Fuglie, K., and Wang, S. L. (2012). Productivity growth in global agriculture shifting to developing countries. *Choices* 27(4), 1–7.

Gibson, R. S. (1994). Content and bioavailability of trace elements in vegetarian diets. *American Journal of Clinical Nutrition* 59, 1123S–1232S.

Girard, A. W., Self, J. L., Mcauliffe, C., and Olude, O. (2012). The effects of household food production strategies on the health and nutrition outcomes of women and young children: a systematic review. *Paediatric and Perinatal Epidemiology* 26, 205–222.

Global Burden of Disease 2015 Obesity Collaborators. (2017). Health effects of overweight and obesity in 195 countries over 25 years. *New England Journal of Medicine* 377(1), 13–27.

Golden, C. D., Fernald, L. C. H., Brashares, J. S., Rasolofoniaina, B. J. R., and Kremen, C. (2011). Benefits of wildlife consumption to child nutrition in a biodiversity hotspot. *Proceedings of the National Academy of Sciences* 108, 19653–19656.

Grace, D. (2015a). Food safety in low and middle income countries. *International Journal of Environmental Research and Public Health* 12, 10490–10507.

Grace, D. (2015b). *Review of Evidence on Antimicrobial Resistance and Animal Agriculture in Developing Countries*. Department for International Development, London. doi:10.12774/eod_cr.june2015.graced

Grace, D., Dominguez-Salas, P., Alonso, S., Lannerstad, M., Muunda, E., Ngwili, N., Omar, A., Khan, M., and Otobo E. (2018). The Influence of Livestock-Derived Foods on Nutrition during the first 1,000 Days of Life. ILRI Research Report 44. International Livestock Research Institute, Nairobi. http://hdl.handle.net/10568/92907 (accessed April 4, 2019).

Grace, D., Gilbert, J., Randolph, T., and Kang'ethe, E. (2012). The multiple burdens of zoonotic disease and an ecohealth approach to their assessment. *Tropical Animal Health and Production* 44, 67–73.

Grace, D., Lindahl, J., Wanyoike, F., Bett, B., Randolph, T., and Rich, K. (2017). Poor livestock keepers: ecosystem-poverty-health interactions. *Philosophical Transactions of the Royal Society B: Biological Sciences* 372, 20160167.

Grenier, L. (1998). *Working with Indigenous Knowledge: A Guide for Researchers*. International Development Research Centre, Ottawa.

Gupta, S. (2016). Brain food: clever eating. *Nature* 531, S12–S13.

Hall, A. G., Ngu, T., Nga, H. T., Quyen, P. N., Anh, P. T. H., and King, J. C. (2017). An animal-source food supplement increases micronutrient intakes and iron status among reproductive-age women in rural Vietnam. *Journal of Nutrition* 147, 1200–1207.

Hallberg, L., and Hulthén, L. (2000). Prediction of dietary iron absorption: an algorithm for calculating absorption and bioavailability of dietary iron. *American Journal of Clinical Nutrition* 71, 1147–1160.

Hallberg, L., and Rossander-Hultén, L. (1991). Iron requirements in menstruating women. *American Journal of Clinical Nutrition* 54(6), 1047–1058.

Havelaar, A. H., Kirk, M. D., Torgerson, P. R., Gibb, H. J., Hald, T., Lake, R. J., Praet N., Bellinger, D. C., de Silva N. R., Gargouri G., Speybroeck, N., Cawthorne, A., Mathers, M., Stein, C., Angulo F. J., and Devleesschauwer, B. (2015). World Health Organization global estimates and regional comparisons of the burden of foodborne disease in 2010. *PLOS Medicine* 12(12), e1001923.

Hawkes, C., Turner, R., Waage, J., Ferguson, E., Johnston, D., Shankar, B., Haseen, F., Homans, H. Y., Hussein, J., Marai, D., and McNeil, G. (2012). *Current and Planned Research on Agriculture for Improved Nutrition: A Mapping and a Gap Analysis*. http://r4d.dfid.gov.uk/pdf/outputs/misc_susag/LCIRAH_mapping_and_gap_analysis_21Aug12.pdf   (accessed April 4, 2019).

Henchion, M., Hayes, M., Mullen, A.M., Fenelon, M., and Tiwari, B. (2017). Future protein supply and demand: strategies and factors influencing a sustainable equilibrium. *Food* 6(7), 53.

Herforth, A., and Ballard, T.J. (2016). Nutrition indicators in agriculture projects: current measurement, priorities, and gaps. *Global Food Security* 10, 1–10.

Herrero, M., and Thornton, P. K. (2013) Livestock and global change: emerging issues for sustainable food systems. *Proceedings of the National Academy of Sciences* 110, 20878–20881.

Higgins, D. A., and Shortridge, K. F. (1988). Newcastle disease in tropical and developing countries. In: Alexander, D. J. (Ed.), *Developments in Veterinary Virology: Newcastle Disease*, pp. 272–302. Kluwer Academic Publishers, Dordrecht.

Hoddinott, J., and Haddad, L. (1994). Women's income and boy-girl anthropometric status in the Côte d'Ivoire. *World Development* 22(4), 543–553.

Hoffmann, S., Devleesschauwer, B., Aspinall, W., Cooke, R., Corrigan, T., Havelaar, A., Angulo, F., Gibb, H., Kirk, M., Lake, R., Speybroeck, N., Togerson, P., and Hald, T. (2017) Attribution of global foodborne disease to specific foods: findings from a World Health Organization structured expert elicitation. *PLoS ONE* 12(9), e0183641.

Hulett, J. L., Weiss, R. E., Bwibo, N. O., Galal, O. M., Drorbaugh, N., and Neumann, C. G. (2014). Animal source foods have a positive impact on the primary school test scores of Kenyan schoolchildren in a cluster-randomised, controlled feeding intervention trial. *British Journal of Nutrition* 111, 875–886.

Hutchison, G. I., Nga, H. H., Kuo, Y. L., and Greenfield, H. (1987). Composition of Australian foods 36: beef, lamb and veal offal. *Food Technology in Australia*. 39(5), 223–227.

Intergovernmental Panel on Climate Change (IPCC). (2014). Climate Change 2014: Synthesis Report. Contribution of working groups I, II and III to the fifth assessment report of the Intergovernmental Panel on Climate Change. Intergovernmental Panel on Climate Change, Geneva.

Isaya, E. L., Agunga, R., and Sanga, C. A. (2016). Sources of agricultural information for women farmers in Tanzania. *Information Development* 34(1), 77–89.

Jones, B. A., Grace, D., Kock, R., Alonso, S., Rushton, J., Said, M. Y., McKeever, D., Mutua, F., Young, J., McDermott, J., and Pfeiffer, D. U. (2013). Zoonosis emergence linked to agricultural intensification and environmental change. *Proceedings of the National Academy of Sciences* 110, 8399–8404.

Kadiyala, S., Morgan, E. H., Cyriac, S., Margolies, A., and Roopnaraine, T. (2016). Adapting agriculture platforms for nutrition: a case study of a participatory, video-based agricultural extension platform in India. *PLoS One* 11(10), e0164002.

Khlangwiset, P., Shephard, G. S., and Wu, F. (2011). Aflatoxins and growth impairment: a review. *Critical Reviews in Toxicology* 41(9), 740–755.

Kimura, S., and Antón, J. (2011). *Risk Management in Agriculture in Australia*. OECD Food, Agriculture and Fisheries Papers, No. 39. OECD Publishing, Paris.

Krasevec, J., An, X. Y., Kumapley, R., Begin, F., and Frongillo, E. A. (2017). Diet quality and risk of stunting among infants and young children in low- and middle-income countries. *Maternal and Child Nutrition* 13, e12430.

Lee, Y. A., Hwang, J. Y., Kim, H., Ha, E. H., Park, H., Ha, M., Kim, Y., Hong, Y. C., and Chang, N. (2011). Relationships of maternal zinc intake from animal foods with fetal growth. *British Journal of Nutrition* 106, 237–242.

Libert, B., and Franceschi, V. R. (1987). Oxalate in crop plants. *Journal of Agricultural and Food Chemistry* 35, 926–938.

Marangoni, F., Cetin, I., Verduci, E., Canzone, G., Giovannini, M., Scollo, P., Corsello, G., and Poli, A. (2016). Maternal diet and nutrient requirements in pregnancy and breastfeeding: an Italian consensus document. *Nutrients,* 8(10), 629.

Marsh, T. L., Yoder, J., Deboch, T., McElwain, T. F., and Palmer, G. H. (2016). Livestock vaccinations translate into increased human capital and school attendance by girls. *Science Advances* 2(12), e1601410M.

Masters, W. A., Rosettie, K., Kranz, S., Pedersen, S. H., Webb, P., Danaei, G., Mozaffarian, D., and Global Nutrition Policy Consortium. (2018). Priority interventions to improve maternal and child diets in Sub-Saharan Africa and South Asia. *Maternal and Child Nutrition* 14(2), e12626.

Meyer-Rochow, V. B. (2009) Food taboos: their origins and purposes. *Journal of Ethnobiology and Ethnomedicine* 5(1), 18.

Murphy, S. P., and Allen, L. H. (2003). Nutritional importance of animal source foods. *Journal of Nutrition* 133, 3932S–3935S.

Neumann, C. G., Bwibo, N. O., Gewa, C. A., and Drorbaugh, N. (2014). Animal source foods as a food-based approach to improve diet and nutrition outcomes. In: Thompson, B., and Amoroso, L. (Eds.), *Improving Diets and Nutrition: Food-Based Approaches*, pp. 157–172. Food and Agriculture Organization of the United Nations and Centre for Agriculture and Bioscience International Rome and Wallingford.

Neumann, C., Harris, D. M., and Rogers, L. M. (2002). Contribution of animal source foods in improving diet quality and function in children in the developing world. *Nutrition Research* 22, 193–220.

Niamir-Fuller, M. (1994). *Women Livestock Managers in the Third World: A Focus on Technical Issues Related to Gender Roles in Livestock Production*. International Fund for Agricultural Development, Rome.

Nigussie, H. (2016). Indigenous communication forms and their potential to convey food security messages in rural Ethiopia. *Indian Journal of Human Development* 10(3), 414–427.

Nu, J., and Bersamin, A. (2017). Collaborating with Alaska native communities to design a cultural food intervention to address nutrition transition. *Progress in Community Health Partnerships* 11(1), 71–80.

Pamphilon, B., and Mikhailovich, K. (2017). Bringing together learning from two worlds: lessons from a gender-inclusive community education approach with smallholder farmers in Papua New Guinea. *Australian Journal of Adult Learning* 57(2), 172–196.

Pandey, V. L., Dev, S. M., and Jayachandran, U. (2016). Impact of agricultural interventions on the nutritional status in South Asia: A review. *Food Policy* 62, 28–40.

Péter, S., Eggersdorfer, M., van Asselt, D., Buskens, E., Detzel, P., Freijer, K., Koletzko, B., Kraemer, K., Kuipers, F., Neufeld, L., Obeid, R., Wieser, S., Zittermann, A., and Weber, P. (2014). Selected nutrients and their implications for health and disease across the lifespan: a roadmap. *Nutrients* 6(12), 6076–6094.

Porter, M. E. (1985). *Competitive Advantage*. Simon and Schuster, New York.

Prescott, S. L., Pawankar, R., Allen, K. J., Campbell, D. E., Sinn, J. K., Fiocchi, A., Ebisawa, M., Sampson, H. A., Beyer, K., and Lee, B. W. (2013). A global survey of changing patterns of food allergy burden in children. *World Allergy Organization Journal* 6, 21.

Pym, R., and Alders, R. G. (2012). *Introduction to Village and Backyard Poultry Production*. Centre for Agriculture and Biosciences International, Wallingford.

Quisumbing, A., Brown, L., Feldstein, H., Haddad, L., and Pena, C. (1995). *Women: The Key to Food Security*. Food Policy Report. http://ebrary.ifpri.org/cdm/ref/collection/p15738coll2/id/125877 (accessed April 4, 2019).

Rahman, S., Rahman, A. S., Alam, N., Ahmed, A. M. S., Ireen, S., Chowdhury, I. A., Chowdhury, F. P., Rahman, S. M. M., and Ahmed, T. (2017). Vitamin A deficiency and determinants of vitamin A status in Bangladeshi children and women: findings of a national survey. *Public Health Nutrition* 20, 1114–1125.

Robinson, T. P., Bu, D. P., Carrique-Mas, J., Fèvre, E. M., Gilbert, M., Grace, D., Hay, S. I., Jiwakanon, J., Kakkar, M., Kariuki, S., Laxminarayan, R., Lubroth, J., Magnusson, U., Thi Ngoc, P. Van Boeckel, T. P., and Woolhouse, M. E. J. (2017). Antibiotic resistance: mitigation opportunities in livestock sector development. *Animal* 11, 1–3.

Rojas-Downing, M. M., Nejadhashemi, A. P., Harrigan, T., and Woznicki, S. A. (2017). Climate change and livestock: impacts, adaptation, and mitigation. *Climate Risk Management* 16, 145–163.

Shrivastava, S. (2012). Comprehensive modeling of communication barriers: a conceptual framework. *IUP Journal of Soft Skills* 6(3), 7–19.

Simopolous, A. P. (2008). The importance of the omega-6/omega-3 fatty acid ratio in cardiovascular disease and other chronic diseases. *Experimental Biology and Medicine* 233, 674–688.

Sobal, J., and Biosogni, C. A. (2009). Constructing food choice decisions. *Annals of Behavioral Medicine* 38(Supp 1), S37–46.

Trant, H. (1954). Food taboos in East Africa. *Lancet* 264(6840), 703–705.

Tucker, C. (2013). Insects, offal, feet and faces: acquiring new tastes in New Zealand? *New Zealand Sociology* 28, 101–122.

United Nations. (2015). *Transforming Our World: the 2030 Agenda for Sustainable Development*. United Nations, New York.

United Nations Economic and Social Council. (2002). *Thematic Issues before the Commission on the Status of Women: Report of the Secretary General*. United Nations, New York.

Vandenplas, Y. (2015). Lactose intolerance. *Asia Pacific Journal of Clinical Nutrition* 24, S9–13.

Viteri, F. E., and Gonzalez, H. (2002). Adverse outcomes of poor micronutrient status in childhood and adolescence. *Nutrition Reviews*, 60(5), S77–S83.

Walker, P., Rhubart-Berg, P., McKenzie, S., Kelling, K., and Lawrence, R. S. (2005). Public health implications of meat production and consumption. *Public Health Nutrition* 8, 348–356.

Wang, Y. Lehane, Ghebremeskel, K., and Crawford, M. A. (2009). Modern organic and broiler chickens sold for human consumption provide more energy from fat than protein. *Public Health Nutrition* 13(3), 400–408.

Williams, P. G., Droulez, V., Levy, G., and Stobaus, T. (2007). Composition of Australian red meat 2002: 3. nutrient profile. *Food Australia* 59(7), 331–341.

Wingett, K., Allman-Farinelli, M., and Alders, R. (2018) Food loss and nutrition security: reviewing pre-consumer loss in Australian sheep meat value chains using a planetary health framework. *CAB Reviews* 13(33), 1–12.

Wolfe, R. R., Baum, J. I., Starck, C., and Moughan, P. J. (2018). Factors contributing to the selection of dietary protein food sources. *Clinical Nutrition* 37, 130–138. doi:10.1016/j.clnu.2017.11.017

Wong, J. T., Bagnol, B., Grieve, H., Jong, J., Li, M., and Alders, R. G. (2018). Factors influencing consumption of animal-source foods in Timor-Leste. *Food Security* 10(3), 741–762.

World Bank. (2007). *From Agriculture to Nutrition: Pathways, Synergies and Outcomes*. Report No. 40196-GLB. http://siteresources.worldbank.org/INTARD/825826-1111134598204/21608903/January2008Final.pdf (accessed April 4, 2019).

World Health Organization (WHO) (2015). *The Global Prevalence of Anaemia in 2011*. http://who.int/nutrition/publications/micronutrients/global_prevalence_anaemia_2011/en (accessed April 8, 2019).

Zambrano, L. D., Levy, K., Menezes, N. P., and Freeman, M. C. (2014). Human diarrhea infections associated with domestic animal husbandry: a systematic review and meta-analysis. *Transactions of the Royal Society of Tropical Medicine and Hygiene* 108(6), 313–325.

Zhang, Z. Y., Goldsmith, P. D., and Winter-Nelson, A. (2016). The importance of animal source foods for nutrient sufficiency in the developing world: the Zambia Scenario. *Food and Nutrition Bulletin* 37, 303–316.

THEME III
AGRICULTURE

18

Optimal Nitrogen Management for Meeting Sustainable Development Goal 2

Kshama Harpankar

1. Introduction

Global cereal production doubled between 1960 and 2000 because of "green revolution" technologies such as new crop varieties, accompanied by increased use of fertilizers and pesticides. This increase in food production made it possible for the global per capita food supply to increase, leading to a decrease in the prevalence of food insecurity (Lassaletta et al. 2014; Tilman et al. 2002). Anthropogenic creation of reactive nitrogen (N_r) played a crucial role in this increase in global food production. In 1850 only about 20% of N_r creation on land was due to cultivation-induced biological nitrogen (N) fixation, whereas by 2010, 75% of the N_r created on land was by human action (Galloway et al. 2014). Use of nitrogen (N) fertilizers has been estimated to be responsible for feeding 48% of the world's population by 2008 (Erisman et al. 2008).

Synthetic N fertilizer inputs in agriculture increased by a factor of nine between 1960 and 2010 globally. Increased use of nitrogen fertilizers has not come without challenges. Evidence suggests that as compared to early 1960s, the proportion of reactive nitrogen added globally onto croplands and converted into harvested products has gone down by 21%. Thus, more than half of the N used for crop fertilization is currently lost into the environment, either in the form of nitrous oxides, which are potent greenhouse gases, or as soluble nitrates that may find their way into aquatic systems (Galloway et al. 2014; Galloway and Leach 2016; Lassaletta et al. 2014). Energy-intensive production of fertilizers is the largest source of fossil fuel consumption in agriculture, with predictions that it will constitute 2% of global energy use by 2050 (Rogers and Oldroyd 2014). With trends in yield growth slowing down and with the world population expected to reach 9.7 billion by 2050, optimizing N management is emerging to be one of the top challenges to ensure global food security and environmental sustainability. According to Galloway et al. (2014), N represents an especially complex challenge because of the way in which it "cascades" between different N_r forms, such as ammonia (NH_3), nitrogen oxides (NO_x), nitrate (NO_3), nitrous oxide (N_2O), and organic N, leading to a plethora of different effects on the environment.

These effects include surface water pollution, groundwater pollution, air pollution, global warming, and loss of ecosystem services and biodiversity. Underutilization of N fertilizer by crops leads to poor product quality and lower yields while underapplication can result in "soil mining," which implies depletion of nutrients from the soil. The dramatic variation in N fertilizer application rates across the world (Table 18.1) illustrates why N is said to be a problem of too much and too little.

Total N fertilization of cropland is mainly the sum of synthetic fertilizers, manure application, symbiotic nitrogen fixation, and atmospheric N deposition (the last one contributing a much smaller share).The proportion of the three main N inputs to overall fertilization varies a great deal among the different world cropping systems (Lassaletta et al. 2014). As shown in Table 18.1, N fertilizer application rates across the world vary considerably. In terms of the relationship between N fertilization and yield, using global data for 1961–2009, Lassaletta et al. (2014) found that some countries, including China, Egypt, and India, presented a simple trajectory with increasing N fertilization over time and gradual reduction in the marginal crop yield response to extra fertilizer. Another group of countries, including the United States, Brazil, and Bangladesh, experienced improved agronomic practices, so that their yield response to nitrogen shifted upward and became steeper, resulting in higher nitrogen usage. A similar pattern occurred in a number of European countries, except that they experienced a reduction in fertilizer usage from the 1980s on, reflecting stronger environmental policies and regulations addressing overusage in the earlier period. Countries such as Morocco, Benin, and Nigeria were characterized by very low N inputs and yields. The nitrogen use variation found in countries around the world demonstrates both challenges as well as opportunities associated with N management. Countries like Egypt, China, and India face the challenge of reducing N use while maintaining yields, while countries like Morocco, Benin, and Nigeria face the challenge of ensuring adequate availability of N from the three sources mentioned earlier.

Seventeen sustainable development goals (SDGs), which replaced the millennium development goals (MDGs) in 2015, acknowledge the multiple demands agriculture faces. SDGs seek to meet these demands based on the joint framework of economic development, environmental sustainability, and social inclusion. By bringing in a

Table 18.1 Variation in N fertilizer application rates around the world

Region	Annual N input kg per hectare of agricultural land (average and range)
Sub-Saharan Africa	8 (0–20)
Latin America	60 (0–120)
Europe and North America	80 (50–300)
South and Central Asia South	40 (10–200)
South East Asia	250 (50–1000)

Source: Sutton et al. (2013).

host of development enablers under the framework of SDGs, these goals collectively aspire to a holistic vision of sustainable development for the future. The goal of this chapter is to discuss the role that science, technology, and innovation (STI) can play in optimal N management as a way to meet SDG2: "end hunger, achieve food security and improved nutrition, and promote sustainable agriculture." Optimal N management will enhance food security by improving yields, nutrition, and promote sustainable agriculture by limiting environmental externalities associated with N.

This chapter aims to address the following two questions:

1. How can new technologies help boost agricultural productivity while also reducing nitrogen pollution?
2. What policy and institutional changes will be needed to encourage innovation and diffusion of these practices and technologies to developing countries in order to achieve SDGs?

To answer these questions, we conducted a review of the literature. To search for relevant articles, we used the following databases: Science Direct, PubMed, Google Scholar, and Web of Science. We focused on peer-reviewed articles/reports published within the last two decades. We used multiple keywords to identify studies relevant to developing countries. Examining the reference lists for the identified articles revealed additional relevant studies. We focused on developing countries in general with some specific references to Africa and Asia. The final section summarizes key findings and frames the discussion in a broader context of economic development in an era of greater complexity and greater urgency due to climate change.

2. Feeding the World Sustainably

The 17 SDGs and 169 targets announced in 2015 build on the MDGs by creating a vision for the world where economic development is driven by practices ensuring environmental sustainability and social inclusion. SDG2 is stated as follows: "end hunger, achieve food security and improved nutrition, and promote sustainable agriculture." It acknowledges the challenge of nourishing people while nurturing the planet. The world has seen remarkable progress when it comes to reducing the number of people undernourished. In 2000, the UN members established the first MDG (MDG1), which included among its targets "cutting by half the proportion of people who suffer from hunger by 2015." According to the State of Food Insecurity in the World report 2015, the commitment to halve the percentage of hungry people has been almost met at the global level. The share of undernourished people in the world population decreased from 18.6% in 1990–1992 to 10.9% in 2014–2016 (FAO 2015).

Despite the impressive progress toward achieving food security, 795 million people (780 million in developing countries) were undernourished in 2014–2016. This represents a reduction from 1011 million in 1990–1992. For sub-Saharan Africa (SSA) and Southern Asia in particular, progress on fighting hunger has been slow.

Table 18.2 Share of population that is undernourished

	1991	2015
World	18.6	10.8
South Asia	24.84	16.23
Sub-Saharan Africa	33.37	18.55
East Asia and Pacific	25.12	9.74
Middle East and North Africa	7.51	8.16
Latin America and Caribbean	14.94	7.39

Source: www.ourworldindata.org.

Over 90 million children under five (one out of seven children worldwide) were underweight as of 2014–2016 and 90% of all underweight children lived in those two regions.

Two recent publications from the FAO (2017; 2018) suggest that the rate of reduction in food insecurity may have slowed significantly. Based on these two reports, the absolute number of people in the world affected by chronic food deprivation is suggested to be on the rise. The reports argue that the global average of the prevalence of undernourishment (PoU) indicator for SDG2 has stagnated from 2013 to 2015. The FAO estimates that in 2016 almost 520 million people in Asia, more than 243 million in Africa, and more than 42 million in Latin America and the Caribbean did not have access to sufficient food energy.

The food insecurity experience scale (FIES) is an experience-based metric of the severity of food insecurity, relying on direct yes/no responses to eight questions regarding access to adequate food. FIES is an indicator for assessing progress on SDG2. Based on the data collected by FAO between 2014 and 2016, 9.3% of people in the world suffered from severe food insecurity, corresponding to about 689 million people. SSA and Latin America are the regions with the highest prevalence of severe food insecurity. The prevalence of stunting, also an indicator for SDG2, is currently highest in Eastern Africa, Middle Africa, Western Africa, Southern Asia, and Oceania (excluding Australia and New Zealand), where more than 30% of children under five years of age are too short for their age.

Agriculture is at the center of the vision for sustainable development implied by the SDGs. Almost 80% of the world's poor live in rural areas, where people depend directly or indirectly on agriculture, fisheries, or forestry as sources of income and food. In low-income and lower-middle-income countries, small family farms are particularly important for achieving food security. In low- and lower-middle-income countries, not only are the majority of farms smaller than five hectares, but they also are responsible for producing a large share of national food output (Lowder et al. 2014). About 70% of all farmland in the low-income countries and about 60% in the lower-middle-income group consists of farms smaller than five hectares. Small family farms dominate rural landscapes across the developing world, accounting for up to 80% of food produced in Asia and SSA, while supporting livelihoods of up to 2.5 billion people (IFAD 2016). In regions faced with severe food insecurity like SSA, family farmers and small-scale food producers dominate the agricultural sector. Thus, growth led by agriculture will be important for ending poverty, food insecurity, and malnutrition.

Agricultural innovation will play a key role in meeting SDG2. Challenges involved in raising the productivity of small farms differ from those of the large farms. STI can be harnessed to introduce new technologies or scale up existing technologies through investment in research and development (R&D) that is aimed toward supporting smallholder farmers. There will also be possibilities for successful adaptation of existing technologies and processes to make them more suitable for small farms. Local knowledge and formal research both will have a role to play. Knowledge of site-specific challenges and opportunities that rests with the local farmers will need to be combined with new/existing technologies to achieve SDG2. Agricultural research will have to seek active engagement with farmers and farming communities to encourage innovation.

The gap between per capita spending on agricultural R&D done by rich countries and poor countries has widened from a 7.7 fold difference in 1980 to an 11.7 fold difference in 2011 (Pardey et al. 2016). Regions experiencing the highest rates of population growth are the regions where per capita agricultural R&D spending is among the lowest in the world. Increased commitment from governments toward R&D spending will need to be coupled with institutional innovation to encourage investments in agriculture in developing economies, to spread the existing successful technologies and make them locally relevant as well as to foster innovation at a local scale (Pardey et al. 2016).

3. Innovations To Boost Productivity and Decrease Pollution

Organized research efforts and policy have played an important role in agriculture for a long time. In the past, these research efforts were mostly concentrated on increasing output produced per unit area of land. Reframing of this objective as increasing output produced per unit area of land while minimizing environmental damages imposed on society is a recent shift. As global population grows, with some of the least-developed and food-insecure countries experiencing more rapid population growth, while negative environmental impacts become more pressing, the role of agricultural innovation becomes ever more important. Climate change adds another layer of complexity to the set of problems described earlier. Successful agricultural technologies will allow food producers to meet the following core challenge: producing food more efficiently with less negative by-products (e.g., fewer GHG emissions, fewer ecosystem services disruptions).

In this section, we will review N-related technologies and innovations that help us to meet this challenge. The focus of the discussion is on developing countries in general with some specific references to Africa and South Asia.

3.1 New Types of Nitrogen Fertilizers

The world used 105 million tonnes of synthetic N fertilizers in 2010. Future consumption of synthetic N fertilizers is expected to be in the range of 80–150 million tonnes/year by 2050 (Bindraban et al. 2015). A path heavily dependent on the use of chemical nitrogen fertilizers manufactured using traditional methods will interfere with plans to mitigate climate change as well as with other aspects of sustainability. STI can contribute in at least two ways. First, STI can help make the fertilizer production process more sustainable. Second, STI can help pair innovative types of N fertilizers with appropriate agronomic practices to minimize pollution in places where use of synthetic N fertilizers is the best way to provide the needed nutrients to crops in order to meet the growing demand for food. Toward that end, improved fertilizer technology will be helpful to minimize N losses through volatilization and leaching, while also increasing the recovery of N fertilizer by crops.

Enhanced-efficiency N fertilizer is one such technology developed to achieve yield increases while minimizing N losses to the environment. Examples of types of enhanced-efficiency fertilizers (EEFs) follow. Urease inhibitors (UI) work by delaying urea hydrolysis, which lowers NH_3 emissions. Nitrification inhibitors (NI) reduce the activities of nitrifying bacteria and therefore decrease both the risks of NO_3 leaching and the level of N_2O emissions. Double inhibitors (DI) are designed to lower NH_3 emissions, NO_3 and N_2O losses by combining urease and nitrification inhibitors. Polymer-coated fertilizers (PCF) use partially permeable coating material to control N release (Herrera et al. 2016; Li et al. 2018).

Based on a meta-analysis of 203 published studies, Li et al. (2018) compared the performance of the aforementioned EEFs on the basis of yield gains, reductions in N_2O emissions, NH_3 volatilization, and NO_3 leaching. The results showed that in general, DI performed the best by increasing yield and N uptake by 11% and 33% respectively, while decreasing aggregated N loss by 47%; the positive effects held under all soil and climate conditions tested. EEFs are likely to match N availability to crop demand better than traditional N fertilizers and thus can contribute to the SDG2 goal of raising agricultural productivity by raising yields ; by increasing yields, this technology also has the potential to address the SDG2 goal of ending hunger and malnutrition.

Reducing the amount of N lost to the environment in a variety of forms will be helpful to achieve a number of SDGs. NO_3 leaching into waterways is a major cause of eutrophication (Chen et al. 2011; Howarth and Marino 2006). Reducing NO_3 leaching will make agriculture more sustainable (SDG2), improve water quality (SDG6.3), and reduce marine nutrient pollution (SDG14.1). The ability to lower N_2O emissions makes the EEF technology relevant for SDG13: taking urgent action to combat climate change. Excessive use of N fertilizer has contributed to anthropogenic soil acidification (Guo et al. 2010; Tian and Niu 2015). Soil acidification (i.e., decrease in pH) caused by leaching of NO_3 (Tian and Niu 2015) has been documented to be a significant threat to species diversity and terrestrial ecosystem functioning. EEFs can therefore contribute to achievement of SDG15 (protect, restore, and promote sustainable use of terrestrial ecosystems).

Nitrogen pollution can also affect human health. Fine particulate matter, formed from nitrogen oxide and ammonia emissions, contributes to respiratory and cardiovascular diseases. Other human health risks associated with N pollution come from increased levels of nitrate in drinking water as well as consumption of seafood contaminated by toxic algae blooms caused by nutrient pollution (Reis et al. 2016). Reducing various forms of N pollution will help with efforts to achieve SDG3.9: "by 2030, substantially reduce the number of deaths and illnesses from hazardous chemicals and air, water and soil pollution and contamination."

The downside of EEFs is the cost of this technology, which can be four to eight times the cost of conventional fertilizers (Herrera et al. 2016). In developing countries where affordability of fertilizers is a pressing concern for farmers, use of traditional N fertilizers in innovative ways paired with well-designed agronomic practices will be helpful. Li et al. (2018) argued that innovative fertilizers are not a panacea for addressing the N challenge without elimination of fertilizer mismanagement and implementation of knowledge-based management practices. Traditional N fertilizers can be used in innovative ways to reduce their negative impacts on the environment. For example, deep placement of urea has been shown to match the N demand of plants in intensive rice cropping systems and also effectively reduce N losses, especially NH_3 volatilization, in field trials in China and Bangladesh (Yao et al. 2018). Similarly, the use of urea super granules (which refers to the local scale conversion of small urea granules or "prills" into larger pellets) 5–7 cm below the soil surface increased yield by 15–18% per ha over surface applied urea using approximately one-third less N fertilizer (Sutton et al. 2013).

Use of synthetic N fertilizer will always need to be guided by the 4R nutrient stewardship principles of 1) the right source of N fertilizer, 2) the right rate, 3) the right timing of application, and 4) the right placement. Agronomic practices used to implement the 4R approach include precision application, deep placement, row application, and split applications (Bindraban et al. 2015).

As traditional fertilizer production process is energy intensive, there is a need for innovations in methods to produce N fertilizers sustainably. Creation of such methods is listed as one of the 50 breakthrough technologies put together by Lawrence Berkeley National Laboratory's Institute for Globally Transformative Technologies (Buluswar et al. 2014).

Pfromm (2017) explored the possibility of retrofitting traditional ammonia synthesis facilities with existing technology for competitive renewable electricity-based feedstock preparation to produce fossil-fuel-free ammonia. Given the decline in the cost of renewable electricity, the study argued that replacing natural gas as the fuel source for hydrogen and nitrogen generation to electrolytic hydrogen and cryogenic nitrogen made using renewable electricity is economically attractive. This innovation offers a path to decouple ammonia synthesis, which claims a major share of energy use for food production, from fossil fuels. This technology has the potential to contribute to SDG7: "increase substantially the share of renewable energy in the global energy mix." The possibility of setting up smaller scale production units of fossil-fuel-free ammonia in developing countries will also be a step toward achieving SDG9.4: "by 2030, upgrade infrastructure and retrofit industries to make them sustainable, with increased resource-use efficiency and greater adoption of clean and environmentally sound technologies and industrial processes, with all countries taking action in accordance with their respective capabilities."

3.2 Integrated Soil and Fertility Management (ISFM)

Soils play an important role in provision of basic human needs like food, clean water, and clean air. They also support biodiversity (Keesstra et al. 2016). Focusing on N management strategies without consideration of the broader contributions soils make will limit our ability to achieve the SDGs. According to Chen et al. (2011), research in the area of development, application, and adaptation of appropriate cropping systems is likely to have the largest payoff in rapidly developing economies. Population growth in SSA and regions of Latin America, Asia, and the Pacific is expected to accelerate soil degradation in those regions. Garrity et al. (2010) argued that restoring soil health is often the first entry point for increasing agricultural productivity, especially for regions like SSA. Toward that end, the principles of integrated soil and fertility management (ISFM) seek to maximize agronomic efficiency by combining a balanced nutrient supply with improved varieties and agronomy adapted to local conditions (Rosegrant et al. 2014; Vanlauwe et al. 2015). ISFM includes practices like use of improved seeds, use of fertilizers, organic inputs, liming materials, water management, and appropriate tillage practices. One goal of ISFM is to develop and promote soil-fertility-replenishment technologies that are suitable for different types of resource-poor farm households (Mutuma et al. 2014). Organic inputs like livestock manures, crop residues, and compost are also sources of nutrients that help increase soil organic matter. Improved soil organic matter increases soil fertility and nutrient cycling capacity. There are opportunities to pair the use of chemical fertilizers with organic inputs in low-input systems like SSA, but often farmers face constraints due to limited availability or high opportunity cost of organic material.

ISFM practices can lead to an increase in agricultural productivity, leading to improvements in incomes and enhanced food security. Thus, ISFM practices can help us tackle SDG2 as well as SDG1 of ending poverty in all its forms everywhere. Based on a 20–year study of the research farm of International Institute of Tropical agriculture in southwestern Nigeria, it was shown that implementing ISFM based farm management generated sustainable increases in crop productivity and input use efficiency which ultimately improved the livelihoods of farmers (Roobroeck et al. 2016). The study also showed that production of maize crops with ISFM practices was less impacted by variations in weather conditions than when fertilizers were used exclusively. Organic inputs played an important role in reducing the climate sensitivity of maize crops in this study. The higher productivity and yield stability achieved in the ISFM system is an indicator of its ability to adapt to and increase resilience to climate-change impacts (SDG13). Combining the use of fertilizers with organic inputs also led to higher soil organic content in this study, thereby lowering carbon emissions from soil, another contribution to achieving SDG13 (Roobroeck et al. 2016). Chen et al. (2016) mentioned that a maize production system in China was able to double the yield with no increases in N fertilizer use in ISFM-based farm-management systems. In another study carried out in Nigeria, average crop yields for maize, sorghum, and soybean increased by more than 200% in the villages covered by the project based on two technology packages using a combined application of inorganic fertilizers and manure with a soybean/maize rotation practice. The results also confirmed the positive correlation between yield increases due to ISFM technologies used and food expenditures of the adopters (Akinola et al. 2009).

3.3 Technologies Helping with Biological Nitrogen Fixation

Use of improved legume varieties and microbial inoculants holds the promise of lowering synthetic N fertilizer applications by relying on biological nitrogen fixation. N fixation at field level depends on the potential of N-fixing plants to establish an effective symbiosis with rhizobia, and the relative ability of the established symbiosis to fix N (Ronner and Franke 2012). The former is a function of a variety of environmental and management factors at the field level, while the latter depends on the genetic potential of the bacteria, the crop, and the symbiosis.

Raising yields and agricultural productivity with a reliance on biological nitrogen fixation is a strategy that contributes to achieving the environmental goals set out (and discussed earlier) in SDGs 6, 13, 14, and 15 by lowering the use of synthetic N fertilizers. In the right circumstances, the inclusion of legumes in crop rotations can also contribute to income goals associated with SDG1 and 2. Grain legumes fix atmospheric nitrogen gas (N_2) that can contribute to the nitrogen economy of fields, provide other rotational benefits to subsequent crops, produce in situ high-quality organic residues and contribute to integrated soil fertility management (Franke et al. 2018; Vanlauwe et al. 2015). Examples of technologies with potential for an increase in biological nitrogen fixation include diversification and intensification via inclusion of grain legumes in cereal, root or tuber based cropping systems (Franke et al. 2018), and climbing bean technology (Ronner 2018).

In a study done in Nigeria, Ronner (2018) found positive yield responses to the use of rhizobium inoculants with Soybean varieties. N2Africa is a large-scale, science-based development-to-research project focused on putting nitrogen fixation to work for smallholder farmers growing legume crops in Africa (Giller et al. 2015). It is reported that over a period of four years, N2Africa's biological nitrogen fixation technology dissemination project realized up to a 15% increases in farm yields of grain legumes and a 17% increase in biological nitrogen fixation (Reis et al. 2016). Ronner (2018) reported that Rhizobial inoculation proved to be a cheap way to increase soybean yields with low financial risks for farmers in Africa. They presented evidence to make the case that use of inoculants in Africa makes economic as well as agronomic sense.

Biological nitrogen fixation can also be facilitated by incorporating local legume tree species into farm management systems. Evergreen agriculture is a set of technology/management practices based on the idea of integration of trees into annual food crop systems. The Zambian Conservation Farming Unit (CFU) used a package of agronomic practices based on the principles of conservation farming (which included minimum tillage methods, mineral fertilizer input application, and crop rotations) along with inclusion of a nitrogen-fixing tree species, *Faidherbia albida*, in the crop fields. This practice led to yield increases in maize and cotton. Similarly, the Malawi Agroforestry Food Security Program assists with the uptake of three types of nitrogen-fixing tree legumes. They include short-term species such as *Tephrosia candida*, *Sesbania sesban*, and pigeon peas, which are planted and incorporated within one year; medium-term solutions such as *Gliricidia*, which can be continuously pruned for organic fertilizer for one to two decades; and long-term full canopy trees of *Faidherbia albida*, which provide benefits for many decades. These species are often combined on the same fields (Garrity et al. 2010). By incorporating a variety of local species in farm management systems, this technology helps address the SDG15 by ensuring healthy ecosystems and biodiversity.

Research has the potential to help by developing elite inoculants like BIOFIX®. A study of soybean farmers in Western Kenya reported improvements in profits and profits for farmers who adopted the BIOFIX® technology (Mutuma et al. 2014).

3.4 Biotechnology Solutions

Opportunities exist for identifying new genetic markers and hybrids for molecular breeding to improve nitrogen uptake by plants. Research is underway into the feasibility of biotechnology solutions to supply nitrogen to crops. The application of novel techniques such as genetically modified organisms (GMOs) and gene editing have the potential to radically change the way we grow our crop plants, and therefore contribute to solutions for hunger, food insecurity, and malnutrition problems, especially in the developing world (Qaim and Kouser 2013; Shrawat et al. 2008; Tester and Langridge 2010).

Oldroyd and Dixon (2014) discuss two such possibilities: genetic engineering of the legume symbiosis into cereals and the introduction of nitrogenase. It might be possible for these technologies to deliver small increases in available nitrogen in the near future, a breakthrough that will be immensely helpful for small-holder farmers in developing countries facing severe nutrient scarcity (Folberth et al. 2013; Oldroyd and Dixon 2014; Rogers and Oldroyd 2014).

Another promising area of research is an approach building on the discovery of a nonrhizobial, naturally occurring nitrogen-fixing bacterium that fixes nitrogen in sugarcane, which is able to intracellularly colonize the root systems of cereals and other major crops. The bacterium, *Gluconacetobacter diazotrophicus* (GD), has the ability to systematically colonize both nonlegumes and legumes intracellularly. Research on GD, which is now at a field-trial evaluation stage, suggests that an adequate level of bacterial intracellular colonization and nitrogen fixation can be established throughout the plant without any need for nodulation (Dent and Cocking 2017).

Altering crops genetically to improve their ability to take up nitrogen while reducing nitrogen pollution is another area where science and technology has potential to play an important role. For example, biological nitrification inhibition (BNI) is the ability of certain plant roots to suppress soil-nitrifier activity, through production and release of nitrification inhibitors. This approach can be harnessed to improve nitrogen-cycling in agricultural systems. Breeding crop varieties with BNI traits is a possibility that would require new investments in agricultural research (Subbarao et al. 2017). Modern biotechnology development has the potential to contribute to the environmental goals of SDGs 2, 6.13, 14, and 15 by lowering our reliance on synthetic N fertilizers. Evidence presented in chapter 24 (Adenle et al.) indicates that the application of modern biotechnology such as GMOs can help address the challenges of SDGs.

3.5 Precision Agriculture

Carefully matching nitrogen supply to crop needs during the growing season has the ability to increase nitrogen uptake by plants while reducing nitrogen pollution. The goal of precision agriculture (PA) is to target the fertilizer application based on yield potential of different parts of a field as opposed to a spatially uniform application of fertilizer. Site-specific nutrient applications have the potential to reduce nitrogen/nutrient pollution. The use of PA technologies will allow us to make progress on SDGs related to environmental management as well as SDG2 of raising agricultural productivity. Aune et al. (2017) found that precision agriculture technologies provided cost-efficient methods to increase yields in semiarid West Africa. Kisinyo et al. (2015) showed that application of micro dosing (application of small amounts of lime and fertilizers P and N) significantly increased maize grain yield. Similarly, Doberman et al. (2004) elucidated lessons learned from a research project in the United Kingdom that set out to maximize profitability and minimize environmental impact of cereal production by using PA. The application of N in a spatially variable manner improved the efficiency of cereal production. It reduced N pollution by approximately one third and created economic benefit per ha that was higher as compared to a standard N management policy (Dobermann et al. 2004).

PA can also help with biological nitrogen fixation technology. Micronutrients are important for the legume–rhizobia symbiosis. Soil micronutrient deficiencies reduce the potential for biological nitrogen fixation. Micro-nutrient deficiencies are increasingly being demonstrated to be influential in crop yield in the unresponsive soils of Africa (Bindraban et al. 2015). Incorporating PA techniques like geospatial analysis based on soil testing and agronomic fertilizer trials would allow for improving the process of biological nitrogen fixation by factoring in micro-nutrient deficiencies.

Technologies used in precision agriculture include remote sensing, geographical information systems, spectral imaging, and information and communication technologies (ICT). ICT can be used to provide farmers with recommendations on fertilizer use customized for different crops and locations. In collaboration with Bangladink and Grameenphone, a mobile-phone-based fertilizer information service was launched in Bangladesh. eGeneration, a local information technology company, developed the software application in the local language (Bangla) with attention to the agricultural users and local context. Since its launch in July 2009, users have reduced their fertilizer costs by up to 25% and have obtained higher crop yields by up to 15% (UNCTAD 2017). Use of ICT for agriculture in developing countries has the potential to address SDGs 1, 2, 6, 13, and 14.

4. Challenges in the Path of STI Solutions to Sustainable Development Goals

In this section, we discuss obstacles inhibiting the widespread uptake of technologies and innovations discussed in earlier sections. Although the use of EEFs can help us make progress toward SDGs 2, 3, 6, 13, 14, and 15, the cost of this technology remains a barrier for resource-poor smallholder farmers in developing countries. Countries like Zambia, Malawi, Mali, and Nigeria have implemented fertilizer-subsidy programs to increase the affordability of nutrients to small-scale farmers (Sutton et al. 2013). Such subsidy programs for resource-poor farmers along with payments for ecosystem services will help to accelerate the transition to productive sustainable systems (Vanlauwe et al. 2014). India introduced the nutrient-based subsidy (NBS) program in 2010 to encourage a balanced use of fertilizers to increase food production while reducing nutrient pollution. This program broadened the scope of the basket of fertilizers farmers can choose from to give farmers the option to choose the right fertilizer for their soil-crop system.

Technologies that make N fertilizer production process sustainable were discussed as a way to achieve SDG2 and SDG7. With the decline in the cost of renewable energy, fossil-fuel-free NH_3 seems likely to become economically profitable, and it may already be so in some cases (Pfromm 2017). Given the interest in using NH_3 for energy storage/transportation, funding research in this area allows us to take advantage of the synergies in SDGs 2, 7, and 9. Being able to produce renewable NH_3 that can be used in food production as well as for energy storage and transportation will not only make our energy mix more sustainable (SDG7), but will also help to build resilient and sustainable infrastructure (SDG9). Research on alternatives to the Haber-Bosch NH_3 production is also called for. Specifically for Africa, very little investment in fertilizer production by governments has caused imports of fertilizers and access to them to become constraining factors for agriculture.

Developing countries like China have a record of using N fertilizer excessively, whereas countries in Africa have low N use on farms. Traditional fertilizers may be more accessible to smallholder farmers, so it is important to invest in initiatives to promote the adoption of technologies like ISFM and PA, as discussed earlier. Adoption of technologies like PA and ISFM will be more likely if they provide the farmers a short-term return on investment at low risk. Profitability will likely be a stronger motivation force for adoption of new practices/technologies by farmers than their sustainability, particularly for off-farm aspects of sustainability. To provide high returns on investments for resource-poor smallholder farmers, cost-effective innovations like urea super granules will be important. Understanding what limits the adoption of technologies like PA/ISFM will be helpful to facilitate their wider adoption. For example, Xia et al. (2017) discussed reasons for lack of popularity of knowledge-based N management systems (built on principles of PA/ISFM) in China. Most Chinese farms are small, so large-scale mechanization required by knowledge based N management systems is viewed as an impediment. In addition, due to increasing urbanization, farming is a part-time occupation for many in China. Thus, the opportunity cost of time to be devoted to learning and adopting N management practices is typically high. Successful technology adaptation will depend on its ability to create a production surplus without requiring extra labor, and the ability to set in a motion a positive feedback loop that provides an incentive to sustain the practice (Aune et al. 2017). Major technical advances are needed to develop low-cost in-situ methods to diagnose field-level variability in relevant micronutrients allowing variable-rate micronutrient application to optimize biological nitrogen fixation technologies (Thilakarathna and Raizada 2018).

Although the technologies we have discussed are largely for use on the farm, their potential to contribute toward SDGs is likely to be hampered or enhanced as a function of a variety of off-farm socioeconomic and institutional variables. Investments in robust distribution networks for agricultural inputs, credit sources, retail outlets, and roads will help in improving access to agricultural inputs. Low quality of agricultural inputs is often a barrier faced by farmers in SSA. Masso et al. (2017) report that of 22 rhizobial inoculants evaluated in SSA, about 40% did not contain the declared active ingredients and, not surprisingly, did not perform as claimed. Other inputs such as animal manures may contain little N due to poor feed quality and poor manure management (Masso et al. 2017). There is a need to improve quality control for commercially supplied agricultural inputs through improved regulatory mechanisms.

Improvements in agronomic efficiency of applied N in SSA have been limited by lack of research capacity, and especially by a lack of long-term experimental trials (Giller et al. 2015; Vanlauwe et al. 2015). The ability to diagnose N-related problems and design solutions for them is hindered by a lack of detailed on-farm nutrient budgets (Vitousek et al. 2009). Quantification of N flows at the system level is rare and a deterrent to knowledge-based N management in many developing countries (Xia et al. 2017). Tools such as N budgets and the N footprint are valuable in providing information to design practices and policies to reduce N losses.

The technology options presented in the earlier sections will need to function under a variety of socioeconomic conditions and different governance structures. Meeting the dual goal of increased food production with lower N pollution will depend on the types of policies used and enforcement mechanisms in place. A developing country may not have a national program/policy to reduce or monitor water quality in rural areas. Industrial and point sources of N pollution have been reduced with the help of policy and technology in developed countries, whereas they remain a challenge in developing countries (Gu et al. 2011). Misguided government policies can hurt the environment by promoting excessive fertilizer use, as in case of China (Li et al. 2018). Developing countries also face particular challenges in reducing nutrient pollution from agricultural sources. A developing country may also lack a culture that expects and supports enforcement with the result that they lack the means or the motivation to enforce regulations (Sutton et al. 2013). Policies that integrate agricultural-support policies with environmental protection policies hold promise for developing countries.

Heterogeneity of smallholder farming landscapes in developing countries suggests that optimal N management pathways to achieve SDGs will need to be flexible and adapted to local farming conditions, cropping patterns, resource endowments, production objectives, and risk preferences of farmers. Continuous collaboration between various stakeholders and the need for learning from experience will be the hallmarks of the participatory development approaches needed. A successful example of such an approach in operation is N2Africa. N2Africa seeks to understand why certain technologies work best for particular farmers; using feedback loops through adaptive research it seeks to refine and improve the technologies through addressing those problems that emerge (Ampadu-Boakye et al. 2017).

5. Conclusion

There has been impressive progress in reducing the prevalence of food insecurity in the last quarter of the twentieth century, yet a significant number of people still suffer from severe food insecurity. Nitrogen plays an important role in food production globally and is responsible for a variety of environmental problems associated with its loss in various forms. SDG2, which aims for zero hunger and sustainable agricultural production, is one of the most difficult of the SDGs to achieve. It will require a mix of interventions to increase N use in some cases and to decrease N use in other cases, or other means to reduce nitrogen losses to the environment.

We considered a variety of technologies and innovations with the potential to contribute to SDGs via the joint goal of lowering N pollution and increasing agricultural productivity. While technologies like ISFM and PA hold great promise for developing countries in meeting the dual objective of increased food production and lower nitrogen pollution, the number of constraints experienced by smallholder farmers in developing countries in adopting these technologies is high. Affordability of technologies, technical expertise to implement them, lack of supporting infrastructure, and lack of market access are some of the limiting factors that will have a bearing on the success of STI solutions outlined earlier.

In order to achieve SDG2 of zero hunger and sustainable agriculture, the STI solutions discussed in the earlier section cannot be considered in isolation. There are important synergies across different SDGs and ignoring them will be counterproductive. For example if poverty is a barrier to adoption of a certain technology that lowers nitrogen pollution, then solutions to lowering nitrogen pollution need to be considered with solutions to the problem of poverty. Echoing the call to take a systems approach made by Keesstra et al. (2016), there is a need for studying the STI solutions discussed in this chapter in an interdisciplinary and transdisciplinary manner to deepen our understanding of the progress made on SDGs and the trade-offs involved along each path.

Science has helped us achieve great feats in food production in the past, but as Barret and Palm (2016) argued, the low-hanging fruit of dealing with relatively homogeneous agroecosystems have largely been harvested (Barrett and Palm 2016). Going forward, the challenge is to apply STI to heterogeneous and often degraded agroecosystems across the world. In some cases, these systems have lagged in agricultural productivity growth in recent decades. Given the complexity of the N problem, there is no one-size-fits-all solution. To provide a context-specific solution, an adaptive learning process based on feedback loops will be needed to determine the suitability of options from the basket of technologies outlined earlier.

References

Akinola, A. A., Alene, A. D., Adeyemo, R., Sanogo, D., and Olanrewaju, A. S. (2009). Impacts of balanced nutrient management systems technologies in the northern Guinea savanna of Nigeria. *Journal of Food Agriculture and Environment* 7(2), 496–504.

Ampadu-Boakye, T., Stadler, M., and Kanampiu, F. (2017). *N2Africa Annual Report 2016.* N2Africa. https://n2africa.org/n2africa-annual-report-2016 (accessed March 5, 2019).

Aune, J. B., Coulibaly, A., and Giller, K. E. (2017). Precision farming for increased land and labour productivity in semi-arid West Africa: a review. *Agronomy for Sustainable Development* 37(3), 16.

Barrett, C. B., and Palm, C. (2016). Meeting the global food security challenge: obstacles and opportunities ahead. *Global Food Security* 11, 1–4.

Bindraban, P. S., Dimkpa, C., Nagarajan, L., Roy, A., and Rabbinge, R. (2015). Revisiting fertilisers and fertilisation strategies for improved nutrient uptake by plants. *Biology and Fertility of Soils* 51(8), 897–911.

Buluswar, S., Friedman, Z., Mehta, P., Mitra, S., and Sathre, R. (2014). 50 Breakthroughs: Critical Scientific and Technological Advances Needed for Sustainable Global Development. LBNL Institute for Globally Transformative Technologies, Berkeley, CA.

Chen, X.-P., Cui, Z.-L., Vitousek, P. M., Cassman, K. G., Matson, P. A., Bai, J.-S., Meng, Q.-F., Hou, P., Yue, S-C, Römheld, V., and Zhang, F-S. (2011). Integrated soil–crop system management for food security. *Proceedings of the National Academy of Sciences* 108(16), 6399–6404.

Dent, D., and Cocking, E. (2017). Establishing symbiotic nitrogen fixation in cereals and other non-legume crops: the greener nitrogen revolution. *Agriculture and Food Security* 6(1), 7, doi: https://doi.org/10.1186/s40066-016-0084-2

Dobermann, A., Blackmore, S., Cook, S. E., and Adamchuk, V. I. (2004). Precision farming: challenges and future directions. In: Fischer, R., Turner, N., Angus, J., McIntyre, L., Robertson, M., Borrell, A., and Lloyd, D. (Eds.), *New Directions for a Diverse Planet: Proceedings of the 4th International Crop Science Congress*, Brisbane, Australia, September 26–October 1, Published on CDROM. www.cropsceince.org.au (accessed April 13, 2020).

Erisman, J. W., Sutton, M. A., Galloway, J., Klimont, Z., and Winiwarter, W. (2008). How a century of ammonia synthesis changed the world. *Nature Geoscience* 1(10), 636–639.

Folberth, C., Yang, H., Gaiser, T., Abbaspour, K. C. and Schulin, R. (2013). Modeling maize yield responses to improvement in nutrient, water and cultivar inputs in sub-Saharan Africa. *Agricultural Systems* 119, 22–34.

Food and Agriculture Organization of the United Nations (FAO) (2017). *The State Of Food Security and Nutrition in the World: Building Resilience for Food and Food Security.* FAO, Rome.

Food and Agriculture Organization of the United Nations (FAO) (2018). *The State of Food Security and Nutrition in the World: Building Climate Resilience for Food Security and Nutrition.* FAO, Rome.

Franke, A., Van den Brand, G., Vanlauwe, B., and Giller, K. (2018). Sustainable intensification through rotations with grain legumes in Sub-Saharan Africa: a review. *Agriculture, Ecosystems and Environment* 261, 172–185.

Galloway, J. N., and Leach, A. M. (2016). Sustainability: your feet's too big. *Nature Geoscience* 9(2), 97–98.

Galloway, J. N., Winiwarter, W., Leip, A., Leach, A. M., Bleeker, A., and Erisman, J. W. (2014). Nitrogen footprints: past, present and future. *Environmental Research Letters* 9(11), https://doi.org/10.1088/1748-9326/9/11/115003.

Garrity, D. P., Akinnifesi, F. K., Ajayi, O. C., Weldesemayat, S. G., Mowo, J. G., Kalinganire, A., Larwanou, M., and Bayala, J. (2010). Evergreen Agriculture: a robust approach to sustainable food security in Africa. *Food Security* 2(3), 197–214.

Giller, K. E., Andersson, J. A., Corbeels, M., Kirkegaard, J., Mortensen, D., Erenstein, O., and Vanlauwe, B. (2015). Beyond conservation agriculture. *Frontiers in Plant Science* 6, 1–14.

Gu, B., Zhu, Y., Chang, J., Peng, C., Liu, D., Min, Y., Luo, W., Howarth, R., and Ge, Y. (2011). The role of technology and policy in mitigating regional nitrogen pollution. *Environmental Research Letters* 6(1), doi: 10.1088/1748-9326/6/1/014011

Guo, J. H., Liu, X. J., Zhang, Y., Shen, J., Han, W., Zhang, W., Christie, P., Goulding, K.W., Vitousek, P. M., and Zhang, F. S. (2010). Significant acidification in major Chinese croplands. *Science* 327(5968), 1008–1010.

Herrera, J. M., Rubio, G., Häner, L. L., Delgado, J. A., Lucho-Constantino, C. A., Islas-Valdez, S., and Pellet, D. (2016). Emerging and established technologies to increase nitrogen use efficiency of cereals. *Agronomy* 6(2), doi: 10.3390/agronomy6020025

Howarth, R. W., and Marino, R. (2006). Nitrogen as the limiting nutrient for eutrophication in coastal marine ecosystems: evolving views over three decades. *Limnology and Oceanography* 51(1, part 2), 364–376.

International Fund for Agricultural Development (IFAD). (2016). *Rural Development Report 2016 Fostering Inclusive Rural Transformation.* IFAD, Rome.

Keesstra, S. D., Bouma, J., Wallinga, J., Tittonell, P., Smith, P., Cerdà, A., Montanarella, L., Quinton, J. N., Pachepsky, Y., van der Putten, W. H., Bardgett, R. D., Moolenaar, S., Mol, G., Jansen, B., and Fresco, L.O. (2016). The significance of soils and soil science towards realization of the United Nations Sustainable Development Goals. *Soil* 2(2), 111–128.

Lassaletta, L., Billen, G., Grizzetti, B., Anglade, J., and Garnier, J. (2014). 50 year trends in nitrogen use efficiency of world cropping systems: the relationship between yield and nitrogen input to cropland. *Environmental Research Letters* 9(10), doi: 10.1088/1748-9326/9/10/105011

Li, T., Zhang, W., Yin, J., Chadwick, D., Norse, D., Lu, Y., Liu, X., Chen, X., Zhang, F., Powlson, D., and Dou, Z. (2018). Enhanced-efficiency fertilizers are not a panacea for resolving the nitrogen problem. *Global Change Biology* 24(2), e511–e521.

Lowder, S. K., Skoet, J., and Singh, S. (2014). What Do We Really Know About the Number and Distribution of Farms and Family Farms Worldwide? Background Paper for The State of Food and Agriculture 2014. ESA Working Paper No. 14-02. FAO, Rome.

Masso, C., Baijukya, F., Ebanyat, P., Bouaziz, S., Wendt, J., Bekunda, M., and Vanlauwe, B. (2017). Dilemma of nitrogen management for future food security in sub-Saharan Africa: a review. *Soil Research* 55(6), 425–434.

Mutuma, S., Okello, J., Karanja, N., and Woomer, P. (2014). Smallholder farmers' use and profitability of legume inoculants in Western Kenya. *African Crop Science Journal* 22(3), 205–214.

Oldroyd, G. E., and Dixon, R. (2014). Biotechnological solutions to the nitrogen problem. *Current Opinion in Biotechnology* 26, 19–24.

Pardey, P. G., Chan-Kang, C., Dehmer, S. P., and Beddow, J. M. (2016). Agricultural R&D is on the move. *Nature News* 537(7620), 301–303.

Pfromm, P. H. (2017). Towards sustainable agriculture: fossil-free ammonia. *Journal of Renewable and Sustainable Energy* 9(3), https://doi.org/10.1063/1.4985090

Qaim, M., and Kouser, S. (2013). Genetically modified crops and food security. *PloS One* 8(6), e64879.

Reis, S., Bekunda, M., Howard, C. M., Karanja, N., Winiwarter, W., Yan, X., Bleeker, A., and Sutton, M. A. (2016). Synthesis and review: tackling the nitrogen management challenge: from global to local scales. *Environmental Research Letters* 11(12), doi: 10.1088/1748-9326/11/12/120205

Rogers, C., and Oldroyd, G. E. (2014). Synthetic biology approaches to engineering the nitrogen symbiosis in cereals. *Journal of Experimental Botany* 65(8), 1939–1946.

Ronner, E. (2018). From targeting to tailoring. PhD diss., Wageningen University.

Ronner, E., and Franke, A. (2012). *Quantifying the Impact of the N2Africa Project on Biological Nitrogen Fixation.* N2Africa, Wageningen.

Roobroeck, D., Van Asten, P. J. A., Jama, B., Harawa, R., and Vanlauwe, B. (2016). Integrated Soil Fertility Management: Contributions of Framework and Practices to Climate-Smart Agriculture. Climate-Smart Agriculture Practice Brief. CGIAR Research Program on Climate Change, Agriculture and Food Security (CCAFS), Copenhagen.

Rosegrant, M. W., Koo, J., Cenacchi, N., Ringler, C., Robertson, R. D., Fisher, M., Cox, C. M., Garrett, K., Perez, N. D., and Sabbagh P.(2014). *Food Security in a World of Natural Resource*

Scarcity: The Role of Agricultural Technologies. International Food Policy Research Institute, Washington, DC.

Shrawat, A. K., Carroll, R. T., DePauw, M., Taylor, G. J., and Good, A. G. (2008). Genetic engineering of improved nitrogen use efficiency in rice by the tissue-specific expression of alanine aminotransferase. *Plant Biotechnology Journal* 6(7), 722–732.

Subbarao, G., Arango, J., Masahiro, K., Hooper, A., Yoshihashi, T. ando, Y., Nakahara, K., Deshpande, S., Ortiz-Monasterio, I., Ishitani, M., Peters, M., Chirinda, N., Wollenberg, L., Lata, J. C., Gerard, B., Tobita, S., Rao, I. M., Braun, H. J., Kommerell, V., Tohme, J., and Iwanaga, M. (2017). Genetic mitigation strategies to tackle agricultural GHG emissions: the case for biological nitrification inhibition technology. *Plant Science* 262, 165–168.

Sutton M. A., Bleeker A., Howard C. M., Bekunda M., Grizzetti B., de Vries W., van Grinsven H. J. M., Abrol Y. P., Adhya T. K., Billen G., Davidson E. A, Datta A., Diaz R., Erisman J. W., Liu X. J., Oenema O., Palm C., Raghuram N., Reis S., Scholz R. W., Sims T., Westhoek H. and Zhang F. S., with contributions from Ayyappan S., Bouwman A. F., Bustamante M., Fowler D., Galloway J. N., Gavito M. E., Garnier J., Greenwood S., Hellums D. T., Holland M., Hoysall C., Jaramillo V. J., Klimont Z., Ometto J. P., Pathak H., Plocq Fichelet V., Powlson D., Ramakrishna K., Roy A., Sanders K., Sharma C., Singh B., Singh U., Yan X. Y., and Zhang Y. (2013) *Our Nutrient World: The Challenge to Produce More Food and Energy with Less Pollution. Global Overview of Nutrient Management. Centre for Ecology and Hydrology.* On behalf of the Global Partnership on Nutrient Management and the International Nitrogen Initiative, Edinburgh.

Tester, M., and Langridge, P. (2010). Breeding technologies to increase crop production in a changing world. *Science* 327(5967), 818–822.

Thilakarathna, M. S., and Raizada, M. N. (2018). Challenges in using precision agriculture to optimize symbiotic nitrogen fixation in legumes: progress, limitations, and future improvements needed in diagnostic testing. *Agronomy* 8(5), https://doi.org/10.3390/agronomy8050078.

Tian, D., and Niu, S. (2015). A global analysis of soil acidification caused by nitrogen addition. *Environmental Research Letters* 10(2), doi: 10.1088/1748-9326/10/2/024019

Tilman, D., Cassman, K. G., Matson, P. A., Naylor, R., and Polasky, S. (2002). Agricultural sustainability and intensive production practices. *Nature* 418(6898), 67, 1–7.

United Nations Conference on Trade and Development (2017). *The Role of Science, Technology and Innovation in Ensuring Food Security by 2030.* UNCTAD, Geneva.

Vanlauwe, B., Descheemaeker, K., Giller, K., Huising, J., Merckx, R., Nziguheba, G., Wendt, J., and Zingore S. (2015). Integrated soil fertility management in sub-Saharan Africa: unravelling local adaptation. *Soil* 1(1), 491–508.

Vanlauwe, B., Coyne, D., Gockowski, J., Hauser, S., Huising, J., Masso, C., Nziguheba, G., Schut, M., and Van Asten, P. (2014). Sustainable intensification and the African smallholder farmer. *Current Opinion in Environmental Sustainability* 8(0), 15–22.

Vitousek, P. M., Naylor, R., Crews, T., David, M., Drinkwater, L., Holland, E., Johnes, P., Martinelli, P. A., Matson, G. Nziguheba, D., Ojima, C. A., Palm, G. P., Robertson, P. A., Sanchez, A. R., Townsend, F. S., and Zhang, F. S. (2009). Nutrient imbalances in agricultural development. *Science* 324(5934), 1519–1520.

Xia, L., Lam, S. K., Chen, D., Wang, J., Tang, Q., and Yan, X. (2017). Can knowledge-based N management produce more staple grain with lower greenhouse gas emission and reactive nitrogen pollution? a meta-analysis. *Global Change Biology* 23(5) 1917–1925.

Yao, Y., Zhang, M., Tian, Y., Zhao, M., Zhang, B., Zhao, M., Zeng, K., and Yin B. (2018). Urea deep placement for minimizing NH3 loss in an intensive rice cropping system. *Field Crops Research* 218, 254–266.

19

Adoption of Integrated Crop Management Technology for Poverty Reduction and Food Security

The Case of Smallholder Rice Production in Timor Leste

Maria Fay Rola-Rubzen, Renato Villano, Marcolino Estevão Fernandes E. Brito, J. Brian Hardaker, and John Dixon

1. Introduction

Timor Leste, also known as East Timor, has been combating poverty since it regained independence from Indonesia in 2002. About two-thirds of the population live in rural areas with the majority relying on agriculture for their main source of livelihood (Ministry of Finance 2008). Given the strong reliance on agriculture, the government has focused on agricultural development as a key pillar of economic development (MAF 2012).

Many farmers in Timor Leste are largely subsistence producers. Agricultural production is typically low-input, low-output, with little adoption of new technology. Hence, the government is seeking to develop the agriculture sector in order to increase food production, to provide sufficient food for the population, and to reduce rural poverty. The plan is to transform subsistence farming to more market-oriented production, without degrading the environment (RDTL 2010; MAF 2012).

The two most important food crops in Timor Leste are maize and rice. Maize is the main staple in upland areas (MAF 2009), while rice is mainly grown in the lowlands. Given its importance to the economy, the government is keen to improve rice production. To this end, the Ministry of Agriculture and Fisheries, in collaboration with the German international cooperation agency GTZ, is encouraging rice farmers to adopt various new rice production technologies, one of which is integrated crop management (ICM) (GTZ-RDP 2008).

ICM refers to the integrated use of technologies that are suited to farmers' needs and designed to increase crop yields and farmers' incomes (Kumar and Shivay 2008; Balasubramanian et al. 2005). ICM is a systems approach to sustainable agriculture, meaning account is taken of agro-ecological, socioeconomic, and environmental factors (CABI 2016). Specifically, ICM takes account of soil health, suitable varieties, crop establishment, water management, and pest control, and an enabling policy and institutional environment (FAO 2011). As such, ICM is potentially a tool that can be used to meet the government's goal of reducing poverty, obtaining food security, and protecting the environment. More broadly, it also may contribute to the sustainable development goals (SDGs) of ending poverty (SDG1), ending hunger (SDG2), economic growth, employment and decent work (SDG8), and combatting climate change (SDG13).

ICM has been promoted by the Food and Agriculture Organization of the United Nations (FAO) as a means to identify and deploy technologies for increasing rice productivity in ways suited to smallholder situations (Balasubramanian et al. 2005; FAO 2011). It is designed to promote sustainability. However, adoption of ICM in Timor Leste has been slower than hoped for. Farmers naturally need to be convinced of the advantages before they will adopt a new technology (Leeuwis 2004; Rogers 1995; Van Den Ban and Hawkins 1990; Feder and Slade 1984; Feder et al. 1985). Moreover, the farmers of Timor Leste face constraints to adoption, such as limited labor, deficient rural infrastructure, lack of information, shortage of credit, and lack of access to markets for new inputs or to sell surplus production (MAF 2012). Given these constraints, it is important to understand whether the adoption of ICM indeed leads to higher production and incomes. Can ICM be a tool to meet the national goals, and more broadly, the SDG goals?

The aims of this study is to examine the effects of ICM for East Timorese farmers who have adopted it and determine whether it can contribute to the achievement of SDGs in Timor Leste. Thus, the research questions we seek to answer are:

- Has adoption brought benefits to farmers in terms of production and income?
- Can ICM contribute to the achievement of the SDGs?
- What factors influence ICM adoption?

Ultimately, we seek to find out whether ICM adoption can contribute to meeting the goals for East Timor of poverty reduction (SDG1), food security (SDG2), and inclusive sustainable economic growth (SDG8), while promoting environmental sustainability (SDG13).

2. Consequences of ICM Technology: Literature Review

As part of efforts to increase rice productivity, the Food and Agriculture Organization of the United Nations adopted ICM in 2004 as a strategy to extend and promote rice technologies worldwide (Balasubramanian et al. 2005). According to FAO (2011), ICM is a holistic management system using best management practices based on objective recommendations oriented toward sustainable agriculture and designed to increase crop production in balance with the environment; thus it is well aligned with SDG8 (economic growth, employment, and decent work) and SDG13 (combat climate change). With ICM, farmers can use a combination of organic inputs (i.e., animal manures and composts) with inorganic inputs (i.e., chemical fertilizers, pesticides, and herbicides) (Kumar and Shivay 2008; Balsubramanian et al. 2005). ICM is closely aligned with SDG13 in that it aims to meet agricultural production objectives in conjunction with ecological and economic imperatives, using local resources rather than a heavy reliance on external inputs (Kumar and Shivay 2008; FAO 2011). By using a combination of sustainable principles, ICM technology improves the resilience of farming systems and thus directly aligns with SDG target 13.1, strengthening the resilience and adaptive capacity to climate-related hazards and natural disasters.

ICM has four components—crop management, integrated pest management, integrated nutrient management, and financial management—taking local conditions into consideration (Kumar and Shivay 2008; Abdulrachman et al. 2005). ICM involves the use of best agronomic techniques such as using good quality seeds and low seeding rates, efficient planting methods, efficient irrigation, mechanical weeding, nutrient-based management, and pest management (Abdulrachman et al. 2005). Since its first introduction, it has been used in many countries around the world, including Brazil, the Philippines, Bangladesh, Vietnam, Indonesia, and Australia (Balasubramanian et al. 2005; Abdulrachman et al. 2005; Cruz et al. 2005; Pham et al. 2005).

ICM technology for rice production has been promoted and adopted at varying degrees in a number of countries. Studies of ICM have generally shown that the technology improves crop yields when compared with traditional methods. For example, Saleque et al. (2005) found that adoption of ICM in Bangladesh led to farmers achieving higher yields. There is evidence that ICM helped farmers integrate the best production and postproduction technologies, which resulted in increased productivity, enhanced grain quality, increased market value of rice, improved farmer incomes, and increased rice production (Kueneman 2006; Balasubramanian et al. 2005). In Indonesia, ICM was found to have increased rice production with just a slight increase in the use of inputs (Abdulrachman et al. 2005). In the Philippines and Vietnam, increases in rice yields, along with reductions in inputs, led to improved farmers' profits (Cruz et al. 2005; Pham et al. 2005). An ICM program in India resulted in increased rice yields and reduced production costs, meaning higher profits for ICM farmers than from traditional rice cultivation methods (Balasubramanian et al. 2005). In Indonesia, adoption of ICM technology resulted in lower seeding rates, lower use of inorganic fertilizers, increased efficiency of irrigation water, and higher yields compared to traditional rice cultivation methods (Abdulrachman et al. 2005). Similarly, in Latin American countries, ICM adoption has generally resulted in increased rice production (Kueneman 2006). The positive effects of ICM on yield, productivity, and incomes demonstrate its potential positive contribution to poverty reduction (SDG1), reducing hunger (SDG2), and improved sustainable economic growth (SDG8), particularly if adopted at a large scale. Similarly, the resulting decreased use of inorganic fertilizers and more efficient water use are contributing to SDG8 and SD13. Given the importance of utilizing resources efficiently in response to climate change, the experience in Indonesia and Latin America confirms ICM's potential contribution to climate action (SDG13).

ICM entails the use of an integrated package of technology and farm management practices. The use of the modified rice mat seedling nursery method in an ICM system has been shown to reduce seed and labor costs and to increase production (Rajendran et al. 2004). These authors reported that application of this system can reduce seed used from 50–80 kg/ha using conventional wet nursery systems to 10–12 kg/ha. They found that robust young seedlings, planted at around 14–15 days, contributed to the production of more tillers, and an increased number of rice panicles per plant.

Among studies investigating the effect of the technology on labor use, research in India found that ICM technology in rice cultivation reduced labor for rice production relative to traditional rice cultivation methods (Balasubramanian et al. 2005). A similar study in Indonesia and Vietnam also found that the technology led to lower labor use in rice production (Abdulrachman et al. 2005; Pham et al. 2005). In a review of ICM in Asia, Balasubramanian et al. (2005) concluded that ICM used less labor, particularly in weeding activities, owing to the use of mechanical weeders. Such a reduction in labor use of ICM is particularly beneficial in areas where farm labor supply is limited.

The effects of ICM technology are not, however, always positive. Outcomes are influenced by other aspects such as socioeconomic conditions and resource endowments of farmers. For instance, rice ICM technology implemented in Senegal had mixed results, with a boost in farmers' productivity and profitability for those farmers who had sufficient resources, whereas, for those who had limited resources, adoption of ICM led to lower rice production and profitability (Kebbeh and Miezan 2003). ICM technology requires some initial investments to purchase inputs (e.g., fertilizers, herbicides, mechanical weeders) and very poor farmers may not have sufficient resources to purchase these inputs. As a result, they may not be able to get the increased yields expected. It is important, therefore, to find out whether ICM will work in the context of Timor Leste where many farmers are smallholders and resource poor.

Generally, however, it appears that the benefits of ICM in rice production, when well-applied, are manifested in three main ways: higher yields, improved household income, and sometimes decreased labor use partly resulting from the labor-saving seeding method or from mechanized weeding. With these advantages of ICM technology in rice cultivation in mind, the Timor Leste government, through the Ministry of Agriculture and Fisheries (MAF), has sought to encourage farmers to apply ICM in their rice farming with the aim of attaining the benefits described earlier. ICM technology was introduced by MAF in conjunction with GTZ to two districts in East Timor, with technical assistance on ICM technology provided to farmers. Farmers in these districts were trained in the principles of ICM using a set of best management practices for Timor Leste conditions (Ogoshi et al. 2008). Given the dependence of a large proportion of the population on agriculture, it was expected that wide adoption of ICM would bring short-term benefits and would contribute to longer term improvements in economic, social, and environmental conditions—all of which are directly related to SDG1, SDG2, SDG8, and SDG13. Consequently, it is important to investigate the effects of ICM in Timor Leste and to find out whether it has brought benefits to ICM farmers and, if so, to determine what factors influence adoption in order to promote further spread of the technology.

3. Methods

The research method used is outlined below.

3.1 Data

Data were collected in 2009 from farmers in two districts in Timor Leste (Baucau and Maliana) where rice was the main agricultural product. Enumerators comprised of staff and students from a university in Timor Leste. Prior to the survey, enumerators were trained in the use of the questionnaire to minimize differences in the interpretation of questions. The targeted population was rice farmers in the two districts. A pretested survey questionnaire was administered to a random sample of rice farmers, stratified by farmer type (i.e., ICM and non-ICM farmers). Originally, the plan was to survey equal numbers of ICM and non-ICM farmers. However, following the introduction of ICM in the two districts, many farmers in these districts had already adopted or were using components of ICM. The total number of ICM farmers surveyed was higher than the non-ICM farmers. Also, some questionnaires were excluded in the analysis due to incomplete information. The final sample comprised 223 farmers from Baucau (120 ICM farmers and 103 non-ICM farmers) and 177 farmers from Bobonaro (126 ICM and 51 non-ICM farmers), for a total of 400 farmers.

3.2 Analytical Framework

We used a multistage approach to examine the research questions posed earlier. In the first stage, we examined the factors affecting decisions to adopt ICM using logit analysis. In the second stage, we examined the effect of ICM adoption. Here, we first conducted propensity score matching to find matched samples of adopters and non-ICM adopters, and then evaluated the effects of ICM technology adoption on production and outcome variables using the "treatment effect" concept (Cameron and Trevide 2005).

If we were to follow a commonly preferred way to examine the effects of an intervention, as described in the evaluation literature, we would randomly select farmers before the intervention, randomly assign individual farmers to treatment (ICM farmers) and control (non-ICM farmers) groups, and then compare data for the same farmers before and after adoption, correcting for any changes occurring among nonadopters, to make causal inferences (White and Florence 2014; Duflo et al. 2007). Randomization enables unbiased estimation of treatment effects. In our case, ICM had already been introduced in the area when the study was conducted; hence, we were unable to randomly assign the treatments to farmers. Comparing results of ICM farmers with non-ICM farmers would have been possible but would be subject to bias if the two groups differed in important ways, such as access to inputs. We therefore decided to use a quasi-experimental method using propensity score matching (Rosenbaum and Robin 1983) to draw common samples which we could use to compare ICM farmers and non-ICM farmers.

Propensity score matching is a statistical matching technique used for observational data. Propensity score matching is used to estimate the effect of a treatment by accounting for the covariates that predict receiving the treatment (Rosenbaum and Robin 1983). To implement propensity score matching, a binary-choice model is used to generate a "propensity score" for each farmer in the sample. These scores represent the probability of being an adopter of ICM, considering both adopter and nonadopter farmers, based on a set of covariates (Cameron and Trivedi 2005; Becker and Ichino 2002; Imbens and Wooldridge 2008). The estimated propensity scores are then used, via an appropriate matching algorithm, to match ICM with non-ICM farmers falling within a *common support region*. To form this region, ICM farmers with a propensity score smaller than the minimum or larger than the maximum for the non-ICM group are removed from the sample (Caliendo and Kopeinig 2008). The common support region ensures that farmers with the same characteristics, Z, have nonzero probabilities of being either an adopter or nonadopter of ICM. To ensure that the samples within the common support area have the same distribution of observable characteristics, regardless of whether the farmer has adopted the technology or not, it is necessary to test for the "balancing property" (Becker and Ichino 2002). Once appropriately matched samples are identified, the impact of adoption of ICM is measured as the average treatment effect on the treated or *ATET* (Khandker et al. 2010). The *ATET* is the average impact of the treatment on those individuals who adopted ICM. Thus, the adoption of a technology and its influence on key production variables (e.g., seeds, labor, fertilizer) and outcome variables (e.g., yield, income) can be evaluated.

3.3 Empirical Application

To estimate the propensities for farmers to adopt ICM, for propensity score matching we defined ICM farmers as those farmers who had adopted at least one of the components of ICM. We assumed that the decision to adopt was a binary choice, and that ICM was adopted when farmers judged that the benefits from adopting the technology would be greater than those of not adopting. We used a logit model to estimate the required propensities to adopt ICM by individual farmers.

The definitions, summary statistics and units of the Z variables used in the logit model are presented in Table 19.1. The demographic characteristics (age, education, household size, experience) are those typically included in technology adoption studies. We also included the variables INTAGEN, GROUPME, and MASLEAF, to examine the importance of the mode of acquisition of information and its significance for technology adoption and dissemination.

Table 19.1 Descriptive statistics for variables used in logit and production function

Variable	Description	Both (N = 400) Mean	SD	Adopter (N = 246) Mean	SD	Nonadopter (N = 154) Mean	SD
GENDER (Z_1)	Dummy: 1 if farmer is male	0.75	0.43	0.80	0.40	0.67 [a]	0.47
AGE (Z_2)	Age of farmer (years)	42.45	11.07	43.06	11.15	41.46	10.91
EDUCATION (Z_3)	Years of schooling of household head	4.06	4.70	3.96	4.42	4.21	5.13
HHSIZE (Z_4)	Number of members in the household	6.72	2.38	6.72	2.29	6.71	2.52
EXPERIENCE (Z_5)	Years of experience in rice farming	19.79	9.99	20.04	10.29	19.40	9.51
FARMSIZE (Z_6)	Total landholding (ha)	1.57	0.97	1.63	1.02	1.47	0.88
NRICINC (Z_7)	Income from nonrice farming ($)	89.05	120.59	91.57	122.42	85.02	117.89
OWNDUM (Z_8)	Dummy: 1 if farm is owner-operated	0.95	0.22	0.96	0.21	0.94	0.25
INTAGEN (Z_9)	Dummy: 1 if source of information on ICM is international agencies	0.58	0.49	0.63	0.48	0.49 [a]	0.50
GROUPME (Z_{10})	Dummy: 1 if source of information is through group meeting	0.62	0.49	0.77	0.42	0.38 [a]	0.49
MASLEAF (Z_{11})	Dummy: 1 if source of information is through leaflet	0.11	0.31	0.17	0.37	0.01 [a]	0.11

[a] denotes significantly different from the means of ICM farmers.

Because matching is based on the assumption of conditional independence, we checked that the variables included in the model satisfied the balancing requirement to ensure that all farmers had comparable propensities to adopt the technology. We performed a comparison of means test before and after matching and also used the pseudo-R^2 test for differences in propensity scores between ICM farmers and non-ICM farmers. The pseudo-R^2 is a measure of how well the regressors, Z, explain the adoption probability. After matching, there should be no systematic differences in the distribution of covariates between the groups and, therefore, the pseudo-R^2 should be acceptably low (Wu et al. 2010). The results of the balancing tests are in Table 19.2.

Finally, we examined the results of adoption of ICM on key production inputs of seeds, fertilizer, and chemicals and on output variables. The standard errors of the estimated effects were computed using bootstrapping and are presented in Table 19.3. All estimation procedures were performed using STATA12.

Table 19.2 Balancing tests for the observed variables used in the logit model specification using unmatched and matched sample data

Variable	Data	Mean			% reduction	t-test	
		Treated	Control	%bias	In bias	T	p>t
GENDER	Unmatched	0.797	0.669	29.1		2.89	0.004
	Matched	0.816	0.864	−10.9	62.7	−1.3	0.193
EXPERIENCE	Unmatched	20.037	19.399	6.4		0.62	0.536
	Matched	20.96	21.196	−2.4	63.1	−0.23	0.821
EDUCATION	Unmatched	3.963	4.208	−5.1		−0.51	0.614
	Matched	3.856	3.578	5.8	−13.6	0.61	0.541
AGE	Unmatched	43.061	41.461	14.5		1.41	0.16
	Matched	43.08	43.524	−4	72.2	−0.4	0.692
HHSIZE	Unmatched	6.715	6.714	0		0	0.996
	Matched	6.682	6.587	3.9	−8038.8	0.38	0.701
FARMSIZE	Unmatched	1.627	1.471	16.4		1.57	0.117
	Matched	1.658	1.658	−0.1	99.7	−0.01	0.996
OWNDUM	Unmatched	0.955	0.935	8.9		0.88	0.379
	Matched	0.95	0.939	4.8	45.9	0.48	0.632
NRICEINC	Unmatched	91.573	85.03	5.4		0.53	0.598
	Matched	93.318	85.261	6.7	−23.1	0.69	0.491
INTAGENC	Unmatched	0.626	0.494	26.9		2.62	0.009
	Matched	0.582	0.544	7.7	71.5	0.76	0.446
GROUPME	Unmatched	0.772	0.383	85.5		8.47	0
	Matched	0.721	0.668	11.8	86.2	1.17	0.243
MASLEAF	Unmatched	0.167	0.013	55.7		4.96	0
	Matched	0.015	0.049	−12.3	78	−1.94	0.054

Table 19.3 Impacts of ICM adoption on key production and outcome variables

Variable	Unit	Nearest-Neighbor matching [§]		Kernel Density Matching [§§]	
		ATET	Std.Error	ATET	Std.Error
Production variables					
Seed	kg/ha	−59.78 ***	8.63	−65.86 ***	9.44
Labor	h/ha	5.24	13.80	8.57	15.33
Chemical	US$/ha	4.34 ***	1.40	3.90 *	2.18
Fertilizer	US$/ha	27.01 ***	4.78	28.62***	5.35
Others	US$/ha	29.28 ***	6.15	26.44 ***	6.37
Total cost of production	US$/ha	33.56 ***	6.64	30.28 ***	6.95
Outcome variables					
Output	kg/ha	1877.54 ***	241.25	1829.77 ***	245.91
Total value	US$/ha	563.26 ***	72.37	548.93 ***	73.77
Rice income	US$/ha	350.77 ***	58.07	286.41 ***	58.17

[§] No. of observations (Treated = 211; Control = 142); the balancing property is satisfied and the region of common support is (0.209, 0.990).

[§§] No. of observations (Treated = 234; Control = 140); Bandwidth of 0.10 was used 150 replications to compute the standard errors of ATET.

***P<0.01, * P<0.10.

4. Results and Discussion

The results of the analysis are as follows.

4.1 Determinants of ICM Adoption

The results of the logit model are presented in Table 19.4.

Adoption of technology was significantly higher for larger farms (at 1% significance level). Male farmers (GENDER) were more likely to adopt the technology than females, while age (AGE) was not found to significantly affect farmers' decisions. Counterintuitively, education (EDUCATION) was found to have a negative effect on adoption, but not significantly so. This result suggests that education is not a major factor in the decision of farmers on whether to adopt ICM in this case study. Only 6% of respondents indicated lack of education as a limiting factor.

The fact that male farmers were more likely than females to adopt the technology does not necessarily imply that the technology is male-biased. Instead, the result is likely because agricultural training sessions are mainly targeted toward male farmers and hence, mostly attended by men. Thus to promote gender equality (SDG5), training and extension of ICM should be made available to both men and women,

Table 19.4 Factors affecting decisions of farmers to adopt ICM, Timor Leste

Variable	Logistic Regression [§]		Marginal Effects		Odds Ratio	
	Coefficient	SD	Coefficient	SD	Coefficient	SD
Constant	−2.132 ***	0.892			0.119 ***	0.106
GENDER	0.833 ***	0.287	0.151 ***	0.05	2.300 ***	0.661
EXPERIEN	0.026	0.021	0.005	0.004	1.026	0.021
EDUCATION	−0.01	0.029	−0.002	0.005	0.99	0.029
AGE	−0.019	0.019	−0.0034	0.003	0.981	0.018
HHSIZE	−0.014	0.05	−0.00263	0.009	0.986	0.049
FARMSIZE	0.330 ***	0.136	0.059 ***	0.024	1.390 ***	0.19
OWNDUM	0.281	0.524	0.051	0.095	1.324	0.694
NRICEINC	0.00114	0.001	0.000206	0.0002	1.001	0.001
INTAGENC	0.732 ***	0.246	0.133 ***	0.043	2.079 ***	0.511
GROUPMEE	1.627 ***	0.248	0.296 ***	0.034	5.090 ***	1.26
MASLEAF	2.463 ***	0.762	0.448 ***	0.134	11.746 ***	8.946

[§] Pseudo-R^2 = 0.196, LLF = −213.28; * significant at 1% level.

with appropriate good-quality extension materials made accessible to them both, which will also have a bearing on SDG4 (quality education).

In the literature, the role of access to information was often found to be an important factor that determines technology adoption (e.g., Wood et al. 2014; Matuschke et al. 2007). In our study, about 54% of farmers indicated that lack of information was a major factor that hampered their adoption decisions. This view by farmers is supported by evidence from the logit model of a strongly positive effect of source of information influencing adoption. Specifically, exposure to an international agency (INTAGEN) disseminating information about the benefits of ICM was found to influence adoption. How information was disseminated was also significant; both attendance at group meetings (GROUPME) and the distribution of leaflets (MASLEAF) had positive effects on adoption. The probability of adoption was found to increase by almost 45% if a farmer had received a leaflet or other printed material detailing the features and benefits of ICM.

These results show the critical importance of providing good information about ICM, including its benefits, to encourage adoption. The manner in which the information is disseminated and who is involved in its dissemination are also important. These are elements that have a bearing in the provision of quality education (SDG4) to rice farmers. Furthermore, given that attendance at group meetings was found to be a significant factor in ICM adoption, it is important to include both men and women farmers in the training sessions to promote gender equality (SDG5). Owing to sociocultural norms, women are not normally targeted for agriculture-related training. Thus, efforts are needed to make ICM trainings more gender-inclusive.

There may be a case for specifically targeting women for ICM-related training and putting in place strategies to encourage the attendance of female farmers (e.g., timing trainings when women are not busy with other household duties, setting quotas, having separate women-only training sessions) to increase their participation and thus promote SDG5.

4.2 Selection of the Matched Groups

By employing propensity score matching, PSM, we obtained matched groups of ICM and non-ICM farmers, with a total of 201 ICM farmers and 152 non-ICM farmers, respectively. To check that the observable characteristics of the adopter and nonadopter groups are equal, on average, we performed balancing tests on the variables used in the logit specification. The results of the tests, reported in Table 19.3, show that the means of the observed characteristics for ICM farmers and non-ICM farmers after matching are not significantly different at the 5% level of significance. The results of these tests imply that the matched samples can serve as a good proxy for the missing counterfactuals, which allows us to measure the effects of using ICM.

4.3 Effects of ICM Adoption

We obtained the estimates of the effects of ICM adoption on key inputs and outcome variables. The corresponding results are presented in Table 19.4. There is evidence of a significant reduction in seed usage per hectare with ICM, most likely related to the low seeding rate recommended in the ICM technology package (Balasubramanian et al. 2005; Rajendran et al. 2004). On the other hand, there was a significant increase in the use of other inputs, including chemicals, fertilizers, and machines as a source of power. Following the recommended practice, ICM farmers applied more fertilizer than non-ICM farmers. The ICM farmers applied fertilizer twice, first during transplanting and then again when rice plants reached a certain growth stage.

The increase in seed utilization efficiency is a positive impact, although the higher use of chemicals, fertilizers, and machines among ICM adopters would have a negative effect on their input costs. Overall, there was an average increase of US\$33.56 per hectare in the cost of production for ICM farmers. ICM farmers used somewhat more labor for planting and for fertilizer application. The modest extra labor for planting was required because ICM technology includes a different, more laborious method of rice planting (constant distance between plants). The finding is in contrast with studies in India (Balasubramanian et al. 2005), Indonesia (Abdulrachman et al. 2005), and Vietnam (Pham et al., 2005) where labor use decreased. The modest difference arises principally because farmers in Timor Leste did not use mechanical transplanters, but might also stem from the lower opportunity cost of family labor and the relatively early stage of adaptation of ICM cropping practices to local farm household circumstances.

As shown in the second part of Table 19.4, there is evidence of significant improvement in quantity and value of rice produced per hectare by ICM farmers. A difference of an additional US$561 per hectare was observed in the value of output from rice for those farmers who adopted the ICM technology, amounting to an additional US$351 net household income for rice farmers practicing ICM. This outcome shows the potential of ICM technology in tackling poverty (SDG1). Given that the majority of the poor in Timor Leste are farmers, wide-scale adoption is likely to lead to poverty reduction in the country.

As Table 19.4 shows, our study confirms the significant effects of ICM adoption on farmer's production practices and well-being.

5. Conclusions and Implications

The results of this study indicate an overall positive outcome from the adoption of ICM in East Timor, confirming ICM's potential to contribute to the SDG goals in East Timor. If these results were to be replicated throughout the rice-growing areas of Timor Leste, the consequence would be an increase in the total amount of rice sold to the market as well as higher rice consumption by producing households, which would have a direct contribution to food security (SDG3). Moreover, increased incomes to rice producers from adoption of ICM would contribute to poverty reduction (SDG1) among Timor Leste farm households. Because ICM is an integrated approach that applies sustainability principles, it can contribute to both SDG8 and SDG13, if appropriate supporting policies are in place. Historically, there has been weak policy support for rice farmers, thus adoption of ICM has been slow and productivity remains low in the country as a whole.

Although the study showed that adoption of even some of the components of ICM can lead to improved yield and income, thereby contributing to SDG1, because ICM technology is a "packaged" technology, the adoption of all or the main components may be critical if farmers are to obtain the potential rice yield. Important components in ICM technology are crop management, including use of good rice varieties, and appropriate application of fertilizers and chemicals, along with integrated nutrient and pest management (Balasubramanian et al. 2005). Thus, availability and affordability of the necessary inputs in rural areas in Timor Leste is important in encouraging farmers to take up ICM and for them to reap the associated benefits.

In terms of determinants of adoption, we found that adoption was significantly higher for larger farms, while male farmers were more likely to adopt the technology than females. Often, men are the ones who attend information sessions or receive training in new technology, which could be the reason for why females are less likely to adopt ICM. Therefore, if the government aims to increase the adoption by female farmers (i.e., female-headed farm households), then it needs to target women in the ICM trainings. This is particularly helpful for promoting gender equality (SDG5).

Information was found to be an important factor influencing farmers' decision to adopt ICM technology. This is consistent with the results of the survey where more than half of farmers in the study indicated that lack of information was a major factor hampering their decisions to adopt ICM. To address this constraint, agricultural extension is important in promoting ICM effectively. Good quality information about the various components of ICM and the benefits of ICM will contribute to quality education of farmers, thus also promoting SDG4. Promotion of ICM can be used as part of a strategy to improve incomes and reduce poverty (SDG1), and improve food security (SDG2) in a sustainable manner (SDG8 and SDG13). We also found that the source of information influenced the adoption of ICM. Contact with international agencies appeared to have been important in the dissemination of information about the value of ICM and in influencing farmers' early adoption decisions, thus demonstrating the important role of international aid in development, particularly if the international program is aligned with the national interest to achieve a common purpose. Also, as attendance at group meetings was found to be a significant determinant of adoption, it is important to include both men and women farmers in these meetings to promote gender equity (SDG5). The type of communication was also found to affect adoption. Leaflets were shown to be effective in communicating to smallholder farmers and so can be used to disseminate information in other areas of Timor Leste. Making extension more effective, with the assistance of the international agencies, would enable the government of Timor Leste to expand ICM adoption in the country, thereby contributing to its national goals and to the SDG goals of poverty reduction, eliminating hunger, gender equality, promoting sustainable economic growth, and climate action.

Finally, the finding in this study have broader implications for sustainable agricultural development (SDG1, SDG2, SDG8) in Asia and Africa. Many developing countries are facing similar challenges of stagnant or declining agricultural productivity in the face of climate change. ICM offers a possible strategy to meet this challenge, as long as mechanisms are put in place to deal with potential barriers to adoption. This could include strategies to make the technology more available to women and to farm households with poor resource endowments (e.g., small landholdings). Such strategies could include targeting training, setting gender quotas, women-friendly timing of training, and improved access to credit. In particular, credit availability and access will allow resource-poor famers to purchase the necessary inputs and machinery (e.g., mechanical transplanter). Likewise, providing an enabling environment for the private sector to provide hiring services for machinery which men and women smallholder farmers may not normally be able to afford to purchase will improve their access to machinery, thus encouraging ICM adoption. Through effective policy mechanisms of these types, ICM will be more likely to spread widely throughout Timor Leste and meet its full potential to contribute to the SDGs.

Acknowledgments

We are grateful to the Timor Leste farmers who participated in the study, for sharing their time and knowledge with us. We also thank the staff and students of the Universidade Nacional de Timor Lososa'e who assisted in the survey. Finally, we thank Professor Harry Bloch (School of Economics and Finance, Curtin University) and Professor Ahmed Mushfiq Mobarak (Economic Growth Center and School of Management, Yale University) for their valuable comments on an earlier draft of this manuscript. Any errors are our own and should not tarnish the reputations of these esteemed colleagues.

References

Abdulrachman, S., Las, I., and Yuliardi, I. (2005). Development and dissemination of integrated crop management for productive and efficient rice production in Indonesia. *International Rice Commission Newsletter* 54, 73–82.

Balasubramanian, V., Makarim, A.K., Kartamadtja, S., Zaini, Z., Nguyen, N. H., Tan, P. S., Heong, K. L., and Buresh, R. J. (2002). Integrated resource management in Asian rice farming for enhanced profitability, efficiency and environmental protection. Poster paper presented at the 1st International Rice Congress, Beijing, China, September 15–21, 2002.

Balasubramanian, V., Rajendran, R., Ravi, V., Chellaiah, N., Castro, E., and Chandrasekaran, B. (2005). Integrated crop management for enhancing yield, factor productivity and profitability in Asian rice farms. *International Rice Commission Newsletter* 54, 63–72.

Becerril, J., and Abdulai, A. (2010). Impact of improved maize varieties on poverty in Mexico: a propensity score-matching approach. *World Development* 38(7), 1024–1035.

Becker, S. O., and Ichino, A. (2002). Estimation of average treatment effects based on propensity scores. *The Stata Journal* 2, 358–377.

Bravo-Ureta, B. E., Almeida, A., Solís, D., and Inestroza, A. (2011). The economic impact of Marena's investment on sustainable agricultural systems in Honduras. *Journal of Agricultural Economics* 62, 429–448.

Caliendo, M., and Kopeinig, S. (2008). Some Practical Guidance for the Implementation of Propensity Score Matching. *Journal of Economic Surveys* 22(1), 31–72.

Cameron, A., and Trivedi, P. (2005). *Microeconometrics: Methods and Applications*. Cambridge University Press, Cambridge.

Centre for Agriculture and Bioscience International (CABI) (2016). Integrated Crop Management, http://www.cabi.org/about-cabi/cabi-centres/switzerland/integrated-crop-management, (accessed March 9, 2019).

Centre for Agriculture and Bioscience International (CABI) (2016). Integrated Crop Management, http://www.cabi.org/about-cabi/cabi-centres/switzerland/integrated-crop-management (accessed March 9, 2019).

Cerdán-Infantes P., Maffioli, A., and Ubfal, D. (2008). *The Impact of Agricultural Extension Services: The Case of Grape Production in Argentina*. Office of Evaluation and Oversight (OVE) Working Papers 0508, Inter-American Development Bank, Washington D.C.

Cruz, R. T., Llanto, G. P., Castro, A. P., Barroga, K. E. T., Bordey, F. H., Redoña, E. D., and Sebastian, L. S. (2005). PalayCheck: the Philippines' rice integrated crop management system, *International Rice Commission Newsletter* 54, 83–90.

Davis, K., Nkonya, E., Kato, E., Mekonnen, D. A., Odendo, M., Miiro, R., and Nkuba, J. (2011). Impact of farmer field schools on agricultural productivity and poverty in east Africa. *World Development* 40(2), 402–413.

Dehejia, R. H., and Wahba, S. (2002). Propensity score-matching methods for nonexperimental causal studies. *Review of Economics and Statistics* 84(1): 151–161.

Duflo, E., Glennerster, R., and Kremer, M. (2007). Using randomization in development economics research: a toolkit. In Rodrik, D., and Rosenzweig M. (Eds.), *Handbook of Development Economics* vol. 4, pp. 3895–3962. Amsterdam, North-Holland.

Feder, G., Just, R. E., and Zilberman, D. (1985). Adoption of agricultural innovations in developing countries: a survey. *Economic Development and Cultural Change* 33(2), 255–298.

Feder, G., and Slade, R. (1984). The acquisition of information and the adoption of new technology. *American Journal of Agricultural Economics* 66(3), 312–320.

Food and Agriculture Organization of the United Nations (FAO) (2011). *Save and Grow: A Policy Makers Guide to the Sustainable Intensification of Smallholder Production*. Rome.

Gesellschaft für Technische Zusammenarbeit- Rural Development Program (GTZ-RDP) (2008). *Timor Leste Farming System Development Integrated Crop Management*. Baucau, Timor Leste.

Imbens, G. W., and Wooldridge, J. M. (2009). Recent developments in the econometrics of program evaluation. *Journal of Economic Literature* 47(1), 5–86.

Kebbeh, M., and Miezan, F. H. (2003). Ex-ante evaluation of integrated crop management options for irrigated rice production in the Senegal River Valley. *Field Crops Research* 81, 87–94.

Khandker, S. R., Koolwal, G. B., and Samad, H. A. (2010). *Handbook on Impact Evaluation: Quantitative Methods and Practices*, pp. 53–103. World Bank, Washington D.C.

Kueneman, E. A. (2006). Improved Rice Production in a Changing Environment: From Concept to Practice. Food and Agriculture Organization of the United Nations, Rome.

Kumar, D., and Shivay, Y. S. (2008). *Modern Concepts of Agriculture: Integrated Crop Management*. Indian Agriculture Research Institute, New Delhi.

Lamda, P., Filson, G., and Adekunle, B. (2008). Factors affecting the adoption of best management practices in southern Ontario. *Environmentalist* 29, 64–77.

Leeuwis, C. (2004). *Communication for Rural Innovation: Rethinking Agricultural Extension*. Blackwell, London.

Matuschke, I., Mishra, R., and Qaim, M. (2007). Adoption and impact of hybrid wheat in India. *World Development* 35(8), 1422–1435.

Ministry of Agriculture and Fisheries (MAF) (2009). *Agriculture in Number, Ministry of Agriculture and Fisheries*. MAF, Dili, Timor Leste.

Ministry of Agriculture and Fisheries (MAF) (2010). *Data Record on Potential Area, Planted Area and Production by 2011 First Crop Season*. MAF, Dili, Timor Leste.

Ministry of Agriculture and Fisheries (MAF) (2012). *Strategic Plan 2014-2020*. MAF, Dili, Timor Leste.

Ministry of Finance (2008). *Final Statistical Abstract: Timor-Leste Survey of Living Standards 2007*. Ministry of Finance, Dili, Timor Leste.

Ogoshi, R., Balasubramanian, V., and Jones, M. (2008). *Integrated Crop Management (ICM): Implementing IM Technology in Timor Leste*. United States Agency for International Development (USAID), Dili, Timor Leste.

Pham, T. S., Trinh, K. Q., and Tran, D. V. (2005). Integrated crop management for intensive irrigated rice in the Mekong Delta of Viet Nam. *International Rice Commission Newsletter* 54, 91–96.

Rajendran, R., Balasubramanian, V., Ravi, V., Valliappan, K., Jayaraj, T., and Ramanathan, S. (2004). Nursery technology for early production of robust rice seedings to transplant under integrated crop management. In: *Proceedings of the 5th International Rice Genetics Symposium*, pp. 73–75.

República Democrática de Timor Leste (RDTL) (2010). *Timor Leste Strategic Development Plan 2011–2030*. RDTL, Dili, Timor Leste.

Rivera, W. M., and Qamar, M. K. (2003). *Agricultural Extension Rural Development and Food Security Challenge*. Food and Agriculture of the United Nations, Rome.

Rogers, E. M. (1995). *Diffusion of Innovations*. Free Press, New York.

Rosenbaum, P. R., and Rubin, D. B. (1983). The central role of the propensity score in observational studies for causal effects. *Biometrica* 701, 41–55.

Saleque, M. A., Abedin, M. J. Bhuiyan, N. I. Zaman, S. K., and Panaullah, G. M. (2005). Long term effects of inorganic and organic fertiliser sources on yield and nutrient accumulation of lowland rice. *Field Crops Research* 86, 53–65.

Thavaneswaran, A., and Lix, L. (2008). Propensity Score Matching in Observational Studies, Manitoba Centre for Health Policy. Monograph, University of Manitoba, Winnipeg.

Van Den Ban, A. W. and Hawkins, H. S. (1990). *Agricultural Extension*. Longman, London.

White, H., and Sabarwal, S. (2014). *Quasi-experimental Design and Methods, Methodological Briefs: Impact Evaluation* 8, UNICEF, Florence.

Wood, S. A., Jina, A. S., Jain, M., Kristjanson, P., and DeFries, R. S. (2014). Smallholder farmer cropping decisions related to climate variability across multiple regions. *Global Environmental Change* 25, 163–172.

World Bank. (2002). *East Timor Policy Challenges for a New Nation*. World Bank, Washington, DC.

Wu, H., Ding, S., Pandey, S. T., and Tao, D. (2010). Assessing the impact of agricultural technology adoption on farmers' well-being using propensity score matching analysis in rural China. *Asian Economic Journal* 24, 141–160.

20

Two Decades of GMOs

How Modern Agricultural Biotechnology Can Help Meet
Sustainable Development Goals

Ademola A. Adenle, Hans De Steur, Kathleen Hefferon,
and Justus Wesseler

1. Introduction

Genetically modified (GM) organisms (GMOs) have been available commercially for
over twenty years. According to the International Service for the Acquisition of Agri-
biotech Applications (ISAAA), in 2017, the global area of crops bred using biotech
increased by 3% or 4.7 million hectares, with over half of these increases coming from
smallholder farmers in developing countries (James 2018). Early GM crops increased
yields and saved costs by bestowing resistance to biotic stresses. With the advent of
climate change, second-generation biotech crops focused more on abiotic stresses
such as tolerance to heat, drought, and flooding. Currently, the world is entering a
stage of third-generation GM crops that provide traits that are highly valued by consumers and the food and manufacturing industries. Examples include crops that are
biofortified, nutritionally-enhanced, or able to reduce food waste (James 2017).

The millennium development goals (MDGs) were originally set up by the United
Nations between 2000 to 2015 to fight poverty by establishing a series of eight measurable objectives that helped to promote political accountability, improved metrics,
and global awareness of the world's poor (Sachs 2012). During this time period, great
strides were made in the fight against poverty; much of this advance was a result of a
boost in economic growth in China. However, although some countries have managed to achieve the majority of the MDGs, others have not. More recently, the advent
of climate change, rising poverty, and other pressing global challenges have necessi-
tated a revision of these goals in order to incorporate overriding sustainability objectives that can be addressed alongside poverty-reduction objectives. Rising incomes,
along with predicted population growth, will increase food demand particularly with
respect to meat consumption, and in turn are expected to exert more pressure on
ecosystems. According to FAO, the number of chronically hungry currently exceeds 1
billion, and the gap between wealthy and poor continues to grow. The sustainable development goals (SDGs) were thus developed in response to these changes in parameters (Fullman et al. 2017). The SDGs and their corresponding indicators represent

a broader range of sustainable development challenges than the MDGs; this enables more appropriate decision-making and implementation to take place. Given these challenges, new approaches, including technological innovation, are required to implement SDGs. In this regard, GMOs have great potential to help meet part of the SDGs, especially in the developing world.

In view of the benefits of GMOs, we conducted an extensive review of literature and focus on its advantages and the challenges that are associated with the application of GMOs over the past two decades. We used a variety of sources and databases including Science Direct, Google Scholar, Web of Science, and PubMed as well as multiple keywords to identify GMO studies. We focus on empirical published studies and recognized annual reports on global adoption of GMOs. Finally, additional studies that were not captured in other searches were located through references identified from the studies.

The following chapter discusses the development of GM crops which have the ability to address several of the SDGs (Table 20.1), while acknowledging that modern biotechnology is also important for animal breeding and using microorganisms. We also acknowledge that it is not the technology itself that is of importance, but rather the institutional environment that allows making use of these technologies (Wesseler et al. 2017). The chapter begins by describing the role of GMOs in achieving environmental goals laid out in the SDGs, including lowering greenhouse gas emissions, conserving biodiversity, and reducing food waste. The chapter then describes potential direct and indirect health advantages of GMOs. Finally, various socioeconomic dimensions and future challenges with respect to the deployment of GMOs, and possible solutions to address the challenges, within the context of implementing SDGs, are described.

2. GMOs and the Environmental Dimension

The correlation between the application of GMOs and achievement of the SDGs will be examined among GM producing countries with focus on SDGs 7, 12, 13, and 15.

2.1 Lowering Greenhouse Gases Emissions

The agricultural sector remains the third largest producer of global GHG emissions. Technological innovation such as GM technology could help reducing GHG emissions particularly with regard to the implementation of SDG13. GM crops have been engineered to lower the application of chemical pesticides, thereby lowering GHG emissions.

The analysis of Brookes and Barfoot showed on a global scale, the agricultural environmental footprint was reduced by 613 million kg of pesticide active ingredient among GM producing countries between 1996 and 2015 including the United States, China, India, South Africa, Brazil, and Argentina (Brookes and Barfoot 2017a). Also,

Table 20.1 Overview of genetically modified crop traits that address SDGs

Sustainable development goals (SDGs)	Description	GM crop example	Reference (examples)
SDG1	End poverty in all its forms everywhere	*Bt* cotton enabled reduced pesticide use, yield increase	Graham Brookes and Barfoot (2017b), Qaim and Zilberman (2003)
SDG2	End hunger, achieve food security and improved nutrition and promote sustainable agriculture	Golden rice, anthocyanin tomatoes, insect-resistant cow peas and corn, bacterial-wilt-resistant banana	Butelli et al. (2008), De Steur et al. (2015), Wesseler et al. (2017)
SDG3	Ensure healthy lives and promote well being for all at all ages	Golden rice Folate biofortified rice	De Steur et al. (2015), Wesseler and Zilberman (2014)
SDG5	Achieve gender equality and empower all women and girls	Herbicide-tolerant plants, *Bt* cotton reduces laborious tasks in the field	Gandhi and Namboodiri (2006)
SDG7	Ensure access to affordable, reliable, sustainable, and modern energy for all	GM sugarcane for biofuel, GM switchgrass	Janda et al. (2012), Shen et al. (2013)
SDG8	Promote sustained, inclusive, and sustainable economic growth, full and productive employment and decent work for all	Maintenance of plant material, GM crops that assist with sustainable growth	Smart et al. (2017), Wesseler et al. (2017)
SDG12	Ensure sustainable production and consumption patterns	Reduction of waste via prolonged shelf life, nonbrowning apples	Bovay and Alston (2018), Ricroch and Henard-Damave (2016)
SDG13	Take urgent action to combat climate change and its impacts	Herbicide-tolerant soybean reduces tillage and greenhouse-gas emission, GE switchgrass for biomass	Shen et al. (2013), Graham Brookes and Barfoot (2015), Barrows et al. (2014), Smyth et al. (2011)
SDG15	Protect, restore, and maintain sustainable use of terrestrial ecosystems, sustainably manage forests, combat desertification, and halt and reverse land degradation and halt biodiversity loss	*Bt* cotton and retention of nontarget insect populations GM maize enhances conservation tillage practices	Marvier et al. (2007), Gouse (2012), Barrows et al. (2014), Carpenter (2011)

Continued

Table 20.1 *Continued*

Sustainable development goals (SDGs)	Description	GM crop example	Reference (examples)
SDG16	Promote peaceful and inclusive societies for sustainable development, provide access to justice for all, and build effective, accountable, and inclusive institutions at all levels	GM can respond to food shortages and contribute to peace GMO governance inclusivity	James (2009), Kettenburg et al. (2018), Wesseler and Zilberman (2014), Wesseler et al. (2017).

these authors estimate that 174 million hectares of land were saved through the adoption of GM crops. Recently, the meta-analysis of 147 studies showed that 37% in reduction of chemical pesticide use was a result of adoption of GM technology, especially in developing countries (Klümper and Qaim 2014).

The use of genetically engineered herbicide-tolerant and insecticide-resistant crops is known to have facilitated the adoption of conservation tillage practices among farmers. For example, according to Brookes and Barfoot (2015), in 2013, the adoption of GM crops such as soybean that encourages conservation tillage practices represents 80% and 89% in terms of area planted in US and Argentina, respectively. According to the FAO (2008), a conservation tillage system offers better recycling of soil nutrients during the first decade of adoption, resulting in a decrease of 1.8 tonnes of CO_2 released per hectare per year. Hence, reduction in GHG emissions is more likely to be encouraged when GM technology is combined with the conservation tillage practices, thereby contributing directly to SDG13.

While first-generation transgenic crops or GMOs have focused on pest resistance and herbicide tolerance, the second generation have been tailored to improved agronomic traits, including nitrogen use efficiency (Han et al. 2015). Nitrogen use efficiency in plants encompasses multiple genes, and includes processes ranging from carbon/nitrogen metabolism to nitrogen uptake, remobilization, and storage (Brauer and Shelp 2010). Concerted efforts in this field will lead to crops that thrive on smaller agricultural inputs, thus contributing to SDGs 13 and 15.

There has also been a great deal of interest in the application of GM technology to improve biofuel production. For example, in the United States, the yield of cellulosic ethanol increased by more than two fold through genetically engineered switchgrass (*Panicum virgatum*) (Shen et al. 2013). Transgenic switchgrass with reduced lignin content and improved ethanol yield have also been developed (Johnson et al. 2017). Moreover, a team of researchers (Brennan and Owende 2010; Singh et al. 2011) have reported that genetic engineering can help develop new types of microalgae strains for biodiesel production by increasing the quality of lipids generated, which in turn can be extracted for energy purposes. According to Brennan and Owende (2010), biofuel production from microalgae has great potential for carbon capture under different conditions. The algae biofuel production research related program, particularly through genetic engineering, continues to emerge and is expected to contribute to a positive environmental impact.

2.2 Conserving Biodiversity

In order to cope with increased demands for agricultural products from growing populations, there will be more competition for land and water, which in turn will contribute to deforestation, depletion of soil, and the loss of biodiversity. The response to this challenge can be partly addressed by the use of GM technology (Barrows et al. 2014), thereby helping to achieve SDG15.

The implication of introducing GM technology for biodiversity can be complex and has generated a debate about the impact of GM technology on biodiversity. A large number of studies have been conducted on the introduction of GM technology for biodiversity. The conservation tillage system forms part of agriculture management practices as described in section 2.1, and positive impact of GM technology on biodiversity has been reported. A study by (Carpenter 2010) analyzed 49-peer reviewed publications on the adoption of GM crops from 12 countries by comparing the yields between adopters and nonadopters. The author showed that out of 168 results (based on yield outcome), 124 results were positive for GM adopters (most of these farmers came from developing countries) when compared to nonadopters. This result is consistent with the ISAAA report that 18 million poor smallholders have benefited from the adoption of GM crops in the past two decades (James 2016). Taken together, these authors argue that fewer lands have been converted to agricultural production through the adoption of GM crops while the yields increased, thereby conserving biodiversity in land and forest areas.

The beneficial role of insects and other arthropods remains important in maintaining soil fertility (Culliney 2013). The potential positive and negative impacts of GM technology on nontarget invertebrates have been discussed multiple times in the literature. In this regard, several studies have revealed the impact of GM technology (*Bacillus thuringiensis-Bt*) on soil biodiversity. Using meta-analysis studies, the results of 42 field experiments of *Bt* maize and *Bt* cotton versus non-*Bt* crops were analyzed based on the abundance of nontarget invertebrates (Marvier et al. 2007). Their results showed that there was greater abundance of nontarget invertebrates in *Bt* crops compared to non-*Bt* crops. In a similar study, another author analyzed the effect of *Bt* on soil biodiversity over a period of 15 years and concluded that little or no effects of *Bt* on soil biodiversity have been recorded, although some differences reported in soil biodiversity are known to be due to differences in temperature, soil type, and crop varieties (Carpenter 2011). Overall, the adoption of GM crops has created more environmentally friendly farming practices, particularly through the enhancement of soil biodiversity.

2.3 GMOs in the Context of SDGs7, -12, -13, and -15

In view of the explanation in the previous section, there is causal relationship between GM technology and some aspects of SDG7, SDG13, and SDG15; therefore, promotion of GM technology can help meet SDGs around the world, especially in developing countries.

One of the key targets for SDG7 is to increase the share of renewable energy mix by the year 2030. Brazil is arguably the only country in the world where ethanol remains an important component of the energy mix since the 1930s, when a 5% ethanol blend with gasoline was introduced (Leite et al. 2009), and now stands at 18–25% for gasoline (Janda et al. 2012). In 2003, the Brazilian government introduced the Flex Fuel Vehicle to cut down on gasoline use in order to reduce GHG emissions. Because of the mandatory blend, 50% of gasoline has been replaced by ethanol for transport (Janda et al. 2012). The energy mix policy is largely driven by huge investment in agricultural research and development (R&D) programs. The main Brazilian agricultural institution EMBRAPA is leading the research effort in developing various GM products including drought resistant sugar-cane for biofuel production. Brazil, being one of the largest producers of GM crops, can be argued to be in an excellent position to achieve the SDG7 goal. Other advanced developing countries such as Argentina and South Africa also have extensive biotechnology R&D programs, which provides opportunities for improving also their renewable energy programs in order to meet SDG7 targets. More importantly, other developing countries will have to follow in Brazil's footsteps in their efforts to achieve SDG7 targets.

The ability to invest in biotechnology R&D, especially among GM producing countries, has been partly attributed to adoption of GM crops (Zilberman et al. 2016). The development of robust strategies for new GM crop traits can help achieve both SDG13 and SDG15 in these countries. Brookes and Barfoot (2017) argue that the reduction of GHG emissions in 2015 alone is equivalent to removing 12 million cars off the road, suggesting that promotion of GM technology in agricultural practices, particularly in developing countries, can help achieve SDG13. According to James (2016) and Brookes and Barfoot (2015), inter alia, the application of GM technology has contributed considerably to a reduction of GHGs into the atmosphere, due to limited spraying of pesticides and herbicides. While industrialized countries have largely benefitted from adoption of GM crops, it is important to emphasize the potential role of other GM crops that are under R&D programs or confined field trials and how these GM crops can help contribute to meeting the SDGs in developing countries. GM technologies such as drought and salinity tolerance traits can play important roles toward implementing SDGs especially in meeting the target of SDG13 of strengthening resilience and adaptive capacity related to the impact of climate change. Drought-tolerant GM crops can help protect high biodiversity areas and reduce soil contamination, thereby resulting in land savings and maintaining soil fertility. Moreover, GM drought tolerant crops could lead to a significant reduction in the use of water with huge environmental benefits, particularly in drought prone regions of Africa. This could potentially meet the target of SDG15 by taking significant action to reduce natural habitat degradation, drought, and desertification. The availability of new GM crop varieties (for example, GM maize, Wesseler et al., 2017) can position many African countries to achieve SDG15 and other relevant SDGs. Taken together, the combined benefits of GM crops can reduce the impact of climate change, improve the health of farmers and put their economy on a sustainable path of growth.

More recently, GM technology has also been used to combat food waste along the chain. According to FAO, up to one third of all food is lost before it is consumed (FAO 2018). Through eliminating cosmetic issues associated with crops, GM technology can contribute to SDG12 and its targets to halve global food waste at retail and consumer levels. For example, several varieties of late-blight disease resistant potato have recently been approved for commercialization in several countries. These transgenic potatoes will not only increase yield and reduce the use of fungicides as much as 45%, but will also reduce food wastage through its nonbrowning trait (James 2017). Similarly, several apple varieties were genetically modified to be resistant to browning and, oftentimes, bruising, two attributes causing retailer and consumer reluctance (Thompson and Kidwell 1998; Waltz 2015a). This new technology is much anticipated, as approximately 31% or 133 billion pounds of food is wasted annually (GMO Answers 2016).

3. GMOs and Health Dimension

Whereas the majority of commercialized GMOs are targeted at increased productivity in order to help ensure food security and end hunger, GM technology also has the potential to reduce the global burden of malnutrition and hidden hunger. Health-oriented GMOs, like vitamin or mineral enriched GM foods (GM biofortified foods), are considered to be part of the third-generation of GMOs, by which output/quality traits are introduced or improved (James 2017). With direct, tangible benefits for the end-user (nutritious, anti-allergic, improved ripening/extended shelf life, and improved taste/appearance) (Parisi et al. 2016), this generation complements farmer-oriented generations, which typically focused on the improvement (first) and stacking of input traits (second-generation). Nutritionally-enhanced GMOs can be considered to be potentially promising means to contribute to the achievement of SDG2. Furthermore, GMOs with agronomic traits can also indirectly improve farmers' health through lowering exposure to insecticides and herbicides (e.g., (Pray et al. 2001), which aligns with SDG3 on health.

3.1 Direct Effects on Public Health

Biofortification is the process of improving the nutritional quality of (staple) food crops through conventional plant breeding (conventional biofortification), fertilizer optimization or soil improvement (agronomic biofortification), or use of biotechnology (GM biofortification) (Adenle et al. 2012; Carvalho and Vasconcelos 2013). Non-GM biofortified crops are widely developed and commercialized (Bouis and Saltzman 2017), but they are inadequate for crops with a low or absent level of a certain micronutrient (Beyer 2010). This is where GM technology comes into play. A recent review has summarized successful R&D efforts in the field of GMOs with increased micronutrient content in staple crops (De Steur et al. 2015). They have identified 35 studies that reported a substantial fold change in one of the following micronutrients: pro-vitamin A (rice, potato, corn, wheat, cassava, sorghum), vitamin B9/folate (rice, corn), vitamin C (corn, potato), iron (rice, corn, barley), zinc (rice, barley), and copper (rice).

Two well-known, advanced examples of GM biofortification are Golden Rice, enriched with pro-vitamin A (β-carotene) (Paine et al. 2005), and vitamin B9 (folate) enhanced rice (Storozhenko et al. 2007). Here, conventional breeding techniques were not applicable given the absence and low content of these micronutrients in rice. For Golden Rice, a combination of transgenes from daffodil and *Pantoea* were inserted to introduce pro-vitamin A levels in the rice endosperm (Paine et al. 2005). Initially developed in 2000, Golden Rice was further improved to contain up to 23-fold increase in carotenoid content as compared to the original, known as GR2 (Paine et al. 2005). Folate biofortified rice was developed through overexpression of transgenes from *Arabidopsis* in rice endosperm. Through transgenic breeding, a four-fold increase of the baseline folate concentrations in rice could be reached (Storozhenko et al. 2007), while folate stability for long term storage could be improved (Blancquaert et al. 2015). Besides, there have been considerable advancements in the use of biotechnology for micronutrient enhancement, as illustrated in reviews citing pro-vitamin A, B1 (thiamine), B6, B9 (folic acid), vitamin C, (ascorbate), vitamin E, iron, and iodine (Giuliano 2017; Lee 2017; Van Der Straeten and Strobbe 2017; Mène-Saffrané and Pellaud 2017).

Furthermore, there has been a recent interest and shift toward research on multibiofortification, that is, the simultaneous enhancement of the level of multiple micronutrients (De Steur et al. 2015). While several researchers targeted rice for the enhancement iron, zinc, and copper (Wu et al. 2018), few researchers successfully accumulated micronutrient levels in corn (pro-vitamin A, folate, vitamin C) and barley (iron, zinc) (Ramesh et al. 2004).

Whereas most applications of GM biofortification aim to elevate vitamin and/or mineral concentrations in staples (such as zinc, iron, and pro-vitamin A), GM biofortification could also aim to elevate macronutrient levels, such as amino acids and fatty acids, with the aim to correct or prevent deficiency (Codex Alimentarius Commission 2015; Zhao and Shewry 2011). Two applications that were approved for commercialization are high-oleic acid soybeans (low in trans-fats; only small-scale commercialization) and high-lauric acid canola (currently not commercialized due to inferior agronomic traits) (Kramer et al. 2001; National Academies of Sciences and Medicine 2016). Examples of ongoing research targeting the enhancement of other macronutrients are omega-3 enhanced oilseed crops (Venegas-Calerón et al. 2010) or protein-rich potato ("protato") (Chakraborty et al. 2010; Changat and Krishna 2006), which can be linked to health improvement, respectively.

While applications of GM technology for health often target the optimization and enhancement of desirable traits, they can also offer health benefits through down regulating undesirable traits. For example, foods have been genetically engineered to reduce anti-nutrients, compounds in foods that impair the nutrient absorption (Katoch and Thakur 2013). Phytic acid, an antinutrient that interferes with the absorption of key minerals like iron and zinc, have been reduced in cereals, like maize, rice, and soybean (Shahzad et al. 2014). GMOs can also provide health benefits through reduction of toxins. In the United States, for example, the recently approved Innate™ potato varieties not only have bacterial blight resistance and nonbrowning traits, but also lower toxin concentrations of acrylamide (Waltz 2015b). Ongoing research have attempted to lower levels of toxic cyanogenic glycosides in cassava (Piero et al. 2015) or aflatoxins in crops (Ostrý et al. 2015).

When it comes to the impact of the aforementioned R&D efforts, researchers have assessed the nutrition/health effects of GM biofortification, either *ex ante* or *ex post*. Regarding the latter, only one clinical trial is available. A randomized trial in the United States resulted in a high bioconversion factor of b-carotene in Golden Rice (3.8:1), by which 100 g of uncooked Golden Rice would provide about 80–100% of the estimated average requirement (EAR) for adult men and women (Tang et al. 2009). Findings from 15 simulation analyses, as reported in economic evaluation studies and R&D studies, confirmed the promising effects of GM biofortified crop consumption on dietary intake and nutritional outcomes in humans (De Steur, Mehta et al. 2017). In nearly all studies, a regular portion of the targeted biofortified crop would provide the daily micronutrient requirements. Golden Rice, for instance, could reduce the prevalence of dietary vitamin A inadequacy by up to 30% (children) and 55–60% (women) in Indonesia and the Philippines, and up to 71% (children) and 78% (women) in Bangladesh, thereby contribute to SDG3 targets (De Moura et al. 2016).

3.2 Indirect Effects on Farmers' Health

GMOs with improved agronomic traits could also generate health effects, though indirectly, through its impacts on the environment and insecticide and herbicide use. A review of Klümper and Qaim (2014) has demonstrated a substantial lower use of chemical pesticides for GMOs, that is, an average reduction of pesticide quantity of 37%. This study argues that considerable reductions are mainly reported for insect-resistant (*Bt*) crops, as herbicide-tolerant crops may both increase and decrease different types of herbicide use. Furthermore, the decline in insecticide treatments is often associated with a lower number of farmers reporting poisonings. This has been illustrated in studies with *Bt* cotton farmers in China (Huang et al. 2002), India (Kouser and Qaim 2011), Pakistan (Kouser and Qaim 2013), and South Africa

(Bennett et al. 2006). Besides, the adoption of *Bt* cotton has been associated with adoption of eco-friendly rather than chemical pesticides (Veettil et al. 2017).

The effects of herbicides that are used in association with herbicide-tolerant plants have also been examined in detail. By far the most well-known herbicide that is used in conjunction with a GM crop is glyphosate. Since 1996, the planting of glyphosate tolerant crops, in combination with the herbicide, greatly simplified weed management (Duke 2015). However, controversial results regarding the link between glyphosate and cancer, as well as its impact on soil, water, and aquatic species have been published. This series of contradictory statements by regulatory bodies have caused much public confusion regarding the safety of glyphosate (Landrigan and Belpoggi 2018).

3.3   GMOs in the Context of SDGs 2 and 3

While GM biofortified crops, once approved, could help meet nutrition targets of SDG2 especially in the developing world, GMOs with agronomic traits could offer indirect health effects and thereby contribute to SDG3.

Despite great progress in the reduction of micronutrient malnutrition and the achievement of the nutrition targets of SDG2 (Fullman et al. 2017), about 2 billion people are still considered to suffer from one or more micronutrient deficiencies, calling for complementary strategies to existing, effective micronutrient interventions, like vitamin A or iron supplementation (Wesseler et al. 2017). As a consequence, considerable progress has been made in conventional biofortification since the turn of the century, with many micronutrient enriched crops being implemented in over 50 priority countries (HarvestPlus 2015). There is considerable and ever-growing evidence showing that such crops are a highly cost-effective strategy to improve micronutrient status among target populations (Haas et al. 2013).

Similar conclusions can be drawn for the future implementation of GM biofortified crops, illustrating its potential to contribute to the nutrition targets of SDG2. In view of the various successful applications of biotechnology for biofortification, and their potential to substantially increase micronutrient intake levels, *ex ante* assessments have demonstrated that GM biofortified crops like Golden Rice (De Moura et al. 2016; Stein et al. 2006) would be highly cost-effective investments to reduce target micronutrient deficiencies, like vitamin A, iron, and zinc (De Steur, Wesana et al. 2017b). This is particularly the case for policy interventions targeting multibiofortified crop varieties, as they would generate economies of scale and, hence, larger benefits at a relatively low cost (De Steur et al. 2012; Fiedler et al. 2013).

Unlike GM crops with agronomic traits, GM foods with health traits are only starting to be introduced in the global pipeline of GMOs (Parisi et al. 2016), of which only applications are already approved for cultivation, that is, low-acrylamide GM potatoes (Emily Waltz 2015b) and GM oilseed crops (National Academies of Sciences and Medicine 2016). Within the context of SDG2, however, GMOs that address highly prevalent vitamin and mineral deficiencies are not (yet) approved. Main reason is the current regulatory climate and anti-GMO lobbying efforts (Moghissi et al. 2016; Ingo Potrykus 2017). Nevertheless, proof of concept has been realized for various nutritionally enhanced GMOs (De Steur et al. 2015; Van Der Straeten et al. 2017), which triggered an increase in the number of nutritional traits in the global GM crops pipeline over the last two decades and is expected to further increase in the near future (Parisi et al. 2016). Another reason behind the growing interest in health traits relates to the consumer. A meta-analysis of studies on consumers' willingness-to-pay revealed general positive reactions toward GM biofortified crops. This is in line with conventional biofortified crops, indicating that consumers' opinion on nutritious crops are hardly affected by the applied technology (De Steur, Wesana et al. 2017a). Worldwide, they are willing to pay on average 24% more for GM biofortified food as compared to regular food (De Steur, Wesana et al. 2017b). Price premiums for a health policy intervention like this should be only interpreted as positive preferences. Consumers' optimism, however, can be significantly influenced by information provision, in both ways. While positive information on the nutritional content/benefits or technology increased consumers' intention to purchase, the opposite was true for negative information on GM technology. This lends support for the significant effect of lobbying which polarized the public opinion, regardless of the scientific basis of given arguments (Wesseler and Zilberman 2014).

As such, GM biofortified crops are still marked by contradiction. On the one hand, evidence underpins their impacts on nutritional crop content and potential effects on nutrient intake, public health, and social welfare. On the other hand, these well accepted GMOs are prevented from meeting the needs for contributing to SDG2 and saving threatened lives of malnourished populations in developing regions.

Given that the indirect health effects of GMOs on farmers' health are generated by GMOs with improved agronomic/input traits that are widely cultivated in the world, like pest resistant GMOs (James 2017), a large body of ex-post evidence is available. While the design of the impact studies is sometimes criticized (Kathage and Qaim 2012), the role of insect-resistant (*Bt*) crops on reducing the use of chemical pesticides is well established (Klümper and Qaim 2014). Although its negative association with lower poisonings needs more underpinning, evidence on farmers' health impacts, particularly in developing countries, is growing (National Academies of Sciences and Medicine 2016). As such, GMOs with input traits could also help to meet SDG3 (ensure healthy lives and promote well-being for all at all ages) and its particular target 3.9 that deals with disease reduction from hazardous chemicals.

4. GMOs and the Socioeconomic Dimension

The adoption of GMOs has improved the well-being of many farm households in developing countries. The health benefits of reduced pesticide use can be substantial (Antle and Pingali 1995). This not only benefits farm households, in particular children and women, but also provides positive benefits for farm laborers (Beckmann and Wesseler 2004). Pesticides saving technologies are not only appreciated for directly reducing time and costs but also for saving the burden of labor. The adoption of broad spectrum herbicides in Africa can be explained by reducing the burden of manual weeding mainly done by children and women (Haggblade et al. 2017). This contributes to empowerment of woman contributing to SDG5: "achieve gender equality and empower all women and girls." In the United States the wide and rapid adoption of herbicide resistant crops such as soybeans has been explained among other factors by reducing the burden of "thinking" how to control weeds. Herbicide resistant weeds have simplified weed control (Marra et al. 2002). The implication for weed control in developed as well as developing countries and the nutritional and food safety contributions support SDG3: "ensure healthy lives and promote well-being for all at all ages."

One of the major advantages of the GMO technology is that poorer households in developing countries in general benefit from the technology. In regions where the use of pesticides is low but pest pressure high, which is more often observed among poorer households, substantial yield increase has been observed benefitting in particularly poor households (Klümper and Qaim 2014).

While cotton is a nonfood crop, using GMOs in food crops has even wider socioeconomic implications for poor households. Corns resistant to lepidopteran pests such as corn borers has lower levels of mycotoxins. This is in particular important for countries where the postharvest storing facilities are not well developed. Insect resistant corn varieties show substantially lower levels of mycotoxins even after longer postharvest periods, generating health benefits for subsistence farm households (Gressel and Polturak 2018). Delaying the introduction of GM crops harms several societal groups. Not all groups assess the introduction of GM crops positively. An urban–rural divide has been observed (Paarlberg 2008). Wealthy and well-educated urban consumers show on average a lower willingness-to pay for GM food than poor rural and less formally educated consumers (Kikulwe et al. 2011).

In developed and developing countries many societal groups have expressed concerns about GMOs (Tosun and Scaub 2017: Paarlberg 2008: Wesseler et al. 2017). They are related to the potential negative effects on the environment and human health and ethical issues, but also the implications for market structure of the seed industry and related intellectual property rights. The environmental and human health effects are related to unknown risks. This has been rejected as a logically inconsistent line of reasoning as the same can be said about plants produced using "conventional" breeding technologies (Wesseler 2014). Furthermore, the properties of transgenic plants are often better known than those of conventional ones. Ethical issues mainly refer to the "unnaturalness" of transgenic plants. Deciding about what is to be considered "natural" or "unnatural" is a difficult issue. The majority of EU citizens prefer a governance system where experts decide based on scientific evidence, while about a third prefers a governance system based on public views considering moral and ethical issues (European Commission 2010). Taking different societal views into consideration and providing opportunities for those interested to get involved in the policy processes contributes to peaceful and inclusive societies for sustainable development, provide access to justice for all and build effective, accountable, and inclusive institutions at all levels (SDG16).

The concerns about intellectual property rights (IPRs) relates to the argument that large international seed companies own the rights over transgenic plants and can charge farmers a higher price and restrict their ability to save seeds, thus undermining farmers' rights. IPRs and plant breeders' rights apply not only to transgenic crops but to other seeds as well. Industry concentration has increased over the past two decades. An analysis of the recent mergers in the seed sector shows an increase in concentration but at levels below measures of concentration that regulators in the European Union or the United States use for intervention. Nevertheless, the recent merger between Bayer and Monsanto required a sale of some of the seed production to maintain a sufficient level of competition (Bonanno et al. 2017). Several authors have related the concentration in the seed industry with the regulation companies face and in particular the more stringent regulation for the approval of transgenic crops (Smart et al. 2017). The Nagoya Protocol has been implemented to protect farmer rights and to ensure benefit sharing of plant genetic resources maintained by farmers mainly in the Global South. The seed industry has complained about the additional costs and negative implications for developing improved plants targeting in particular agriculture production in developing countries. Addressing this problem at an international level will be important for the achievement of several SDGs (SDG8, SDG12, SDG16).

Several countries such as the European Union and Japan have introduced mandatory labeling policies for food derived from GMOs (Adenle 2017; Carter and Gruere 2003). As a result, there are hardly any GMO labeled food products to be found in those countries (Venus et al. 2018). In Europe and the United States, a voluntary market for "GMO-free" labeled food products has emerged, driven by private sector initiatives. Several retailers in the European Union have a GMO-free policy for their own brands (Wesseler 2014). This is partly a response to consumer demands and also serves retailers' own interests in product differentiation and can increase revenues, but also results in an increase in food prices, negatively affecting low income households in particular (Bovay and Alston 2018). Depending on the country, policies can have negative implications for consumption and production patterns (SDG12) as well as the achievement of sustainable economic growth (SDG8).

In summary, the socioeconomic impacts of GMOs can contribute to a number of the SDGs. The adoption of GMOs in developing countries has already contributed to alleviate poverty in many regions including China, India, and South Africa. They have great potential to further reduce poverty in many other regions of the world. Poor as well as better-off farm households can benefit. In addition they increase food availability and food safety at farm household level, and in particular among poor farm households. Hence, the main contributions of GMOs are to achieving sustainable and inclusive economic growth including full and productive employment and decent work for all (SDG8).

5. Challenges Facing Modern Agricultural Biotechnology: Obstacles to the Implementation of SDGs

The problems facing modern agricultural biotechnology, especially GMOs, are enormous. These obstacles include limited biotechnology R&D programs, politics and regulatory delays (Smyth 2017), and the perception on the potential risks of GMOs (Morris 2011), among others. They present a major challenge to the application of GMOs to meeting SDGs, especially in developing countries.

Many developing countries especially those in Africa have limited biotechnology R&D programs, weak laboratory facilities, and a lack of well-trained experts. This problem continues to undermine the application of GMOs in developing countries in addition to the controversy surrounding the use of GMOs (Adenle et al. 2012; Paarlberg 2008). Over the past decade, a number of new GM crop varieties have been in the R&D pipeline, while some have undergone confined field trials in different parts of the continent. However, there has been little or no progress in finalizing many of these GM products, mainly because of the institutional capacity problem and government policy failure. A three-year intensive study conducted by Adenle and his team concluded that lack of scientific capacity and poorly equipped laboratories remain a huge constraint in developing new GMO products in Africa (Adenle et al. 2013). For example, super GM cassava (beta-carotene biofortified GM cassava has undergone confined field trials in Nigeria) projects have been abandoned, partly due to lack of expertise required in conducting further lab research (Adenle et al. 2012). Other GM products, including GM banana and GM maize (under R&D programs and confined field trials), will have to be sent to developed countries (e.g., the United States, Australia) for laboratory analysis where there is an ongoing collaboration (Adenle 2017; Wesseler et al. 2017). Apart from South Africa, Tunisia, Egypt, Kenya, and Nigeria, which have relatively advanced biotechnology R&D programs, many African countries still lack the capacity to regulate and develop GMOs. Further to this, several African scholars (including one of the authors of this chapter) argue that the development of GMOs will have to be led by the African scientists from design phase to the final stage in order to facilitate its adoption in the continent. As a result of this problem, several pro-poor GM products continue to suffer delays at the various stages of R&D programs. Given this challenge, it may be difficult to meet the relevant SDGs where GM technology could play a part.

Another problem many developing countries face is related to the political economy of using modern biotechnology, for example, with the United States and the European Union taking different approaches in regulating GMOs, continue to affect the adoption of GMOs around the world (Paarlbarg 2008).

Over the past twenty years, differing European and US perspectives on the application of GMOs have constituted a substantial obstacle for accepting GMOs. In fact, it has been reported that environmental nongovernmental organization and the European countries have spent over \$10 billion in 2015 to prevent the use of GMOs around the world (Byrne 2015), when compared with \$8.6 billion spent on agricultural biotechnology R&D by the United States led multinational companies (Hobbs et al. 2014). This overwhelming difference continues to dominate the debate, particularly at the international arena. The national GMO policy is often influenced by the international GMO regulatory framework. The European factor as reflected in the precautionary principle continues to affect the public R&D programs of local GMO products in developing countries (Paarlberg 2008). After two decades, Golden Rice, one of the flagships of GM biofortified crops, has been approved in Bangladesh. However, technical and regulatory challenges continue to undermine the adoption of Golden Rice in developing regions (Potrykus 2010).

Given the enormous health benefits it could generate, delaying a valuable micronutrient intervention comes at a cost that are difficult to justify (Zilberman et al. 2016). In Asia alone, the annual economic loss of its delayed adoption was estimated at about \$15.6 billion (Anderson et al. 2005). A more recent analysis in India estimated the annual losses to be about 1.4 million life years lost over the past decade, which translates into a cost of about US\$199 million per year (Wesseler and Zilberman 2014). Not surprisingly, 107 Nobel Laureates have recently advocated for GM biofortified foods (Roberts 2018; Washington Post 2016).

6. Conclusions

While GMOs can contribute to reaching the SDGs, policy controversies remain a significant hurdle to overcome. That countries have adopted current international regulatory processes for the approval of GM crops has hindered their progression into the international marketplace. Yet the acreage of GM crops continues to increase with every new year, illustrating that farmers desire and take advantage of new traits that are offered. In the face of climate change, improved crops that can withstand drought yet maintain high yields will become even more important. Furthermore, crops with improved micronutrient content that can be readily absorbed through human consumption will become even more necessary in meeting targets of SDG3, particularly in developing countries. Other new breeding techniques, including genome editing, will also play an important role.[1] Genome edited crops could thus be a venue by which improved nutritional traits can be introduced to crops of the future (Ma et al. 2018). The fact that many of these crops do not retain foreign DNA sequences in their final products makes them exceedingly difficult to distinguish from their conventional crop counterparts. Although this feature might have enabled genome edited crops to avoid the public perception problems that have dogged GM crop acceptance

---

[1] Genome editing is a group of technologies that give scientists the ability to change an organism's DNA. Several approaches to genome editing have been developed, with recent one called CRISPR Cas9. The CRISPR-Cas9 system has generated a lot of excitement in the scientific community because it is faster, cheaper, more accurate, and more efficient than other existing genome editing methods.

for decades, the European Court of Justice has recently ruled them to follow the same regulatory guidelines as GMOs (Purnhagen et al. 2018).

Given rapid adoption of GM crops, since over 20 years of commercialization, the potential role of GMOs in contributing toward meeting SDGs cannot be over-emphasized. Indeed, GM technology is one of the modern innovations that should be encouraged in view of recorded benefits (social, economic, and environment) in addressing sustainable development challenges over the past two decades. However, this would require an overhaul of the current international GMO regulatory framework, which remains one of the most significant challenges to the application of GMO in developing world (Adenle et al. 2017). According to Adenle et al. (2017), the Cartagena Protocol on Biosafety (CPB) that guides the use of GMOs is no longer fit for purpose because it is outmoded and largely influenced by the European Union. Therefore, there is a need for the international community, especially the UN Convention on Biological Diversity, the UN Food and Agriculture Organization, and the World Trade Organization, to review the controversial CPB. Also, a recent article by Adenle et al. (2018) argue that the continued implementation of CPB will create more barriers rather than solutions in developing countries, thereby limiting access to the innovation which can play a part in meeting SDGs. They argue further that the current CPB should be abandoned and that risk-assessment models focusing on local agricultural and environmental practices in developing countries rather than a conventional model should be encouraged. By so doing, it will reduce burdens of poverty and hunger on millions of people.

References

Abbott, A. (2015). Europe's genetically edited plants stuck in legal limbo. *Nature* 528, 319–320.

Adenle, A. A. (2011). Response to issues on GM agriculture in Africa: are transgenic crops safe? *BMC Research Notes*, 4, 388 doi:10.1186/1756-0500-4-388.

Adenle, A. A. (2017). Modern biotechnology for innovation of agricultural development in the developing world: what role can Japan play? *AgBioForum* 20(1), 54–66.

Adenle, A. A., Aworh, G. C., Akromah, R., and, Parayil, G. (2012). Developing GM super cassava for improved health and food security: future challenges in Africa. *Agriculture & Food Security*, 1(1), 1–11.

Adenle, A. A., Morris, E. J., and Govindan, P. (2013). Status of development, regulation and adoption of GM agriculture in Africa: views and positions of stakeholder groups. *Food Policy* 42, 159–166.

Adenle, A. A., Morris, E. J., and, Murphy, D. J. (2017). *Genetically Modified Organisms in Developing Countries: Risk Analysis and Governance.* Cambridge University Press, UK.

Adenle, A. A., Morris, E. J., Murphy, D. J., Phillips, P. W. B, Trigo, E., Peter, K., Li, Y.-H., Quemada, H., Falck-Zepeda, J., and, John, K. (2018). Rationalizing governance of genetically modified products in developing countries. *Nature Biotechnology* 36, 137–139.

Anderson, K., Jackson, L. A., and, Nielsen, C. P. (2005). Genetically modified rice adoption: implications for welfare and poverty alleviation. *Journal of Economic Integration* 20(4),771–788.

Antle, J. M., and, Pingali, L. P. (1995). Pesticides, productivity, and farmer health: a Philippine case study. In: Prabhu, L. P., and Pierre, A. R. (Eds.), *Impact of Pesticides on Farmer Health and the Rice Environment*, pp. 361–387. Springer Netherlands, Dordrecht.

Barrows, G., Sexton, S., and Zilberman, D. (2014). The impact of agricultural biotechnology on supply and land-use. *Environment and Development Economics* 19(6), 676–703.

Beckmann, V., and Wesseler, J. (2004). How labour organisation may affect technology adoption: an analytical framework analysing the case of integrated pest management. *Environment and Development Economics* 8(3), 437–450.

Bennett, R., Morse, S., and Ismael, Y. (2006). The economic impact of genetically modified cotton on South African smallholders: yield, profit and health effects. *The Journal of Development Studies* 42(4), 662–677.

Beyer, P. (2010). Golden Rice and "golden" crops for human nutrition. *New Biotechnology* 27(5), 478–481.

Blancquaert, D.,Van Daele, J. Strobbe, S. Kiekens, F. Storozhenko, S. De Steur, H. Gellynck, X. Lambert, W. Stove, C., and, Van Der Straeten, D.(2015). Improving folate (vitamin B9) stability in biofortified rice through metabolic engineering. *Nature Biotechnology* 33(10), 1076–1078.

Bonanno, A., Materia, V. C., Venus, T., and, Wesseler, J. (2017). The plant protection products (PPP) sector in the European Union: a special view on herbicides. *European Journal of Development Research* 29(3), 575–595.

Bouis, H. E., and Saltzman, A. (2017). Improving nutrition through biofortification: a review of evidence from HarvestPlus, 2003 through 2016. *Global Food Security* 12, 49–58.

Bovay, J., and, Alston, J. M. (2018). GMO food labels in the United States: economic implications of the new law. *Food Policy* 78, 14–25.

Brauer, E. K., and, Shelp, B. J. (2010). Nitrogen use efficiency: re-consideration of the bioengineering approach. *Botany* 88(2), 103–109.

Brennan, L., and, Owende, P. (2010). Biofuels from microalgae: a review of technologies for production, processing, and extractions of biofuels and co-products. *Renewable and Sustainable Energy Reviews* 14(2), 557–577.

Brookes, G., and, Barfoot, P. (2015). Environmental impacts of genetically modified (GM) crop use 1996–2013: impacts on pesticide use and carbon emissions. *GM Crops and Food* 6(2), 103–133.

Brookes, G., and, Barfoot, P. (2017a). Environmental impacts of genetically modified (GM) crop use 1996–2015: impacts on pesticide use and carbon emissions.*GM Crops and Food* 8(2),117–147.

Brookes, G., and, Barfoot, P. (2017b). Farm income and production impacts of using GM crop technology 1996–2015. *GM Crops and Food* 8(3), 156–193.

Byrne, J. (2015). Presentation by President of v-Fluence Interactive at the Agricultural Bioscience International Conference. Melbourne, Australia, September 8.

Carpenter, J. E. (2010). Peer-reviewed surveys indicate positive impact of commercialized GM crops. *Nature Biotechnology* 28(4), 319–321.

Carpenter, J. E. (2011). Impacts of GM crops on biodiversity. *Landes Bioscience* 2(1), 1–17.

Carter, C. A., and Gruere, G. P. (2003). Mandatory labeling of genetically modified foods: does it really provide consumer choice? *AgBioForum* 6(1–2), 68–70.

Carvalho, M. P., and Vasconcelos, M. W. (2013). Producing more with less: strategies and novel technologies for plant-based food biofortification. *Food Research International* 54(1), 961–971.

Cerdeira, A. L., and Duke, S. O. (2006). The current status and environmental impacts of glyphosate-resistant crops: a review. *Journal of Environmental Quality* 35(5), 1633–1658.

Chakraborty. S., Chakraborty, N., Agrawal, L., Ghosh, S., Narula, K., Shekhar, S., Prakash S. Naik, P. C., S.K. Chakrborti, S. K. P., and, Datta, A. (2010). Next-generation protein-rich potato expressing the seed protein gene AmA1 is a result of proteome rebalancing in transgenic tuber. *Proceedings of the National Academy of Sciences* 107(41), 17533–17538.

Changat, D. P., and Krishna, V. V. (2006). An ex-ante economic evaluation of nutritional impacts of transgenic-biofortified potato in India. Presentation at Tropentag 2006: Prosperity and Poverty in a Globalized World: Challenges for Agricultural Research University of Bonn, Germany, October 11–13, 2006

Codex Alimentarius Commission (2015). *Joint Food Agrilcuture and Organization of United Nations (FAO)/World Health Organization (WHO) Food Standards Programme.* Codex Committee on Nutrition and Foods for Special Dietary Uses, 37th Session. Proposed Draft Definition For Biofortification. FAO, Rome.

Culliney, T., (2013). Role of arthropods in maintaining soil fertility. *Agriculture* 3 (4), 629–659.

De Moura, F., Moursi, M., Angel, M. M., Angeles-Agdeppa, I., Atmarita, A., Gironella, G. M., Muslimatum, S., and, Carrquiry, A. (2016). Biofortified β-carotene rice improves vitamin A intake and reduces the prevalence of inadequacy among women and young children in a simulated analysis in Bangladesh, Indonesia, and the Philippines. *American Journal of Clinical Nutrition* 104(3), 769–775.

De Steur, H., Blancquaert, D. Strobbe, S. Lambert, W. Gellynck, X., and Van Der S. D. (2015). Status and market potential of transgenic biofortified crops. *Nature Biotechnology* 33(1), 25–29.

De Steur, H., Gellynck, X., Blancquaert, D., Lambert, W., Van Der Straeten, D., and Qaim, M. (2012). Potential impact and cost-effectiveness of multi-biofortified rice in China. *New Biotechnology* 29(3), 432–442.

De Steur, H., Mehta, S., Gellynck, X., and, Finkelstein J. L. (2017). GM biofortified crops: potential effects on targeting the micronutrient intake gap in human populations. *Current Opinion in Biotechnology* 44, 181–188.

De Steur, H., Wesana, J., Blancquaert, D., Van Der Straeten, D., and Gellynck, X. (2017a). Methods matter: a meta-regression on the determinants of willingness-to-pay studies on biofortified foods. *Annals of the New York Academy of Sciences* 1390(1), 34–46.

De Steur, H., Wesana, J., Blancquaert, D., Van Der Straeten, D., and Gellynck, X. (2017b). The socio-economics of genetically modified biofortified crops: a systematic review and meta-analysis. *Annals of the New York Academy of Sciences* 1390(1), 14–33.

Duke, S. O. (2015). Perspectives on transgenic, herbicide-resistant crops in the United States almost 20 years after introduction. *Pest Manag Sci* 71(5), 652–657.

European Commission (2010). *Biotechnology.* Eurobarometer 73.1. Brussels.

Fiedler, J. L., Kikulwe, E. M., and Birol, E. (2013). An ex ante analysis of the impact and cost-effectiveness of biofortified high-provitamin A and high-iron banana in Uganda. IFPRI Discussion Paper 1277, International Food Policy Research Institute, Washington, DC.

Food and Agriculture Organization of United Nations (FAO) (2008). Conservation Agriculture Carbon Offset Consultation. West Lafayette, Indiana, USA, October 28–30.

Food and Agriculture Organization of United Nations (FAO) (2018). Food loss and food waste. http://www.fao.org/food-loss-and-food-waste/en/ (accessed April 5, 2019).

Fullman, N., Barber, R. M., Abajobir, A. A., Abate, K. H., Abbafati, C., and Abba, K. M. (2017). Measuring progress and projecting attainment on the basis of past trends of the health-related sustainable development goals in 188 countries: an analysis from the Global Burden of Disease Study 2016. *The Lancet* 390(10100), 1423–1459.

Gandhi, V. P., and Namboodiri, N. V. (2006). *The Adoption and Economics of Bt Cotton in India: Preliminary Results from a Study.* Indian Institute of Management, Ahmedabad, Gujarat.

Gianessi, L. P. (2005). Economic and herbicide use impacts of glyphosate-resistant crops. *Pest Management Science* 61, 241–245.

Giuliano, G. (2017). Provitamin A biofortification of crop plants: a gold rush with many miners. *Current Opinion in Biotechnology* 44, 169–180.

Gouse, M. (2012). GM maize as subsistence crop: the South African smallholder experience. *AgBioForum* 15(2), 163–174.

Goyer, A. (2017). Thiamin biofortification of crops. *Current Opinion in Biotechnology* 44, 1–7.

Gressel, J., and Polturak, G. (2018). Suppressing aflatoxin biosynthesis is not a breakthrough if not useful. *Pest Management Science* 74(1), 17–21.

Haas, J. D, Finkelstein, J. L., Udipi, S. A., Ghugre, P., and Mehta, S. (2013). Iron biofortified pearl millet improves iron status in Indian school children: results of a feeding trial. *The FASEB Journal*, 27 (1 Supplement), 355.2.

Haggblade, S., Smale, M., Kergna, A., Theriault, V., and Assima, A. (2017). Causes and consequences of increasing herbicide use in Mali. *European Journal of Developmental Research* 29(3), 648–674.

Han, M., Okamoto, M., Beatty, P. H., Rothstein, S. J., and Good, A. G., (2015). The genetics of nitrogen use efficiency in crop plants. *Annual Review of Genetics* 49, 269–289.

HarvestPlus (2015). Nutritious staple food crops: who is growing what? https://www.harvestplus.org/sites/default/files/HarvestPlus_BiofortifiedCropMap_2015.pdf (accessed April 5, 2019).

Hobbs, J. E., Kerr, W. A., and Smyth, S. J. (2014), The perils of zero tolerance: technology management, supply chains and thwarted globalisation. *International Journal of Technology and Globalisation* 7, 203–216.

Huang, J., Pray, C., and Roszelle, S. (2002). Bt cotton benefits, costs, and impacts in China. *AgBioForum* 5(4), 153–166.

James, C. (2009). *Global Status of Commercialized Biotech/GM Crops: 2009* (ISAAA Brief No. 41). International Service for the Acquisition of Agri-biotech Applications, New York.

James, C. (2010). *Global Status of Commercialized Biotech/GM Crops: 2010* (ISAAA Brief No.42). International Service for the Acquisition of Agri-biotech Applications, New York.

James, C. (2016). *Global Status of Commercialized Biotech/GM Crops: 2016* (ISAAA Brief No. 52). International Service for the Acquisition of Agri-biotech Applications, New York.

James, C. (2017). *Global Status of Commercialized Biotech/GM Crops: 2016* (ISAAA briefs: No. 49-199). International Service for the Acquisition of Agri-biotech Applications, New York.

James, C. (2018). *Global Status of Commercialized GM/Biotech Crops: 2017* (ISAAA) Brief No. 53-2017). International Service for the Acquisition of Agri-biotech Applications, New York.

Janda, K., Kristoufek, L. and Zilberman, D. (2012). Biofuels: policies and impacts. *Agricultural Economics* 58(8), 367–371.

Johnson, Chelsea R., Millwood, Reginald J., Tang, Y., Gou, J., Sykes, Robert W., Turner, Geoffrey B., Davis, Mark F., Sang, Y., Wang, Z-Y., and C. Neal Stewart. (2017). Field-grown miR156 transgenic switchgrass reproduction, yield, global gene expression analysis, and bioconfinement. *Biotechnology for Biofuels* 10(1), 255.

Kathage, J., and Qaim, M. (2012). Economic impacts and impact dynamics of Bt (Bacillus thuringiensis) cotton in India. *Proceedings of the National Academy of Sciences of the United States of America* 109(29), 11652–11656.

Katoch, R., and Thakur, N. (2013). Advances in RNA interference technology and its impact on nutritional improvement, disease and insect control in plants. *Applied Biochemistry and Biotechnology* 169(5), 1579–605.

Kettenburg, A. J., Hanspach, J., Abson, D. J., and Fischer, J. (2018). From disagreements to dialogue: unpacking the Golden Rice debate. *Sustainability Science* 13(5), 1469–1482.

Kikulwe, E. M., Wesseler, J., and Falck-Zepeda, J. (2011). Attitudes, perceptions, and trust: insights from a consumer survey regarding genetically modified banana in Uganda. *Appetite* 57(2), 401–413.

Klümper, W., and Qaim, M. (2014). A meta-analysis of the impacts of genetically modified crops. *PLoS ONE* 9(11): e111629. doi.org/10.1371/journal.pone.011162.

Kouser, S., and Qaim, M. (2011). Impact of Bt cotton on pesticide poisoning in smallholder agriculture: a panel data analysis. *Ecological Economics* 70(11), 2105–2113.

Kouser, S., and Qiam, M. (2013). Valuing financial, health, and environmental benefits of Bt cotton in Pakistan. *Agricultural Economics* 44(3), 323–335.

Kramer, K., Hoppe, P. P., and Packer, L. (2001). *Nutraceuticals in Health and Disease Prevention.* CRC Press, Ohio, USA.

Landrigan, P. J., and Belpoggi, F. (2018). The need for independent research on the health effects of glyphosate-based herbicides. *Environmental Health* 17, 51.

Lee, H. (2017). Transgenic pro-vitamin a biofortified crops for improving vitamin a deficiency and their challenges. *The Open Agriculture Journal* 11(1).

Leite, R. C., Leal, M. R. V., Cortez, L. A. B., Griffin, W. M., and Scandiffio, M. I. G. (2009). Can Brazil replace 5% of the 2025 gasoline world demand with ethanol? *Energy* 34(5), 655–661.

Ma, X., Mau, M., and Sharbel, T. F. (2018). Genome editing for global food security. *Trends in Biotechnology* 36(2), 123–127.

Marra, M. P., Pardey, P., and Alston, J. (2002). The payoffs to transgenic field crops: an assessment of the evidence. *AgBioForum* 5(2), 43–50.

Marvier, M., McCreedy, C., Regetz, J., and Kareiva, P. (2007). A meta-analysis of effects of Bt cotton and maize on non-target invertebrates. *Science* 316, 1475–1477.

Mène-Saffrané, L., and Pellaud, S. (2017). Current strategies for vitamin E biofortification of crops. *Current Opinion in Biotechnology* 44, 189–197.

Moghissi, A. A., Pei, S., and Liu, Y. (2016). Golden rice: scientific, regulatory and public information processes of a genetically modified organism. *Critical Reviews in Biotechnology* 36(3), 535–541.

Morris, E. J. (2011). A semi-quantitative approach to GMO risk-benefit analysis. *Transgenic Research* 20(5):1055–1071.

National Academies of Sciences, Engineering and Medicine (2016). *Genetically Engineered Crops: Experiences And Prospects.* National Academies Press, Washington DC, USA.

Ostrý, V., František, M., and Pfohl-Leszkowicz, A. (2015). Comparative data concerning aflatoxin contents in Bt maize and non-Bt isogenic maize in relation to human and animal health: a review. *Acta Veterinaria Brno* 84(1), 47–53.

Paarlberg, R. (2008). Africa's organic farms. *New York Times.* http://www.nytimes.com/2008/02/29/opinion/29iht-edpaarlberg.1.10576543.html (accessed April 5, 2019).

Paine, J. A., Shipton, C. A., Chaggar, S., Howells, R. M., Kennedy, M. J., Vernon, G., Wright, S. Y., Hinchliffe, E., Adams, J. L., Silverstone, A. L., and Drake, R. (2005). Improving the nutritional value of Golden Rice through increased pro-vitamin A content. *Nature Biotechnology* 23(4), 482–487.

Parisi, C., Tillie, P., and Rodríguez-Cerezo, E. (2016). The global pipeline of GM crops out to 2020. *Nature biotechnology* 34(1), 31–36.

Piero, N. M., Joan, M. N., Richard, O. O., Jalemba, M. A., Omwoyo, O. R., and Chelule, C. R. (2015). Determination of cyanogenic compounds content in transgenic acyanogenic Kenyan cassava (Manihot esculenta Crantz) genotypes: linking molecular analysis to biochemical analysis. *Journal of Analytical and Bioanalytical Techniques* 6(5), 264.

Potrykus, I. (2010). Lessons from the "Humanitarian Golden Rice" project: regulation prevents development of public good genetically engineered crop products. *New Biotechnology* 27(5), 466–472.

Potrykus, I. (2017). The GMO-crop potential for more, and more nutritious food is blocked by unjustified regulation. *Journal of Innovation and Knowledge* 2(2), 90–96.

Pray, C., Ma, D., Huang, J., and Qiao, F. (2001). Impact of Bt cotton in China. *World Development* 29(5), 813–825.

Purnhagen, K., Esther K., Gijs K., Hanna S., Richard V., and Wesseler, J. (2018). EU Court casts new plant breeding techniques into regulatory limbo. *Nature Biotechnology* 36(9), 799–800.

Qaim, M., and Zilberman, D. (2003). Yield effects of genetically modified crops in developing countries. *Science* 299(5608), 900–902.

Ramesh, S. A., Choimes, S., and Schachtman, D. P. (2004). Over-expression of an Arabidopsis zinc transporter in Hordeum vulgare increases short-term zinc uptake after zinc deprivation and seed zinc content. *Plant Molecular Biology* 54(3), 373–385.

Ricroch, A. E., and Henard-Damave, M. C. (2016). Next biotech plants: new traits, crops, developers and technologies for addressing global challenges. *Critical Reviews in Biotechnology* 36(4), 675–690.

Roberts, R. J. (2018). The Nobel laureates campaign supporting GMOs. *Journal of Innovation and Knowledge* 3(2), 61–65.

Sachs, J. D. (2012). From millennium development goals to sustainable development goals. *The Lancet* 379, 2206–2211.

Shahzad, Z., Rouached, H., and Rakha, A., (2014). Combating mineral malnutrition through iron and zinc biofortification of cereals. *Comprehensive Reviews in Food Science and Food Safety* 13(3), 329–346.

Shen, H, Poovaiah, C.R., Ziebell, A., Tschaplinski, T. J., Pattathil, S., Gjersing, E., Engle, N. L., Katahira, R., Pu, Y.,Sykes, R., Chen, F., Ragauskas, A. J., Mielenz, J. R., Hahn, M. G., Davis, M., Stewart, C. N., and Dixon, R. A. (2013). Enhanced characteristics of genetically modified switchgrass (Panicum virgatum L.) for high biofuel production. *Biotechnology for Biofuels* 6(1), 71.

Singh, A., Nigam, P. S., and Murphy, J. D. (2011). Renewable fuels from algae: an answer to debatable land based fuels. *Bioresource Technology* 102(1), 10–16.

Smart, R., Blum, M., and Wesseler, J. (2017). Trends in genetically engineered crops' approval times in the United States and the European Union. *Journal of Agricultural Economics* 68(1), 182–198.

Smyth, S. J. (2017). Genetically modified crops, regulatory delays, and international trade. *Food and Energy Security* 6(2), 78–86.

Stein, A. J, Sachdev, H. P. S., and Qaim, M. (2006). Potential impact and cost-effectiveness of Golden Rice. *Nature Biotechnology* 24(10), 1200–1201.

Storozhenko, S., De Brouwer, V. Volckaert, M. Navarrete, O. Blancquaert, D. Zhang, G. Lambert, W., and Van Der Straeten, D. (2007). Folate fortification of rice by metabolic engineering. *Nature Biotechnology* 25(11), 1277–1279.

Tang, G., Qin, J., Dolnikowski, G. G., Russell, R. M., and Grusak, M. A. (2009). Golden Rice is an effective source of vitamin A. *American Journal of Clinical Nutrition* 89(6), 1776–1783.

Washington Post (June 30, 2016). 107 Nobel laureates sign letter blasting Greenpeace over GMOs. https://www.washingtonpost.com/news/speaking-of-science/wp/2016/06/29/more-than-100-nobel-laureates-take-on-greenpeace-over-gmo-stance/?utm_term=.fa8114f70df3 (accessed April 5, 2019).

Thompson, G. D., and Kidwell, J. (1998). Explaining the choice of organic produce: cosmetic defects, prices, and consumer preferences. *American Journal of Agricultural Economics* 80(2), 277–287.

Tosun, Y., and Scaub, S. (2017). Mobilization in the european public sphere: the struggle over genetically modified organisms. *Review of Policy Research* 34(3), 310–330.

Van Der Straeten, D., Fitzpatrick, T. B., and De Steur, H. (2017). Editorial overview: biofortification of crops: achievements, future challenges, socio-economic, health and ethical aspects. *Current Opinion in Biotechnology* 44, vii–x. doi: 10.1016/j.copbio.2017.03.007.

Van Der Straeten, D., and Strobbe, S. (2017). Folate biofortification in food crops. *Current Biotechnology*, 44, 202–211.

Veettil, P. C., Krishna, V. V., and Qaim, M. (2017). Ecosystem impacts of pesticide reductions through Bt cotton adoption. *Australian Journal of Agricultural and Resource Economics* 61 (1), 115–134.

Venegas-Calerón, M., Sayanova, O., and Napier, J. A. (2010). An alternative to fish oils: metabolic engineering of oil-seed crops to produce omega-3 long chain polyunsaturated fatty acids. *Progress in lipid research* 49(2), 108–119.

Venus, T. J., Drabik, D., and Wesseler, J. (2018). The role of a German multi-stakeholder standard for livestock products derived from non-GMO feed. *Food Policy* 78, 58–67.

Waltz, E. (2015a). Nonbrowning GM apple cleared for market. *Nature Biotechnology* http:// blogs.nature.com/tradesecrets/2015/03/30/nonbrowning-gm-apple-cleared-for-market (accessed April 5, 2019).

Waltz, E. (2015b). The United States Department of Agriculture (USDA) approves next-generation GM potato. *Nature Biotechnology* 33(1), 12–13.

Waltz, E. (2016). Gene-edited CRISPR mushroom escapes United States (US) regulation. *Nature* 532(7599), 293.

Wesseler, J. (2014). Biotechnologies and agrifood strategies: opportunities, threats and economic implications. *Bio-based and Applied Economics* 3(3), 187–204.

Wesseler, J., Smart, R. D., Thomson, J., and Zilberman, D. (2017). Foregone benefits of important food crop improvements in Sub-Saharan Africa. *PLoS ONE* 12(7): e0181353.

Wesseler, J., and Zilberman, D. (2014). The economic power of the Golden Rice opposition. *Environment and Development Economics* 19(6): 724–742.

Wu, T. Y., Gruissem, W., and Bhullar, N. K. (2018). Targeting intra-cellular transport combined with efficient uptake and storage significantly increases grain iron and zinc levels in rice. *Plant Biotechnol Journal* 17(1), 9–20.

Zhao, F.-J., and Shewry, P. R. (2011). Recent developments in modifying crops and agronomic practice to improve human health. *Food Policy* 36, S94–S101.

Zilberman, D., Kaplan, S., and Wesseler, J. (2016). The loss from underutilizing GM technologies. *AgBioForum* 18(3), 312–319.

21

Farmer-Prioritized Climate-Smart Agriculture Technologies

Implications for Achieving Sustainable Development Goals in East Africa

Caroline Mwongera, Chris M. Mwungu, Mercy Lungaho, and Steve Twomlow

1. Introduction

The sustainable development goals' (SDGs) triple-bottom-line approach—targeting economic development, environmental sustainability, and social inclusion—provide a focus on addressing current global challenges such as population growth, demographic change, resource competition and scarcity, degraded agricultural soils, loss of habitat, pollution, greenhouse gas emissions, and climate change (Godfray and Garnett 2014; Sachs 2012). Agriculture lies at the heart of delivering the SDGs, playing a pivotal role in providing access to adequate food and nutrition, advancing more productive and resilient livelihoods and ecosystems, poverty reduction, and fostering secure, sustainable and safe food systems (Beddington et al. 2012; Schmidt-Traub and Sachs 2015). The SDGs succeeded the millennium development goals (MDGs), for the period 2015 to 2030, with the aim of being applicable to all countries and a guide toward sustainable development (Le Blanc 2015).

For the majority of food producers, especially smallholder farmers in low-income countries, significant concerns for agriculture consist of adapting and building resilience to climate change, reducing food losses, increasing agricultural productivity and incomes, improving nutrition outcomes, and sustaining ecosystem services (Godfray et al. 2010). Since agriculture is one of the most important economic sectors, making up 20 to 40% of gross domestic product for many sub-Saharan Africa (SSA) countries (FAO 2015; 2014), it is not surprising that development and diffusion of appropriate agricultural technologies is one of the critical priorities for rural transformation (Barrett et al. 2017).

CSA, as promoted in developing countries, has three aims: i) increasing agricultural productivity to support equitable increases in incomes, food security, and development; ii) enhancing adaptive capacity and resilience; and iii) mitigating climate change by reducing greenhouse gas emissions and increasing carbon sequestration (Campbell et al. 2014; FAO 2013; 2010). In their Intended Nationally Determined

Contributions (INDCs) under the United Nations Framework Convention on Climate Change (UNFCCC), more than 30 countries specifically refer to CSA, most of them in sub-Saharan Africa (FAO 2016). Despite the various benefits of CSA technologies, the current rate of awareness and adoption by farmers remains low (Palanisami et al. 2015). Africa is the only region in the world where food crop productivity has remained stagnant since the 1970s (Jama and Pizarro 2008; Jayne et al. 2010), highlighting the need for interventions to support a green revolution.

A challenge with CSA is to identify approaches that allow for spatially explicit and contextually relevant targeting of approaches for diverse national and subnational contexts. The diversity of contexts is illustrated in East Africa, which has a complex matrix of land-cover typologies, governance structures, and social-economic realities, all of which produce multiple livelihood priorities, challenges, and opportunities for scaling CSA technologies. Recognizing this variability, Mwongera et al. (2017) developed the Climate-Smart Agriculture Rapid Appraisal (CSA-RA) tool to prioritize CSA technologies for specific contexts across diverse landscapes.

The CSA-RA is a rapid and comprehensive mixed-method approach that integrates participatory, qualitative, and quantitative tools with a combination of gender-disaggregated methods and climate analysis. The tool also assesses multiple aspects of the farming system, such as land health, agriculture production, agriculture practices, perceptions of climate impacts on farming systems and rural livelihoods, and incorporating multiple scales and actors, to identify context-specific CSA options (Mwongera et al. 2017).

In contrast to the SDGs, CSA is framed differently for global, developing, and developed country scales, without a specific set of criteria for integration at various levels (Chandra et al. 2018). This chapter aims to show whether prioritized climate-smart technologies, can contribute to achieving SDGs for smallholder farmers, as well as identify potential barriers and trade-offs.

2. Methodology

This section describes the study area, study design, data collection strategies, and models.

2.1 Study Area

We conducted the study in Bagamoyo, Kilolo, Kilosa, and Mbarali districts in the Southern Agriculture Growth Corridor of Tanzania (SAGCOT), Bungoma, Kakamega, and Siaya districts in western Kenya, and Nwoya district in Northern Uganda, between March 2014 and May 2016.

The SAGCOT region covers about one-third of the Tanzanian mainland. The area experiences tropical climate with marked seasonal temperature and rainfall variations. The rainfall season lasts from November to May. Subsistence farming is the main economic activity characterized by livestock production, crop farming, and fishing. The most common crops in the region are maize, paddy rice, pulses, and vegetables. The SAGCOT land area is about 31 million ha, of which approximately 27% are arable lands. There is a growing demand for large-scale agricultural investment in the region (Hart et al. 2014).

In Kenya, small-scale farmers operating on land below 0.8 hectares occupy most of the agricultural area of western Kenya. The most-widely cultivated crops include maize, banana, sorghum, millet, cassava, sweet potatoes, and pulses. Cattle and poultry farming are the main livestock production systems. In Bungoma County, rainfall is bimodal, with an average of 1,100 mm. Kakamega County also has two rainy seasons with an average range of 1,300 to 2,200 mm annually. Siaya County receives an average rainfall of between 1,800 and 2,000 mm annually. Livestock husbandry and subsistence farming are the main economic activities in Nwoya district, northern Uganda. Crop production is primarily rain-fed, favoring multicropping. The most important crops in Nwoya are beans, cassava, and rice. Brick making and charcoal burning provide off-farm income for farmers. Rainfall is bimodal, with the first rainy season lasting from March to June and the second rainy season lasting from July to November. Mean annual precipitation is about 1500 mm (Mwongera et al. 2014).

2.2 Data Collection

We conducted stakeholder workshops and household surveys to collect quantitative and qualitative data for this study. Participatory rural appraisal tools, including pairwise ranking, were also employed in the data collection (Mwongera et al. 2017).

2.3 Stakeholder Workshops

We carried out the identification and prioritization of context-specific CSA technologies in the participatory workshops. There was a total of five farmers' workshops in Uganda, four in Tanzania, and two in western Kenya, to represent the agro-ecological variability across the sites. Each farmers' workshop had approximately 30 to 50 participants. The participants included farmers, agricultural officers, climate change and environmental experts, and representatives from local NGOs and the private sector. There were efforts to generate a representative sample of workshop participants based on gender, agro-ecologies, and age groups. Each stakeholder workshop followed a four-stage process.

In stage one, farmers selected the most important CSA practices from a long list, prepared earlier through literature review and key informant interviews. Practices that participants were aware of or were implementing were included to generate a master list of potential CSA options for each site.

In stage two, participants identified and agreed on a set of indicators that were relevant to the local context, to guide the evaluation of different practices. This included characteristics considered the most important to foster adoption by farmers. Discussions on the choice and ranking of indicators took place in the plenary session.

In stage three, farmers in different groups by gender used a pairwise matrix comparison to rank the long list of practices in order of importance. Scoring of the indicators followed in step four. Each group assigned a scale of 1 to 5, lower values denoting low importance.

In stage four, a single practice's potential impact was evaluated against indicators related to the CSA pillars (productivity, adaptation, and mitigation), to conduct the climate-smartness assessment. The scoring was from −10 to +10, where −10 = very high negative change (or 100% negative change); −1 = very low negative change (or 10% negative change); 0 = No change; 1 = very low positive change (or 10% positive change), 10 = very high positive change (or 100% positive change); N/A = Not applicable. For each practice, we obtained the sum of individually assigned scores for indicators corresponding to each pillar, and then calculate the means. Each respondent was required to evaluate only the practices for which they have experience implementing at the study site.

2.4 Household Surveys

Sampled farming households interviews consist of structured questionnaires consisting of both closed and open-ended questions. In all sites, we selected household heads (Mwungu et al. 2017; Ng'ang'a et al. 2017). In all sites, the selection of households for the interview was through random sampling. Ultimately, we interviewed a total of 1281households (608 from Tanzania, 585 from Uganda and 88 from Kenya).

2.5 Econometric Analysis

In this study, we applied a multinomial endogenous switching regression (MESR) to evaluate the impacts of adopting CSA technologies on farm income and labor demand using the Southern Tanzania dataset. We based the analysis on three practices: minimum tillage, use of improved varieties, and integrated pest management (IPM). There were eight potential combinations of these practices (2^3) as shown in Table 21.1. The approach allowed for the possibility of there being a biased sample of adopters. For example, it may a relatively large proportion of well-educated farmers who are likely to take up new agriculture technologies. In this case, it is not

Table 21.1 Prioritized indicators for adopting climate-smart agriculture technologies by farmers in Uganda and Tanzania by gender

Farmer indicator	Sustainable Development Goals indicator	Farmer indicator scores			
		Tanzania		Uganda	
		Male	Female	Male	Female
Land size	1.4.2 Secure tenure rights to land	5.0	5.0	3.0	1.0
Food security	2.1.1 Prevalence of food insecurity	4.0	5.0	4.0	5.0
Yield	2.3.1 Volume of production per labor unit	4.5	5.0	5.0	5.0
Cost	2.3.2 Average income	5.0	5.0	5.0	4.0
Income	2.3.2 Average income	4.0	5.0	5.0	5.0
Labor availability	2.3.1 Volume of production per labor unit	3.0	4.0	3.0	3.5
Availability of inputs	2.3.1 Volume of production per labor unit	3.0	5.0	4.0	4.0
Weed infestation	2.4.1 Agriculture area under sustainable agriculture	3.0	1.0	4.0	4.0
Sustainability of land	2.4.1 Agriculture land under sustainable agriculture	3.0	5.0	N/A	N/A
Pests and Diseases	2.4.1 Agriculture area under sustainable agriculture	5.0	5.0	4.0	4.0
Availability of seed	2.5.1 Number of plants in conservation facilities	4.0	3.0	4.0	4.0
Health	3.3.5 Interventions against neglected tropical diseases	2.0	2.0	5.0	5.0
Availability of water	6.4.1 Bodies of water with good ambient water quality	5.0	5.0	N/A	N/A
Total rainfall	13.b.1 Capacity building for climate change management	5.0	5.0	5.0	5.0

Continued

Table 21.1 *Continued*

Farmer indicator	Sustainable Development Goals indicator	Farmer indicator scores			
		Tanzania		Uganda	
		Male	Female	Male	Female
Knowledge/skills	13.b.1 Capacity building for climate change management	5.0	5.0	5.0	5.0
Indigenous knowledge	13.b.1 Capacity building for climate change management	N/A	N/A	2.0	2.0
Adapting to climate change	13.b.1 Capacity building for climate change management	2.0	3.0	2.0	3.0
Soil fertility	15.3.1 Degraded land over total land area	5.0	5.0	1.0	1.0
Soil moisture retention	15.3.1 Degraded land over total land area	N/A	N/A	1.0	1.0
Topography	15.3.1 Degraded land over total land area	N/A	N/A	1.0	1.0

Low scores indicate least importance; N/A = Not relevant. Numbers in the second column refer to the proposed SDG goal indicator according to the global indicator framework developed by the Inter-Agency and Expert Group on SDG Indicators. Source: authors.

possible to estimate the impact of a technology without constructing a valid and justifiable counterfactual. The MESR model allows us to compute observed and potential (counterfactual) outcomes, which further enables one to calculate the treatment effects of adopting a particular technology by getting the difference between the observed and potential outcomes. In other words, the model computes average treatment effects if a farmer has adopted a given technology and an estimated counterfactual outcome corresponding to what value the outcome would take if the farmer had not adopted the technology. The difference will show whether a given technology has a positive or negative impact on the outcomes of interest, otherwise referred to as the treatment effect.

3. Results and Discussion

In this section we discuss factors that farmers consider important when choosing CSA technologies and we link the outcomes of adoption to the SDGs.

3.1 Farmers' Prioritization of CSA Options

Table 21.1 presents indicators farmers consider essential when considering what CSA technologies to adopt in Uganda and Tanzania. The total number of indicators selected by farmers was 17 in Tanzania and 18 in Uganda. Results show that the importance assigned by farmers to each indicator when evaluating CSA. There is variation in the scores depending on geographic location and gender. For example, results indicate that land size, land sustainability, health, and availability of water are considered more important by farmers in Tanzania that Uganda. In addition, weed infestation and indigenous knowledge are relatively highly rated indicators in Uganda. This information illustrates the variation in stakeholder preferences which may reflect differences in biophysical characteristics of farms (agro-ecology, climate, topography, soil health) or differences in socioeconomic factors (income, land sizes, labor availability, access to infrastructures such as markets and roads, availability of agriculture information and advice).

Malhotra and Schuler (2005) have demonstrated the relevance of identifying context-specific indicators. The process provides both quantitative and qualitative information that can be used to design and implement programs at a local level (Khatri-Chhetri et al. 2017). Using an indicator development process with extensive stakeholder participation and consultation gives legitimacy and credibility to the outcome (Jónssona et al. 2016). The indicators reflect the essential qualities considered by farmers for evaluating the agricultural technology. The process may also benefit farmers through the unfolding of greater understanding of the issues surrounding agriculture and climate change management at a local level.

In Tanzania, farmers ranked CSA practices according to farm types. Inorganic fertilizer, intercropping, and irrigation were the most preferred as shown in Figure 21.1. Figure 21.1 shows how preferred practices can vary within a country, in this case Tanzania. The most preferred technologies are the use of inorganic fertilizer and intercropping in Bagamoyo; inorganic fertilizer, irrigation, crop rotation, and intercropping in Kilolo; intercropping, mulching, and zero grazing in Kilosa; and irrigation and inorganic fertilizer application in Mbarali (Figure 21.1).

When considering adoption of CSA technologies, farmers focus on income, yields, food security, nutrition, health, land size, labor, and time demands. Adoption is hampered by cost, low soil fertility, rainfall variability, pests and diseases, weed infestations, lack of inputs, lack of knowledge, and lack of relevant skills (Table 21.1).

The results show that the practices preferred by farmers vary with site, highlighting the context-specific nature of climate-smart agriculture, and the need for context-specific targeting of CSA practices. Agricultural practices and technologies do not necessarily have universal applicability; they have to be selected to suit each context,

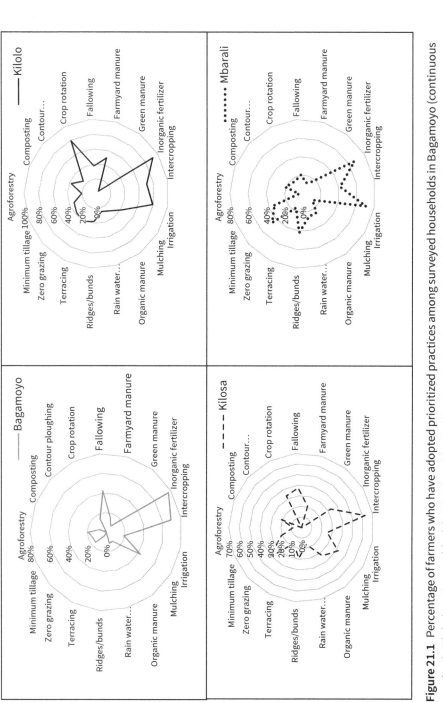

Figure 21.1 Percentage of farmers who have adopted prioritized practices among surveyed households in Bagamoyo (continuous grey line), Kilolo (continuous black line), Kilosa (dashed line), and Mbarali (dotted line) districts in Tanzania

Source: authors

including agro-ecological zones and farming systems, as well as cultural and socioeconomic circumstances (Dinesh 2016).

3.2 Prioritized Climate-Smart Technologies and Climate Action (SDG13)

To evaluate the link between CSA and SDG13, Figure 21.2 shows the results of the assessment by farmers of selected climate-smart practices for different farm types in western Kenya. Indicators selected by stakeholders for each of the three CSA pillars (productivity, adaptation, and mitigation) represent the outcomes that they are most interested in achieving and for which data are available.

Inorganic fertilizer, certified seed, and crop rotation were rated most highly for productivity; farmyard manure, conservation agriculture, and agroforestry for adaptation; and agroforestry and composting for mitigation (Figure 21.2). For the productivity pillar, indicators chosen by the stakeholders are yield and income. Indicators selected for the adaptation pillar are the quantity of water available for production, water used per unit of product, soil capacity to retain water, level of soil disturbance, the ability of farmers to manage climate risks, and the ability of farmers to limit the production system exposure to climate risks. Others included diversification of income sources on the farm, use of local and traditional knowledge, the content of soil organic matter in soils, quality of animal diet, household income spent on food per month, and amount of soil lost through erosion.

For mitigation potential, the quantity of greenhouse gas (GHG) released in terms of carbon dioxide equivalent (CO_2eq) per unit of production, the amount of biomass aboveground, and below-ground amount of biomass were the indicators used for the assessment. Conservation agriculture (CA) contributes to mitigation by avoiding soil compaction, reducing water run-off, increasing aeration of the soils, reducing emissions, and increasing soil carbon (Rochecouste et al. 2015; Stagnari et al. 2010). Composting and farmyard manure (FYM) are considered as having a potential for soil carbon sequestration (Lal 2004; Smith et al. 2008). For inorganic fertilizer, energy consumption in manufacturing and transportation generate GHG emissions (Snyder et al. 2009). Crop rotations can have benefits in terms of biomass production, quality of the organic matter released to the soil, and soil carbon sequestration, especially rotations with nitrogen-fixing legumes that can substantially reduce the nitrogen input through chemical fertilizers (Isbell et al. 2017; Stagnari et al. 2017). Increasing crop intensity or cropping frequency by reducing the frequency of bare land in the crop rotation is another useful approach to improve biomass production and soil carbon sequestration, as well as returning more crop residues to the soil, compared to monoculture. In the right conditions, certain crop rotations can also help in reducing pest and pathogen loads, resisting climate extremes, and increasing overall system productivity (Bedoussac et al. 2015; Dias et al. 2015; Gan et al. 2015; Godfray et al. 2010; Harrington, 1959).

Farm type	Practice	Productivity	Adaptation	Mitigation
Small scale mixed	Farm yard manure	5	2.5	0.9
Small scale mixed	Conservation agriculture	5.2	2.7	1.3
Small scale mixed	Composting	4.7	2.3	1.6
Medium-mixed-hort	Intercropping	3.5	1.3	1.1
Medium-mixed-hort	Certified seed	6.8	3.2	N/A
Medium-mixed-hort	Crop rotation	3.5	2.2	1
Medium-mixed-dairy	Farm yard manure	3.8	3.7	1.5
Medium-mixed-dairy	Agroforestry	4.5	2.6	2.3
Medium-mixed-dairy	Inorganic fertilizer	3.5	1.5	-3.3
Medium-mixed-cereal	Crop rotation	6.5	0.5	N/A
Medium-mixed-cereal	Herbicide	5	2.2	-0.3
Medium-mixed-cereal	Inorganic fertilizer	7	1.2	-4.3
Large scale commercial	Conservation agriculture	5.5	3.6	1.2
Large scale commercial	Liming	2.5	0.7	N/A
Large scale commercial	Agroforestry	5.5	3.5	2.7

Figure 21.2 Assessment of the performance of climate-smart practices against the three CSA pillars (productivity, adaptation, and mitigation) for western Kenya. 1 = very low positive change (or 10% positive change), 10 = very high positive change (or 100% positive change), N/A = not applicable.

Source: authors

Overall, for western Kenya, conservation agriculture provides the highest potential impacts on climate-smart agriculture outcomes among small-scale mixed farm types. For the medium-sized mixed dairy and large-scale commercial farms, agroforestry has the greatest potential benefits. Intercropping and crop rotation have the highest climate-smartness scores for medium-sized mixed horticulture and medium-sized mixed cereal farm types respectively.

The climate smartness assessment considers productivity, adaptation, and mitigation as the critical outcomes of sustainable development in addressing food security and climate concerns (FAO 2013). Comparably, the goals of SDG13 integrate both adaptation and mitigation (Campbell et al. 2018). Similar to the target represented by the SDGs, CSA is conceptualized as agriculture that brings us closer to safe operating spaces for agricultural and food systems (Neufeldt et al. 2013). Achieving adaptation and mitigation in food systems will require success in other SDGs as enabling conditions, such as sustainable consumption and production (12), poverty reduction (1), food security (2), education (4), gender equity (5), water (6), and life on land (15) (Campbell et al. 2018).

3.3 Prioritized Climate-Smart Technologies and Ending Poverty (SDG1)

Results shown are for the eight possible combinations of the three CSA practices selected as most relevant by farmers in Tanzania (Table 21.2). The practices are improved varieties, minimum tillage and IPM. IPM is a sustainable and broad-based approach that combines practices for the economic prevention and control of pests

Table 21.2 Frequency of implementation of the three prioritized Climate-smart agriculture practices on farms in Kilolo and Mbarali districts, Tanzania

Climate-smart agriculture technologies	Frequency (%)
$M_0V_0I_0$ (no adoption)	28.9
$M_1V_0I_0$ (adoption of minimum tillage)	8.2
$M_0V_1I_0$ (adoption of improved varieties)	16.7
$M_0V_0I_1$ (adoption of integrated pest management (IPM)	15.0
$M_1V_1I_0$ (adoption of both minimum tillage and improved varieties)	9.5
$M_1V_0I_1$ (adoption of both minimum tillage and IPM)	7.0
$M_0V_1I_1$ (adoption of both improved varieties and IPM)	12.0
$M_1V_1I_1$ (adoption of all three strategies, minimum tillage, improved varieties, IPM)	7.8

Note: The abbreviation M refers to minimum tillage; V refers to drought-tolerant varieties while I refers to integrated pest management. Source: authors.

and diseases in crops. It seeks to use natural predators or parasites to control pests, using selective pesticides for backup only when natural means fail.

The adoption of CSA technologies was low among the sampled households. Twenty-nine percent of the respondents did not adopt any of the three prioritized technologies. The highest adoption rate was recorded for improved varieties (17%), followed by the use of integrated pest management at 15%. Minimum tillage had the lowest frequency of implementation in the study, at 8%. Farms that adopted a combination of the technologies ranged from 7% for both minimum tillage and IPM, and 9% for minimum tillage and improved varieties. The highest combination was the use of improved varieties at 12%. Seven percent of the farms adopted all three CSA practices (Table 21.2).

Of the various practices assessed, the ones with the greatest impacts on income were the adoption of improved varieties, followed by the adoption of IPM (Table 21.3). Minimum tillage has the lowest effect on farm incomes. Farmer priorities include increasing farm incomes (Bulte et al. 2014), and therefore are more likely to adopt technologies that have a positive impact on profit. However, Noltze et al. (2013) argue that adoption patterns are context-specific and variation is expected depending on agro-ecological and socioeconomic features at a particular location.

Table 21.3 also presents the average effect of CSA adoption on labor demand, measured in man days. We compare the level of income and labor use among the adopters (observed farm income and observed labor use) with the level for the nonadopters. The results indicate that gross farm incomes are increased by the adoption of all three CSA practices, individually or in any combination (Table 21.3). The highest increases

Table 21.3 Impact of adopting various climate-smart agriculture practices on gross farm income and labor use in Kilolo and Mbarali districts, Tanzania

	Gross farm income (per season, per hectare in USD)			Farm labor use (per season, per hectare in man days)		
	Actual farm income	Counterfactual farm income	Treatment effect	Actual labor use	Counterfactual labor use	Treatment effect
$M_1V_0I_0$	376	338	38	90.27	108.49	-18.20***
$M_0V_1I_0$	569	381	188***	103.58	101.82	1.76***
$M_0V_0I_1$	292	180	112***	118.93	103.88	15.05***
$M_1V_1I_0$	474	378	96***	82.04	99.70	-17.66***
$M_1V_0I_1$	371	323	48***	91.63	99.70	-8.07***
$M_0V_1I_1$	539	492	47	97.99	96.97	1.02
$M_1V_1I_1$	624	471	153***	83.99	102.89	-18.90***

Notes: Statistical significance at 0.01(***), 0.05(**) and 0.1 (*). The abbreviation M refers to minimum tillage, V refers to drought tolerant varieties, and I refers to integrated pest management. Source: authors.

in farm incomes result from adopting improved varieties (188 USD in one season for each hectare cultivated), followed by adoption of all the three practices (112 USD).

Adoption of all the three practices has the most substantial benefits in terms of labor savings: 19 man days in one season for each hectare cultivated. Minimum tillage has the benefit of reduced farm labor requirements (18 man days). Adoption of IPM and improved varieties increases the requirement for farm labor by 15- and two-man days, respectively.

Our results are similar to others regarding the benefit of the adoption of minimum tillage practices for farm profitability, arising from higher yields (Ngoma 2018; Ngoma et al. 2015; Sijtsma et al. 1998). Several previous studies have shown the positive impact of CSA technologies on farm incomes. Adoption of improved maize varieties helped raise household per capita expenditure by farmers, thereby reducing their probability of falling below the poverty line in Mexico (Becerril and Abdulai 2010) and Zambia (Khonje et al. 2015). Agricultural technologies that reduce labor can free up time for other activities, thereby contributing to increasing income. However, it is important to determine whose labor is saved and at what point during the agricultural season (Doss 2001; Noltze et al. 2012). Since adopting a new technology often implies a need for additional labor, labor availability determines successful adoption (Doss and Morris 2001). Smallholder farmers can implement a range of climate-smart agricultural practices and technologies, to minimize the adverse effects of climate change and variability, but their adoption largely depends on economic benefits associated with the practices (Khatri-Chhetri et al. 2016). The pathway to fostering sustainable development includes finding technologies that provide adequate farmers' income with positive effects on the household's labor demand.

3.4 Prioritized Climate-Smart Technologies and Ending Hunger (SDG2)

The link between nutrition and the SDGs has been documented by the 2016 Global Nutrition Report (IFPRI 2016). There is growing global recognition that agriculture offers several opportunities for improved nutrition outcomes, especially among mothers and children. In general, there are three opportunities: 1) as a source of food, which can affect food availability and utilization for the household; 2) as a source of agricultural income for expenditure on food and healthcare; and 3) through women's empowerment, which affects access to income, improves maternal and child-caring capacity and practices, and increases female expenditure on energy (Herforth and Harris 2014). By affecting these pathways and improving nutrition outcomes, climate-smart agricultural technologies offer a strategic opportunity to achieve multiple SDGs.

The data presented in this chapter shows that it is plausible that adopting CSA options improved nutrition outcomes through increased incomes and reduced labor demand. Table 21.3 gives evidence that adoption of all three prioritized CSA practices did improve agricultural incomes by a significant margin. Improving food security, yield, and health are among the priorities for adopting agriculture interventions for smallholder farmers (Table 21.1). Besides, Table 21.3 shows some CSA options that present the opportunity to reduce labor demand thus allowing better time use and time saving. When activities introduce a more equitable workload, they can encourage men and women to use the newly available time for childcare and self-care. Time availability is especially important for pregnant and lactating women, infants, and young children, as this is the period when nutrition and healthcare needs are more significant and can prevent malnutrition.

Agriculture is the primary livelihood for most rural households in the developing world (Cobbinah et al. 2015; Dixon, 1990). We know that families dependent on agriculture to make a living are highly vulnerable to environmental shocks and stresses (Harvey et al. 2014; Perez et al. 2015). Climate change disrupts agricultural productivity and food markets, thereby posing risks to the food supply (Lipper et al. 2014; Wheeler and Braun 2013). By increasing income, agricultural activities can reduce the potential impacts of future risks. With additional revenue, families are better able to make informed, joint decisions about household purchases that can contribute to healthier diets and better childcare (Danton 2016). Studies linking agriculture to nutrition and household diets have also shown that most agricultural interventions have positive impacts on food production (Berti et al. 2004; Bhutta et al. 2008; Leroy and Frongillo 2007). Climate-smart agriculture increases resilience and resource-use efficiency in agricultural production systems, helping to adapt food systems to the threats of climate change.

3.5 Prioritized Climate-Smart Technologies and Life on Land (SDG15)

Figure 21.3 shows the results of the assessment of prioritized practices against farmers' preferred indicators in western Kenya. Small-scale subsistence farmers considered yield and income as the most desired benefits. Of the practices, farmyard manure was most preferred, followed by crop rotation and intercropping. Small-scale farms have problems of low soil fertility on their farms and low resource endowment. There could be a challenge obtaining adequate manure due to the small numbers of livestock and poultry on most farms. There are also trade-offs involved with reallocating crop residues from livestock feed (a short-term benefit) to recycling back to crop fields as soil amendment (a long-term benefit). On medium-scale farms with commercial cereal production, farmers ranked certified seeds as most beneficial for improving yield, income, and yield variability reduction. Composting and agroforestry rank high on enhancing soil organic matter, while terracing is useful for controlling erosion.

On medium-scale farms with commercial horticulture, farmers believe that intercropping and the use of certified seeds has the potential to increase yield and income. Intercropping contributes to yield variability reduction and improving the soil

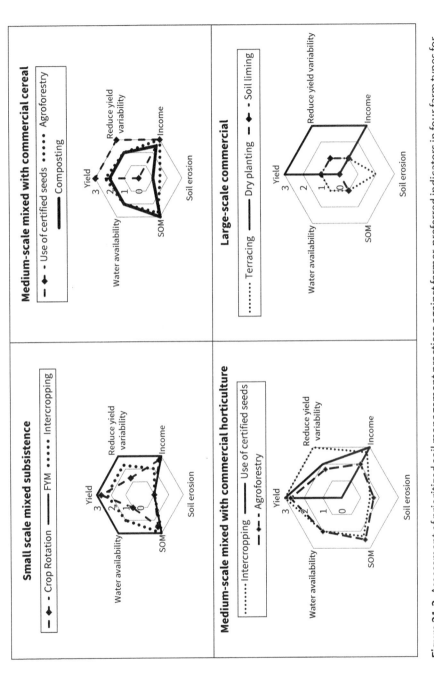

Figure 21.3 Assessment of prioritized soil management practices against farmer-preferred indicators in four farm types for Western Kenya

Source: authors

organic matter content. Agroforestry and intercropping are useful for increasing the amount of water available in the soil. In regards to soil erosion benefits, both agroforestry and intercropping have some advantages.

On large-scale commercial farms, farmers assessed early planting as being favorable for increasing yield and income and reducing yield variability. Terracing is seen as better for controlling soil erosion, whereas soil liming has low benefits in terms of yield and soil organic matter (Figure 21.3).

Soils contribute approximately 37% of greenhouse gas emissions from these farms (Tubiello et al. 2015), and soil carbon sequestration can be useful in climate mitigation, offsetting human emissions of greenhouse gases. Prioritized climate-smart practices include the application of farmyard manure, residue management, composting and agroforestry (Figure 21.3), each of which can increase soil organic matter, thereby buffering plants from climate change and enhancing soil carbon (C) stocks. Further benefits accrue when organic fertilizers replace synthetic fertilizers. Practices to increase productivity in low-yielding, nutrient-deficient systems provide greater C inputs to the soil (Tubiello et al. 2015). Intercropping and agroforestry help to protect or enhance ecosystem services while producing food, feed, and wood products. Intercrop systems often produces more biomass than monocultures, potentially enhancing soil carbon sequestration. Additional mitigation benefits result from reduced deforestation by substituting wood fuels from forests with fast-growing wood fuels from agroforestry (Bogdanski 2012). The Intergovernmental Panel on Climate Change estimates that 17.4% of global greenhouse gas emissions comes from the forest sector, in large part from deforestation in developing countries (IPCC 2007). Tropical deforestation globally resulted in the release of an estimated 1.1 to 2.2 Gt/year in the decade prior to 2009 (Sasaki and Putz 2009).

3.6 Trade-Offs and Challenges

Stakeholders highlighted several factors that contribute to the low adoption of improved technologies across the three regions: western Kenya, Northern Uganda, and the SAGCOT. Small herd sizes mean that the quantity of manure is less than ideal; there is a lack of the required knowledge and skills among farmers; inputs such as seed, fertilizer, and lime are costly or unavailable; and there is a perception that some practices are slow to deliver benefits (e.g., manure releases nutrients slowly as compared to inorganic fertilizers). Farmers also mentioned small land size as a barrier to implementation of agroforestry, crop rotation, and fallowing. For agroforestry, barriers to adoption are difficulty in accessing the seedlings, lack of knowledge on the most locally appropriate types of trees to plant, and land tenure systems that discourages tree planting among farmers who do not have land ownership documents. It is clear that social and economic factors such as profitability and risk implications of agriculture practices, input and output price, and climate variability influence adoption (Kassie et al. 2015).

Several factors such as agro-ecological suitability, gender priorities, labor availability, and income outcomes affect the achievement of multiple SDGs. Unfortunately, promising CSA solutions are sometimes relatively narrowly focused—their developers may not consider other factors such as nutrition. Mainstreaming nutrition into CSA requires research to identify specific technologies and leverage opportunities to improve nutrition in particular contexts. Nutritional outcomes could be served if CSA projects incorporated explicit nutrition objectives and indicators into their designs. Efforts to improve households' dietary practices guide project design and implementation among farming communities could be highly beneficial. For example, improving yield is important, particularly in places where there is insufficient food or where households depend on agriculture for their livelihood. However, merely increasing or diversifying production does not automatically lead to better nutrition (Leroy and Frongillo 2007; Webb and Kennedy 2014). Thus, a project design would first set out the causal path linking CSA adoption to nutrition outcomes. Then impact assessment would go beyond the proximal indicator of yield and measure the distal indicators for nutrition outcome.

In Tanzania, we noted trade-offs between desirable outcomes following adoption of CSA practices. For example, improved varieties had the highest increases in income but required additional labor use on the farm. Similarly, IPM increases gross income but requires a substantial increase in labor use. Farmer priorities in western Kenya reveal a trade-off in the achievement of the three goals of CSA, with the highest proportion of prioritized indicators focusing on adaptation, followed by productivity and lastly mitigation. Based on this result, stakeholders feel that consideration of climate change is important when assessing agricultural technologies. The climate assessment also reveals that indicators selected by the stakeholders do not provide a balanced assessment within the three pillars of CSA. Some farmers have a low awareness of the full range of benefits that new technologies can provide and they may focus on the most tangible and familiar benefits. For instance, end users in developing countries may consider productivity to be most important. However, from a global perspective, the challenges linked to climate change mitigation may outweigh such local priorities.

The western Kenya assessment demonstrates that further increases in fertilizer application to meet SDG1 and SDG2 may undermine mitigation goals to support SDG13. Fertilizer N is responsible for more than 30% of agriculture-related N_2O emissions. Fertilizer production relies on significant use of fossil fuels, contributing to greenhouse gas emissions (Bogdanski 2012). In the Netherlands, sustainable fertilizer application has reduced fertilizer use while doubling yields (Campbell et al. 2018). Responsible fertilizer N use also helps increase biomass production necessary to help restore and maintain soil organic carbon (SOC) levels (Snyder et al. 2009)

The interactions among SDGs, as discussed in this chapter, are important but introduce challenges. For example, achieving effective mitigation of greenhouse gas emissions from agriculture will require success in food security, poverty reduction and life on land (Campbell et al. 2018). On the other hand, SDG13 fosters the stability of food markets and food supply (Lipper et al. 2014). Likewise, household and individual incomes affect food access and utilization indirectly. Lack of access to drinking water could impair food utilization and cause damage to health (Wheeler and Braun 2013). At the same time, climate change and increased climate variability will enhance the pressure on agricultural production systems and forests, and thus essential sources of energy for many of the world's poor, making people more vulnerable without adaptation (Bogdanski 2012).

4. Conclusion and Recommendation

The CSA-RA is a mixed-method approach that aims to facilitate the prioritization of locally specific climate-smart agriculture technologies. This approach collects qualitative and quantitative data from a broad range of stakeholders, allowing extensive analysis, triangulation, and validation. The process reveals stakeholder priorities, their preferences for different indicators, and suitable CSA practices at the local level.

This study demonstrates the value of evaluating preferred indicators and practices for a local context. Farmers' priorities for CSA technologies are dependent on the geographical context and vary between genders. We have presented a set of indicators for CSA prioritization that have been developed with stakeholder participation. The indicator selection reveals gender differences in perceptions. For instance, relative to men, women in both Tanzania and Uganda assigned greater importance to food security, labor availability, and adapting to climate change (Table 21.1). The results show the need for interventions to embrace an approach that addresses gender differences (Beuchelt and Badstue 2013). The potential for technologies to generate benefits related to gender equality is most significant where these technologies contribute to sustainable agricultural intensification when they are adopted by women to improve their situations (Mathews et al. 2018).

Improving incomes, productivity, food security, nutrition, health, labor availability, climate adaptation, pest, diseases, and weed management were of high-interest to the stakeholders. These indicators are useful to assist decision-making regarding agriculture production and represent what stakeholders consider as being relevant for sustainable development in their context. This selection process lends credibility to the selected core set of indicators and gives us confidence that technologies that meet the criteria will be widely adopted (Mwongera et al. 2017).

Farmer-prioritized CSA technologies to integrate climate change and agriculture development have the potential to contribute to several SDGs—those concerned with ending poverty (SDG1), zero hunger (SDG2), climate action (SDG13), and life on land (SDG15). The chapter provides compelling results to facilitate the effective promotion of CSA practices to support agriculture and rural development.

Economic assessments are highly useful to evaluate impacts of adopting CSA technologies for specific target groups within specific agricultural systems. A significant challenge is to minimize CSA-related trade-offs identified in this study such as the increase in labor demand arising from the adoption of certain technologies.

A key concerns among policymakers, planners, and implementers is whether adoption of new agricultural practices yields social and economic benefits without undue risk, and to what subset of farms. Assessing agricultural practices against the SDGs can assist to generate benefits, and manage trade-offs.

This study indicates that farmers prefer productivity- and adaptation-enhancing technologies such as agroforestry, irrigation, and conservation agriculture. A significant challenge mentioned by farmers in adopting these practices is their high capital investments and long payback periods. Financial support from the government and other agencies can enhance the uptake of practices to promote the achievement of SDGs. Identifying appropriate ways to incentivize the uptake of climate-smart alternatives that are both more sustainable and more productive is a priority (Campbell et al. 2014). A CSA prioritization approach provides information for climate development planning at a local level, ensuring that selected practices are relevant for a particular location, and have the highest adoption potential, and therefore are likely to contribute to sustainable development.

Acknowledgments

The International Center for Tropical Agriculture (CIAT) supported this study through the project Increasing Food Security and Farming System Resilience in East Africa through Wide-Scale Adoption of Climate-Smart Agricultural Practices. The project funding was provided by the International Fund for Agriculture Development, and the project Climate-Smart Soil Protection and Rehabilitation in Western Kenya was funded by Deutsche Gesellschaft für Internationale Zusammenarbeit (GIZ). The work was carried out under the Climate Change, Agriculture and Food Security (CCAFS), and Water, Land and Ecosystems (WLE) Research programs of the Consultative Group on International Agricultural Research (CGIAR). We thank the anonymous reviewers for their careful reading of our chapter and their many insightful comments and suggestions.

References

Barrett, C. B., Christiaensen, L., Sheahan, M., and Shimeles, A. (2017). On the Structural Transformation of Rural Africa, Policy Research Working Papers. The World Bank. doi.org/ doi:10.1596/1813-9450-7938

Becerril, J., and Abdulai, A. (2010). The impact of improved maize varieties on poverty in Mexico: a propensity score-matching approach. *World Development* 38, 1024–1035.

Beddington, J., Asaduzzaman, M., Clark, M., Fernández, A., Guillou, M., M, J., Erda, L., Mamo, T., Bo, N. Van, Nobre, C. A., Scholes, R., Sharma, R., and Wakhungu, J. (2012). *Achieving Food Security in the Face of Climate Change: Final Report from the Commission on Sustainable Agriculture and Climate Change*. Commission on Sustainable Agriculture and Climate Change, Copenhagen.

Bedoussac, L., Journet, E.-P., Hauggaard-Nielsen, H., Naudin, C., Corre-Hellou, G., Jensen, E. S., Prieur, L., and Justes, E. (2015). Ecological principles underlying the increase of productivity achieved by cereal-grain legume intercrops in organic farming: a review. *Agronomy for Sustainable Development* 35, 911–935.

Berti, P. R., Krasevec, J., and FitzGerald, S. (2004). A review of the effectiveness of agriculture interventions in improving nutrition outcomes. *Public Health Nutrition* 7, 599–609.

Beuchelt, T. D., and Badstue, L. (2013). Gender, nutrition- and climate-smart food production: opportunities and trade-offs. *Food Security* 5, 709–721.

Bhutta, Z. A., Ahmed, T., Black, R. E., Cousens, S., Dewey, K., Giugliani, E., Haider, B. A., Kirkwood, B., Morris, S. S., Sachdev, H. P. S., and Shekar, M. (2008). What works? Interventions for maternal and child undernutrition and survival. *The Lancet* 371, 417–440.

Bogdanski, A. (2012). Integrated food–energy systems for climate-smart agriculture. *Agriculture and Food Security* 1(9), 1–10.

Bulte, E., Beekman, G., Falco, S. Di, Hella, J., and Lei, P. (2014). Behavioral responses and the impact of new agricultural technologies: evidence from a double-blind field experiment in tanzania. *American Journal of Agricultural Economics* 96, 813–830.

Campbell, B. M., Hansen, J., Rious, J., Stirling, C. M., Twomlow, S., and Wollenberg, E. (2018). Urgent action to combat climate change and its impacts (SDG 13): transforming agriculture and food systems. *Current Opinion in Environmental Sustainability* 34, 13–20.

Campbell, B. M., Thornton, P., Zougmoré, R., van Asten, P., and Lipper, L. (2014). Sustainable intensification: what is its role in climate smart agriculture? *Current Opinion in Environmental Sustainability* 8, 39–43.

Chandra, A., McNamara, K. E., and Dargusch, P. (2018). Climate-smart agriculture: perspectives and framings. *Climate Policy* 18, 526–541.

Cobbinah, P. B., Erdiaw-Kwasie, M. O., and Amoateng, P. (2015). Rethinking sustainable development within the framework of poverty and urbanisation in developing countries. *Environmental Development* 13, 18–32.

Danton, H. (2016). Five ways agriculture can improve nutrition, https://www.agrilinks.org/blog/five-ways-agriculture-can-improve-nutrition (accessed March 5, 2019).

Dias, T., Dukes, A., and Antunes, P. M. (2015). Accounting for soil biotic effects on soil health and crop productivity in the design of crop rotations. *Journal of the Science of Food and Agriculture* 95, 447–454.

Dinesh, D. (ed.) (2016). Agricultural practices and technologies to enhance food security, resilience and productivity in a sustainable manner (No. 146), Messages to SBSTA 44 agriculture workshops. Copenhagen, Denmark.

Dixon, C. (1990). *Rural Development in the Third World*, 1st ed. Routledge, London.

Doss, C. R. (2001). Designing agricultural technology for African women farmers: lessons from 25 years of experience. *World Development* 29, 2075–2092.

Doss, C. R., and Morris, M. L. (2001). How does gender affect the adoption of agricultural innovations? the case of improved maize technology in Ghana. *Agricultural Economics* 25, 27–39.

Food and Agriculture Organization of the United Nations (FAO) (2010). *"Climate Smart" Agriculture: Policies, Practices and Financing for Food Security*. FAO, Rome.

Food and Agriculture Organization of the United Nations (FAO) (2013). *Climate Smart Agriculture Sourcebook*. FAO, Rome.

Food and Agriculture Organization of the United Nations (FAO) (2014). *FAO Statistical Yearbook 2014*. FAO, Rome.

Food and Agriculture Organization of the United Nations (FAO) (2015). *FAO Statistical Pocketbook: World Food and Agriculture 2015*. FAO, Rome.

Food and Agriculture Organization of the United Nations (FAO) (2016). *The State of Food and Agriculture: Climate Change, Agriculture and Food Security*. FAO, Rome.

Gan, Y., Hamel, C., O'Donovan, J. T., Cutforth, H., Zentner, R. P., Campbell, C. A., Niu, Y., and Poppy, L. (2015). Diversifying crop rotations with pulses enhances system productivity. *Scientific Reports* 5, 14625.

Godfray, H. C. J., Beddington, J. R., Crute, I. R., Haddad, L., Lawrence, D., Muir, J. F., Pretty, J., Robinson, S., Thomas, S. M., and Toulmin, C. (2010). Food security: the challenge of feeding 9 billion people. *Science* 327(5967), 812–818.

Godfray, H. C. J., and Garnett, T. (2014). Food security and sustainable intensification. *Philosophical Transactions of the Royal Society B Biological Sciences* 369(1639), doi.org/10.1098/rstb.2012.0273.

Harrington, J. F. (1959). *The Value of Moisture Resistant Containers in Vegetable Seed Packaging, Bulletin 7*. California Agricultural Experiment Station, Berkeley, CA.

Hart, A., Tumsifu, E., Nguni, W., Recha, J., Malley, Z., Masha, R., and Buck, L. (2014). *Participatory Land Use Planning to Support Tanzanian Farmer and Pastoralist Investment: Experiences from Mbarali District, Mbeya Region, Tanzania*. International Land Coalition, Rome.

Harvey, C. A., Rakotobe, Z. L., Rao, N. S., Dave, R., Razafimahatratra, H., Rabarijohn, H., Rajaofara, H., and Mackinnon, J. L. (2014). Extreme vulnerability of smallholder farmers to agricultural risks and climate change in Madagascar. *Philosophical Transactions of the Royal Society B Biological Sciences* 369(1639), doi.org/10.1098/rstb.2013.0089

Herforth, A., and Harris, J. (2014). *Understanding and Applying Primary Pathways and Principles, Brief 1, Improving Nutrition through Agriculture, USAID Co Improving Nutrition through Agriculture Technical Brief Series*. USAID/Strengthening Partnerships, Results, and Innovations in Nutrition Globally (SPRING) Project, Arlington, VA.

Intergovernmental Panel on Climate Change (2007). Synthesis Report: Contribution of Working Groups I, II and III to the Fourth Assessment Report of the Intergovernmental Panel on Climate Change. IPCC, Geneva.

International Food Policy Research Institute (2016). *Global Nutrition Report 2016: From Promise to Impact: Ending Malnutrition by 2030, Global Nutrition Report*. International Food Policy Research Institute, Washington, DC.

Isbell, F., Adler, P. R., Eisenhauer, N., Fornara, D., Kimmel, K., Kremen, C., Letourneau, D. K., Liebman, M., Polley, H. W., Quijas, S., and Scherer-Lorenzen, M. (2017). Benefits of increasing plant diversity in sustainable agroecosystems. *Journal of Ecology* 105, 871–879.

Jama, B., and Pizarro, G. (2008). Agriculture in Africa: strategies to improve and sustain smallholder production systems. *Annals of the New York Academy of Science* 1136, 218–232.

Jayne, T. S., Mather, D., and Mghenyi, E. (2010). Principal challenges confronting smallholder agriculture in Sub-Saharan Africa. *World Development* 38, 1384–1398.

Jónssona, J. Ö. G., Davíðsdóttir, B., Jónsdóttir, E. M., Kristinsdóttir, S. M., and Ragnarsdóttir, K. V. (2016). Soil indicators for sustainable development:aA transdisciplinary approach for indicator development using expert stakeholders. *Agriculture, Ecosystems and Environment* 232, 179–189.

Kassie, M., Teklewold, H., Jaleta, M., Marenya, P., and Erenstein, O. (2015). Understanding the adoption of a portfolio of sustainable intensification practices in eastern and southern Africa. *Land Use Policy* 42, 400–411.

Khatri-Chhetri, A., Aggarwal, P. K., Joshi, P., and Vyas, S. (2017). Farmers' prioritization of climate-smart agriculture (CSA) technologies. *Agricultural Systems* 151, 184–191.

Khatri-Chhetri, A., Aryal, J. P., Sapkota, T. B., and Khurana, R. (2016). Economic benefits of climate-smart agricultural practices to smallholder farmers in the Indo-Gangetic Plains of India. *Current Science* 110, 1251–1256.

Khonje, M., Manda, J., Alene, A. D., and Kassie, M. (2015). Analysis of adoption and impacts of improved maize varieties in eastern Zambia. *World Development* 66, 695–706.

Lal, R. (2004). Soil carbon sequestration impacts on global climate change and food security. *Science* 304, 1623–1627.

Le Blanc, D. (2015). Towards integration at last? the sustainable development goals as a network of targets. *Sustainable Development* 23, 176–187.

Leroy, J. L., and Frongillo, E. A. (2007). Can interventions to promote animal production ameliorate undernutrition? *Journal of Nutrition* 137, 2311–2316.

Lipper, L., Thornton, P., Campbell, B. M., Baedeker, T., Braimoh, A., Bwalya, M., Caron, P., Cattaneo, A., Garrity, D., Henry, K., Hottle, R., Jackson, L., Jarvis, A., Kossam, F., Mann, W., McCarthy, N., Meybeck, A., Neufeldt, H., Remington, T., Sen, P. T., Sessa, R., Shula, R., Tibu, A., and Torquebiau, E. F. (2014). Climate-smart agriculture for food security. *Nature Climate Change* 4, 1068–1072.

Malhotra, A., and Schuler, S. R. (2005). Women's empowerment as a variable in international development. In: Narayan, D. (Ed.), *Measuring Empowerment: Cross-Disciplinary Perspectives*, pp. 71–88. World Bank, Washington, DC.

Mwongera, C., Shikuku, K. M., Twyman, J., Läderach, P., Ampaire, E., Van Asten, P., Twomlow, S., and Winowiecki, L. A. (2017). Climate smart agriculture rapid appraisal (CSA-RA): a tool for prioritizing context-specific climate smart agriculture technologies. *Agricultural Systems* 151, 192–203.

Mwongera, C., Shikuku, K. M., Twyman, J., Winowiecki, L., Ampaire, E., Koningstein, M., and Twomlow, S. (2014). *Climate Smart Agriculture Rapid Appraisal (CSA-RA) Report for Northern Uganda.* Nairobi, Kenya.

Mwungu, C. M., Mwongera, C., Shikuku, K. M., N. Nyakundi, F., Twyman, J., Winowiecki, L. A., L. Ampaire, E., Acosta, M., and Läderach, P. (2017). Survey data of intra-household decision making and smallholder agricultural production in Northern Uganda and Southern Tanzania. *Data in Brief* 14, 302–306.

Neufeldt, H., Jahn, M., Campbell, B. M., Beddington, J. R., DeClerck, F., De Pinto, A., Gulledge, J., Hellin, J., Herrero, M., Jarvis, A., LeZaks, D., Meinke, H., Rosenstock, T., Scholes, M., Scholes, R., Vermeulen, S., Wollenberg, E., and Zougmoré, R. (2013). Beyond climate-smart agriculture: toward safe operating spaces for global food systems. *Agriculture and Food Security* 2(12), 1–6.

Ng'ang'a, S. K., Notenbaert, A., Mwungu, C.M., Mwongera, C., and Girvetz, E. (2017). Cost and Benefit Analysis for Climate-Smart Soil Practices in Western Kenya. Working Paper, CIAT Publication No. 439, Kampala, Uganda.

Ngoma, H. (2018). Does minimum tillage improve the livelihood outcomes of smallholder farmers in Zambia? *Food Security* 10, 381–396.

Ngoma, H., Mason, N. M., and Sitko, N. J. (2015). Does minimum tillage with planting basins or ripping raise maize yields? meso-panel data evidence from Zambia. *Agriculture, Ecosystems and Environment* 212, 21–29.

Noltze, M., Schwarze, S., and Qaim, M. (2012). Understanding the adoption of system technologies in smallholder agriculture: the system of rice intensification (SRI) in Timor Leste. *Agricultural Systems* 108, 64–73.

Noltze, M., Schwarze, S., and Qaim, M. (2013). Impacts of natural resource management technologies on agricultural yield and household income: the system of rice intensification in Timor Leste. *Ecological Economics* 85, 59–68.

Palanisami, K., Kumar, D. S., Malik, R. P. S., Raman, S., Kar, G., and Monhan, K. (2015). Managing water management research: analysis of four decades of research and outreach programmes in India. *Economic and Political Weekly* 50(26&27), 33–43.

Perez, C., Jones, E. M., Kristjanson, P., Cramer, L., Thornton, P. K., Förch, W., and Barahona, C. (2015). How resilient are farming households and communities to a changing climate in Africa? a gender-based perspective. *Global Environmental Change* 34, 95–107.

Rochecouste, J.-F., Dargusch, P., Cameron, D., and Smith, C. (2015). An analysis of the socio-economic factors influencing the adoption of conservation agriculture as a climate change mitigation activity in Australian dryland grain production. *Agricultural Systems* 135, 20–30.

Sachs, J. D. (2012). From millennium development goals to sustainable development goals. *The Lancet* 379, 2206–2211.

Sasaki, N., and Putz, F. E. (2009). Critical need for new definitions of "forest" and "forest degradation" in global climate change agreements. *Conservation Letters* 2, 226–232.

Schmidt-Traub, G., and Sachs, J. D. (2015). Financing Sustainable Development: Implementing the SDGs through Effective Investment Strategies and Partnerships. Working Paper. Sustainable Development Solutions Network, New York.

Sijtsma, C., Campbell, A., McLaughlin, N., and Carter, M. (1998). Comparative tillage costs for crop rotations utilizing minimum tillage on a farm scale. *Soil Tillage Research* 49, 223–231.

Smith, P., Martino,D., Cai, Z., Gwary, D., Janzen, H., Kumar, P., McCarl, B., Ogle, S., O'Mara, F., Rice, C., Scholes, B., Sirotenko, O., Howden, M., McAllister, T., Pan, G., Romanenkov, V., Schneider, U., Towprayoon, S., Wattenbach, M., and Smith, J. (2008). Greenhouse gas mitigation in agriculture. *Philosophical Transactions of the Royal Society B Biological Sciences* 363, 789–813.

Snyder, C. S., Bruulsema, T. W., Jensen, T. L., and Fixen, P. E. (2009). Review of greenhouse gas emissions from crop production systems and fertilizer management effects. *Agriculture, Ecosystems and Environment* 133, 247–266.

Stagnari, F., Maggio, A., Galieni, A., and Pisante, M. (2017). Multiple benefits of legumes for agriculture sustainability: an overview. *Chemical and Biological Technologies in Agriculture* 4(2), 1–13.

Stagnari, F., Ramazzotti, S., and Pisante, M. (2010). Conservation agriculture: a different approach for crop production through sustainable soil and water management: a review. In: Lichtfouse, E. (Ed.), *Organic Farming, Pest Control and Remediation of Soil Pollutants*, pp. 55–83. Springer, Dordrecht.

Tubiello, F. N., Salvatore, M., Ferrara, A. F., House, J., Federici, S., Rossi, S., Biancalani, R., Condor Golec, R. D., Jacobs, H., Flammini, A., Prosperi, P., Cardenas-Galindo, P., Schmidhuber, J., Sanz Sanchez, M. J., Srivastava, N., and Smith, P. (2015). The contribution of agriculture, forestry and other land use activities to global warming, 1990–2012. *Global Change Biology* 21, 2655–2660.

Webb, P., and Kennedy, E. (2014). Impacts of agriculture on nutrition: nature of the evidence and research gaps. *Food Nutrition Bulletin* 35, 126–132.

Wheeler, T., and von Braun, J. (2013). Climate change impacts on global food security. *Science* 341(6145), 508–513.

22

Toward Sustainable Agri-Food Systems in Brazil

The Soybean Production Complex as a Case Study

Cecilia G. Flocco

1. Introduction

An increasing global demand for agricultural products, driven by population growth, is producing environmental pressure in commodity-producing countries and regions such as the United States and South America, a situation that can be exacerbated by climate-change-related events (Fernandes et al. 2012). The major challenge and, at the same time, the window of opportunity for establishing reliable and stable food production systems is to avoid, mitigate, or reverse the degradation of ecosystems while meeting the increasing demand for agricultural products in sustainable ways.

The United Nations' sustainable development agenda has put in place a set of seventeen sustainable development goals (SDGs) to end poverty and ensure a sustainable future for all, to be attained by the year 2030 (UN 2015). Engineering sustainable agri-food supply chains is part of the solution to some of the issues addressed in the ambitious sustainability agenda. The challenge requires both contributions from science, technology, and innovation (STI) to develop sustainable production systems and improve or scale up existing ones, and the establishment of harmonized regulatory frameworks supporting those efforts. Such innovative strategies and policies should target all actors along the multistakeholder agricultural products and food supply chain, encompassing producing and importing countries, consumers, and global organizations.

This chapter aims to identify key leverage points in agricultural and food production systems, in which the contributions of STI can help balance the trade-offs bound to the expansion of agriculture in food-and-raw-material-producing regions in South America. To illustrate the analysis undertaken, the soybean production complex in Brazil is presented as a case study, through which cross-sectoral efforts to develop integrated sustainable production strategies are analyzed. Emphasis is placed on exploring STI contributions and policy approaches that help preserve land ecosystems and soil functions (SDG15), given that a healthy biophysical environment (SDGs 12, 13, 15) is critical to the achievement of many interlinked SDGs focusing on economic development and societal well-being (SDGs 1, 2, 3, 6, 8) at large (Keesstra et al. 2016). Also, examples of successful partnerships and initiatives (SDG17) supporting interlinked SDGs targeting different stakeholders in the supply chain, as well as opportunities to further develop effective policy recommendations, are discussed. A detailed description of the chapter's contents and the research methodology employed are presented in the next section.

2. Chapter organization and methodological approach

This chapter delves into the contributions of the STI sector and policy interventions supporting a sustainable agriculture enterprise in Brazil, with a focus on the soybean production complex. Also, the connections of agriculture to the natural and socioeconomic environments are discussed. Due to the availability of historical records of land use change and the magnitude of the impacts produced, the chapter focuses on the Amazon and Cerrado biomes, taking the Mato Grosso state as proxy to the situation in the region. The state, comprising large sections of both biomes, constitutes a focal point of promising agricultural innovations aimed at engineering sustainable agricultural practices (Richards et al. 2015; Galford et al. 2013).

The first part of this chapter summarizes the role of STI in the development of the agriculture sector in Brazil and its contribution toward the achievement of the millennium development goals (MDGs) (UN n.d.) and their successors, the SDGs. Also, the impacts of the agricultural development on land ecosystems and soil functions are discussed, given the important services that they provide and their tight relation with human and environmental well-being (Wall et al. 2015; Nielsen et al. 2015; Keesstra et al. 2016). In the second part, a selection of STI contributions and policy leverage points are presented, aiming to further advance the SDGs and harmonize those with agricultural and food supply chains interventions, both locally and in a global context.

The methodology used for this case study included a survey of the scientific literature reviewing the contributions of STI to the development of the agricultural enterprise in Brazil (starting from the period of the agricultural revolution in the 1970s onward), with a focus on the soybean production sector. The research approach also considered the role of the agricultural sector in the country's socioeconomic development, its impacts on the natural environment, and key policy and regulatory frameworks. In some cases, the information was obtained from reports produced by nongovernmental (NGOs) and private-sector organizations and by conducting interviews with leaders at institutions fostering the sustainable development of agriculture in Latin-America (EMBRAPA, FONTAGRO, IADB, IICA).

Research and development contributions were defined as creative and systematic work undertaken to increase the stock of knowledge and to devise new applications of available knowledge (according to guidelines given in OECD 2015). In order to identify robust STI contributions to sustainable agriculture and craft the recommendations proposed in this chapter, the different sources of information surveyed were evaluated in an integrated fashion (that is, considering the viability of the intervention together with supporting policies and under given social, political, and geographical frameworks), in order to select the most promising interventions. Those were defined as innovative approaches that have shown success in a middle-scale range or in a regional setting and that have the potential of being scaled up and/or replicated. To facilitate the visualization, the selected STI contributions and supporting policy recommendations applying to different levels and sectors were also listed in a tabular format, highlighting those that are considered pillars of an integrated strategy aiming to bolster sustainable agriculture and food supply chains.

3. The Evolution of the Brazilian Agricultural Sector

3.1 South America, a Key Player in the Global Food Supply Scenario

As noted earlier, the increasing global demand for food and agricultural products, driven by population and macroeconomic growth, is exerting pressure on natural resources and on the integrity of the environment (Fernandes et al. 2012). The phenomenon is mostly affecting major producers of agricultural commodities, such as the United States and countries in South America (Zeigler and Truitt Nakata 2014), with various economic and social impacts (Vera-Diaz et al. 2009; VanWey et al. 2013; Richards et al. 2015). With an estimated global population over 9 billion by 2050 (UN 2017), it is foreseen that the demand for food will increase by 59 to 98% from 2005 to 2050, with a socioeconomic development model that considers climate change and bioenergy scenarios (Valin et al. 2014). This situation, in turn, will stimulate an expansion of arable lands in agricultural countries, exacerbating existing environmental challenges and potentially threatening food security in vulnerable regions and economies. In addition, the economic development of middle-class sectors in some regions, particularly in Asia, has caused a change of consumption patterns and preferences. For example, in China, a shift from traditional diets to a consumption pattern with a higher proportion of meat has been observed during the last two decades (Wang et al. 2015). With 20% of the world's population, China holds only 8% of the world's arable land and very limited possibilities to expand it. Once a self-sufficient country in terms of food, China must rely nowadays on imported agricultural products since the increased demand surpassed the local production capacity (Schneider 2011). An example is the import of soy, an annual crop that produces an edible bean with a high oil and protein content with multiple uses for human food, animal feed, and industrial applications. China is importing more than 70% of their domestic demand (Hansen and Gale 2014). The impacts of the aforementioned consumption trends among

Asian countries reverberate beyond their borders, since they catalyze socioeconomic and environmental impacts in other crop-, meat-, and feed-producing countries and regions. For example, more than a half of the global soy exports are produced in the fertile lands of South America (Zeigler and Truitt Nakata 2014).

3.2 Pivotal STI contributions to Brazil's agricultural development

Among the South American countries producing food and other agricultural products, Brazil constitutes a representative case study depicting the linkages between global food demand and agriculture-driven challenges to sustainable production systems.

The Brazilian agriculture sector has experienced dramatic changes over the last five decades: from being a net food importer of agricultural and food products in the early seventies to constituting one of the main contributors of agricultural commodities to global markets nowadays (Arraes Pereira et al. 2012; Zeigler and Truitt Nakata 2014). The sector moved progressively from an extensive agriculture model (increasing production by expanding croplands) to a more efficient and productive one, supported by the outcomes of the STI sector. The key early contributions to the agricultural enterprise were the development of adapted crop and forage varieties and soil fertilization schemes. These pivotal STI contributions were bound to the efforts to adapt temperate environments crops to tropical regions, thus expanding the agricultural frontier (Arraes Pereira et al. 2012; Freitas and Landers 2014; Martha and Alves 2018). The mechanization of farming practices, enabling the cultivation of larger agricultural plots, was another crucial driver of the agricultural development. However, intensive tillage practices, aggravated by a lack of soil cover during noncropping seasons, led to soil erosion, with the inherent loss of the fertile topsoil (Freitas and Landers 2014, and references therein). This prompted pioneering farmers in the early seventies to adopt innovative soil-management practices and concepts of conservation agriculture (Saturnino 1998; Freitas and Landers 2014). Those soil-management innovations were the minimum tillage and zero-tillage farming practices which, after overcoming initial implementation hurdles resulting from a lack of appropriated equipment and know-how, were widely adopted by farmers in Brazil (Freitas and Landers 2014 provide a historical review of the subject) and neighboring agricultural countries such as Paraguay and Argentina (Peiretti and Dumanski 2014).

The agricultural development in Brazil experienced another boost in the late 1990s with the adoption of genetically modified (GM) crops, in particular soybeans engineered to tolerate the application of herbicides. The cultivation of glyphosate-tolerant soy, starting in 1998, allowed farmers to reduce the number of agrochemical applications and labor input and increase their crop yields (Qaim 2005). The biotechnological pack of GM soy seeds plus herbicide (and later other GM crops, such as cotton and maize designed to resist insect attacks) was rapidly adopted by Brazilian producers and nowadays more than 96% of the soybeans produced in Brazil are transgenic (Companhia Nacional de Abastecimento 2017). However, heated debates around the technology are fueled by several controversial environmental and socioeconomic issues such as: (i) the promotion of mono-cropping (planting extensive areas with one crop variety; Mendonça 2011), thus reducing the diversity of plants and microorganisms and creating concerns about the landscape's resilience to shocks such as pests outbreaks or severe weather events (Kremen and Miles 2012; Kremen and Merenlender 2018); (ii) the potential negative impacts of the applied agrochemicals to nontarget organisms and habitats (Nicolopoulou-Stamati et al. 2016 and references therein); and (iii) rising concerns around monopolist practices, since the technology pack (seeds plus agrochemicals) is mainly commercialized by one multinational company. A central dispute is the right of farmers to save seeds produced in their farms with the aforementioned technology (protected under the developing company's intellectual property rights) for their next planting season (Filomeno 2013 and references therein).

Regarding the role of institutions and government agencies in Brazil's agricultural revolution, the creation of the state-owned Brazilian Agricultural Research Corporation (EMBRAPA, under the Ministry of Agriculture) in the early 1970s constituted a key stepping stone supporting the development of the sector. The institutional framework supporting agricultural research reflected the recognition of the important contribution of the sector to the country's economy and societal development (Dias Avila et al. 2010). As such, EMBRAPA's nationwide reach supports and coordinates the activities of Brazilian research centers and associated institutions and promotes collaborative international partnerships. Its main pillars are the continued and balanced promotion of basic and applied research, a strong focus on human resources development, and an effective technology transfer strategy, working directly with farmers at the agricultural extension units throughout the country (a detailed description of EMBRAPA´s development, structure and impact is provided by Martha and Alves 2018).

As a result, the agricultural sector has greatly contributed to the country´s stability in front of fluctuating global markets (Dias Avila et al. 2010; Martha and Alves 2018). By harnessing this positive momentum, and harmonizing it with policies and national programs targeting other areas key to societal development (UNDP 2012), Brazil achieved many of the eight MDGs (established by the United Nations following its Millennium Summit in the year 2000), by their deadline in the year 2015, including eradicating extreme poverty, achieving universal primary education, reducing child mortality, and combating malaria and other diseases (UN n.d; VanWey et al. 2013; Richards et al. 2015). However, these positive outcomes did involve tradeoffs in terms of costs imposed to the natural environment, as discussed in the following sections.

3.3 The Soybean Production Complex

Brazil has seen a major expansion of the soybean production during the period 2001 to 2006, with one million hectares replacing cattle grasslands, plus large areas of conversion from forest, peaking toward the end of that period (Morton et al. 2006). The adoption of GM crops in the 1990s, as commented in previous sections, has greatly contributed to the increase of the soybean production (Companhia Nacional de Abastecimento 2017). To counter some of the adverse environmental consequences of the agricultural expansion, a multistakeholder initiative was formed, the Soy Moratorium (Soy M), aiming to curb the commercialization of soybeans grown in deforested lands (Gibbs et al. 2015; Box 22.1). The initiative has been very successful, as evidenced by the reduction of soybean-induced deforestation in the Amazon after its implementation in the year 2006. The percentage of new soybean plots created by clearing native vegetation dropped from 30% to approximately 1% by the year 2014 (Macedo et al. 2012; Gibbs et al. 2015). Nevertheless, steady efforts are needed in order to contain deforestation (Box 22.1), given that the production of soybeans in Brazilian lands is constantly expanding in response to the growing global demand for agricultural products (Soterroni et al. 2019). Currently, Brazil is the world's second-largest soy meal exporter, after the United States (Companhia Nacional de Abastecimento 2017). Its current export market share is expected to set an all-time record (around 114 million tonnes), fueled by relatively low-cost production in Brazil and stricter exporting rules imposed on shipments from the United States. China imports 60% of the soybeans globally traded, buying 50.9 million tonnes from Brazil and 32.9 tonnes from the United States, representing approximately 53% and 34% of China´s purchases, respectively (USDA 2018).

3.4 The Trade-Offs Bound to Agricultural Expansion

As mentioned earlier, the expansion of the agricultural frontier in Brazil has brought economic and infrastructural development to some regions of the country (Vera-Diaz et al. 2009; Garret and Rausch 2016). In the case of the Mato Grosso region, it was estimated that each square kilometer of soybeans contributed approximately USD (US dollar) 150,000 to the gross domestic product (GDP) through the creation of formal jobs outside the agricultural sector, such has commerce, services, construction, education, and public health (Richards et al. 2015), hence contributing to advance several SDGs related to socioeconomic development (mainly SDGs 1, 2, 3, 6 and 8).

However, this positive development and the associated expansion of urban settlements have generated significant challenges to the sustainable use of natural resources. It is well documented that unbalanced land-use change and the expansion of the agricultural frontier challenge the well-being of natural ecosystems (Fearnside

Box 22.1 Regulatory Framework and Certification Initiatives: A Chronology

The use of forest lands in private properties is regulated by the Brazilian Forest Code (FC, established in 1965, with subsequent reforms). It mandates maintenance of 80% of forests as legal reserves while prohibiting the clearing of land in sensitive areas, such as hilltops, slopes, and along river margins (Nepstadt et al. 2014). A relatively recent reform to the Code mandates registration of private lands in the online Rural Environmental Registry (Portuguese acronym: CAR). The purpose of this requirement is to increase transparency and compliance with the FC. The same reform created an amnesty for illegal deforestation in small rural properties ("small" defined as properties up to 440 ha in the Amazon region) and established a market to trade surplus forest (Nepstadt et al. 2014). These reforms were supported by the agribusiness sector while criticized by environmental protection advocates (Soares Filho et al. 2014). In spite of the efforts to increase transparency and compliance, it was shown that registration in CAR alone did not contain deforestation: for example, 25 to 30% of deforestation in the states of Mato Grosso and Pará took place in legal reserves (Gibbs et al. 2015).

In 2006, in response to increasing pressure from consumers and NGOs concerned about soybean driven deforestation, the major stakeholders in the soy agribusiness in Brazil, food supply chain companies, and European alliances signed the Soy Moratorium (Soy M; Gibbs et al. 2015), a voluntary pact pledging not to trade and finance soy grown on lands deforested after July 2006 in the Brazilian Amazon or produced under unfair labor conditions. The agreement monitors an area that produces 98% of the soybeans grown in the Amazon biome. The initiative has been very successful as evidenced by the reduction of soybean induced deforestation, ranging from 30% in the two years prior to its implementation to approximately 1% in 2014 (Macedo et al. 2012; Gibbs et al. 2015). The Soy M mandate has been extended eight times since its introduction and after the last extension deadline (May 2016), the commitment has been extended indefinitely (World Wide Fund for Nature 2016). The argument to discontinue its implementation relied on the expectation that government mechanisms mandated by the FC and enforcing efforts will be mature enough to contain deforestation (Gibbs et al. 2015). Unfortunately, the periods of political instability and associated budgets cuts that the country has been facing over the last years conspire against the continuity of successful policies and environmental protection programs (Angelo 2016; Tollefson 2016b) and jeopardize monitoring and enforcement of the FC mandates. In fact, a new spike in illegal deforestation was observed in 2015 and 2016, reaching its 2008 levels (7,989 km^2) (Tollefson 2016a). Due to the extension and heterogeneity of the areas under surveillance (Richards et al. 2014; 2017), monitoring and enforcing efforts are needed to contain deforestation and avoid losing decades of progress.

2005; Vera-Diaz et al. 2009; Gibbs et al. 2015; Richards et al. 2015). In particular, agriculture driven-deforestation and the associated degradation of the biophysical environment (impacting biodiversity, the provision of ecosystems services and ultimately, human well-being) spark international concern (Zeigler and Truitt Nakata 2014; Amundson et al. 2015; KRBG 2016; Tollefson 2016a; 2018). Due to the extension of farmed lands and land-use changes that had been ramping up over recent decades (Schneider 2011; Richards et al. 2014; 2015), the Amazon biome (tropical forest) represents the initial focal point of agriculture driven-deforestation, which is currently spilling over into the neighboring Cerrado biome (savanna/grasslands), catalogued as a biodiversity hotspot (Françoso et al. 2015, KRBG 2016; Soterroni et al. 2019). In particular, the Mato Grosso state, an extension of 900,000 square kilometers situated across the Amazon and the Cerrado biomes, represents the nation's soy and beef production center (Macedo et al. 2012; Richards and VanWey 2016). Unfortunately, 40% of the deforestation in the Amazon biome during the first half of the past decade took place within its boundaries (Macedo et al. 2012). For those reasons, the state constitutes a good proxy to analyze the development of agriculture and linked environmental issues in the region. In an effort to monetize the value of the forest, Richards and collaborators (2015) estimated that the loss of forest in the Mato Grosso up to the year 2006 (300.000 km^2) represented 4.8 billion metric tons of carbon, which translated into USD 48 billion, assuming a price of 10 USD per ton of carbon, which is broadly consistent with European and Californian carbon markets. The authors pointed out that the forest loss value was still below the calculated direct and indirect economic benefits of the agricultural expansion, when considered over a period of time, but indicated that the estimations did not consider other functions and services that land ecosystems and soils provide.

In addition to deforestation, poor land management practices contribute to the degradation of soils and land habitats (Gubler 2002; Bauch et al. 2015; Terrazas et al. 2015 and references therein), exacerbating the current challenges to ecosystems and potentially impacting human health (Wall et al. 2015). For example, the loss of vegetation cover due to deforestation leads to a steady degradation of the fertile topsoil with a loss of the below-ground biodiversity, progressing toward desertification (Fearnside et al. 2005; Thomaz and Luiz 2012; Chimini Sobral et al. 2015). The tillage of dry, bare soils increases the amount of particles released into the air, which are transported by the wind to distant regions, raising the incidence of respiratory diseases on-site and off-site (Schenker 2005); their deposition affects the quality of surface water bodies (Miranda et al. 2015; Chimini Sobral et al. 2015). In some cases, those particles transport fungal spores, contributing to environmental contamination and the spread of severe fungal diseases beyond endemic areas (Sprigg et al. 2014). Also, a reduced soil quality and microbial biodiversity translates into a decreased biocontrol of pests for plants, animals, and humans, facilitating the outbreak of diseases (Sehgal 2010; Nielsen et al. 2015). These situations represent some examples illustrating the tight links between soil quality and environmental and human well-being, as shown in Figure 22.1.

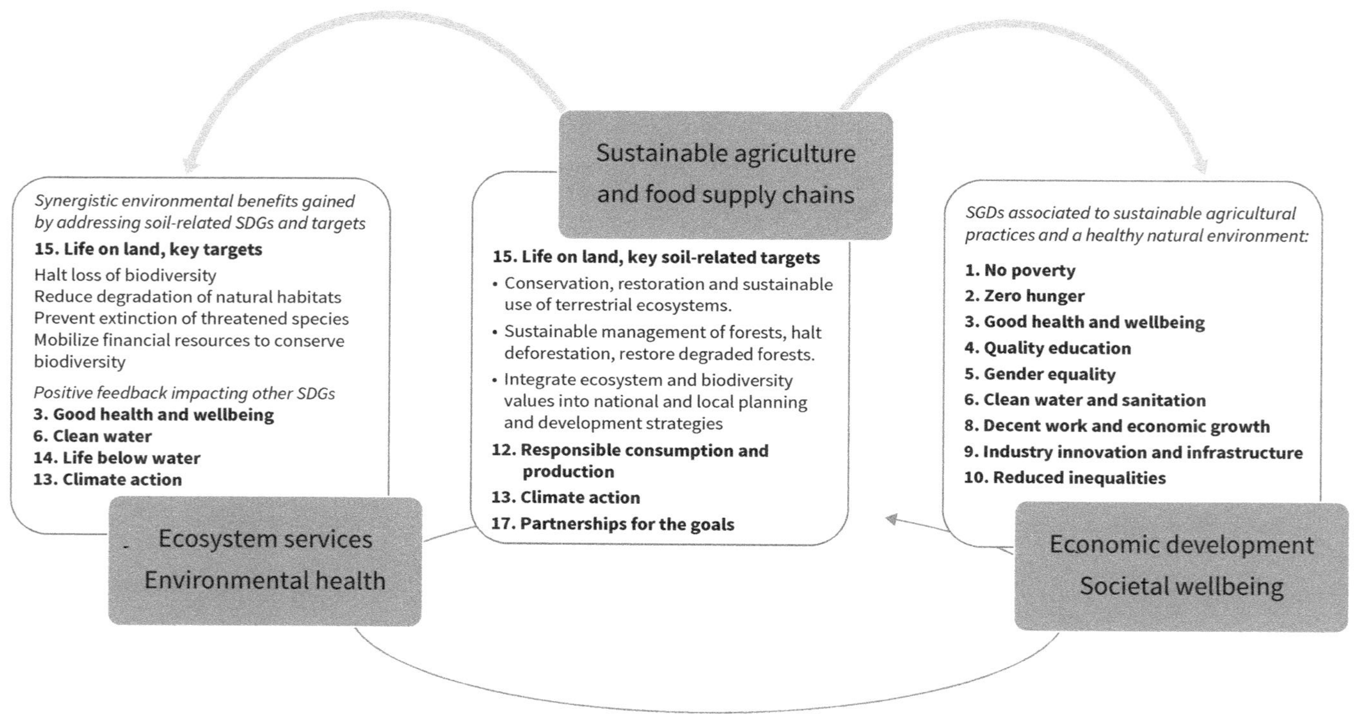

Figure 22.1 Schematic representation of the interlinked nature of the SDGs defined by the United Nations (n.d). The arrows indicate the links between soil-centered actions and sustainable practices with SDGs related to the biophysical and socioeconomic environment and the interdependencies among those domains.

Source: authors

4. Further Steps Toward a Sustainable Agriculture

4.1 Healthy Soils at the Center of the Sustainability Agenda

Although preserving natural ecosystems and their key provisioning, regulating and cultural services (for a comprehensive review see Millennium Ecosystems Assessment 2015) is central to the sustainability agenda, the pressure on them is not likely to recede, given the anticipated increasing demand for food and agricultural products (Schneider 2011; FAO 2013; UN 2017). Hence, concerted efforts to avoid, reduce or mitigate the environmental costs bound to agriculture and food supply chains are needed (Galford et al. 2013). The significant but oftentimes overlooked contributions of soils to ecosystems services (Amundson et al. 2015; Nielsen et al. 2015; Wall et al. 2015; Zhu and Meharg 2015; Keesstra et al. 2016) are brought to light when dissecting the numerous functions of soils, as defined by the European Commission (EC 2006), encompassing: (i) biomass production (including agriculture and forestry), determining the quantity and quality of food fiber and other bioresources; (ii) storage, transformation, and filtration of nutrients and water, playing a key role in biogeochemical cycles and water quality; (iii) biodiversity pool (comprising habitats, species, and genes), supporting a complex web of life contributing to the biogeochemical cycling of nutrients, the suppression of disease and pest control, and constituting a source of useful biomolecules; (iv) carbon pool, key to global climate regulation; (v) source of raw materials; (vi) physical and cultural environment for humans; and (vii) archive of geological and archaeological heritage. Hence, soils constitute both the foundation and a key target for action, guiding the design of sustainable agricultural alternatives presented in this chapter, since actions aimed at conserving this natural resource have direct links to SDGs related to both the biophysical and socioeconomic environments (Figure 22.1).

4.2 Systems Approach to a Sustainable Agri-Food Supply Chain

From the agricultural perspective, sustainability is defined as the capacity to maintain the production (food, fiber, bioresources) through time in a given set of environmental and socioeconomic conditions (Pretty 2008). In other words, it aims to meet the needs of the present without compromising those of future generations. Although some of the different factors affecting agricultural sustainability are site- and situation-specific (and addressing those requires a case-based analysis), defining an overarching framework based on a set of agricultural best management practices and supporting policy recommendations can constitute the scaffold that facilitates the construction of regional (or even crop-specific) approaches to sustainable agricultural practices. In this sense, the following sections describe a holistic strategy for a sustainable agri-food supply chain, taking the Brazilian soy sector as model system.

Sustainable soy production requires concerted actions among key stakeholders in the supply chain, a supportive governance framework and best management practices in the field. The soy supply chain is a complex, multilevel, and multisectorial transboundary system. The main stakeholders can be enumerated as follows: (i) farmers (smallholders and large-scale enterprises); (ii) local communities (families practicing subsistence agriculture and indigenous people); (iii) private sector (agribusiness and food companies providing resources and goods, buying production from farmers, exporting, producing food and other soy-derived products); (iv) commerce, transport, trading institutions, and organizations; (v) banks and financing institutions; (vi) producer's associations and commercial chambers; (vi) government and certifying institutions (national, state, and regional authorities); (vii) multistakeholders ́ dialogue tables (such as the Round Table on Responsible Soy n.d); (viii) nongovernmental organizations (local and international NGOs); (ix) academic and research institutions; and (x) consumers (at the local level and transboundary). Integrated interventions aimed at improving its sustainability must identify critical leverage points across this complex system.

Based on the analysis of the supply-chain's structure and interactions and the assessment of the most pressing environmental issues, national regulatory frameworks, international trade perspectives and previously successful initiatives and interventions, a systems strategy for a sustainable soy supply chain is presented and discussed in the next sections of this chapter. In a nutshell, a market-based strategy backed by the governance of the soy and food supply chain is proposed (summarized in Figure 22.2). It recognizes the central role of the biophysical environment (SDG15, life on land and key soil-bound targets) and stands on four pillars that support the achievement of several socioeconomic SDGs linked to conserving healthy soils (as shown in Figure 22.1), in particular, goals 1, 2, 3, (aiming at combatting poverty and hunger and promoting good health and well-being), 6 (water quality), 12 (responsible consumption and production), 13 (climate action), and 14 (protection of life below water). This systems strategy also seeks to harness the momentum of the sustainability agenda at a global level (UNDP 2012; Zhu and Mehrag 2015; Keesstra et al. 2016) and the growing environmental awareness and public pressure on key players in the agri-food corporate sector (Gibbs et al. 2015; Donofrio 2018; Gross 2018) to foster the formation of "partnerships for the goals" (SDG17), since those alliances are pivotal driving forces boosting the shift to business and development models that factor in the conservation of the natural capital (World Resources Institute 2017; The Nature Conservancy 2020; Soterroni et al. 2019). Key stepping stones and STI outcomes helping advance the proposed integrated strategy for a sustainable agri-food system are discussed in the next section.

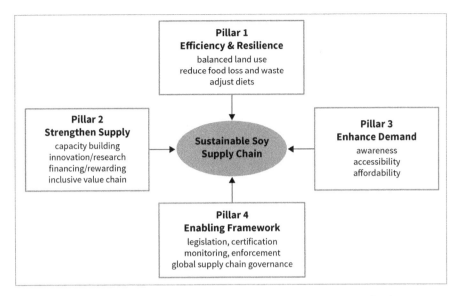

Figure 22.2 Market-based integrated strategy for a sustainable soy supply chain, as a representative agri-food supply chain example. Its four pillars constitute key leverage points across the different sectors in the soy supply chain and its multistakeholder governance.

Source: authors

4.3 Key STI contributions, management practices and policy recommendations

Engineering sustainable agricultural food-supply chains requires contributions from the science and technology sector to develop innovative production systems and improve or scale up existing sustainable models and a supportive market and policy framework. The key actions embedded in the pillars supporting sustainable agricultural production systems presented in Figure 22.2 embrace harmonized production, market, environmental protection, and policy actions aiming to: (i) strengthen efforts to monitor and enforce the compliance with laws and policies supporting the sustainable use of natural resources and avoid further conversion of forest land to agriculture; (ii) optimize the use of resources and increase efficiency in processes along the food supply chain; (iii) increase the availability and affordability of sustainable products; and (iv) foster associations of private agribusiness companies with NGOs supporting sustainable agri-food supply chains.

In order to avoid counteracting actions and increase the chances of sustained success, it is necessary to coordinate efforts along the whole supply chain, including the food industry and the consumers, which comprises a transboundary and multistakeholder scenario, as described earlier. Hence, the STI contributions and supporting policy recommendations proposed in this section are aimed at the production regions, consumers, and global organizations and were classified according to those categories. The key STI contributions are discussed in detail next, together with examples of successful cases that substantiate the rationale for the proposed actions or policy recommendations. For enhanced clarity and faster access, those recommendations and guidelines are summarized in Table 22.1.

A continued flow of STI outcomes is pivotal to advancing the aforementioned actions along the agri-food supply chain. These outcomes arise from both research groups and institutions dedicated to agricultural research, such as Brazil´s EMBRAPA (Martha and Alves 2018) and from an array of scientific disciplines and innovation sectors not related to agriculture, a priori. For example, geographical information systems (GIS) and satellite surveillance contribute to advance SDG goals related to environmental protection, primarily SDG15 (terrestrial ecosystems and biodiversity) and SDG13 (combat climate change), and SDGs linked to those, as previously discussed (Figure 22.1). Its application to the remote observation of forested areas and farms is essential to monitor compliance with the mandates of the Forest Code and international agreements (Box 22.1) aiming to combat deforestation in agricultural producing countries (Gibbs et al. 2015; Nepstadt et al. 2014). GIS software also provides a key tool for designing zoning areas for forest protection, detecting degraded lands that could be reforested or recovered for agriculture by means of sustainable land management practices (Thomaz and Luiz 2012; Kremen and Merenlender 2018) and classifying and monitoring farms that qualify for economic incentives designed to foster environmental and climate protection (Table 22.1, section A). In fact, the technology has been crucial to the implementation of the REDD+ (Reducing Emissions from Deforestation and Forest Degradation) program in Brazil (Duchelle et al. 2014). Notwithstanding, illegal deforestation spots, escaping the official control systems, proliferate (Wilkinson 2017). Hence, persistent efforts and constantly improving monitoring technologies (Richards et al. 2017) are needed to counteract that trend.

Another well-known application of GIS technology is precision agriculture, which is a methodology aiming to optimize the application of agricultural inputs (seeds, fertilizers, water) according to the temporal and spatial needs of the agricultural plots (contrasting with traditional uniform, non–site-specific input applications). This agricultural management approach should be considered complementary to other sustainable management practices (such as those aiming to preserve soil's properties, presented in Table 22.1, section A.5), rather than an alternative to those. Precision agriculture has been documented as an effective management measure to reduce both, input costs, and the potential environmental burden associated to fertilizer and pesticide applications (Walter et al. 2017), supporting SDGs 13, 14, and 15 (protection of land, water and climate) and tightly linked to SDG3 (healthy lives). Such contributions gain particular relevance when taking into account the vast extensions of cultivated lands in the Amazon and Cerrado regions (Richards et al. 2015). Collaborative work and partnerships (SDG17) with rural extension services help overcome the initial hurdles related to knowledge acquisition and technology implementation (Kutter

Table 22.1 Proposed management practices, interventions and complementary policy recommendations supporting a sustainable soy and agri-food supply chain.

A. Soy Production Regions

A.1. Policies and commitments

- Apply regulatory and market-based mechanisms to halt deforestation and reduce overuse of fertilizers and pesticides
- Integrateestablished sustainable agricultural practices in financing mechanisms, commercialization channels, and global trade policies
- Maintain successful initiatives to halt deforestation such as the Soy Moratorium (Soy M) in the Amazon; avoid deforestation leakage to other regions (see B.1 and B.2); apply similar moratorium to neighboring regions (Cerrado) to avoid spillover effects

A.2. Enforcement

- Enhance and develop capacity for the effective monitoring and enforcing compliance with agreed policies and regulations

A.3. Incentives and regulations

- Reward individuals and entitites committed to sustainable practices. Penalize those that do not comply (fines, embargoes)
- Empower and support local communities applying sustainable agriculture initiatives

A.4. Zoning

- Apply the mandates of the Forest Code to register rural properties and mitigate/recover deforested land
- Observe the conservation of endangered and critical zones (fragile ecosystems, biodiversity hotspots)

A.5. Land use planning and best agricultural management practices

- Balance land use, integrating soy with production forest and cattle (conservation agriculture)
- Promote strategies to avoid conversion: establish new soy fields in degraded and unused land
- Improve productivity, minimize agrochemical and energy use, reduce environmental impact, improve or maintain soil quality

B. Soy supply chain: global

Including: producers, retailers from the soy industry, environmental and social NGOs, commodity traders, goods manufactures, consumers, feed industry, and financial institutions

B.1. Commitments and market incentives

- Produce globally applicable standards to help build a market for responsible soy (flagship initiative: Soy Moratorium, Box 22.1)
- Engineer business models to facilitate compliance with sustainability standards (see B.2)

B.2. Engagement

- Foster commitment of agribusiness corporations owning a global presence to participate in sustainable programs, promote their success stories, and establish those as a model

C. Consumers

C.1. Change diet/perceptions

- Reassess diet recommendations based on science-based standards; reduce consumption of animal products, promote sustainable sources of protein

C.2. Increase demand for sustainable products

- Exert consumer's pressure on retailers and brands to commit to responsible soy throughout their supply chains; identify NGOs, communication tools and education campaigns as key resources for this action

C.3. Reduce food loss and waste

- Foster strategic alliances to reduce food loss and waste and related environmental and economic impacts

Continued

Table 22.1 *Continued*

D. Research and Academic Sector

D.1. Basic and applied research and innovation
• Invest and promote agricultural and food technology development supporting sustainable production and consumption patterns
• Balance support to both applied and basic research, enabling development of breakthrough technologies

D.2. Assessment and evaluation
• Design and implementat holistic evaluation tools to assess the impacts of proposed sustainable agriculture innovations considering different scenarios (integrated impact assessment and life-cycle assessment methods)

The proposed actions and policies are aimed at soy production regions, consumers and global organizations and grouped accordingly. Examples of key STI contributions and successful cases that substantiate the rationale for the proposed actions or policy recommendations are discussed in detail under Section 4.3.

Source: the authors. Examples supporting each proposal included in the table are provided in the text of Section 4.3. The text indicates the corresponding location in the table for the item discussed.

et al. 2011; Lindblom et al. 2017). Small rural communities (oftentimes located either in remote or not-well-connected areas) can benefit as well from GIS applications, communication technology, and the increasing availability of smart telephones. For example, it is possible to quickly identify an emerging agricultural pest on the field (that could damage crops overnight) through optical information (such as a picture taken with a smart telephone) and promptly receive management advice (Pongnumkul et al. 2015). At the same time, the technology reduces the isolation of rural communities and inequalities (SDGs 5, 10), facilitating access to health services (SDGs 3), information, and educational resources (SDG4), the stepping stones for societal development and well-being.

Agricultural research is crucial to the design and evaluation of best management practices, which embrace production, conservation of natural resources (SDG14, SDG15), and climate protection (SDG13). Such agricultural management approaches (Table 22.1, section A5) support a balanced land use (Kremen and Merenlender 2018) by integrating crops with forest and cattle production. For example Nielsen et al. (2015) evaluated cases of the effective implementation of crop-livestock rotations to increase soil organic matter content. Thomaz and Luiz (2102) described the results of a seven-year field scale experiment reporting the effective use of palisades and enclosures to retain sediment and promote soil recovery and plant succession. In front of a scenario encompassing a fast growing demand for natural resources and agricultural products (FAO 2013; UN 2017), implementing agricultural practices and strategies that avoid further forest conversion is fundamental (Barreto and Gibbs 2015). This can be achieved by establishing crops in unused lands and recovering degraded ones (Kremen and Miles 2012; Thomaz and Luiz 2012; Galford et al. 2013). On this avenue, Brazil launched in the year 2016 an ambitious strategy to recover 22 million hectares of degraded land for restoration and low-carbon agriculture (including integrated crop, livestock, and forest management) by the year 2030. The strategy aligns with global efforts to protect natural resources and the climate, such as the Bonn Challenge (n.d) and initiatives promoted by NGOs, such as the World Resources Institute (-2016), facilitating interactions with the public and private sectors and providing data and analysis tools to identify opportunities for sustainability commitments (SDG17). The research carried out by Macedo et al. (2012) provides evidence showing that reduced deforestation and increased agricultural production can occur simultaneously in tropical forest frontiers, provided that land is available and that policies promoting the efficient use of already cleared lands (intensification) and restricting deforestation are enforced. This analysis highlights the importance of aligning market and policy frameworks (Figure 22.2) that support sustainable actions on-site (Table 22.1, section A) and globally (Table 22.1, sections B and C), to balance the growing pressure on natural resources. An additional strategy to recover degraded lands is to harness the natural capacity of microorganisms to degrade remaining pesticides and recalcitrant compounds (bioremediation) present in highly stressed agricultural soils (Merini et al. 2007; Cuadrado et al. 2008; Flocco 2016 and references threin) due to previous nonsustainable management practices. It has been calculated that approximately 50% of Brazilian pastures are degraded (Dias-Filho 2014), hence the importance of developing innovative methods to recover degraded lands.

As mentioned, commitments and market incentives along the supply chain are central to the continuity of sustainable agriculture practices and enterprises. The Soy M initiative (Box 22.1 and Table 22.1, section B1) is another remarkable example of partnerships committing to sustainability goals (SDG17) through responsible production and consumption (SDG12). The research carried out by Gibbs et al. (2015) identifies a large reduction in agriculture-driven deforestation after the Soy M inception in 2006, decreasing from 30% to ~1% in a ten years period. The Soy M success inspired the application of its principles and governance system to the Cerrado Region, which has faced a *spillover* effect due to deforestation moving away from the protected Amazon (Soterroni et al. 2019). The Cerrado Manifesto, pledging to halt deforestation and loss of the native Brazilian savannah vegetation, has been recently signed by key local and international players in the food processing and retail industry, but the crucial support of the main commodities traders in the region is still missing (Gross 2018). The engagement of agribusiness corporations, which have a worldwide corporate presence and market influence, is crucial to the long-term success of global sustainability initiatives (SDG17). As mentioned earlier, associations with NGOs catalyze the corporate commitment to sustainable practices, help promote success stories, and foster further agreements. For example, in 2004 the NGO The Nature Conservancy (2020) partnered with one of the major soybean traders (Cargill), pioneering a business approach committed to reducing deforestation in Brazil's Amazon biome, coupled to enforcement efforts by the Brazilian government implementing the Forest Code. They worked with other groups and NGOs to put in place the Soy Moratorium. Currently, there are both, mixed signals and growing expectations that the agribusiness company (together with other four major commodity traders in the region) will extend the commitment to halt deforestation in the Cerrado region (Donofrio 2018), where substantial progress on the matter is yet to be seen.

Research and education institutions and NGOs can help promote healthier and climate-smart diets and consumption habits based on science-supported standards (Table 22.1, section C), contributing primarily to SDGs 3 (health) and 12 (responsible consumption and production), and linked to SDGs 13, 14, and 15 (protecting the environment and climate). For example, reducing the consumption of natural resources- and energy-intensive foods, such as beef, can produce a significant impact on the demand for natural resources at global scale (Adamkeiwicz 2016). Also, consumers and stakeholders in the producing regions and globally, can contribute to a more efficient use of resources by reducing food loss and waste (SDGs 12), a hidden leaking pipe in the food production and supply chains. The Food and Agriculture Organization of the United Nations (FAO n.d) estimates that 32% of all food produced in the world (on weight basis) was lost or wasted in 2009, causing approximately $940 billion per year in economic losses. The global initiative Champions 12.3 Coalition (n.d) tackles these issues, aiming to fulfill the SDG target 12.3: to halve food waste and reduce food loss globally by 2030.

Investment in research should support a balance of both, applied and basic research, since the latter constitutes the stepping stone for developing promising innovative and disruptive technologies (SDG9, industry, innovation, and infrastructure) that can boost progress toward sustainability goals (Table 22.1, section D). A remarkable example of such synergy is represented by the agricultural application of the powerful gene editing technology CRISPR/Cas9 (Clustered Regularly Interspaced Short Palindromic Repeats /CRISPR-associated protein9; Lander 2016). It can be used as a very precise molecular editing tool to modify the genetic sequence of plants (and other organisms) to obtain desired phenotypic traits (for example, increased yield, improved nutritional quality or resilience to biotic and abiotic stress), without requiring the insertion of external genetic (transgenic) material (Haque et al. 2018). Although the technology has yet to overcome regulatory and technical barriers to fully develop (Gao 2018), it is headed to mark the next biotechnological revolution in agriculture, comparable to the introduction of GM soybean in the nineties, as described earlier.

Although the application of gene technology in agriculture constitutes a contentious arena, it is deemed as a game-changing tool to bridge the yield gap in global grain production (Hermans et al. 2010), a target that could not be achieved by other methods for agricultural intensification. A careful and transparent assessment of the benefits and trade-offs bound to agricultural technology innovations is crucial to guide informed decisions and policies that guarantee safety and reduce uncertainties around their applications.

Other examples of innovative and breakthrough technologies (SDG9) advancing sustainability in agricultural and food systems (Table 22.1, sections A5, C1, and D) include: (i) the re-assessment of animal feed composition and search for novel protein sources (alternatives to soybean meal) for animal (Shabat et al. 2016) and human consumption (Henchion et al. 2017); (ii) the application of modern molecular biology analysis methods, such as high throughput sequencing technologies, to gain insights into ruminants' microbiological milieu, and guide the introduction of diet modifications to reduce methane production and the footprint of agriculture on the global climate (Tapio et al. 2017); (iii) the bioengineering of soil microbial communities to recover degraded lands (Neumann et al. 2014; Flocco, 2016), as mentioned earlier; and (iv) the development of lab-designed food, such as plant-based meat, or even meat grown in vitro. Furthermore, a growing food start-up community is embracing the power of artificial intelligence to dissect the molecular structure of animal food, with the aim of obtaining those components from sustainable sources and recreating the organoleptic and texture qualities in laboratory designed food (Card 2017).

Finally, STI outcomes are necessary to assess and evaluate the impacts of proposed sustainable agricultural production systems (Table 22.1, section D.2) considering different aspects, including region of origin and use (da Silva, van der Werf et al. 2010), production practices at the farm level (Grenz et al. 2009), and how the whole cycle of production and transport affects the environmental footprint of food products, from farm to fork (Adamkeiwicz 2016). Furthermore, current research efforts to develop assessment and evaluation tools are focusing on integrated models of environmental sustainability, health, and economic development, which enable in-depth explorations of food production and trade scenarios (Garnett n.d.)

This discussion makes apparent that a continued flow of STI outcomes, harmonized with policies and initiatives that support their effective implementation across the supply chain, is crucial to advancing the sustainability agenda, at the local and global levels.

5. Concluding Remarks and Outlook

This chapter focused on the analysis of the soybean production sector in Brazil, one of the largest food and agricultural product providers of the world, taking the Mato Grosso state as a case study. Much progress on the path to sustainability has been made over the last four decades, from the inception of the MDGs to the successors, the ambitious SDGs, aiming to lift millions out of poverty, improve education and healthcare, and curb agriculture-driven negative impacts on the environment (da Silva, Del Grossi, and Galvão de França 2010; World Bank 2014). However, underlying adverse factors play against the implementation of the country's sustainability agenda. Political and economic instability (Escobar 2018; Tollefson 2016b; 2018) threaten to reverse the progress made so far, in particular in matters related to environmental protection and poverty alleviation. In spite of this threat, Brazil's recent development history has demonstrated that the country is able to pull together the needed determination and the entrepreneurial capacity of its citizens (Henriquez and Li Pun 2013) to come out of adverse situations. Furthermore, the successful outcomes from the country's flagship programs to combat hunger and poverty, such as the Zero Hunger and Family Grant programs, have been presented on the global stage as the models to follow (UNDP 2012; World Bank 2014).

Due to the complex and interlinked nature of pressing global issues, in particular the sustainable use of natural resources and food security and the well-documented long-term impacts on the natural capital driven by agricultural expansion, a progressive shift to proactive approaches (rather than remedial ones) is required, together with concerted actions among all stakeholders and transboundary cooperation. From the perspective of Brazil's development over recent decades, it is apparent that the path to reaching the SDGs pertaining to environmental protection is not a straightforward one, but the solid progress achieved so far indicates that it is achievable. Fostering and harnessing the contributions that STI can deliver to the agricultural sector, learning from successful cases and initiatives and crafting policy approaches harmonized with socioeconomic development goals and environmental protection targets, constitute the pillars supporting sustainable global agri-food systems and the overarching goal of achieving long-term societal *and* environmental well-being.

Acknowledgments

Part of this research was carried out within the framework of the Professional Program in Environmental Policy and International Development of Harvard University Extension School (HES) Cambridge, MA. The author is thankful to: Dr. J. Hunt, L. Tomasso, and H. Silver at HES for their insightful comments and exchanges; Dr. H. Li Pun, Executive Secretary, FONTAGRO (Regional Fund for Agricultural Technology) at the Inter-American Development Bank, Washington, DC; and Dr. P. Henriquez, International Specialist on Management of Technological Innovation in Agriculture, Inter-American Institute for Cooperation on Agriculture (IICA), who kindly discussed successful innovations and sustainable agriculture strategies implemented by rural communities in Latin America; Dr. G.B. Martha Jr. at the Brazilian Agricultural Research Corporation (EMBRAPA), Brasilia, for sharing his knowledge of the role of STI in Brazil's agricultural development; Dr H.K. Gibbs at Nelson Institute for Environmental Studies, Center for Sustainability and the Global Environment (SAGE), University of Wisconsin-Madison, for sharing her insight into the Soy Moratorium and its implementation; and to the editors of this volume and three anonymous reviewers for their thoughtful comments, which contributed to improving the manuscript.

References

Adamkeiwicz, G. (2016). Buying local, do food miles matter? http://www.extension.harvard.edu/inside-extension/buying-local-do-food-miles-matter (accessed April 6, 2019).

Amundson, R., Berhe, A. A., Hopmans, J. W., Olson, C., Sztein, E., and Sparks, D. L. (2015). Soil and human security in the 21st century. *Science* 348(6235), 1261071.

Angelo, C. (2016). Demotion of science ministry angers beleaguered Brazilian researchers. *Nature* 533, 301. doi: 10.1038/nature.2016.19910

Arraes Pereira, P. A., Martha Jr, G. B., Santana, C. A. M., and Alves, E. (2012). The development of Brazilian agriculture: future technological challenges and opportunities. *Agriculture and Food Security* 1, 1–12.

Barreto, P. and Gibbs, H. K. (2015). Imazon Brief: Como melhorar a eficácia dos acordos contra o desmatamento associado à pecuária na Amazônia? (How to improve the effectiveness of agreements against deforestation associated with ranching in the Amazon?). Imazon, Belem Brazil and University of Wisconsin-Madison (in Portuguese language). http://imazon.org.br/PDFimazon/Portugues/livros/TACPecuaria_WEB.pdf (accessed March 21, 2019)

Bauch, S. C., Birkenbach, A. M., Pattanayak, S. K., and Sills, E. O. (2015). Public health impacts of ecosystem changes in the Brazilian Amazon. *Proceedings of the National Academy of Science* 112, 7414–7419.

Bonn Challenge (n.d.). http://www.bonnchallenge.org (accessed November 6, 2018).

Card, J. (2017). Lab-grown food: the goal is to remove the animal from meat production. The disruptors. *Guardian Small Business Network.* https://www.theguardian.com/small-business-network/2017/jul/24/lab-grown-food-indiebio-artificial-intelligence-walmart-vegetarian (accessed October 28, 2018).

Champions 12.3 Coalition. (n.d). https://champions123.org (accessed October 4, 2018).

Chimini Sobral, A., Palcheco Peixoto, A. S., Fernandez Nacsimento, V., Rodgers, J., and da Silva, A. M. (2015). Natural and anthropogenic influence on soil erosion in a rural watershed in the Brazilian southeastern region. *Regional Environmental Change* 15, 709–720.

Companhia Nacional de Abastecimento (2017). Séries Históricas. http://www.conab.gov.br/conteudos.php?a=1252 (accessed October 31, 2018).

Cuadrado, V., Merini, L., Flocco, C. G., and Giulietti, A. M. (2008). Degradation of 2,4-DB in Argentinean agricultural soils with high humic matter content. *Applied Microbiology and Biotechnology* 77, 1371–1378.

da Silva, V. P., van der Werf, H. M., Spies, A., and Soares, S. R. (2010). Variability in environmental impacts of Brazilian soybean according to crop production and transport scenarios. *Journal of Environmental Management* 91, 1831–1839.

Dias Avila, A. F., Romano, L., and Garagorry, F. (2010). Agricultural productivity in Latin America and the Caribbean and Sources of Growth. In: Pingali, P. L., and Evenson, R. E. (Eds.), *Handbook of Agricultural Economics*, pp. 3713–3768. Elsevier, Amsterdam.

Dias-Filho, M. B. (2014). Diagnóstico da Pastagens no Brasil. In: Melo Borges, L. C. (Ed.), *Embrapa Documents 402*, pp 1–36. Embrapa Amazônia Oriental, Belém.

Donofrio, S. (2018). Major soy traders get low grades for Cerrado Sourcing. Ecosystem Market Place, October 15. https://www.forest-trends.org/ecosystem_marketplace/major-soy-traders-get-low-grades-for-cerrado-sourcing (accessed October 29, 2018).

Duchelle, A. E., Cromberg, M., Gebara, M. F., Guerra, R., Melo, T., Larson, A., Cronkleton, P., Börner, J., Sills, E., Wunder., S, Bauch, S., May, P., Selaya, G., and Sunderlin, W. D. (2014) Linking forest tenure reform, environmental compliance, and incentives: lessons from REDD+ initiatives in the Brazilian Amazon. *World Development* 55, 53–67.

Escobar, H. (2018). Scientists, environmentalists brace for Brazil's right turn. *Science* 3, 273–274.

European Commission (EC). (2006). *Soil Protection: The Story Behind the Strategy.* Office for Official Publications of the European Communities, Luxembourg. http://ec.europa.eu/environment/archives/soil/pdf/soillight.pdf (accessed March 5, 2019).

da Silva, J. G., Del Grossi, M. E., and Galvão de França, C. (Eds.) *The Fome Zero (Zero Hunger Program): The Brazilian Experience*. NEAD Special Series 13. Food and Agriculture Organization of the United Nations, Rome. http://www.fao.org/3/i3023e/i3023e.pdf.

Food and Agriculture Organization of the United Nations (FAO) (2013). Feeding nine billion in 2050. http://www.fao.org/news/story/en/item/174172/icode (accessed October 4, 2018).

Food and Agriculture Organization of the United Nations (FAO) (n.d). Save food: global initiative on food loss and waste reduction. http://www.fao.org/save-food/en (accessed October 4, 2018).

Fearnside, P. M. (2005). Deforestation in Brazilian Amazonia: history, rates, and consequences. *Conservation Biology* 19, 680–688.

Fernandes, E. C. M., Soliman, A., Confalonieri, R., Donatelli, M., and Tubiello, F. (2012). Climate change and agriculture in Latin America, 2020–2050: projected impacts and response to adaptation strategies. World Bank, Washington, DC. https://openknowledge. worldbank.org/handle/10986/12582 (accessed March 5, 2019).

Filomeno, A. F. (2013). State capacity and intellectual property regimes: lessons from South American soybean agriculture. *Technology in Society* 35, 139–152.

Flocco, C. G. (2016). Eco-engineering sustainable agricultural practices: an international research cooperation between Europe and Latin America. https://www.linkedin.com/pulse/ international-cooperation-between-europe-latin-flocco-ph-d- (accessed October 4, 2018).

Françoso, R. D., Brandão, R., Nogueira, C. C., Salmona, Y. B., Bomfim Machado, R., and Colli, G. C. (2015). Habitat loss and the effectiveness of protected areas in the Cerrado biodiversity hotspot. *Natureza and Conservação* 13, 35–40.

Freitas, P. L., and Landers, J. N. (2014). The transformation of agriculture in Brazil through development and adoption of zero tillage conservation agriculture. *International Soil and Water Conservation Research* 2, 35–46.

Galford, G. L., Soares-Filho, B., and Cerri, C. E. P. (2013). Prospects for land-use sustainability on the agricultural frontier of the Brazilian Amazon. *Philosophical Transactions of the Royal Society Series B: Biological Sciences* 368(1619), doi.org/10.1098/rstb.2012.0171

Gao, C. (2018). The future of CRISPR technologies in agriculture. *Nature Reviews Molecular Cell Biology* 19, 275–276.

Garnett, T. (n.d.). Modeling the relationship between the food system and health, development and the environment. http://www.futureoffood.ox.ac.uk/project/modelling-relationship-between- food-system-and-health-development-and-environment (accessed October 28, 2018).

Garret, R. D., and Rausch, L. L. (2016). Green for gold: social and ecological tradeoffs influencing the sustainability of the Brazilian soy industry. *Journal of Peasant Studies* 43, 461–493.

Gibbs, H. K., Rausch, L., Munger, J., Schelly, I., Morton, D.C., Noojipady, P., Soares-Filho, B., Barreto, P. Micol, L., and Walker, N. F. (2015). Brazil's Soy Moratorium: supply chain governance is needed to avoid deforestation. *Science* 347, 377–378.

Grenz, J., Thalmann, C., Stämpfli, A., Studer, C., and Häni, F. (2009). RISE: method for assessing the sustainability of agricultural production at farm level. *Rural Development News* 1, 5–9.

Gross, A. S. (2018). The Cerrado manifesto could curb deforestation, but needs support: experts. Mongabay Series. https://news.mongabay.com/2018/03/cerrado-manifesto- could-curb-deforestation-but-needs-support-experts (accessed October 29, 2018).

Gubler, D. J. (2002). The global emergence/resurgence of arboviral diseases as public health problems. *Archives of Medical Research*, 33: 330–342.

Hansen, J., and Gale, F. (2014). China in the next decade: rising meat demand and growing imports of feed. USDA Economic Reference Center. https://www.ers.usda.gov/amber- waves/2014/april/china-in-the-next-decade-rising-meat-demand-and-growing-imports- of-feed (accessed October 4, 2018).

Haque, E., Taniguchi, H., Hassan, M. M., Bhowmik, P., Karim, M. R., Śmiech, M., Zhao, K., Rahman, M., and Islam, T. (2018). Application of CRISPR/Cas9 genome editing technology for the improvement of crops cultivated in tropical climates: recent progress, prospects, and challenges. *Frontiers in Plant Science* 9, 617. doi:10.3389/fpls.2018.00617

Henchion, M., Hayes, M., Mullen, A. M., Fenelon, M., and Tiwari, B. (2017). Future protein supply and demand: strategies and factors influencing a sustainable equilibrium. *Foods* 6, 53. doi:10.3390/foods6070053

Henriquez, P., and Li Pun, H. (eds.) (2013). *Innovaciones de impacto: Lecciones de la Agricultura Familiar en America Latina y el Caribe.* Banco Interamericano de Desarrollo and Inter-American Institute for Cooperation on Agriculture, San Jose, Costa Rica.

Hermans, K., Verburg, P., Stehfest, E., and Müller, C. (2010). The yield gap of global grain production: a spatial analysis. *Agricultural Systems* 103, 316–326.

Keesstra, S. D., Bouma, J., Wallinga, J., Tittonell, P., Smith, P., Cerdà, A., Montanarella, L., Quinton, J. N., Pachepsky, Y., van der Putten, W. H., Bardgett, R. D., Moolenaar, S., Mol, G., Jansen, B., and Fresco, L.O. (2016). The significance of soils and soil science towards realization of the United Nations Sustainable Development Goals. *Soil* 2, 111–128.

Kew Royal Botanical Gardens (KRBG). (2016). State of the world's plants. https://stateoftheworldsplants.org/2016/report/sotwp_2016.pdf (accessed April 8, 2019).

Kremen, C., and Merenlender A. M. (2018). Landscapes that work for biodiversity and people. *Science* 362(6412), eaau6020. doi:10.1126/science.aau6020

Kremen, C., and Miles, A. (2012). Ecosystem services in biologically diversified versus conventional farming systems: benefits, externalities, and trade-offs. *Ecology and Society* 17, 40. doi:10.5751/ES-05035-170440

Kutter, T., Tiemann, S., Siebert, R., and Fountas, S. (2011). The role of communication and cooperation in the adoption of precision farming. *Precision Agriculture* 12, 2–17.

Lander, E. S. (2016). The heroes of CRISPR. *Cell*, 164, 19–28.

Lindblom, J., Lundström, C., Ljung, M., and Jonsson, A. (2017). Promoting sustainable intensification in precision agriculture: review of decision support systems development and strategies. *Precision Agriculture* 18, 309–331.

Macedo M. N., DeFries, R. S., Morton, D. C., Stickler, C. M., Gallford, G. L., and Shimabukuro, Y. E. (2012). Decoupling of deforestation and soy production in the southern Amazon during the late 2000s. *Proceedings of the National Academy of Science* 109, 1341–1346.

Martha Jr., G. B., and Alves E. (2018). The role and impact of public research and technology transfer in Brazilian agriculture. In: Kalaitzandonakes, N., Carayannis, E., Grigoroudis, E. and Rozakis, S. (Eds.), *From Agriscience to Agribusiness. Innovation, Technology, and Management,* pp. 429–444. Springer, Cham.

Mendonça, M. L. (2011). Monocropping for agrofuels: the case of Brazil. *Development* 54, 98–103.

Merini, L., Cuadrado, V., Flocco, C. G., and Giulietti, A. M. (2007). Dissipation of 2,4-D in soils of the humid pampa region, Argentina: a microcosm study. *Chemosphere* 68, 259–265.

Millennium Ecosystems Assessment (2015). *Ecosystems and Human Well-Being: Synthesis.* Millennium Assessment Board of Editors. Island Press, Washington, DC.

Miranda, R. B., D'Almeida Scarpinella, G., Siloto da Silva, R., and Mauad, F. F. (2015). Water erosion in Brazil and in the world: a brief review. *Modern Environmental Science and Engineering* 1, 17–26.

Morton, D. C., DeFries, R. S., Shimabukuru, Y. E., Anderson, L. O., Arai, E., del Bon Spiritu-Santo, F., Freitas, R., and Morisette, J. (2006). Cropland expansion changes deforestation dynamics in the southern Brazilian Amazon. *Proceedings of the National Academy of Science* 103, 14637–14641.

Nepstadt D., Mc Grath D., Stickler, C., Alenear, A., Azevedo, A., Swette, B., Bezerra, T., DiGiano, M., Shimada, J., Seroa da Motta, R., Armijo, E., Castello, L., Brando, P., Hansen,

M. C., McGrath-Horn, M., Carvalho, O., and Hess, L. (2014). Slowing Amazon deforestation through public policy and interventions in beef and soy supply chains. *Science* 344, 1118–1123.

Neumann, D., Heuer, A., Hemkemeyer, M., Martens, R., and Tebbe, C. C. (2014). Importance of soil organic matter for the diversity of microorganisms involved in the degradation of organic pollutants. *The International Society for Microbial Ecology Journal* 8, 1289–1300.

Nicolopoulou-Stamati, P., Maipas, S., Kotampasi, C., Stamatis, P. and Hens, L. (2016). Chemical pesticides and human health: the urgent need for a new concept in agriculture. *Frontiers in Public Health* 4, 418. doi:10.3389/fpubh.2016.00148

Nielsen, U. N., Wall, D. H., and Six, J. (2015). Soil biodiversity and the environment. *Annual Reviews of Environmental Resources* 40, 63–90.

Organisation for Economic Co-operation and Development (OECD). (2015). *Frascati Manual 2015: Guidelines for Collecting and Reporting Data on Research and Experimental Development: The Measurement of Scientific, Technological and Innovation Activities.* OECD Publishing, Paris.

Peiretti, R., and Dumanski, J. (2014). The transformation of agriculture in Argentina through soil conservation. *International Soil and Water Conservation Research* 2, 14–20.

Pongnumkul, S., Chaovalit, P., and Surasvadi, N. (2015) Applications of smartphone-based sensors in agriculture: a systematic review of research. *Journal of Sensors* 2015(195308), doi.org/10.1155/2015/195308.

Pretty, J. (2008). Agricultural sustainability: concepts, principles and evidence. *Philosophical Transactions of the Royal Society of London B: Biological Sciences* 363, 447–465.

Qaim, M. (2005). Agricultural biotechnology adoption in developing countries. *American Journal of Agricultural Economics* 87, 1317–1324.

Richards, P., Arima, E., VanWey, L., Cohn, A., and Bhattarai, N. (2017). Are Brazil's deforesters avoiding detection? *Conservation Letters* 10: 470–476. doi:10.1111/conl.12310

Richards P., Pellegrina, H., VanWey, L., and Spera S. (2015). Soybean development: the impact of a decade of agricultural change on urban and economic growth in Mato Grosso, Brazil. *PLoS ONE* 10, e0122510. doi.org/10.1371/journal.pone.0122510

Richards, P. D., and VanWey, L. (2016). Farm-scale distribution of deforestation and remaining forest cover in Mato Grosso. *Nature Climate Change* 6, 418–426.

Richards P. D., Walker, R. T., and Arima, E. Y. (2014). Spatially complex land change: the indirect effect of Brazil's agricultural sector on land use in Amazonia. *Global Environmental Change* 29, 1–9.

Round Table on Responsible Soy (n.d.). http://www.responsiblesoy.org (accessed October 4, 2018).

Saturnino, H. M. (1998). Sustentabilidade do Agronegócio: contribuição do sistema de Plantio Direto. In: de Caldas, A., Pinheiro, L. E. L., de Medeiros, J. X., Mizuta, K., da Gama, G. B. M. N., Cunha, P. R. D. L., Kuabara, M. Y., and Blumenschein, A. (Eds.), *Agronegócio Brasileiro: Ciência, Tecnologia e Competitividade*, pp. 215–224. National Council for Scientific and Technological Development (CNPq), Brasília.

Schneider, M. (2011). *Feeding China's Pigs: Implications for the Environment, China's Smallholder Farms and Food Security.* Institute of Agriculture and Trade Policy. http://www.iatp.org/files/2011_04_25_FeedingChinasPigs_0.pdf (accessed March 21, 2019)

Schenker M. B. (2005). Farming and asthma. *Occupational and Environmental Medicine* 62, 211–212.

Sehgal, R. N. M. (2010). Deforestation and avian infectious diseases. *Journal of Experimental Biology* 213, 955–960.

Shabat, S. K., Sasson, G., Doron-Faigenboim, A., Durman, T., Yaacoby, S., Berg Miller, M. E., White, B. A., Shterzer, N., and Mizrahi, I. (2016). Specific microbiome-dependent mechanisms underlie the energy harvest efficiency of ruminants. *The International Society for Microbial Ecology Journal* 10, 2958–2972.

Soares-Filho, B., Rajão, R., Macedo, M., Carneiro, A., Costa W., Coe. M., Rodrigues, H., and Alencar, A. (2014). Cracking Brazil's forest code. *Science* 344, 363–364.

Soterroni, A. C., Ramos, F. M., Mosnier, A., Fargione, J., Andrade, P. R., Baumgarten, L., Pirker, J., Obersteiner, M., Kraxner, F., Câmara, G., Carvalho, A. X. Y., and Polasky, S. (2019). Expanding the soy moratorium to Brazil's Cerrado. *Science Advances* 5 (7), eaav7336.

Sprigg, W. A., Nickovic, S., Pejanivic, G., Petkovic, S., Vujadinovic, M., Vukovic, A., Dacic, M., DiBiase, S., Prasad, A., and El-Askary, H. (2014). Regional dust storm modeling for health services: the case of valley fever. *Aeolian Research* 14, 53–73.

Tapio, I., Snelling, T. J., Strozzi, F., and Wallace, R. J. (2017). The ruminal microbiome associated with methane emissions from ruminant livestock. *Journal of Animal Science and Biotechnology* 8, 7. doi:10.1186/s40104-017-0141-0

Terrazas, V. C. M., de Souza Sampaio, V., Barros de Castro. D., Costa Pinto, R., de Albuquerque, B.C., Sadahiro, M., Dos Passos, R. A., and Braga, J. U. (2015). Deforestation, drainage network, indigenous status, and geographical differences of malaria in the State of Amazonas. *Malaria Journal* 14, 379–387.

The Nature Conservancy (2020). Companies investing in nature. Protecting the forests and grasslands of Brazil. https://www.nature.org/en-us/about-us/who-we-are/how-we-work/working-with-companies/companies-investing-in-nature1/cargill/ (accessed April 17, 2020).

Thomaz, E. L., and Luiz, J. C. (2012). Soil loss, soil degradation and rehabilitation in a degraded land area in Guarapuava (Brazil). *Land Degradation and Development* 23, 72–81.

Tollefson, J. (2016a). Deforestation spikes in Brazilian Amazon. *Nature*, 540, 182.

Tollefson, J. (2016b). Political upheaval threatens Brazil's environmental protections. *Nature* 539, 147–148.

Tollefson, J. (2018). Brazil´s lawmakers renew push to weaken environmental rules. *Nature*, 557, 17.

United Nations (2015). *Transforming Our World: the 2030 Agenda for Sustainable Development, A/RES/70/1*. United Nations, New York.

United Nations (2017). World Population Prospects: The 2017 Revision, Key Findings and Advance Tables. Working Paper No. ESA/P/WP/248. United Nations, Department of Economic and Social Affairs, Population Division, New York.

United Nations (n.d.). Millennium development goals indicators. Country Progress Snapshot: Brazil. http://mdgs.un.org/unsd/mdg/Data.aspx (accessed November 5, 2018).

United Nations (n.d.). Sustainable Development Goals. http://www.un.org/sustainabledevelopment/sustainable-development-goals (accessed October 4, 2018).

United Nations Development Programme (2012). Brazil MDG Award salutes innovation in development. http://www.undp.org/content/undp/en/home/presscenter/articles/2012/06/01/un-development-chief-president-of-brazil-present-mdg-awards.html (accessed October 4, 2018).

United States Department of Agriculture (2018). Oilseeds: world markets and trade. https://apps.fas.usda.gov/psdonline/circulars/oilseeds.pdf (accessed March 21, 2019)

Valin, H, Sands, R. D., van der Mensbrugghe, D., Nelson, G.C., Ahammad, H., Blanc, E., Bodirsky, B., Fujimori, S., Hasegawa, T., Havlik, P., Heyhoe, E., Kyle, P., Mason-D'Croz, D., Paltsev, S., Rolinski, S., Tabeau, A., van Meijl, H., von Lampe, M., and Willenbockel, D. (2014). The future of food demand: understanding differences in global economic models. *Agricultural Economics* 45, 51–67.

VanWey, L. K., Spera, S. A., Sa, R. D., Mahr, D., and Mustard, J. F. (2013). Socioeconomic development and agricultural intensification in Mato Grosso. *Philosophical Transactions of the Royal Society of London. Series B, Biological Sciences* 368(1619), doi.org/10.1098/rstb.2012.0168

Vera-Diaz, M. d. C., Kaufmann, R. K., and Nepstad, D. C. (2009). The Environmental Impacts of Soybean Expansion and Infrastructure Development in Brazil's Amazon Basin. GDAE Working Paper Series 09-05. Tufts University, Medford, MA.

Wahabzada, M., Mahlein, A-K., Bauckhage, C., Steiner, U., Oerke, E. C., and Kersting, K. (2016). Plant phenotyping using probabilistic topic models: uncovering the hyperspectral language of plants. *Scientific Reports* 6(22482). doi.org/10.1038/srep22482

Wall, D. H., Nielsen, U. N., and Six, J. (2015). Soil biodiversity and human health. *Nature* 528, 69–76.

Walter, A., Finger, R., Huber, R., and Buchmann, N. (2017). Smart farming is key to developing sustainable agriculture. *Proceedings of the National Academy of Sciences* 114, 6148–6150.

Wang, Z. H., Zhai, F. Y., Wang, H. J., Zhang, J. G., Du, W. W. Su, C., Zhang, J., Jiang, H. R., and Zhang, B. (2015). Secular trends in meat and seafood consumption patterns among Chinese adults, 1991–2011. *European Journal of Clinical Nutrition* 69, 227–233.

Wilkinson, A. (2017). Brazilian Amazon still plagued by illegal use of natural resources. *Nature* doi:10.1038/nature.2017.22830.

World Bank. (2014). How to reduce poverty: a new lesson from Brazil for the world? http://www.worldbank.org/en/news/feature/2014/03/22/mundo-sin-pobreza-leccion-brasil-mundo-bolsa-familia (accessed April 6, 2019).

World Resources Institute (2016). Brazil announces restoration and low-carbon agriculture target: 22 million hectares by 2030. https://wribrasil.org.br/en/blog/2016/12/brazil-announces-restoration-and-low-carbon-agriculture-target-22-million-hectares-2030 (accessed November 5, 2018).

World Resources Institute (2017). P4G initiative launches to accelerate sustainable growth through innovative partnerships. Press release, September 20. https://www.wri.org/news/2017/09/release-p4g-initiative-launches-accelerate-sustainable-growth-through-innovative (accessed November 1, 2018).

World Wide Fund for Nature (2016). Soybean Moratorium is extended for indefinite period. https://www.wwf.org.br/informacoes/english/?52122/soybean-moratorium-is-extended-for-indefinite-period (accessed October 5, 2018).

Zeigler, M., and Truitt Nakata, G. (2014). The Next Global Breadbasket: How Latin America Can Feed the World: A Call to Action for Addressing Challenges and Developing Solutions. Inter-American Development Bank (IDB) publications, monograph: 202. IDB, Washington, DC.

Zhu, Y. G., and Meharg, A. (2015). Protecting global soil resource for ecosystem services. *Ecosystem Health and Sustainability* 1, 1–4.

23
Transformation Governance for Sustainable Development

Making Science, Technology, and Innovation Work for Small-Scale Fisheries

Karin Wedig

1. Introduction

Worldwide, small-scale fisheries (SSFs) provide income and nutrition (Béné et al. 2007; FAO 2003; Heck et al. 2007) and thus contribute to the sustainable development goals (SDGs): SDG1 (end poverty) and SDG2 (end hunger). At the same time, wild-capture fisheries, including SSFs, play a significant role in the degradation and depletion of marine and inland water resources, mainly through overfishing, thereby potentially threatening the achievement of SDG14 (conserve and sustainably use the oceans, seas and marine resources). As the man-made reduction in wild fish stocks is putting unprecedented pressures on aquatic ecosystems (Pauly and Zeller 2016) and on the coastal and riparian societies that depend on them (UN 2012), the expansion of large-scale cage aquaculture has emerged as a response to the challenge of securing the availability of fish (World Bank 2013; HLPE 2014; AFSPAN 2015).

Advantages of aquaculture as an instrument for the sustainable growth of fish production include its efficiency as an animal protein production system: fish converts significantly more feed into body mass than do terrestrial animals, and fish production generates far lower nitrogen and phosphorus emissions than does red meat production (HLPE 2014). However, large-scale fish farming also has negative ecological effects, including eutrophication (Ottinger et al. 2016), chemical discharge (Pillay 2008), transfer of diseases to wild fish (Pillay 2008), and overfishing of wild fish that are used as fish feed (Boyd and McNevin 2015). As a result, aquaculture can create new pressures on wild fish stocks instead of relieving existing ones. If negative ecological effects degrade the quality of fishing grounds that small fishers depend on, aquaculture can jeopardize local income sources, thereby compromising the achievement of SDGs 1 and 2. In addition, cages tend to be installed in highly productive water areas, thereby creating competition with small fishers' access to fishing grounds (Wedig 2018). In both cases, there can be negative effects on local welfare, especially if small fishers lack access to new economic opportunities created by aquaculture (Wedig 2018).

Whether aquaculture will contribute to or threaten the achievement of SDGs 1, 2, and 14 will, to a large extent, depend on how its expansion is governed. Science, technology, and innovation (STI) can play a key role in enabling the ecologically, economically, and socially sustainable governance of fisheries resources by creating knowledge and fostering a greater understanding of the ecological effects of economic development strategies, increasing the efficiency of natural resource use, and providing solutions to some of the challenges that arise from conflicting interests between different groups of resource users. In the context of declining wild fish stocks and the degradation of aquatic systems that small fishers depend on—combined with a continuously increasing global demand for fish and persistent poverty among small fishers in the Global South—STI needs to support the transformation of consumption and production structures, and a redistribution of access to productive assets and inputs in favor of primary producers. While an extensive analysis of current fisheries resource governance approaches is beyond the scope of this chapter, the following discussion illuminates the key characteristics of a socially and ecologically sustainable approach to the governance of aquaculture expansion, which protects the natural resources that small fishers depend on and maximizes their ability to benefit from fish-farming. Such an approach, I argue, needs to advance a utilization of STI that puts the economic and social-ecological needs of marginalized, small-scale natural resources users at center stage.

In agriculture, the longstanding insight that technologies and innovative approaches for sustainable intensification need to be accessible to and designed for small producers (Schumacher 1973) is currently receiving renewed attention (Peters 2013; Rauch et al. 2016), especially in the context of small producers' persistent poverty and an increasing concentration of corporate power in agri-food chains worldwide. The effects of climate change on agriculture are further intensifying calls for a "shift from conventional, monoculture-based and high-external-input-dependent industrial production toward mosaics of sustainable, regenerative production systems that also considerably improve the productivity of small-scale farmers" (UNCTAD 2013, p. i). In fisheries, however, the current agenda for the management of natural resources seems to favor the *expansion* of high-external-input-dependent production by promoting a private property based approach (World Bank 2014) that has been criticized for disregarding the needs and capacities of SSFs worldwide (WFFP and WFF 2013; TNI 2014; Wedig 2018).

This chapter argues for a transformation governance approach to achieve sustainable and inclusive management of fisheries resources and aquaculture expansion and thereby enable the sector's contributions to the SDGs 1, 2, and 14. The discussion starts out from the key role of the fisheries and aquaculture sector in improving food security and providing incomes to large numbers of people worldwide, and it recognizes the unprecedented challenges to sustainable development that the sector faces due to overexploitation, an increasingly unequal distribution of access to resources, and our still-limited knowledge of aquatic ecosystems. Using new evidence from Lake Victoria, which harbors Africa's largest inland SSFs and a rapidly-growing cage aquaculture industry, the chapter illuminates how small fishers fail to benefit from the introduction of large-scale aquaculture and how ecological pressures on fisheries resources—unabated or potentially aggravated by aquaculture expansion—may create or intensify poverty traps that lock small fishers in unsustainable resource-use practices. The analysis indicates that STI-based instruments for a sustainable expansion of aquaculture that addresses the needs of small fishers are largely absent in the observed setting. Instead, the rapid growth of cages on the lake seems to create additional economic and social-ecological challenges for the SSF. The chapter concludes with suggestions for STI-based solutions which, if integrated in a transformative governance approach, could support the achievement of SDGs 1, 2, and 14 on the basis of a sustainable development of both, capture fisheries and aquaculture.

The findings are based on a face-to-face socioeconomic survey conducted in 2016 among small fishers, fish-workers and fish traders ("fisherfolks," henceforth) in the Lake Victoria basin (Wedig and Bbosa 2016, unpublished data) and 66 semistructured interviews conducted in 2017 with fisherfolks, including committee members of local comanagement systems for fisheries resources, as well as local government officials and aquaculture investors and workers. An analysis of regional policy documents on the lake's fisheries and aquaculture sector provided information about the riparian states' common regulatory framework that guides, more or less closely, the existing governance practices. The survey included 1,304 interviews at 80 landing sites in the lake's three riparian countries: Kenya (22 of the landing sites), Tanzania (28 sites), and Uganda (30 sites). The sample was designed to cover similar numbers of fisherfolks in each country's SSFs, and sample sizes were adjusted to accommodate variations in landing site size (measured by the numbers of boats that actually use the site). The survey and semistructured interviews generated data on fisherfolks' i) income sources and employment relations; ii) economic organization in formal and informal institutions to manage fisheries resources; iii) poverty proxies including nutritional security, access to financial and other support services, and control over assets and facilities; iv) patterns of access to fish; and v) local environment knowledge about the lake. The semistructured interviews further focused on existing forms of comanagement and governance practices in the fisheries and aquaculture sector. The data was generated as part of a larger transdisciplinary research project that develops scenarios for ecologically, socially, and economically sustainable cage-aquaculture on Lake Victoria (Glaser et al. 2015).

2. A Transformative Governance for Fisheries Resources

Transformation governance as a conceptual framework for the organization of fisheries resource management and aquaculture expansion follows the principles of sustainable governance by giving equal weight to the three pillars of sustainability: social/human development, economic viability and inclusiveness, and ecological responsibility. Furthermore, it concentrates on guiding wider societal change and transformation processes with the intention of reducing economic and social inequality in the context of global environmental change. As is discussed in more detail, it aims to go beyond the promotion of greater *overall* eco-efficiency by addressing issues of inequality in the distribution of benefits from a more efficient use of resources. Furthermore, it proposes a more ambitious conceptual framework for the key issues of social inclusion and economic equality by contextualizing participatory and adaptive governance approaches, such as comanagement systems, within a broader social-ecological context. This context is defined first, by the goal of redistributing access to natural resources in favor of marginalized resource users,[1] and second, by a recognition of the limits of the Earth system's ecological capacities and the associated total limits to natural resource use (eco-sufficiency). The overall goal of transformative governance approaches is fully consistent with that of the SDGs: to mitigate and adapt to earth system transformations in equitable ways while improving the well-being of marginalized groups and populations. The approach, however, de-emphasizes the technocratic management of natural resources in favor of democratic transformation processes that are more akin to the 'organic sustainability revolution' imagined by Meadows et al. (2004).

Given the significant role of capture fisheries and aquaculture in natural resource degradation, achieving greater eco-efficiency in production processes is fundamental to ensuring that the expansion of both sectors in response to increasing human demand for fish products is sustainable. In fisheries, improved eco-efficiency metrics are particularly crucial for by-catch, lost nets and debris, and habitat damage caused by mobile fishing gears. In aquaculture, water pollution and damage to marine and other aquatic habitats caused by cages are major factors of environmental degradation. While the validity of aiming to ensure greater eco-efficiency in food production is undisputed, the difficulty lies in defining the monetary value of the affected ecosystem services or, put differently, the cost of the damages that are inflicted upon them. Willison and Côté (2009) treat eco-efficiencies—the concept of creating more goods and services while using fewer resources and creating less waste and pollution—as situations in which the costs of ecological damages are lower than the economic benefits obtained by the production activity in question. For the purpose of industry regulation, this definition is useful, but its limitations are obvious: our still incomplete knowledge of many aquatic systems and the difficulties in measuring ecological damage, especially if it occurs time-delayed and/or geographically removed from its source, incorporate a significant error margin into existing abilities to achieve eco-efficiency.

Furthermore, the monetary valuation of ecosystems services—necessary to assess the cost of damages to biodiversity or to the functionality of ecosystems more

---

[1] It is acknowledged that comanagement systems often fail to integrate marginalized resource users into decision-making in the context of natural resource governance in ways that enable such groups to shape the outcomes of policy processes (Evans et al. 2011; Nunan 2010).

generally—is influenced by cultural perceptions of utility and risk that reflect dynamic and historically-specific human-nature relations. Consequently, the definition of eco-efficiency may be less straightforward than is often assumed, indicating the need for an improved transdisciplinary understanding of social-ecological systems, as developed by Biggs et al. (2015). Another challenge to the realization of eco-efficiency confirms this point: access to STI-based instruments that help to achieve greater eco-efficiency is determined by socioeconomic and political relations that may create barriers to the equitable implementation of eco-efficiency. Thus, if STI-based solutions for eco-efficiency exclusively serve the interests of a relatively small group of actors, such as aquaculture investors, large numbers of asset-poor small fishers may be driven into using increasingly destructive fishing methods to make a living, because they lack alternative employment opportunities and the resources to invest into fish-farming. Similarly, if STI-based solutions for eco-efficiency in capture fisheries are inaccessible to small fishers, an ecologically sustainable development of the sector is unlikely in countries with substantial SSFs. The mandatory introduction of technologies and innovations that are designed for industrial-scale fishers may also push small fishers out of legal fishing, and seriously undermine inclusiveness.

I argue that in addition to eco-efficiency, a redistribution of access to productive assets and inputs in favor of small-scale primary producers is necessary to promote and protect production structures that seek to maintain the long-term viability of natural resources. As an example of a redistributive approach, the regulation of aquaculture expansion could be used to protect small fishers' third-generation human rights to natural resources. This could help to prevent poverty-driven social-ecological traps. Such traps cause SSFs to become stuck in unsustainable practices (including illegal and undocumented fisheries), which are originally entered into often as a survivalist strategy. SSFs who benefit from STI-based solutions to fisheries development are more likely to grow into medium-scale actors who can more easily be held accountable for potentially unsustainable resource use, thus improving eco-efficiency.

The last pillar of transformation governance, which complements eco-efficiency and redistribution, is eco-sufficiency. This concept expresses the self-imposed limitation of consumption and production in absolute terms, the principle of "living well on less." From a governance perspective it should be noted that restrictions on total resource use need to honor equal consumption rights within existing planetary boundaries, as defined by Steffen et al. (2015) and recognized by the Paris 2015 agreement. Following the recognition of the biophysical limits of growth that relies on finite resources (Meadows et al. 2004), this third factor responds to the temporary and potentially absolute limitations to eco-efficiency in the context of rapid global environmental change. One critique of eco-efficiency points to the incompatibility of unregulated capitalism and environmental sustainability (Naess 2011). Assuming that a true decoupling of growth in consumption and production from environmental degradation is impossible, capitalism, which depends on and tends to result in growth, has an in-built tendency toward crisis. Instead of addressing capitalism's internal contradiction, eco-efficiency sidesteps the apparent need for fundamental changes in consumption and production and attempts to provide solutions that minimize demands on human behavioral change.

The most recent data on overfishing (Pauly and Zeller 2016) indicates that the basic economic goal of indefinite growth needs to be reconsidered to achieve sustainable development in the fisheries sector. Even if a true de-coupling could be achieved in fisheries, it would possibly come too late to prevent large-scale and potentially irreversible damage to the world's fisheries resources on which large numbers of marginalized people depend. Furthermore, eco-efficiency gains would need to successfully transform all parts of the global economy that affect the world's aquatic systems. Lastly, a rebound effect may occur, in which the positive environmental effects achieved through eco-efficiency gains are smaller than the income effect that leads to increased consumption (Figge et al. 2014). If eco-efficiency is achieved for a product that is complementary to another more polluting product, eco-efficiency gains may ultimately lead to increased use of both products, resulting in an overall increase in environmental degradation.

3. Policy Shortcomings for Realizing Transformation Governance

Mitigating and adapting to global environmental change has become fundamental to international development efforts across economic and social sectors, and for agriculture and fisheries, the effects of climate change, resource depletion and biodiversity loss are today widely recognized by governments and international donors as major challenges to sustainable development. The rapid degradation of the world's fishing resources, despite their fundamental importance for nutritional security, has been labeled a worldwide crisis of fisheries and has given rise to increasing efforts to secure sustainable SSFs in coastal and freshwater areas (FAO 2015). The Food and Agriculture Organization's (FAO) Voluntary Guidelines for Securing Sustainable Small-Scale Fisheries in the Context of Food Security and Poverty Eradication, which complement the 1995 FAO Code of Conduct for Responsible Fisheries (FAO 2015) is an instrument for promoting social, economic and ecological sustainability principles in national fisheries development structures. The guidelines recognize that customary practices in SSFs, which have promoted sustainable fishing efforts for centuries, are being lost due to nonparticipatory fisheries management systems, the promotion of inappropriate technologies, and demographic changes that are often driven by a rapid expansion of industrial fisheries.

Recognizing the importance of SSFs for a sustainable future in fisheries, the FAO guidelines promote the protection of small fishers' human rights and explicitly mention the responsibility of business enterprises whose operations affect SSFs to comply with international human-rights standards and the duty of states to ensure that they do. They further promote respect of cultures, including customary practices of small fishers, nondiscrimination and gender equity, and equality in policies and practices.

They express the need to ensure consultation and participation of small-scale fishing communities "in the whole decision-making process related to fishery resources and areas where small-scale fisheries operate as well as adjacent land areas" (FAO 2015, p. 2), and they encourage the adoption of institutions and regulatory frameworks that are designed to represent the interests of small-scale fishing communities.

The launch of these guidelines has sent an important signal to policymakers about the key role of SSFs for the sustainable development of the fisheries sector. The role of appropriate technologies, including "existing traditional and local cost-efficient technologies, local innovations and culturally appropriate technology transfers" (FAO 2015, p. 11) is emphasized and the use of state investment for identifying, developing, supporting and distributing such technologies is promoted. However, the guidelines are silent on the challenges that are posed to sustainable fisheries development by inappropriate technologies. Similarly, the guidelines encourage the adoption of integrated and holistic ecosystem approaches to the management of SSF, but little attention is paid to the often unsustainable operational strategies that are pursued by large investors, protected by national governments, and which directly threaten or eliminate sustainable customary practices in SSFs. Unsurprisingly, the statements about the responsibility of states and large-scale actors to protect and respect the rights of SSF communities to the natural resources that they depend on remain vague and are not supported by concrete policy suggestions.

As a result of these shortcomings, the FAO guidelines, despite their important function of highlighting the importance of SSFs, are unlikely to promote governance structures in the fisheries sector that are capable of forming an adequate and timely response to ongoing processes of environmental changes that affect the world's aquatic systems. The rapid loss of biodiversity in marine, coastal, riparian, and freshwater ecosystems requires social and economic transformation processes at local, national and global levels to mitigate and adapt to simultaneous and in some cases mutually reinforcing changes in equitable ways. There are limitations to eco-efficiency: in some cases, technological progress is too slow to halt negative environmental effects in time to avoid irreversible reductions in human well-being. In other cases, the limitations relate to damaging levels of waste, or costs and risks created by new technologies that are larger than their benefits. Further, the possibility of a rebound effect requires the continuous sensitization and education of both local and global consumers about environmental changes and how they relate to consumption and production patterns. Governance responses that support the transition to sustainable consumption and productions patterns need to include an agenda of redistribution and targeted eco-sufficiency—a transformative agenda that requires paradigm shifts in human behavior.

Such transformative governance structures, as described in the previous section, aim to integrate local-level needs into global sustainability goals and vice versa. They require diverse, multilevel institutions, networks, and multidirectional processes of cooperation and negotiation to translate the challenges arising from global environmental changes into regional and national regulations, and conversely local socioeconomic and social-ecological needs into regional and global responses. Mechanisms of democratic control are necessary to identify unequal power relations in value chains and production processes (e.g., Daviron and Ponte 2005; Wedig 2019). Like small-scale agricultural systems, SSFs need democratic, permanent forms of economic organization that can connect them to policymakers and also help to hold large-scale actors accountable if they violate the rights of small-scale fishing communities to sustainably harvest the natural resources they have traditionally depended on. At the same time, such organizations can help policymakers to enforce restrictions on unsustainable resource use among SSFs, but top-down control should not become a dominant function of such organizations, as was the case with comanagement systems in Uganda's fisheries (Lawrence and Watkins 2011). The approach is one of empowerment, rather than control, based on the assumption that SSFs will largely operate in environmentally responsible ways if they can expect to secure decent living standards for themselves and their dependents by doing so. While both small and large actors need to be held accountable for overexploitation, the drivers of unsustainable resource use need to be better understood to allow for more effective and adequate responses.

Following from this, transformation governance needs to be participatory and democratic to reflect local as well as broader regional needs and interests. The principle of participation is here understood to go beyond activities of contribution to or being involved in fisheries resource management to shaping political settlements and exerting democratic control over the outcomes of political negotiation processes that determine relevant natural resource management frameworks. Transformation governance also needs to be ethical and normative, because it requires a foundation in responsibility, foresight, and solidarity to ensure distributive justice and the maintenance of the conditions for human societies to thrive in a healthy natural environment. Lastly, it needs global and regional institutions that maintain open, two-way communication channels with local organizations and national institutional frameworks. The focus of international donors on community-based organizations to support sustainable development in agriculture and fisheries is insufficient, because local self-help groups in poor rural areas are often unable to challenge structural inequalities (Pattenden 2010; Wedig 2019). Economic organizations with democratic control mechanisms that operate at local and national levels tend to be better positioned to represent the interests of small producers vis-à-vis stronger economic actors and the state (Wedig 2019).

4. Poverty and Inequality in Lake Victoria's SSF

This section discusses new evidence from small-scale fishing communities in the Lake Victoria basin, confirming the need for investments into technologies and innovative approaches that are accessible to and address small-fishers' and fish-workers' economic needs. Highlighting widespread destitution among small-scale fishers and fish-workers, the data points to the limited choices in the SSF over the fishing practices used and the associated barriers to the sustainable development of small-scale fishing operations. The ongoing and significant decline in Nile perch[2] fish stocks in Lake Victoria (Glaser 2017) seems to create pressure on small boat owners and their boat crews to engage in unsustainable practices, making poverty-driven social-ecological traps increasingly likely. Such traps can cause a mutually reinforcing downward spiral of wild fish availability and levels of human well-being in the lake basin. In this context, the data indicates that STI-based instruments for the governance of fisheries resources and aquaculture expansion could contribute to the sustainable development of the sector if they enabled the identification and utilization of local environmental knowledge from SSF to inform the design of natural resource use strategies.

Evidence from the fisheries and aquaculture sector on Lake Victoria illustrates the importance of redistribution in addition to eco-efficiency to foster sustainable development. Shared by Uganda (43%), Tanzania (51%), and Kenya (6%), the lake has a densely populated basin with an estimated 33 million people who depend at least partially on fisheries and over three million people who are directly engaged in the sector (Linard et al. 2012; Glaser et al. 2015). The lake's fish stock is a significant source of export revenue for all three riparian countries, with Uganda and Kenya hosting the bulk of export-oriented fish-processing infrastructure. The sector is characterized by large numbers of small-scale gear-owners and boat-owners and even larger numbers of casual workers who supply tilapia and mukene (a native cyprinid fish) to local and regional markets, and Nile perch to domestic factories for export. Male workers are most often boat crew members and are subject to poor and unsafe working conditions, while women tend to engage in mainly survivalist activities at the landing sites, which include cleaning, drying, and trading fish.

The evidence from the survey, which represents the first large appraisal of poverty in Lake Victoria's fisheries since Geheb et al. (2008), shows that destitution is still widespread among both fish-workers and small-scale gear- and boat-owners, and high malnutrition rates persist particularly among vulnerable groups, most notably women.[3] Primary education was the highest level of education for 76% of respondents and only 1.3% had attended a college or university. Given the nature of the fisheries sector, which is heavily male-dominated, 69% of survey respondents were male. The 31% female respondents mainly worked as fishmongers and/or were engaged in small-scale trading as well as handling and primary-processing at the landing sites (see Table 23.1 for the sample distribution by age).

The most common types of economic activities among respondents were self-employment and wage employment in capture fisheries and fish mongering (with many respondents depending on several of these activities), while aquaculture as a

---

[2] Nile perch is the major export fish and the fish species with the highest economic value in Lake Victoria.

[3] Semistructured interviews indicated that malnutrition also exists among children; however, no data was generated among children.

Table 23.1 Total sample distribution by age group (percentage)

18 to 29 years	30 to 49 years	Above 50 years	Unsure of age
24.6%	59.8%	9.4%	6.2%

source of income was negligible (see Table 23.2). The survey showed that most small-scale actors in Lake Victoria's fisheries sector catch, buy, and sell Nile perch. Thus, despite significantly reduced Nile-perch-based incomes due to the decline in Nile perch stock since the late 1990s, few people are capable of switching to fish-farming. The widespread assumption made by policymakers and international donors that aquaculture expansion creates new income opportunities for small fishers is not supported by the data from Lake Victoria. This is significant, because aquaculture operations on the lake have grown rapidly, with Uganda alone reporting a 120-fold growth between 2000 and 2013 (FAO 2016), and investments in Tanzania have started to increase, too.

For people who were self-employed or employed in capture fisheries, living conditions were poor: only 32% of total respondents had electricity at their homes, 53% relied on water from a river/stream/lake and access to sanitation was divided into 80% pit latrines, 10% flush toilets, and 10% bush/surroundings. More than half (55%) reported earth/sand/dung flooring in their homes and corrugated iron was the most common roofing material. Despite an average of 28% self-reporting as traders,

Table 23.2 Type of economic activities among survey respondents (percentage, by country)

	Percentage of respondents in Tanzania	Percentage of respondents in Kenya	Percentage of respondents in Uganda
Self-employed in capture fisheries	29.0%	33.1%	25.8%
Employed in capture fisheries	25.9%	38.9%	34.1%
Self-employed in aquaculture	0.0%	0.3%	2.6%
Employed in aquaculture (low-skilled)	0.0%	2.0%	13.5%
Handling and processing	12.4%	5.7%	12.7%
Ancillary sector	3.1%	0.9%	6.9%
Trader	18.0%	23.7%	22.8%
Fishmonger	27.7%	17.7%	18.1%
Boat build/repair	2.9%	1.4%	1.8%
Nets repair	2.4%	2.0%	1.2%
Employed in aquaculture (semiskilled)	0.0%	0.0%	0.8%

Source: survey data from author.

vehicular ownership was limited: only 0.6% owned a truck, 2.2% a car, 10% a motorcycle, and 20% a bicycle. The difference in poverty indicators between the employed and self-employed was negligible, except for the few larger boat owners who did not depend on additional wage work as boat crew members. For most poverty indicators, there was a tendency, albeit weak, for poverty to reduce with age. Thus, respondents who were 50 or older were most likely to own bicycles, cars, or trucks and to have access to electricity, a flush toilet, and a tiled or cemented floor. However, the living standards assessment indicated widespread and deep poverty throughout all age brackets. Only 46% owned any equipment, including minor and nonspecialized objects with very low value, such as plastic/wooden rakes for spreading and turning fish for drying, plastic containers for carrying fish, ice drums for fish storage, or wooden "sawing needles," used by artisanal fish gear makers and repairer.

Similarly, nutrition security, measured by the variety of foods accessed, was potentially compromised among all respondents, but especially among women. The gender gap, which was also pronounced with regard to ownership of transport vehicles, was particularly pronounced for access to animal proteins other than fish. Semistructured interviews also indicated that women were more likely than men to eat less nutritious parts of fish, such as the tail and head. Unsurprisingly, fresh and dried fish were the most common sources of animal protein, with 79% and 52% of men and women taken together reporting consumption more than once per week, but for most respondents, fish was also the only source of animal protein. Thus, large numbers of respondents reported that they never or only on rare special occasions ("never," henceforth) ate chicken (85%), beef or pork (52%), eggs (60%), or milk/yoghurt (41%), despite all of these being part of the traditionally common diet in Africa's wider Great Lakes region. Among women, these figures rose to 93% for chicken, 62% for beef/pork, 67% for eggs, and 48% for milk/yoghurt.

Given the extent and depth of poverty among fisherfolks, appropriate technologies and innovative approaches to support sustainable fishing practices in SSF need to be accompanied by significant improvements in basic social services and the rural transport and production infrastructure that is available to the population in the Lake Victoria basin. If nutritional security is not stable and housing quality is too low to ensure healthy living conditions, small-scale fishers, traders, and fishmongers (the latter also engage in primary processing) are unlikely to be able to widely adopt new technologies. In the case of wage workers, the establishment of permanent organizations to fight for better working conditions is equally unlikely under conditions of sustained poverty. The next section takes a closer look at the kind of STI-based approaches that could contribute to a sustainable development of the fisheries and aquaculture sector in the Lake Victoria context.

5. STI-Based Approaches for a Sustainable Expansion of Aquaculture

STI-based instruments that work for small fishers are primarily those that can help to ensure that the ongoing growth of aquaculture operations on the lake contributes to, rather than challenges, the achievement of the SDGs 1 and 2 for the lake basin population. However, given the direct dependence of fisherfolks and their household members on the fisheries resources provided by the lake, the achievements of SDGs 1 and 2 are closely linked to significant contributions to SDG14. An ecologically sustainable use of fisheries resources is needed to help secure the economic future of large numbers of people in the lake basin and protect existing livelihoods, which includes ways of living and working that form part of local social structures. Throughout the lake basin, fisherfolks reported not only significantly reduced Nile perch catches, but also changes in the lake's water quality and reductions in the size and quality of existing wetland areas. Bodies of local environmental knowledge, which could entail important information about sustainable fishing practices and adaptation strategies in the context of environmental changes, are already diminished and may be lost without adequate stock-taking exercises in due time.

Local environment knowledge can play an important role in promoting the sustainable governance of fisheries resources and aquaculture expansion by fostering participatory policy design processes that address the social-ecological needs of local resource users, in this case small-scale fishers and fish workers (see Table 23.3). Using participatory mapping, which combines ethnographic methods with film as well as GPS technologies, such bodies of local environment knowledge can be identified, visualized, and communicated to policymakers and external economic stakeholders to inform and guide the design of regulations that manage natural resource use in locally-specific contexts (Reichel and Frömming 2014; Reichel 2019). Such transdisciplinary and participatory methods can form the basis for effective STI-based instruments for inclusive natural resource management. In the case of SSF, participatory mapping can inform local government officials and aquaculture investors about the social-ecological needs and natural resource use strategies of fisherfolks, including in remote areas, where quantitative monitoring of fishing efforts is difficult. If integrated into fisheries resource management approaches, this knowledge can contribute to sustainable governance systems that have a high chance of being accepted and adhered to by SSF.

New technologies and innovations that contribute to greater eco-efficiency need to be accessible to SSFs without creating stifling economic dependencies and new risks. If such developments are designed to address SSFs' economic needs and capacities, they can empower large numbers of small-scale actors to sustainably harvest the natural resources they depend on, even in the context of spatial competition on the lake due to increased aquaculture operations. For example, the development of low-cost boats that use solar panels, wind turbines, and/or hydrogen fuel cells to power batteries that run electric boat motors could help small fishers to reach more remote fishing grounds and avoiding areas of intense aquaculture (essentially cage constructions in the water) that occupy water areas near the shore line (see Table 23.3). Access to such boats would need to be made possible by financial services offered to asset-poor individuals and groups in SSF. Currently, many small boat owners are unable to reach fishing grounds, because they cannot afford the additional fuel needed due to

Table 23.3 Examples for STI-based approaches for sustainable fisheries and aquaculture

STI-based approach	Contributions to SDGs
Participatory mapping of local environmental knowledge about the lake and its wetlands, which uses ethnographic methods, film, and GPS technology.	SDG1: Protect the economic prospects of fisherfolks who have no or limited access to modern fishing gear and fleets by allowing for an inclusion of traditional fishing practices in regulations for fisheries resource management. Protect the economic prospects of all small-scale fishers by integrating the social-ecological and economic needs more effectively into natural resource management. SDG14: Foster sustainable fisheries by expanding existing knowledge about sustainable fishing and natural resource use practices in and around the lake. Integration of results into sustainable fisheries resource governance regulations can help to make the latter more acceptable to local fisherfolks. Evaluation of local ecosystem services for local resource users to support natural stewardship approaches. Develop effective adaptation strategies in the context of environmental changes.
Low-cost boats with electric motors that are powered by renewable energy sources.	SDG1: Protect the economic prospects of fisherfolks who are forced to fish further from shore due to the decline in wild fish stocks and the occupation of shore areas by aquaculture parks. SDG14: Protect the lake's water quality and local air quality by avoiding pollution.
Development of plant-based fish-feed production systems that can operate sustainably, using local inputs.	SDG1: Creation of employment opportunities in other sectors, such as agriculture, that are accessible to fisherfolks. SDG2: Avoidance of pressures on small fish species that contribute to local food security. SDG14: Avoidance of additional pressures on wild fish stocks through the operation of aquaculture farms.

Source: based on survey data from author.

the occupation of fishing grounds near the shore by aquaculture parks. The reduction in Nile perch is also forcing fishers to search for fish further and further ashore.

Likewise, the key goals of social and economic sustainability—equality, health, and human rights, as well as welfare and equitable access to natural resources and to the benefits from growth (economic)—depend on the inclusion of small fishers and fish-workers as beneficiaries of any strategy for the sector's development. This entails the creation of new employment opportunities for fisherfolks inside and outside the fishing industry, as well as improved access to fish and cash income for women and other vulnerable groups to eliminate malnutrition. An example of an STI-based contribution to agricultural employment creation outside fisheries, which also protects the economic interests of those who continue to depend on SSF, is the development of plant-based fish-feed production systems that can operate sustainably (see Table 23.3). Currently, every kilo of farmed fish production relies on equal or greater amounts of wild fish that is captured to feed the farmed fish. While the economic value (to the aquaculture investor) of a unit of farmed fish is greater than that of a unit of wild fish used for feed, neither the ecological cost nor the economic cost (of reduced wild fish stocks) for small fishers are internalized in this calculation. Plant-based fish-feed production can have negative ecological effects, too, but these can arguably be better managed than those that are produced by the largely unregulated capture of wild fish species that have important (but possibly not fully-understood) ecological functions in the lake.

6. Conclusion

Against the background of the examples discussed in this chapter, the question arises how the transformation governance framework can support the resulting "mosaics of sustainable, regenerative production systems" or at least strategies to ensure a coherent and comprehensive approach to sustainable aquaculture expansion and management of fisheries resources in SSF. Current development strategies, especially in Uganda and Kenya, focus on attracting large-scale investments in industrial fishing, fish-processing and—in the face of declining wild fish stocks—commercial aquaculture (Wedig 2018). Environmental impact assessments are required for large investments, but the enormous importance of such projects for macro-economic balance sheets and the relatively weak regulatory frameworks increase the likelihood that large investors can renegotiate investment conditions and secure regulatory incentives. Simultaneously, the jobs created by such investments are few, require few skills (and hence are low-paid, and provide few opportunities for further training), and often have poor working conditions. Development strategies that are predominantly based on large-scale investment projects are therefore unlikely to be socially and economically sustainable. Instead, small fishers may be driven out of the sector or into illegal and unsustainable fishing practices, such as immature fish catch or fishing in protected areas. Large-scale cage aquaculture can have similar effects, for example, if wild fish stocks are used as fish feed, if inadequate waste management creates water pollution, or if aquaculture parks produce spatial competition by occupying areas of the lake that are particularly productive for wild fish populations.

Applying a threefold approach to the governance of fisheries on Lake Victoria could help to introduce a stronger focus on sustainable SSF into current policy thinking and integrate STI-based instruments into natural resource management strategies that systematically address the social, environmental, and economic dimensions of sustainability in closely interlinked ways. Thus, the development and diffusion of appropriate technologies and innovative approaches would be firmly integrated with the interlinked goals of eco-efficiency and redistribution to allow SSFs to operate sustainably and help small boat owners grow into medium-sized enterprises. Such enterprises play an important role for sustainable fisheries, because they are less likely—and less capable—of pursuing a deplete-and-move-on strategy. With lower financial mobility and greater integration into local cultures and domestic development strategies than transnational corporations, they can contribute substantially to sustainable and inclusive growth. A macro-economic assessment of the net income effects of export earnings, such as profits that are not expatriated to shareholders abroad, and of the net improvements in economic well-being that are achieved with such earnings, can help to reevaluate the current emphasis on foreign direct investment and promote a more balanced approach between large- and small-scale fisheries and aquaculture.

Lastly, a recognition of the ecological limits to growth requires transformation governance strategies to include biophysical thresholds and the possibility that some production methods are no longer viable if a transgression of thresholds is to be avoided. The concept of eco-sufficiency in fisheries relates to both the consumption patterns in highly-industrialized countries and the production patterns in Lake Victoria's fisheries. Key questions that need to be answered relate to the conditions for ecologically sustainable aquaculture as an alternative to fishing efforts for declining wild fish, the quantities of wild fish that can be captured for export without threatening local nutrition security, and the ways in which local environmental knowledge and customary fishing practices can contribute to the achievement of a sustainable fisheries sector.

This chapter has argued for an integrated approach to sustainable development in fisheries, which combines eco-efficiency and eco-sufficiency to promote an improved ratio between production value and resource use while restricting total resource use to prevent rebound effects and safeguard against the limitations of eco-efficiency. Eco-efficiency, which is largely STI-driven, is essential to ecologically sustainable intensification, but it may be ultimately ineffective if it excludes large numbers of small actors who will continue to depend on fisheries' natural resources. The inclusion of small-scale fishers and fish workers into a sustainable future of fisheries is essential to prevent poverty-driven social-ecological traps that can trigger a downward spiral of unsustainable fisheries resource use and poverty.

References

Aquaculture for Food Security, Poverty Alleviation and Nutrition (AFSPAN) (2015). *Final Technical Report.* https://cordis.europa.eu/docs/results/289/289760/final1-afspan-final-technical-report.pdf (accessed March 7, 2019)

Béné, C., Hersoug, B., and Allison, E. H. (2010). Not by rent alone: analysing the pro-poor functions of small-scale fisheries in developing countries. *Development Policy Review* 28(3), 325–358.

Béné C., Macfadyen, G., and Allison, E. (2007). Increasing the Contribution of Small- Scale Fisheries to Poverty Alleviation and Food Security. FAO Fisheries Technical Paper. No. 481. FAO, Rome.

Biggs, R., Schlüter, M., and Schoon, M. L. (Eds.) (2015). *Principles for Building Resilience: Sustaining Ecosystem Services in Social-Ecological Systems.* Cambridge University Press, Cambridge.

Boyd, C. E., and McNevin, A. A. (2015). *Aquaculture, Resource Use, and the Environment.* John Wiley & Sons, Hoboken, NJ.

Evans, L., Cherrett, N., and Pemsl, D. (2011). Assessing the impact of fisheries co-management interventions in developing countries: a meta-analysis. *Journal of Environmental Management* 92(8), 1938–1949.

Figge, F., Young, W., and Barkemeyer, R. (2014). Sufficiency or efficiency to achieve lower resource consumption and emissions? the role of the rebound effect. *Journal of Cleaner Production* 69, 216–224.

Food and Agricultural Organization of the United Nations (FAO) (2003). *Strategies for Increasing the Sustainable Contribution of Small-scale Fisheries to Food Security and Poverty Alleviation.* Committee on Fisheries, 25th Session. FAO, Rome.

Food and Agricultural Organization of the United Nations (FAO) (2015). *Securing Sustainable Small-Scale Fisheries: Update on the Development of the Voluntary Guidelines for Securing Sustainable Small-Scale Fisheries in the Context of Food Security and Poverty Eradication SSF Guidelines.* FAO, Rome.

Food and Agricultural Organization of the United Nations (FAO) (2016). Uganda Aquaculture Production 1950–2014 (online query). FAO Fisheries and Aquaculture Department.

Geheb, K., Kalloch, S., Medard, M., Nyapendi, A.-T., Lwenya, C., and Kyangwa, M. (2008). Nile perch and the hungry of Lake Victoria: gender, status and food in an East African fishery. *Food Policy* 33, 85–98.

Glaser, S. M. (2017). The future of Lake Victoria: a looming conflict over fisheries resources. https://securefisheries.org/blog/future-lake-victoria-part-1 (accessed March 7, 2019).

Glaser, S. M., Hamilton, S. E., Kaufman, L. S., Rothman, D., and Wedig, K. (2015). The potential for aquaculture in Lake Victoria and implications for wild fisheries and fish commodity markets. Uganda, Tanzania, Kenya, Report. USA National Science Foundation.

Heck, S., Béné, C., and Reyes-Gaskin, R. (2007). Investing in African fisheries: building links to the millennium development goals. *Fish and Fisheries* 8 (3), 211–226.

High Level Panel of Experts (2014). *Sustainable Fisheries and Aquaculture for Food Security and Nutrition.* A Report by the High Level Panel of Experts Report, UN Committee on World Food Security, Rome.

International Fund for Agricultural Development (IFAD) (2013). *Smallholders, Food Security, and the Environment.* IFAD, Rome.

Lawrence, T. J., and Watkins, C. (2011). It takes more than a village: the challenges of co-management in Uganda's fishery and forestry sectors. *International Journal of Sustainable Development and World Ecology* 19(2), 144–154.

Linard, C., Marius, G., Robert, S. W., Abdisalan, N. M., and Andrew, T. J. (2012). Population distribution, settlement patterns and accessibility across Africa in 2010. *PLoS ONE* 7(2), e31743.

Meadows, D., Randers, J., and Meadows, D. (2004). *Limits To Growth, 30-Year-Update*. Chelsea Green Publishing, White River Junction, VT.

Naess, P. (2011). Unsustainable growth, unsustainable capitalism. *Journal of Critical Realism* 5(2), 192–227.

Nunan, F. (2010). Governance and fisheries co-management on Lake Victoria: challenges to the adaptive governance approach. *Maritime Studies* 9(1), 103–125.

Ottinger, M., Clauss, K., and Kuenzer, C. (2016). Aquaculture: relevance, distribution, impacts and spatial assessments: a review. *Ocean and Coastal Management* 119, 244–266.

Pattenden, J. (2010). A neoliberalisation of civil society? self-help groups and the labouring class poor in rural South India. *Journal of Peasant Studies* 37(3), 485–512.

Pauly, D., and Zeller, D. (2016). Catch reconstructions reveal that global marine fisheries catches are higher than reported and declining. *Nature Communications* 7, 10244.

Peters, P. E. (2013). Land appropriation, surplus people and a battle over visions of agrarian futures in Africa. *Journal of Peasant Studies* 40(3), 537–562.

Pillay, T. V. R. (2008). *Aquaculture and the Environment*, 2nd ed. Wiley-Blackwell, Oxford.

Rauch, T. G., Beckmann, S., Neubert, S., and Rettberg, S. (2016). *Rural Transformation in Sub-Saharan Africa*. Conceptual Study. Centre for Rural Development (SLE), Berlin.

Reichel, C. (2019). *Mensch—Umwelt—Klimawandel: Trans-lokales Wissen und Resilienz im Schweizer Hochgebirge*. Transcript Verlag, Bielefeld.

Reichel, C., and Frömming, U. U. (2014). Participatory mapping of local disaster risk reduction knowledge: an example from Switzerland. *International Journal of Disaster Risk Science* 5(1), 41–54.

Schumacher, E. F. (1973). *Small Is Beautiful: A Study of Economics As If People Mattered*. Blond and Briggs, London.

Steffen, W., Richardson, K., Rockström, J., Cornell, S. E., Fetzer, I., Bennett, E. M., Biggs, R., Carpenter, S. R., de Vries, W., de Wit, A., Folke, C., Gerten, D., Heinke, J., Mace, G. M., Persson, L. M., Ramanathan, V., Reyers, B., and Sörlin, S. (2015). Sustainability. planetary boundaries: guiding human development on a changing planet. *Science* 347(6223), 1259855.

Transnational Institute (2014). *The Global Water Grab: A Primer*. Transnational Institute (TNI), Amsterdam.

United Nations (2012). *Fisheries and the Right to Food*. United Nations Special Rapporteur on the Right to Food, Report presented at the 67th Session of the United Nations General Assembly.

United Nations Conference on Trade and Development (UNCTAD) (2013). *Wake Up Before It's Too Late: Make Agriculture Sustainable Now For Food Security In A Changing Climate*. Trade and Environment Review 2013. United Nations, Geneva.

Wedig, K. (2018). Water grabbing or sustainable development? effects of aquaculture growth. In: Wiegratz, J., Greco, E., and Martiniello, G. (Eds.), *Uganda: The Dynamics of Neoliberal Transformation*, pp. 249–265. Zed Books, London.

Wedig, K. (2019). *Cooperatives, the State, and Corporate Power in African Export Agriculture: The Case of Uganda's Coffee Sector*. Routledge, London.

Willison, J. H. M., and Côté, R. P. (2009). Counting biodiversity waste in industrial eco-efficiency: fisheries case study. *Journal of Cleaner Production* 17(3), 348–353.

World Bank (2013). *World Development Report 2014: Risk and Opportunity—Managing Risk for Development*. World Bank, Washington, DC.

World Bank (2014). *Global Partnership for Oceans:* Framework Document for a Global Partnership for Oceans. World Bank, Washington, DC.

World Forum of Fisher Peoples (WFFP) and World Forum of Fish Harvesters and Fish Workers (WFF). (2013). A Call for Governments to Stop Supporting the Global Partnership for Oceans (GPO) and Rights-Based Fishing (RBF) Reforms. http://worldfishers.org/wp-content/uploads/2015/11/WFFP-WFF-Call-on-Governments_GPO_200313.pdf (accessed March 7, 2019)

24
Value Network Analysis for (Re) Organizing Business Models Toward the Sustainable Development Goals

The Case of the Agricultural Commodity Exchange in Malawi

Domenico Dentoni, Laurens Klerkx, and Felix Krussmann

1. Introduction

Stemming from an ambitious evolution of the millennium development goals (MDGs) (Sachs 2012), the sustainable development goals (SDGs) provide a set of 17 interrelated, measurable goals that overall provide guidance and space for any actor in the global society to take action under a common sustainable development vision (Nilsson et al. 2016). The achievement of these SDGs would require processes of inclusive social, technological, and institutional innovation in multiple realms and including multiple societal actors (Clifford and Zaman 2016; Pansera and Sarkar 2016; Sengupta 2016) through cross-sectoral and multistakeholder partnerships (Boas et al. 2016). Such cross-sectoral partnerships would then form a key element in the enactment of science, technology, and innovation (STI) policies for inclusive development (Foster and Heeks 2013a; Foster and Heeks 2013b), functioning as what have been called "systemic instruments" which foster inclusion and coordination (Wieczorek and Hekkert 2012).

A number of diagnostic tools are already in use to support and coordinate actors in a system to reach sustainable outcomes and inform cross-sectoral multistakeholder partnerships (which are represented as an SDG on their own in SDG17, in target 17.17), including value-chain analysis, social-network analysis, rapid system appraisals, and net-mapping, among others. Nevertheless, we argue that none of the currently used diagnostic tools assess the configuration of the actors in a system along with the resources associated with them (see section 2 for a thorough discussion). This is a remarkable shortcoming of current diagnostics, because the heterogeneous distribution of resources in poorly communicating subsystems (in other words, their modularity; Newman 2006) underlies key systemic constraints that limit the pursuit of transformational changes in society (Niinimäki and Hassi 2011), such as those that the SDGs require.

To address this methodological gap in the use of current diagnostics in relation to the SDGs, this chapter assesses the use of value network analysis (VNA) as a qualitative diagnostic tool for decision-makers seeking to strategically (re-)organize their sustainable business models in support of the SDGs. In view of the theme of this book, the broader question of this chapter is: What is the value of VNA for actors to use to build strategic cross-sectoral partnerships that effectively support transitions toward achieving the SDGs? From a descriptive standpoint, VNA maps the complex interrelationships among actors and their associated tangible or intangible resources (e.g., money, information, knowledge, commodities, technology, hierarchy) in a system (for example, an economic sector or two tightly interrelated economic sectors at a national level). On the basis of these descriptive maps, the interpretation of VNA allows diagnosis of the current modularity of the system and how this modularity underlies the systemic constraints that limit transitions toward the SDGs. Stemming from this interpretation of the modularity of the system, decision-makers can identify and prioritize the potential of new strategic cross-sector partnerships that help addressing the aforementioned systemic constraints to achieving specific SDG targets. Given these features, this use of the VNA targets decision-makers in sustainable business models (Dahan et al. 2010; Dentoni and Peterson 2011; Schaltegger et al. 2016), whose tasks inherently involve the establishment of new strategic cross-sector partnerships to reach the SDGs. While this chapter aims to explain and evaluate the use and value of VNA for business models seeking to achieve SDGs, other literature provides deeper discussion on the step-by-step process of implementing VNA in engagement with the involved stakeholders (Bocken et al. 2013; 2015).

The case of the Agricultural Commodity Exchange (ACE) in Malawi, embedded in legume and maize systems at a national level, provides a rich empirical case for experimenting and assessing the use of VNA as a diagnostic tool in this domain. Since 2006, ACE represents a business model striving to support sustainability transitions toward the reduction of poverty and hunger (SDGs 1 and 2) and the enhancement of gender equality, economic growth, and public infrastructures (SDGs 5, 8 and 9). Over more than a decade, ACE developed its business model to bridge previously disconnected actors across the system to pool finance, storage, seed, and information resources (Dentoni and Dries 2015). Nevertheless, systemic constraints in and around the Malawian legume (soybean, groundnut, common bean, pigeon pea) and maize systems keep constraining its potential in supporting transitions toward achieving the SDGs (Dentoni and Krussmann 2015). The Malawian legume and maize sectors, in which ACE is embedded, provides a suitable setting for this study for two reasons. First, decision-makers in Malawian business models recognize the importance of organizational and systemic innovation to generate sustainable growth while enhancing poverty reduction, soil fertility, and food security in the Southern African region (Kamanga et al. 2010; Rusike et al. 2013; Waddington et al. 2015), embracing several SDGs. Second, decision-makers realize that strengthening coordination and coherence among several innovation-support initiatives is essential to reach these goals effectively (Van Rooyen and Homann 2007), acknowledging the importance of SDG17 (the global partnership for sustainable development). These case conditions provided fertile ground to experiment with how a VNA can inform local stakeholders on entry points for prioritizing and establishing new cross-sector partnerships to effectively support transitions toward achieving the SDGs.

The remainder of this chapter is organized as follows. After discussing the foundations and complementarities of VNA vis-à-vis other systems diagnostic tools in section 2, section 3 provides background on the case of the Malawian legume systems and an overview of the methods used to implement VNA in Malawi. An illustration of the descriptive VNA findings is provided in section 4. Interpretation of the descriptive findings with specific attention to modularity in the analyzed system follows in section 5, and then we conclude in section 6 drawing an assessment of VNA as a diagnostic tool for re-organizing sustainable business models toward the SDGs.

2. VNA as a Diagnostic Tool Toward the SDGs

In this section, we discuss how VNA adds and complements other diagnostic tools supporting sustainable business models in agricultural systems.

2.1 Diagnostics for Sustainable Business Models in Agriculture

Generally speaking, sustainable business models are mechanisms that either for-profit or not-for-profit organizations use to create economic, environmental, and societal value while capturing a part of the created value, that is, some financial reward to keep the organization going over time (Schaltegger et al. 2016). According to this definition, many recently thriving collaborative organizational structures such as innovation platforms (Kilelu et al. 2013) and rural hubs (Kilelu et al. 2017), as well as cross-sector, public–private, or multistakeholder partnerships (Bitzer and Glasbergen 2010; Dentoni et al. 2016), may also be considered partnerships supporting sustainable business models. Through these partnerships, sustainable business models seek to improve inclusion of a diversity of actors in innovation systems and to enhance coordination among them (Wieczorek and Hekkert 2012), while seeking to generate rewards to each of them (often in the form of knowledge, networking, or reputation). By building bridges across multiple actors, such as public agency officers, private actors in agricultural commodity, IT, finance and energy sectors, or nonprofit entities such as universities, nongovernmental organizations (NGOs), or civil society organizations (Dentoni et al. 2012; Murphy et al. 2012; Schut et al. 2016; Kilelu et al. 2017), these sustainable business models have potential to support sustainability transitions. Yet to effectively support sustainability transitions, several authors have argued that these business models need to be more strategic in fostering bridges among actors who effectively address systemic constraints that impede sustainability transitions

(Hounkonnou et al. 2012; Dentoni and Ross 2013; Struik et al. 2014). Examples of these systemic constraints may involve policies, regulations, standards, and markets interconnected with food and agriculture, such as labor, finance, or energy markets, among others.

To inform the design, establishment, or re-organization of these sustainable business models (in pursuit of the SDGs), diagnostic tools are critical to identify what needs to be changed (Rich et al. 2009; Alvarez et al. 2010; Amankwah et al. 2012; Ilukor et al. 2015; Schut et al. 2015). Through diagnostic tools, for example, decision-makers may understand if they need to intervene at a production system level, at a regulatory level, or at a market level, as well as which key actors and resources need to be involved to effectively support sustainability transitions. Several diagnostic tools are already in use in agricultural settings to reach these objectives, including: rapid systemic analyses based on innovation systems theories (Amankwah et al. 2012; Totin et al. 2012; Schut et al. 2015); institutional analysis to understand and compare the governance structure of institutions and its impacts (Amankwah et al. 2014; Struik et al. 2014); value-chain analysis (Rich et al. 2009; Trienekens 2011); social-network analysis (Spielman et al. 2011); and stakeholder analysis (Prell et al. 2009). Nevertheless, none of these diagnostics provides at once a map interrelating the actors (both within and outside value chains) with the resources that they share or exchange with each other. This is an important gap, because a map that indirectly connects all the actors within a system—either within or outside a specific value chain—may support decision-makers to think creatively about new potential partners that they may need to join with to address systemic constraints and thus support their desired transitions toward the SDGs. To fill this gap, this chapter applies a novel application of VNA (Allee 2000; Peppard and Rylander 2006; Biem and Caswell 2008). The following section illustrates the foundations of VNA in detail as well as its complementary with other diagnostic tools in informing transitions toward achieving the SDGs.

2.2 VNA: Foundations and Complementarity with Other Tools

Initially applied in sectors other than agriculture (Fjeldstad and Ketels 2006; Basole 2009), the general aim of VNA is to provide a comprehensive description of where value lies in a network and how the resource exchange or sharing among interconnected actors creates value (Peppard and Rylander 2006; Allee 2008). Value networks are composed of sets of nodes and links whose distinctive feature is the complementarity of resources associated with these nodes and links to create value to the engaging actors (Peppard and Rylander 2006). As such, value networks have three key elements: first, the nodes as autonomous units (e.g., individuals or organizations) that operate in a system; second, the links indicating the relationships (or ties) among nodes; and third, the resources (e.g., information, knowledge, capital, reputation,

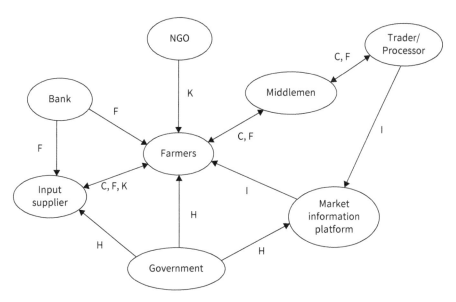

Figure 24.1 Illustrative example of VNA

Legend: Nodes represent actors in a simplified system; arrows represent relationships among actors; letters associated to each relationship represent the following resources: C = commodity; F = funding; H = hierarchy or rules; K = knowledge or advice; I = information; arrows direction represents the direction of transfer of the associated resources.

Source: Authors.

hierarchy) associated with these relationships (Figure 24.1). By mapping these three elements simultaneously, VNA provides a tool to understand how actors in a system can complement and integrate their resources to cocreate (or, potentially, destroy) value through purposive networks.

The qualitative use of VNA reveals relevant features of actors in the system, including their reciprocity, distance, channel diversification, and resource diversification (Allee 2000; 2008). *Reciprocity* involves the extent to which an actor shares resources bi-directionally; for example, farmers and government in Figure 24.1 display low reciprocity. *Distance* represents the degree of separation from an actor to another; for example, in Figure 24.1, banks and input suppliers show high distance in relation to traders/processors. *Resource diversification* entails the heterogeneity of resources provided or received from other actors and *channel diversification* is the wideness of relationships through which resources are shared; for example, farmers in Figure 24.1 have high resource and channel diversification (i.e., indicated by the fact that they receive or give many resources from multiple actors). These features of actors in a system can be interpreted as the challenges and opportunities that they face in coordinating effectively with others in supporting innovation (see sections 5 and 6). While the assessment of these features is mostly qualitative, recent attempts have been made to quantify these features emerging from the VNA (Barzola Iza et al. 2019).

Given its foundations and essential elements, VNA provides a qualitative methodology, empirical data, and implications that are different and complementary to other common analytical tools in use to map food and agricultural systems. First of all, similar to value-chain analysis (Fitter 2001; Kaplinsky 2004; Rich et al. 2009; Horton et al. 2010), VNA tackles the broad question of how value is created among multiple actors. Yet while value-chain analysis assumes that value lies mainly within the chain, VNA presuppose that the key to value creation lies in relationships among actors within and outside the chain (Peppard and Rylander 2006). The focal point of the value chain is the end product and the chain is designed around the activities required to produce it. Instead, from a value network perspective, relationships are viewed as part of a larger interconnected whole.

Second, similar to social-network analysis (Conley and Udry 2003; Cassidy and Barnes 2012; Spielman et al. 2011; Wu and Guclu 2013), VNA allows mapping a system of interconnected actors beyond the value chain, including nodes in the network and strong and weak ties between actors, to show where centrality and power lie in the network. Yet while social-network analysis only maps the set of relationships among actors, VNA also maps the key resources transferred, exchanged, or shared among them. Mapping the key resources exchanged among interconnected actors helps identifying what they gain from their existing relationships, or what they could potentially gain if a potential relationship would be in place (Das and Teng 1998; Zott and Amit 2010; Dentoni and Peterson 2011). By assessing the resources available in the system, the existing links as well as the missing links, through VNA change agents can identify opportunities to develop cross-sector partnerships that build resource complementarities and may represent entry points for innovation.

Third, by combining value-chain and social-network analysis, Lazzarini et al. (2001) introduce "netchain analysis" to assess how tangible and intangible resources are exchanged and pooled along product chains. Yet sustainable transitions need the involvement also of agents that are not directly engaged in value chains (e.g., farmers, processors, banks, manufacturers, retailers, consumers); for example, the engagement also of NGOs, government agencies, social movements, research institutes, and media is necessary to support or accelerate the transitions (Levy 2008; Mair and Marti 2009; Kulve 2010; Klerkx et al. 2010). Therefore, VNA allows mapping and acting upon the agents that are indirectly connected to value chains—which is outside the scope of netchain analysis.

Finally, similar to VNA, net-mapping techniques map networks among stakeholders and their power relationships in and around agricultural and food chains (Aberman et al. 2012; Solomon et al. 2014; Ilukor et al. 2015). Different from net mapping, VNA focuses on their resource exchange and pooling among actors. Thus, VNA complements net-mapping techniques because the type and amount of resource shared among actors can be considered as a determinant of their power relationships. For the purpose of providing implications to agents undertaking innovation-support interventions, mapping the resource exchange and sharing among actors (as VNA does) may be more informative than mapping the power relationships, as it provides a strategy for building relationships among actors in a system to recombine and catalyze their complementary resources in ways that create value.

3. Methods

This section illustrates how VNA has been empirically used to diagnose constraints that limit transitions towards the SDGs, thus suggesting entry points for re-organizing sustainable business models.

3.1 Case and Sample Selection

The case of the Malawian maize and legume sectors, and more specifically of the Agricultural Commodity Exchange operating within them, provides a suitable setting to illustrate the application and use of VNA as a diagnostic tool for supporting transitions toward attaining the SDGs. A wide literature informs us that the maize and legume sector need urgent transitions to sustainably meet the needs of farming communities in peripheral areas of the country (Kamanga et al. 2010) and of the rising urban middle-class (Rusike et al. 2013). In urban areas, legumes represent an affordable, nutritious food for consumers and they can be exported to other growing urban centers in the region (Steyn et al. 2012). Yet legume chains still suffer major inefficiencies that constrain safe and high-quality supply, increase coordination costs, and limit the growth of market demand of Malawian legumes (Rusike et al. 2013) similar to other African contexts (Fafchamps 2004; Eifert et al. 2008; Collier and Dercon 2014; Struik et al. 2014). While the involved actors are aware of these inefficiencies, they have so far collectively failed to achieve coherence in coordinating with each other (Sartorius and Kirsten 2007; Ricker-Gilbert et al. 2013; Collier and Dercon 2014). Part of the challenge of coordination lies in the fact that the level of complexity of the system is too high for any individual actor, or even groups of actors collaborating with each other, to understand and act upon. In this situation, a mapping tool such as VNA helps analyzing these coordination problems in terms of existing (and missing) relationship among actors and related resource exchange.

3.2 Data Collection and Analysis

Interviews took place in two rounds between May and December 2014. In the first round, interviewees were first asked broad questions on the key problems that they were facing upstream and downstream from their position in the supply chain (i.e., respectively on the supply or procurement side and on the demand or market side).

Interviewees outside the supply chain were asked similar questions depending on the position of the chain where they wished to intervene (e.g., farmers' organizations, extension officers and NGOs at farmer level). Moreover, interviewees were asked to identify the key policy and institutional issues that underpinned the key problems along their supply chains. For each problem, the interviewee was asked to mention which actors or organizations were involved in the problem and/or seeking a solution to it, as well as how they were involved (i.e., which key resources were they providing or receiving).

After this first round of interviews, the notes were transcribed and coded by the research team. Respondents' individual problems (e.g., "petrol is expensive, so legume supply from farm to processing is limited") were logically connected to the emerging institutional and policy constraints (e.g., "poor coordination on investments in public infrastructures"). Furthermore, information on the key actors mentioned during the interviews, their relationships with each other and their associated resources were triangulated with secondary data from online reports, websites, and professional publications to confirm the nature of the relationships among the mentioned actors. In total, interviewees mentioned 110 organizations involved in the key problems that they are collectively facing and these organizations are exchanging five key types of resources (information: I, advice: A, funding: F, commodities: C, and hierarchy or rules: H). After this triangulation, the relationships among the mentioned 110 actors, the associated resources and the direction of the resource flow was transposed first to a matrix and then to a map (see Box 24.1).

After producing the VNA maps, a second round of interviews took place with the same interviewees. In this second round, the research team shared the VNA outputs for them to assess and verify the relationships among actors in the VNA map and to gather further insights when needed. Then the research team finalized the VNA map and moved to the interpretation phase. VNA displays the strategic resources that actors exchange or pool as part of their operations. Information on exchanged resources among actors is presented qualitatively (e.g., the maps illustrate that two actors exchange money and information, but not how much money or information, which is a potential limitation of this diagnostic tool). Even if only considered qualitatively, the mapped key resources in the VNA are strategic (Das and Teng 2000), since interviewees consider them as the underlying valuable resources achieved from the relationship (Allee 2000). In this implementation of the VNA, the research undertook the interpretation of the final map (explained in section 5) in interaction with the four key informants and the 20 experts at the Lilongwe University of Agricultural and Natural Resources, yet without the interviewees themselves. A way to overcome this limitation and to achieve a participatory interpretation of the VNA maps with the respondents would be the organization of a workshop where actors in the represented system can reach a synthesis of interpretations (see Burns 2014; Lie et al. 2017). While the results of VNA inherently require interpretative work, they could provide a suitable starting point for a discussion with multiple stakeholders involved in a future interpretation.

Box 24.1 How to set system boundaries for the Value Network Analysis

To establish a sample selection protocol for the VNA implementation, the research team has first defined the boundaries to the analysis. Given its inherent complexity, mapping a complete set of interrelationships would be impossible to represent in any system (Allee 2000). In other words, without set boundaries the VNA would be too broad to be useful to support interventions such as innovation platforms. To narrow down the focus of investigation, the research team focused exclusively on the institutional and policy issues that constrain innovation in the Malawian legume value chains at a market and supplier–buyer relationship level. As other applications of diagnostic tools to support cross-sectoral partnerships for attaining the SDGs illustrate (Burns 2014; Lie et al. 2017), this issue-boundary-setting stage could be implemented more from the bottom up. For example, the actors may determine the boundaries of the analysis by themselves as the interview process goes on, or negotiate these boundaries in a facilitated space (e.g., focus group). Given its focus, the illustrated analysis included managers of organizations within and around the Malawian legume supply chains, as well as key informants on the current status of the sector. In particular, the research team interviewed 39 stakeholders in and around the Malawian value chains and 20 experts in legume systems at the Lilongwe University of Agricultural and Natural Resources (LUANAR). Within the described boundaries, the selection of the sample considered the multiple stakeholders' perspectives around these institutional and policy issues. Among the 39 interviewed stakeholders, representatives of farmers' organizations, input suppliers, traders and processors, NGOs, research institutes, national development agencies, Ministry of Agriculture and Food Security, Ministry of Economic Planning, and Ministry of Industry and Trade were interviewed, as well as four key informants.

Putting boundaries to VNA—similarly to any other diagnostic tool—is a necessary step to give enough focus to analysis and implementation. Yet this inherently involves two key limitations. First, this illustration of VNA captures information about two value chains (maize and legumes), but farmers and other stakeholders actually participate in additional value chains (e.g., rice, livestock), which may influence each other. Second, this application of VNA does not represent relationships within each organization (e.g., NGO, government agency, or company) represented in the systems diagnosis, nor relationships between these organizations and those not represented (for example because they operate outside the Malawian system boundaries). Instead, this VNA representation only captures the relationships among these organizations. Given these limitations, the analysts and users of VNA need to interpret the findings with the awareness that this diagnostic tool represents only a part of a more complex system when drawing implications on how to manage different levels of the innovation systems.

4. Findings: Value Networks in Malawian Legume Systems

In order to identify the entry points for business model collaboration to effectively support transitions toward the SDGs, findings of the VNA first diagnose the set of interrelationships among actors and the resource exchange and pooling constituting the Malawian maize and legume systems around ACE. The actors in the Malawian maize and legume systems are first described based on their key relationships and associated resources (i.e., information, advice, funding, rules and product, which could be either agricultural commodities or inputs). The complete map with the results of the VNA is represented as a diagram and summarized in Tables 24.1 to 24.5. In each table, actors are located in the map quadrant: for example, smallholder farmers are represented in D2 and the Malawian government is in G3. From these findings, interpretation in terms of identifying modularity in the system and thus the strategic entry points for collaboration to effectively support transitions toward the SDGs will follow in section 5.

VNA findings first illustrate the current multiple roles played by smallholder farmers, their two farmers' associations and their supply chain partners in coordination with other actors in the system in relation to their implications for specific SDGs (Tables 24.1 and 24.2).

The VNA illustrates that smallholders depend on information, knowledge, funding, and inputs received from other actors and are indirectly connected with their resource providers, with negative effects on the reaching of SDG1.4 and SDG2.3 (Table 24.1; UNDESA 2018). The two farmers' associations (the National Association of Smallholder Farmers of Malawi, or NASFAM, and the Farmers' Union of Malawi, or FUM) play a critical coordination role in providing farmers with the needed resources. Nevertheless, the complexity of their structure increases their distance from farmers with the necessary resources, thus limiting their potential in achieving SDGs 8.2 and 9.3 (Table 24.1). Traders and processors reciprocally exchange inputs and commodities but do not usually share information and advice with other supply chain partners, with negative implications for SDG8.3 and SDG17.9 (Table 24.2). Input suppliers instead do exchange information and advice with farmers, as they are their customers, but are constrained in their early-generation seed supply from government, with which they do not fluently exchange information and advice (thus affecting SDGs 2.3 and 2.5). In this context, relatively new actors are emerging, such as Esoko Limited, as an information technology provider, and ACE, as a commodity exchange providing a bundle of services to farmers and supply-chain actors (Tables 24.1 and 24.2). These broaden the resource channels from supply-chain actors to farmers.

Descriptive VNA findings illustrate the current role that actors in the Malawian system outside the legume supply chain play to support innovations at farm or market level. First of all, international organizations and national donors still mainly play the role of funders and rule-setting agencies, such as the Food and Agriculture

Table 24.1 Description of the value network analysis by actor: farmers and their associations

Actors (map quadrant: see Dentoni and Krussmann 2015, Appendix 2)	Key relationships and associated resources	Actor features in the system	Key implications related to the SDGs
Smallholder farmers (D2)	1) Receive information (I) and advice (A) from national farmer associations through several intermediaries (see farmers' associations); 2) Receive information and advice from extension officers connected with the Ministry of Agriculture; 3) Receive information through Esoko Ltd. and advice the ACE. ACE is a key intermediary of farmers with traders/processors and input suppliers; 4) Exchange commodities (C) for money (F) with processors through local trading centers and middlemen; and trade commodities with input suppliers and receive advice (A). 5) Receive rules (H) from village heads through their communities and rules from the Ministry Industry and Trade through the farmers' associations.	*Low reciprocity*: farmers receive more resources (I, A, F) than they give to others (C); *High distance*: farmers receive resources (I, A) through multiple degrees of separation (for example, the National Association of Smallholder Farmers of Malawi, or NASFAM through its departments); *High channel diversification*: farmers receive resources (I, A, F, C, H) through a wide number of actors; *High resource diversification*: farmers receive a widely diverse range of resources (I, A, F, C, H).	Given their low reciprocity and high distance from several other actors, smallholders depend on other actors to receive most of the resources and thus are one of the weakest actors in the system → **This is a critical gap toward the achievement of SDG1.4 and SDG2.3.** Given their high channel diversification, smallholders have higher chances to mitigate risks and be resilient to socioeconomic and natural shocks → **This is of critical support toward the achievement of SDG1.a and SDG2.a**

Continued

Table 24.1 *Continued*

Actors (map quadrant: see Dentoni and Krussmann 2015, Appendix 2)	Key relationships and associated resources	Actor features in the system	Key implications related to the SDGs
Farmers' associations: Farmers' Union of Malawi (FUM) (C1) and NASFAM (B1)	1. FUM provide information and advice to farmers directly and through cooperatives and local farmers' associations; 2. NASFAM provide information and advice to farmers through NASFAM Extension and "lead farmers" and through NASFAM Development offices; and provide commodities and funding to farmers through NASFAM outlets for inputs and NASFAM Commercial; 3. FUM and NASFAM are members of ACE and through it they exchange information and advice with traders/processors, input suppliers and banks; 4. FUM and NASFAM receive information from research institutes such as Lilongwe University of Agriculture and Natural Resources (LUANAR) and nongovernmental organizations (NGOs); 5. FUM also receives advice from traders/processors since some traders/processors also own farms and therefore they are FUM members.	*High reciprocity:* FUM and NASFAM receive resources from research institutes, NGOs and traders and provide resources to farmers; *Low distance* (FUM): FUM receives and provides advice and information directly from traders and to farmers with minimal degrees of separation; *High distance* (NASFAM): NASFAM receives and provides advice and information through multiple degrees of separation across its departments (e.g., NASFAM Development, Commercial, Extension, outlets for inputs); *High channel diversification:* FUM and NASFAM receive and provide advice and information through multiple channels; *Low resource diversification:* FUM and NASFAM receive and provide mainly advice and information, and commodities and money to a minimal extent.	Given their high reciprocity and channel diversification, farmers' associations have potential to further catalyze advice and information from multiple sources in ways that create value to farmers. → **This is of critical support toward the achievement of SDG8.3 and SDG17.17.** Given its high distance (in the case of NASFAM) and low resource diversification, farmers' associations depend heavily on other actors for the provision of funding, commodities, and rules. → **This is a critical gap toward the achievement of SDG8.2 and SDG9.3.**

Table 24.2 Description of the value network analysis by actor: private supply chain actors

Actors (map quadrant: see Dentoni and Krussmann 2015, Appendix 2)	Key relationships and associated resources	Actor features in the system	Key implications related to the SDGs
Commercial seed suppliers/ breeders	1. Receive early generation seeds from government breeders. 2. Provide seeds, advice, and information to input suppliers or farmers' associations. 3. Multinational seed suppliers (e.g., Monsanto) owns (H) seed brands (e.g., Dekalb Maize) sold to farmers. 4. Exchange information and advice with the International Crop Research Institute for the Semi-Arid Tropics (ICRISAT).	*High reciprocity* in exchanging seeds from government breeders to input suppliers and farmers' associations. *High distance* from farmers to provide them with seeds, information and advice. *Low resource diversification* since they share mainly seed with related information and advice. *Low channel diversification* since they receive early generation seeds only from government breeders, and give seeds only to input suppliers and farmers' associations.	The high reciprocity in exchanging seeds shows that commercial seed breeders are important players in the legume seed chain, although they play a limited role in less profitable crops (e.g., pigeon pea, beans). → **This is of critical support toward the achievement of SDG2.a.** The high distance show that accessing and informing farmers on seed features is a serious challenge. The low resource and channel diversification shows that receiving early generation seeds only from government is inherently risky → **This is a critical gap toward the achievement of SDG2.3 and SDG2.5.**
Input suppliers and traders (i.e., agrodealers, C 4/5) and processors	1. Input suppliers provide products to farmers and farmers' cooperatives. 2. Input suppliers receive information and advice from farmers' associations and Rural Market Development Trust (RUMARK). 3. Some input suppliers (e.g., Farmers' World) and banks (e.g., Merchant Ban) provide funding and advice to ACE and receive commodities either as purchase (as traders) or collateral (banks) from ACE. 4. Traders and processors receive commodities from farmers, middlemen, and ACE and sell commodities to retailers either in Malawi or abroad. They also sell commodities to each other. 5. Traders and processors receive information from Ministry of Agriculture and Food Security and rules from Ministry of Industry and Trade.	*High reciprocity* in exchanging commodities along the supply chain, from seed breeders to farmers, traders, processors, and retailers. *Low reciprocity* in exchanging advice and information with supply chain partners. *High resource and channel diversification* in terms of commodity, advice, information, funding, and rules from/to actors within and outside the legume supply chain. *Low distance* from farmers to provide inputs, advice, and information.	Given their high reciprocity (in exchanging commodities), high agility, High resource and channel diversification, input suppliers, traders, and processors play a critical role as brokers among multiple actors in the system → **This is of critical support toward the achievement of SDG1.a and SDG2.c.** Given their low reciprocity in exchanging advice and information with supply chain partners, input suppliers, traders, and processors are limited in exchanging advice and formulate joint strategies → **This is a critical gap toward the achievement of SDG8.3 and SDG17.9.**

Source: authors.

Table 24.3 Description of the value network analysis by actor: international donors

Actors (map quadrant: see Dentoni and Krussmann 2015, Appendix 2)	Key relationships and associated resources	Actor features in the system	Key implications related to the SDGs
International organizations: United Nations agencies (H2 and F5)); World Bank (G3/4); International Financial Corporation (IFC) (H3)	1. Food and Agriculture Organizations (FAO) provides regulations to the Malawian Bureau of Standards, who in turn authorizes the export of Malawian legumes. 2. World Health Organization (WHO) provides regulation and knowledge on the safety of Malawian legumes for export; 3. World Bank funds the Agriculture System Wide Approach Program (ASWAP) with other international donors and provide regulations to the Ministry of Agriculture on its management.	*Low reciprocity:* FAO, WHO and World Bank provide more resources (F, H) than what they receive from other actors. *Low distance from the government* in giving rules: FAO and WHO gives rules directly to Ministry of Agriculture and its Bureau of Standards. *High distance* in giving knowledge and information to government and farmers' associations: ASWAP and Bureau of Standards mediate between international organizations, government, and farmers' associations. *Low resource and channel diversification:* they mainly give mainly regulation and funding, but limited other resources, and to only few actors.	Given their low reciprocity and low resource/channel diversification, international organizations do not play a role of resource catalyzers in the system around farmers and supply chain actors → **This is a critical gap toward the achievement of SDG1.a and SDG17.17.** Given their low reciprocity and low distance, international organizations provide complementary resources to the government and farmer associations → **This is of critical support toward the achievement of SDG17.3 and SDG17.9.**

| National development agencies: USAID (A3) and DFID (A3/4) | 4. Exchange information and knowledge with each other; through the Civil Society Agriculture Network (CISANET) and legume trade development platform.
5. Provide funding to implementing NGOs and consulting companies and exchange advice.
6. Provide funding to CISANET and to ACE and through them exchange knowledge with Government and legume trade platforms.
7. Provide funding to the Malawian Oil Seed Transformation (MOST) platform. | *Low reciprocity* in giving funding: USAID and DFID provide more funding than what they receive from other actors.
High reciprocity in giving information and knowledge with NGOs and government through MOST and legume trade development platform.
High distance in exchanging knowledge and information from other sources: USAID and DFID exchange knowledge and information with government, supply chain actors and farmers through CISANET, ACE, NGOs and platforms that they fund (or have funded in the past).
High resource and channel diversification: USAID and DFID exchange and give a wide range of resources (F, A, I) to other actors. | The high reciprocity in exchanging multiple resources (except funding), high distance in exchanging advice and information, high resource and channel diversification of these national development agencies show that international donors support networks that share knowledge with many actors → **This is of critical support toward the achievement of SDG17.3 and SDG9.5.**
The low reciprocity in giving funding signals that the NGOs and market-support institutions (e.g., CISANET, ACE) funded by these agencies may struggle to become financially independent and sustainable from this donor's funding → **This is a critical gap toward the achievement of SDG8.10 and SDG9.3.** |

Source: authors.

Table 24.4 Description of the value network analysis by actor: government actors

Actors (map quadrant: see Dentoni and Krussmann 2015, Appendix 2)	Key relationships and associated resources	Actor features in the system	Key implications related to the SDGs
Ministry of Agriculture and Food Security (E1/2)	1. Receives funding from government through treasurer in exchange to advice and information 2. Provides funding to NGOs 3. Provides information and rules to processors and traders 4. Through its departments (Extension, Crop Development, Seed service) provides information to farmers through media and field officers.	*Low reciprocity* of funding from government and *low reciprocity* of advice and information to actors. *High distance* in giving information to farmers: mediation through Ministry departments, media and field officers. *High resource and channel diversification* including funding, advice, information, and rules to/from multiple actors in the system.	Given the high distance and low reciprocity in giving information to farmers, Ministry of Agriculture and Food Security struggles to provide farmers with coherent information and advice about inputs, production, markets, and changing regulations → This is **a critical gap toward the achievement of SDG2.3, 2.a and 2.c.** Ministry of Agriculture and Food Security and Ministry of Industry and Trade are very distant from each other → This is **a critical gap toward the achievement of SDG2.c and SDG8.3.** Given the high resource/ channel diversification and low distance in giving information and rules to supply chain actors, Ministry of Industry and Trade has potential to act as information providers with many actors → **This is a critical opportunity toward the achievement of SDG9.3 and SDG17.3.**
Ministry of Industry and Trade (E2)	1. Receives advice from CISANET, FUM, and Commodity Exchanges. 2. Gives rules to banks, farmer associations, and traders.	*Low reciprocity of funding* (from government), but *high reciprocity of advice and information* (to actors) *Low distance from* banks, farmers and traders to give them rules and information. *High resource and channel diversification* including funding, advice, information, and rules to/from multiple actors in the system.	

Source: authors.

Table 24.5 Description of the value network analysis by actor: research institutes and NGOs

Actors (map quadrant: see Dentoni and Krussmann 2015, Appendix 2)	Key relationships and associated resources	Actor features in the system	Key implications related to the SDGs
Research institutes (C/D 1; F4)	1. The Lilongwe University of Agriculture and Natural Resources (LUANAR) provides FUM and CISANET with information. 2. LUANAR provides advice to traders (e.g., Export Trading Group) and Ministry of Agricultural and Food Security departments. 3. ICRISAT provides advice and information to seed breeders. 4. IFPRI receives funding from donors and gives advice and information to the Oil Seed Product Technical Working Group (OSP TWG).	*Low reciprocity* of information and advice from farmers' associations, NGOs, traders, and government to LUANAR. *High distance* in providing information and advice to farmers and farmers' associations through traders, seed breeders, FUM or CISANET. *Low resource and channel diversification* as research institutes provide only information and advice to few actors (government, Export Trading Group, seed breeders).	The low reciprocity and high distance in providing information and advice to farmers and their associations are typical features of university and research institutes: they provide knowledge rather than conveying it. → **This is a critical gap toward the achievement of SDG9.5 and SDG17.9.** The *low resource and channel diversification* show that research institutions play an expert role by exchanging funding with advice and information with only few actors in the system. → **This is a critical opportunity toward the achievement of SDG9.5.**
NGOs (A3/C5)	1. CISANET and the Agricultural Cooperative Development International/ Volunteers in Overseas Cooperative Assistance (ACDI/VOCA) receives funding and information from donors and Ministry of Agriculture. 2. CISANET provides advice and information to African Institute of Corporate Citizenship and advice to legume trade development platform, which in turn receives advice from NASFAM and provides advice to the government. 3. ACDI/VOCA exchanges information and advice with farmers' associations, Esoko, and ACE.	*High reciprocity* of information and advice from donors to farmers' associations also through legume trade development platforms *High distance* in exchanging information with farmers, government, and supply chain actors as mediated through trade development platforms and farmers' associations. *High resource and channel diversification* including information, advice, and funding from/to multiple actors.	Given the *high reciprocity and high distance*, NGOs such as CISANET and ACDI/VOCA play a mediating role among farmers and their associations, government, international donors, and supply chain actors. Furthermore, the high resource and channel diversification shows that NGOs play an important role in facilitating trade development platforms or collaborating with ACE in ways that allows them to interact simultaneously with multiple actors in the system → **This is of critical support toward the achievement of SDG1.a, SDG8.3 and SDG17.9.** High distance from Malawian Ministries and supply chain actors may limit their financial sustainability and independence from international development agencies in the future → **This is a critical gap toward the achievement of SDG8.10 and SDG9.3.**

Source: authors.

Organization, the Bureau of Standards, the World Health Organization, the sanitary export regulations, the World Bank, or the Agricultural Sector-Wide Approach Program fund. Yet a persisting challenge for international organizations is how to catalyze information and advice in complementarity with the other resources among the multiple actors in the legume chain (thus affecting the achievement of SDGs 1.a and 17.17; Table 24.3). Second, Malawian ministries play their traditional role as funders and regulators, but they lack coordination across ministries, posing serious challenges for the achievement of SDGs 8, 9, and 17 (Table 24.4). Third, national agencies of donors, such as the UK Department for International Development (DFID) and the US Agency for International Development (USAID), fund NGOs (e.g., Civil Society Agriculture Network, or CISANET) or other market-support platforms (e.g., ACE) to coordinate efforts to catalyze multiple resources, including knowledge and information, around farmers (Table 24.3). With financial support from these national donors, their sharing of information and advice takes place mainly through relatively new institutions such as the legume trade development platforms (e.g., the Malawian Oil Seed Transformation platform, or MOST, and MAlawian Platform for Aflotoxin Control, or MAPAC), facilitated by NGOs (Tables 24.3 and 24.4). Nevertheless, their funding to these emerging institutions raises questions about the ability of the institutions to be financially independent and sustainable in the future, with possible negative effects on SDG8.10 and SDG9.3. Fourth, NGOs act as coordinators among multiple resources in the system; for example, CISANET mediates between Malawian ministries, supply chain actors and farmers' associations while NGOs coordinate resources among supply chain actors, farmers' associations, and their members (Table 24.5). Finally, research institutes such as the International Crops Research Institute for the Semi-Arid Tropics (ICRISAT) or the International Food Policy Research Institute (IFPRI) play a traditional role as knowledge providers, while Lilongwe University of Agricultural and Natural Resources share information and advice with a larger network of stakeholders (Malawian office of agricultural extension, FUM, CISANET), yet without playing coordination roles among these actors. This position of research institutes in the value network may act as a double-edged sword (i.e., possibly both supporting and constraining the process) toward SDG9.5 and may affect SDG17.9.

5. Discussion: Addressing Constraints of Transitions towards SDGs

Based on these descriptive results of VNA, Malawian decision-makers in sustainable business models seeking to support the SDGs 1, 2, 8, 9, and 17 and the specific targets within them can interpret the current interconnectedness among actors to 1) identify the current systemic constraints in terms of modular distribution of strategic resources through the networks; and 2) identify the entry points for new potential collaborations that, through new resource complementarities, can tackle or circumvent the existing systemic constraints. Through these two dimensions of interpreting the VNA diagnostics, we finally assess the value added of VNA vis-à-vis other diagnostic tools (see section 2) in providing deeper insights on how to build new strategic cross-sector partnerships that effectively support transitions toward attaining particular SDG targets.

5.1 Addressing Constraints Around Knowledge Brokering

First of all, descriptive VNA findings show that the Agricultural Commodity Exchange (ACE)—given its low distance vis-à-vis actors in the supply chain and the resource and channel differentiation—currently plays an important role in catalyzing multiple resources from traders, processors, input suppliers, banks, and international donors (funding, information, and knowledge) and sharing them to farmers and supply chain actors through Esoko Ltd. (information) and banks (loans). Therefore, given its embeddedness in the value network, ACE provides critical support toward SDGs 1.a and 2.c. This interconnectedness of actors already addresses (to some extent) the systemic constraints of uncoordinated information and of poor infrastructure systems available to farmers, specifically in terms of transport and storage facilities. The issues for farmers persist, though, because ACE currently cannot provide market knowledge but only uncoordinated information that farmers struggle to organize and use, in combination with loans provided from banks through ACE, to seize market opportunities. This represents a critical limitation toward the achievement of SDGs 1.4 and SDG2.3.

The current status of the interconnectedness among actors also illustrates the modularity existing around these issues. A first concentration of financial and hierarchical resources in a small subsystem involves traders/processors, input suppliers, and banks, which fund and own ACE and, at the same time, are their major suppliers (as storage owners, commodity brokers, or loan providers) and clients (as commodity purchases or loan receivers), while farmers depend on supply chain actors to receive market knowledge. A second concentration of knowledge and financial resources lies with the government, which through extension officers, farmers' associations (NASFAM and FUM), and universities provide knowledge, advice, and input subsidies to farmers. These two concentrations of resources are remarkably modular or, in other words, disconnected from each other in separate subsystems. This modularity among actors and resources across the subsystems of supply chain actors and publicly-funded knowledge providers highlights a first entry point of new potential collaboration given the resource complementarity across the two subsystems, which would be necessary to achieve SDGs 1.4 and SDG2.3. This resource complementary would involve funding and storage space provided by the private sector in the ACE subsystem, while the capacity to reach out to farmers pertains to farmer associations and universities supported by the government. For example, supply chain actors around ACE have the opportunity to connect with farmers' associations, extension officers, and universities to provide market knowledge to farmers (with a market orientation and potential), while receiving funding from banks, government, and international donors (for example through incubators for farmers[1]). By developing resource-sharing processes, these types of public-private cross-sector partnerships would help closing the gap between the two existing concentrations of power. Furthermore, other actors outside the supply chain, including universities, NGOs, and international organizations, have the opportunity to support or complement farmers' associations in providing knowledge to farmers that combines a market orientation with other social and environmental aspects of farming that markets do not usually consider.

5.2 Addressing Constraints Around Accessing Technology and Credit

Second, the descriptive VNA results reveal that commercial seed suppliers/breeders and other input suppliers are directly connected with each other, with breeders from the Ministry of Agriculture receiving early-generation seeds, and with farmers' associations (FUM and NASFAM) supplying inputs and advice on their use. This interconnectedness among seed/input suppliers and farmers' associations relates to the systemic constraints around weak agricultural input and credit markets and to farming as a business and cooperative formation (thus, they play an important role in reaching SDG1.a and SDG2.a). Despite their high reciprocity in exchanging resources with government breeders, farmers' associations, and traders/processors, these input suppliers still have high distance from farmers, especially those growing low-profit commodities (e.g., pigeon peas, common beans) and in peripheral regions (i.e., distant from main markets of Lilongwe and Blantyre), and the offices of crop development, extension, and seed service at the Ministry of Agriculture. The existing lack of interconnectedness among the commercial seed/input companies and the offices at the Ministry of Agriculture do not help with coordination of the government supply of early-generation seeds. Specifically, the government uses subsidies to distribute seeds and fertilizers to farmers, which distort the market by making the demand and price of inputs uncertain for companies seeking to invest.

This high distance of commercial input suppliers from the Ministry of Agriculture and farmers' associations causes a climate of the persisting market uncertainty, with negative repercussions on the credit markets. Specifically, since the role of government subsidies on agricultural seeds and fertilizer remains uncertain, farmers do not need to risk asking for credit from banks (in fact, farmers and banks are connected mainly through ACE, not directly) to purchase inputs. Furthermore, due to subsidies

---

[1] Early examples of farmers' incubators that catalyze complementary resources in a similar actor system include Equity Group Foundation's Group Accelerator Program in Kenya and the African Agribusiness Incubation Network.

from Ministry of Agriculture and bureaucratic constraints in formally registering their activities as a business or as part of a cooperative (for which the Ministry of Industry and Trade is responsible), farming of legumes—with exceptions only for soybean and groundnut farmers close to Lilongwe and Blantyre—remains mainly an informal low-input and low-output activity, with limited opportunities along the value chain. These underlying issues create serious bottlenecks in the achievement of several goals under the SDG2, as well as SDG8.3.

On the basis of the VNA description, this modularity between seed companies/ early-generation seed providers and Ministry of Agriculture/farmer associations represents a second entry point to address these underlying systemic constraints. As in the prior entry point, sustainable business models can tackle these constraints by connecting currently disconnected actors with complementary resources. For example, in this case, farmers' associations, NGOs, and universities may support the formation of an alternative channel of early-generation seed producers and seed breeders—for example by combining knowledge from farmers engaging in seed entrepreneurship and new graduates from universities[2]—to create a more competitive seed market for both seed suppliers and farmers. Furthermore, as a second example entry point in relation to this modularity, supply chain actors could coordinate knowledge-exchange and rule-setting with the Ministry of Agriculture and the other stakeholders (farmers' associations, NGOs, and universities) for a long-term plan to match early-generation seed supply and demand and a strategy to stabilize or decrease the fertilizer subsidy program. Exploiting these entry points for establishing new strategic partners would facilitate the achievement of SDGs 2.5 and 9.5.

5.3 Addressing Constraints Around Standards Enforcement

The VNA findings demonstrate that supply-chain actors and farmers' associations are interconnected to the Ministry of Agriculture and the Ministry of Industry and Trade through the legume trade development platform and the MAPAC, with advice from NGOs and research institutes (ACDI/VOCA and CISANET) and funding from the international donors (DFID, USAID, and the European Union Platform for African European Partnership on Agricultural Research for Development, or PAEPARD). These trade development platforms seek to act as knowledge-exchange mechanisms in response to the current systemic constraints of poor public enforcement of quality and safety standards. So far, these platforms have been playing an important role toward the achievement of SDGs 1.a, 8.3 and 17.9. Nevertheless, while the government is the most involved in these legume-specific platforms, the VNA still reveals

---

[2] An example of intervention to bridge complementary resources across similarly disconnected actors in the system is the Integrated Seed Sector Development (ISSD) program with four agricultural universities in Ethiopia.

a poor interconnectedness between the Ministry of Agriculture and the Ministry of Industry and Trade. The Ministry of Agriculture struggles to enforce legume quality and safety standards, while supply chain actors do not have financial incentives to develop private standards without the guarantee that the sector is profitable. In particular, profitability is undermined by the uncertain structures of trade taxation, business licensing, and export bans, which are under the control of the Ministry of Trade and Industry. Funds from multiple international organizations, channeled through the Agricultural Sector-Wide Approach Program, may have the function to unlock investments on quality and safety controls in the legume supple chain. Yet the gap with the actions of the Ministry of Trade and Industry does not provide the private actors with the necessary confidence to invest. As our findings reveal, these gaps in the value networks constrain the achievement of SDGs 2.3, 2.a, 2.c, and 8.3.

This finding from the VNA description reveals modularity in the Malawian legume and maize system in terms of separating actors around the Ministry of Agriculture from those around the Ministry of Trade and Industry. On the one hand, the private sector, development agencies, farmers' associations, and the Ministry of Agriculture share complementary resources across the emerging MAPAC and trade development platforms. Furthermore, the Ministry of Trade and Industry has complementary yet currently un-coordinated resources and rules that influence the outcomes of the mentioned platforms. This modularity separating the subsystems around respectively the Ministry of Agriculture and the Ministry of Trade and Industry also reveals entry points for a new sustainable business model with cross-sector partnerships. A first entry point may lie in a stronger linkage across these emerging platforms (MAPAC and the legume trade development platform), also at a regional or African scale, to stimulate policy dialogue and place stronger pressure on the government to increase coherence in the public investment strategy and regulation.[3] A second relevant entry point may involve direct linkages between farmers and these platforms, when possible even without the intermediation of farmers' association representatives to reduce their distance. This would ultimately ensure that farmers could share knowledge and contribute to decisions that influence the regulations in the legume supply chain.[4] While supporting the achievement of SDGs 2.3, 2.a, 2.c, and 8.3, these entry points would also make the aforementioned platforms more inclusive to farmers, thus with likely positive effects also on SDGs 1.4 and 2.3.

---

[3] For example, the NGO Agriprofocus facilitates an open virtual and physical platform that stimulates policy dialogue across stakeholders working on multiple change interventions and at multiple levels (district, national, and African/regional level).

[4] For example, taking a human rights perspective, the civil society organization FIAN and the NGO Oxfam International support processes of farmers' inclusion in policy debate and highlight the negative consequences of farmers' exclusion from these decision-making processes.

6. Conclusion: Assessing VNA for Implementation of SDGs

This chapter introduces VNA as a diagnostic tool for policymakers and managers of business models that seek to support sustainability transitions. This diagnostic tool can be helpful in supporting SDG target 17.7 (cross-sectoral partnerships) and, through this SDG target, enabling the lifting of constraints toward achieving other SDG targets. So far mostly applied in the field of technological innovation, VNA has potential also for application in the field of social and organizational innovation in the way it diagnoses how actors in a system relate (or fail to relate) to each other and share resources. Conceptually, VNA supports the functioning of so-called systemic instruments in STI policy (Wieczorek and Hekkert 2012) needed for tackling several interconnected system barriers that prevent achieving SDG targets.

The empirical case of ACE in Malawi illustrates that the first unique feature of VNA entails the identification of the modularity among actors in a system which underlies the systemic constraints in transitions toward sustainability. Modularity refers to subsystems of actors that are 1) poorly connected across each other yet strongly connected within each other; and that 2) have different resource endowments with a potential to be connected toward mutual visions (such as the SDGs). As the case of Malawi illustrates, the modularity—between the maize and legume traders, processors, and publicly-funded knowledge institutions; between seed and input companies/early-generation seed companies and the Ministry of Agriculture; and between the Ministry of Agriculture and the Ministry of Trade and Industry—underlies a number of systemic constraints that affect the reduction of poverty and hunger (SDGs 1 and 2) and the enhancement of economic growth and public infrastructure (SDGs 8 and 9). This modularity in the system is therefore essential to understand, because it may keep systems in "lock-in" (Avelino et al. 2009; Deneke et al. 2011; Kieft et al. 2017; Smink et al. 2015; Turner et al. 2016). VNA deepens system diagnostics—such as RAAIS (Schut et al. 2015)—by providing clarity on the resource configurations and exchange patterns, thus understanding the modularity taking place in a complex system of actors.

The second interrelated feature of VNA is that it provides the necessary information to scan and identify the entry points that can address the key modularity that underlies these systemic constraints. By understanding the interconnectedness and the modularity of actors through VNA, innovation support agents can grasp where resources and their associated value lies in the network; which actors provide resources and which are in need of resources; and, in turn, where opportunities and risks for value creation lie when connecting with different actors. The case of the Agricultural Commodity Exchange in Malawi illustrates that farmer accelerator or incubator programs, as well as work-with-farmer field schools, may represent entry points to bridge the modularity between the maize and legume traders and processors and the publicly-funded knowledge institutions. Similarly, investments in a private market of small-scale early-generation seeds may unlock the modularity be-tween the seed companies and the Ministry of Agriculture. Finally, stronger linkages between farmer organizations and the emerging trade development platforms may act as entry points to address modularity surrounding the Ministry of Agriculture and the Ministry of Trade and Industry subsystems. Hence, VNA effectively supports decision-makers with a strategic direction to organize collaborative cross-sector partnerships (SDG17).

The current application of VNA has limitations that ought to be addressed in future research. First, the analytical rigor involved makes VNA findings more time-consuming to read and interpret by decision-makers. Second, the analytical procedure used in this version of the VNA does not allow the participatory involve-ment of several actors in the system to simultaneously provide inputs to the maps (for example through focus groups) and to go through a process of joint interpretation (for example, through a facilitated process among multiple decision-makers in sus-tainable business models). Third, as discussed in the methods section (see Box 24.1), a common challenge in relation to VNA involves setting a boundary to the analysis of a complex system. To address these shortcomings, it is important to communicate to decision-makers in sustainable business models that VNA involves interpretative work: that is, it is the decision-maker herself who needs to define the boundaries of the diagnosis—if possible, in interaction with others—in relation to the specific systemic constraint or SDG she aims to address. In this light, the decision-maker(s) in sustain-able business models can also decide to focus on a specific level of analysis, ranging, for example, from a village/community level to a national or even global level.

Acknowledgments

The research has been implemented as part of the Global Center for Food Systems Innovation (GCFSI), funded by the US Agency for International Development/Higher Education Solutions Network program.

References

Aberman, N. L., Johnson, M. E., Droppelmann, K., Schiffer, E., Birner, R., and Gaff, P. (2012). Mapping the Contemporary Fertilizer Policy Landscape in Malawi: A Guide for Policy Researchers. Working Paper series 1204. International Food Policy Research Institute (IFPRI), Washington, D.C.

Allee, V. (2000). Reconfiguring the value network. *Journal of Business Strategy* 21(4), 36–39.

Allee, V. (2008). Value Network Analysis and value conversion of tangible and intangible as-sets. *Journal of Intellectual Capital* 9 (1), 5–24.

Alvarez, S., Douthwaite, B., Thiele, G., Mackay, R., Córdoba, D., and Tehelen, K. (2010). Participatory impact pathways analysis: a practical method for project planning and evalua-tion. *Development in Practice* 20, 946–958.

Amankwah, K., Klerkx, L., Oosting, S. J., Sakyi-Dawson, O., van der Zijpp, A. J., and Millar, D. (2012). Diagnosing constraints to market participation of small ruminant producers in northern Ghana: an innovation systems analysis. *Wageningen Journal of Life Sciences* 60, 37–47.

Amankwah, K., Klerkx, L., Sakyi-Dawson, O., Karbo, N., Oosting, S. J., Leeuwis, C., and van der Zijpp, A. J. (2014). Institutional dimensions of veterinary services reforms: responses to structural adjustment in Northern Ghana. *International Journal of Agricultural Sustainability* 12, 296–315.

Avelino, F., and Rotmans, J. (2009). Power in transition: an interdisciplinary framework to study power in relation to structural change. *European Journal of Social Theory* 12, 543–569.

Barzola Iza, C., Dentoni, D., Allievi, F., van der Slikke, T., Isubukalu, P., Oduol, J., and Omta, S. W. F. (2019). Challenges in youth involvement for sustainable food systems: assessing farmers' embeddedness in coffee value networks in Ugandan multi-stakeholder platforms. In: Valentini, R., Sievenpiper, S., Antonelli, M., and Dembska, K. (Eds.), *Achieving the Sustainable Development Goals Through Sustainable Food Systems*, pp. 113–129. Springer, Berlin.

Basole, R. C. (2009). Visualization of interfirm relations in a converging mobile ecosystem. *Journal of Information Technology* 24(2), 144–159.

Biem, A., and N. Caswell (2008). A Value Network Model for Strategic Analysis. Proceedings of the Proceedings of the 41st Annual Hawaii International Conference on System Sciences 361-368, IEEE Computer Society, Washington, DC.

Bitzer, V., and Glasbergen, P. (2010). Partnerships for sustainable change in cotton: an institutional analysis of African cases. *Journal of Business Ethics* 93, 223–240.

Boas, I., Biermann, F., and Kanie, N. (2016). Cross-sectoral strategies in global sustainability governance: towards a nexus approach. *International Environmental Agreements: Politics, Law and Economics* 16, 429–446.

Bocken, N., Short, S., Rana, P., and Evans, S. (2013). A value mapping tool for sustainable business modelling. *Corporate Governance* 13(5), 482–497.

Bocken, N. M. P., Rana, P., and Short, S. W. (2015). Value mapping for sustainable business thinking. *Journal of Industrial and Production Engineering* 32(1), 67–81.

Borrás, S., and Edquist, C. (2013). The choice of innovation policy instruments. *Technological Forecasting and Social Change* 80, 1513–1522.

Burns, D. (2014). Systemic action research: changing system dynamics to support sustainable change. *Action Research* 12(1), 3–18.

Cassidy, L., and Barnes, G. D. (2012). Understanding household connectivity and resilience in marginal rural communities through social network analysis in the village of Habu, Botswana. *Ecology and Society* 17(4), 11.

Clifford, K. L., and Zaman, M. H. (2016). Engineering, global health, and inclusive innovation: focus on partnership, system strengthening, and local impact for SDGs. *Global Health Action* 9, 30175–30181.

Collier, P., and Dercon, S. (2014). African agriculture in 50 years: smallholders in a rapidly changing world? *World Development,* 63, 92–101.

Conley, T., and Udry, C. (2001). Social learning through networks: the adoption of new agricultural technologies in Ghana. *American Journal of Agricultural Economics*, 668–673.

Dahan, N. M., Doh, J. P., Oetzel, J., and Yaziji, M. (2010). Corporate-NGO collaboration: co-creating new business models for developing markets. *Long range planning* 43(2–3), 326–342.

Das, T. K., and Teng, B. S. (2000). A resource-based theory of strategic alliances. *Journal of Management* 26(1), 31–61.

Deneke, T. T., Mapedza, E., and Amede, T. (2011). Institutional implications of governance of local common pool resources on livestock water productivity in Ethiopia. *Experimental Agriculture* 47(1), 99–111.

Dentoni, D., Bitzer, V., and, Pascucci, S. (2016). Cross-sector partnerships and the co-creation of dynamic capabilities for stakeholder orientation. *Journal of Business Ethics* 135(1), 35–53.

Dentoni, D., and Dries, L. (2015). Private sector investments to create market-supporting institutions: the case of Malawian Agricultural Commodity Exchange. Selected Paper prepared for presentation for the 2015 Agricultural and Applied Economics Association and Western Agricultural Economics Association Annual Meeting, San Francisco, CA, July 26–28, 2015. https://ageconsearch.umn.edu/bitstream/205709/2/AAEA%202015%20ACE%20final.pdf (accessed March 7, 2019).

Dentoni, D., Hospes, O., and Ross, R. B. (2012). Managing wicked problems in agribusiness: the role of multi-stakeholder engagements in value creation. *International Food and Agribusiness Management Review*, 15(B), 1–12.

Dentoni, D., and Krussmann, F. (2015). Value network analysis of Malawian legume systems: implications for institutional entrepreneurship. Paper presented at the Conference on "Complex-Systems Dynamics Principles Applied to Food Systems," Food and Agriculture Organization (FAO), Rome, June 1–6, 2015.

Dentoni, D., and Peterson, H. C. (2011). Multi-stakeholder sustainability alliances in agri-food chains: a framework for multi-disciplinary research. *International Food and Agribusiness Management Review* 14(5), 83–108.

Dentoni, D., and Ross, R. B. (2013). Towards a theory of managing wicked problems through multi-stakeholder engagements: evidence from the agribusiness sector. *International Food and Agribusiness Management Review* 16(A), 1–10.

Eifert, B., Gelb, A., and Ramachandran, V. (2008). The cost of doing business in Africa: evidence from enterprise survey data. *World Development* 36(9), 1531–1546.

Fafchamps, M. (2004). *Market Institutions in Sub-Saharan Africa: Theory and Evidence.* MIT Press, Cambridge, MA.

Fjeldstad, Ø. D., and Ketels, C. H. (2006). Competitive advantage and the value network configuration: making decisions at a Swedish life insurance company. *Long Range Planning* 39(2), 109–131.

Fitter, R. (2001). Who gains from product rents as the coffee market becomes more differentiated? a value chain analysis. *International of Development Studies (IDS) Bulletin*, 32(3), 69–82.

Foran, T., Butler, J. R. A., Williams, L. J., Wanjura, W. J., Hall, A., Carter, L., and Carberry, P. S. (2014). Taking complexity in food systems seriously: An interdisciplinary analysis. *World Development* 61, 85–101.

Foster, C., and Heeks, R. (2013a). Analyzing policy for inclusive innovation: the mobile sector and base-of-the-pyramid markets in Kenya. *Innovation and Development* 3, 103–119.

Foster, C., and Heeks, R. (2013b). Conceptualising inclusive innovation: modifying systems of innovation frameworks to understand diffusion of new technology to low-income consumers. *European Journal of Development Research* 25, 333–355.

Horton, D., Akello, B., Aliguma, L., Bernet, T., Devaux, A., Lemaga, B., Magala, D., Mayanja, S., Sekitto, I., Thiele, G., and Velasco, C. (2010). Developing capacity for agricultural market chain innovation: experience with the "PMCA" in Uganda. *Journal of International Development* 22, 367–389.

Hounkonnou, D., Kossou, D., Kuyper, T. W., Leeuwis, C., Nederlof, E. S., Röling, N., and van Huis, A. (2012). An innovation systems approach to institutional change: smallholder development in West Africa. *Agricultural Systems* 108, 74–83.

Ilukor, J., Birner, R., Rwamigisa, P. B., and Nantima, N. (2015). The provision of veterinary services: who are the influential actors and what are the governance challenges? A case study of Uganda. *Experimental Agriculture* 51(3), 408–434.

Kamanga, B. C. G., Waddington, S. R., Robertson, M. J., and Giller, K. E. (2010). Risk analysis of maize-legume crop combinations with smallholder farmers varying in resource endowment in central Malawi. *Experimental agriculture* 46(01), 1–21.

Kaplinsky, R. (2004). Spreading the gains from globalization: what can be learned from value-chain analysis? *Problems of Economic Transition* 47(2), 74–115.

Kieft, A., Harmsen, R., and, Hekkert, M.P. (2017). Interactions between systemic problems in innovation systems: the case of energy-efficient houses in the Netherlands. *Environmental Innovation and Societal Transitions* 24, 32–44.

Kilelu, C. W., Klerkx, L., and Leeuwis, C. (2013). Unravelling the role of innovation platforms in supporting co-evolution of innovation: contributions and tensions in a smallholder dairy development programme. *Agricultural Systems* 118, 65–77.

Kilelu, C. W., Klerkx, L., and Leeuwis, C. (2017). Supporting smallholder commercialization by enhancing integrated coordination in agri-food value chains: experiences with dairy hubs in Kenya. *Experimental Agriculture* 53 (2), 269–287

Klerkx, L., Aarts, N., and, Leeuwis, C. (2010). Adaptive management in agricultural innovation systems: the interactions between innovation networks and their environment. *Agricultural Systems* 103(6), 390–400.

Kulve, H. T. (2010). Emerging technologies and waiting games. *Science, Technology and Innovation Studies* 6(1), 7–31.

Lazzarini, S. G., Chaddad, F. R., and Cook, M. L. (2001). Integrating supply chain and network analyses: the study of netchains. *Journal on Chain and Network Science* 1(1), 7–22.

Levy, D. L. (2008). Political contestation in global production networks. *Academy of Management Review* 33(4), 943–963.

Lie, H., Rich, K. M., and Burkart, S. (2017). Participatory system dynamics modelling for dairy value chain development in Nicaragua. *Development in Practice* 27(6), 785–800.

Mair, J., and Marti, I. (2009). Entrepreneurship in and around institutional voids: a case study from Bangladesh. *Journal of business venturing* 24(5), 419–435.

Murphy, M., Perrot, F., and Rivera-Santos, M. (2012). New perspectives on learning and innovation in cross-sector collaborations. *Journal of Business Research* 65(12), 1700–1709.

Newman, M. E. (2006). Modularity and community structure in networks. *Proceedings of the National Academy of Sciences* 103(23), 8577–8582.

Niinimäki, K., and Hassi, L. (2011). Emerging design strategies in sustainable production and consumption of textiles and clothing. *Journal of Cleaner Production* 19(16), 1876–1883.

Nilsson, M., Griggs, D., and Visbeck, M. (2016). Map the interactions between sustainable development goals. *Nature* 534(7607), 320–323.

Pansera, M., and Sarkar, S. (2016). Crafting sustainable development solutions: frugal innovations of grassroots entrepreneurs. *Sustainability* 8, 1–51.

Peppard, J., and Rylander, A. (2006). From value chain to value network: insights for mobile operators. *European Management Journal* 24(2), 128–141.

Prell, C., Hubacek, K., and Reed, M. (2009). Stakeholder analysis and social network analysis in natural resource management. *Society and Natural Resources* 22(6), 501–518.

Rich, K. M., Baker, D., Negassa, A., and Ross, R. B. (2009). Concepts, applications, and extensions of value chain analysis to livestock systems in developing countries. International Association of Agricultural Economists Conference, Beijing, August 22–24, 2009.

Ricker-Gilbert, J., Jayne, T., and, Shively, G. (2013). Addressing the wicked problem of input subsidy programs in Africa. *Applied Economic Perspectives and Policy* 35(2), 322–340.

Rusike, J., van den Brand, G., Boahen, S., Dashiell, K., Kantengwa, S., Ongoma, J., and Mongane, D. J. (2013). *Value Chain Analyses of Grain Legumes in N2Africa: Kenya, Rwanda, Eastern DRC, Ghana, Nigeria, Mozambique, Malawi and Zimbabwe.* N2Africa, Wageningen University, Wageningen, The Netherlands.

Sachs, J. D. (2012). From millennium development goals to sustainable development goals. *The Lancet* 379(9832), 2206–2211.

Sartorius, K., and Kirsten, J. (2007). A framework to facilitate institutional arrangements for smallholder supply in developing countries: an agribusiness perspective. *Food Policy* 32(5–6), 640–655.

Schaltegger, S., Hansen, E. G., and Lüdeke-Freund, F. (2016). Business models for sustainability: origins, present research, and future avenues. *Organization and Environment* 29(1), 3–10.

Schut, M., Klerkx, L., Rodenburg, J., Kayeke, J., Hinnou, L. C., Raboanarielina, C. M., and Bastiaans, L. (2015). RAAIS: Rapid Appraisal of Agricultural Innovation Systems (Part I): a diagnostic tool for integrated analysis of complex problems and innovation capacity. *Agricultural Systems* 132, 1–11.

Schut, M., Klerkx, L., Sartas, M., Lamers, D., McCampbell, M., Ogbonna, I., and Leeuwis, C. (2016). Innovation platforms: experiences with their institutional embedding in agricultural research for development. *Experimental Agriculture* 52 (4), 537–561.

Sengupta, P. (2016). How effective is inclusive innovation without participation? *Geoforum* 75, 12–15.

Smink, M. M., Hekkert, M. P., and Negro, S. O. (2015). Keeping sustainable innovation on a leash? exploring incumbents' institutional strategies. *Business Strategy and the Environment* 24(2), 86–101.

Solomon, B. O., Ebegba, R., and Gidado, R. S. M. (2014). Influencing politicians and policy makers for a viable biotechnology sector: a case study of the Nigerian biosafety bill drafting, and passage process at the parliament. In: Wambugu, F., and Kamanga, D. (Eds.), *Biotechnology in Africa: Emergence, Initiatives and Future*, pp. 271–288. Springer, Berlin.

Spielman, D., Davis, K., Negash, M., and Ayele, G. (2011). Rural innovation systems and networks: findings from a study of Ethiopian smallholders. *Agriculture and Human Values* 28, 195–212.

Steyn, N. P., Nel, J. H., Parker, W., Ayah, R., and Mbithe, D. (2012). Urbanisation and the nutrition transition: a comparison of diet and weight status of South African and Kenyan women. *Scandinavian Journal of Public Health* 40(3), 229–238.

Struik, P. C., Klerkx, L., and Hounkonnou, D. (2014). Unravelling institutional determinants affecting change in agriculture in West Africa. *International Journal of Agricultural Sustainability* 12, 370–382.

Totin, E., van Mierlo, B., Saïdou, A., Mongbo, R., Agbossou, E., Stroosnijder, L., and Leeuwis, C., (2012). Barriers and opportunities for innovation in rice production in the inland valleys of Benin. *Wageningen Journal of Life Sciences* 60–63, 57–66.

Trienekens, J. H. (2011). Agricultural value chains in developing countries a framework for analysis. *International Food and Agribusiness Management Review* 14, 51–82.

Turner, J., Klerkx, L., Rijswijk, K., Williams, T., and Barnard, T. (2016). Systemic problems affecting co-innovation in the New Zealand Agricultural Innovation System: identification of blocking mechanisms and underlying institutional logics. *Wageningen Journal of Life Sciences* 76, 99–112.

United Nations Department of Economic and Social Affairs (UNDESA) (2018). Final list of proposed sustainable development goals indicators. United Nations Department of Economic and Social Affairs. https://sustainabledevelopment.un.org/content/documents/11803Official-List-of-Proposed-SDG-Indicators.pdf (accessed August 7, 2018).

Van Rooyen, A., and Homann, S. (2007). Innovation platforms: a new approach for market development and technology uptake in southern Africa. International Crops Research Institute for the Semi-Arid Tropics (ICRISAT). http://www.icrisat.org/locations/esa/esa-publications/Innovation-platform.pdf (accessed March 7, 2019).

Waddock, S., Meszoely, G. M., Waddell, S., and Dentoni, D. (2015). The complexity of wicked problems in large scale change. *Journal of Organizational Change Management* 28(6), 993–1012.

Wieczorek, A. J., and Hekkert, M. P. (2012). Systemic instruments for systemic innovation problems: a framework for policy makers and innovation scholars. *Science and Public Policy* 39, 74–87.

Wu, F., and Guclu, H. (2013). Global maize trade and food security: implications from a social network model. *Risk Analysis* 33(12), 2168–2178.

Zott, C., and Amit, R. (2010). Business model design: an activity system perspective. *Long Range Planning* 43(2), 216–226.

25

Making Scale Work
for Sustainable Development

A Framework for Responsible Scaling of Agricultural Innovations

Seerp Wigboldus, Laurens Klerkx, and Cees Leeuwis

1. Introduction

The use of the vocabulary of scaling solutions delivered through science, technology, and innovation (STI) is gaining momentum (e.g., Africa Union Commission 2014; UNDESA, UNCTAD and UNOSD 2017; National Academies of Sciences, Engineering, and Medicine 2017; United Nations Industrial Development Organization 2017). Adoption of innovations (framed as "solutions") at large scales is commonly argued as needed given the scale of challenges facing humanity (Wigboldus and Jochemsen 2018) and the ambitions of the sustainable development goals (SDGs). Hence the common approach of finding solutions to specific problems, in specific contexts and treating those as generalizable solutions (Schurman 2018). However, scaling such solutions does not automatically help meet societal goals. Related assumptions need to be made explicit to test their validity.

"Theories of change" have become widely used in the context of international development by private sector enterprises—for example to define what makes for sustainable farming (UTZ 2017)—and in agricultural research (Balmann and Valentinov 2016; CGIAR 2012; Maru et al. in press; Mayne and Johnson 2015; Thornton et al. 2017). The purpose of such theories of change is to capture and express the way in which aspired change in agricultural systems and value chains is thought to be possible (Maru et al. 2018) and to identify key assumptions upon which these expectations are based (Archibald et al. 2016). The articulation of theories of change and related impact pathways has become a more common practice in agricultural research and innovation design over the past decade, especially within the concept of agricultural research for development (AR4D). The aim is to support assessment of the appropriateness of proposed research and innovation strategies (Thornton et al. 2017). The process of articulating a theory of change (ToC) creates opportunities for interaction between stakeholders, elucidating stakeholders' assumptions regarding exactly what change is needed and their potential roles in effectuating change (Grygoruk and Rannow 2017; Tavella 2016).

A ToC in the context of research efforts thus aims to identify impact pathways, in a continuum from planned research outputs to innovations and finally to impact at scale relating to local, national, and global public goods (e.g., Douthwaite et al. 2003; Gaunand et al. 2015; Thornton et al. 2017), including impacts related to the SDGs, as stated in the Ag4SDGs initiative (CGIAR 2017). In this context, "scaling up" and "scaling out" are key terms. Scaling up means increasing the usage of particular practices within an existing region. Scaling out usually refers to wider application. In this chapter, we use the generic term "scaling" unless quoted literature phrases it differently.

Research outputs may have an indirect relation to outcomes and impact (e.g., dissemination of knowledge through communication channels such as articles, briefs, media messages) or a more direct relationship (e.g., delivery of new technologies or improved practices that can be used more widely). If agricultural research aims to connect to impact at scale through knowledge, technologies, and practices that it generates, related theories of change are required to express how scaling processes are expected to take place (Passioura 2010). However, as Matt et al. (2017) argue in relation to the impact dimension, much of the question of how scaling happens tends to remain a black box in theories of change and related impact pathways in the context of research programs, but also in wider development initiatives (Figure 25.1). Darbas et al. (2015) call this the output–outcome gap. Those who do address this gap almost always do so from a purely instrumentalist perspective of how to make scaling happen and rarely do a more comprehensive exploration of, for example, potentially negative side effects (see, e.g., Gillespie et al. 2015; Oddsdóttir 2014).

Related complexities tend to be left mostly unexplored and unanticipated (see also Apgar et al. 2016; Ely et al. 2014; Wigboldus et al. 2016). As many project and program proposals include a significant scaling phase, the implication may be that decision-makers are not appropriately informed about options for, and implications of, scaling processes. Contributions of research to development impact at scale can be assessed through *ex post* impact evaluation (e.g., Douthwaite et al. 2003; EIARD 2003; Maredia et al. 2014; Matt et al. 2017). However, it is also useful to enhance the ability for *ex ante* assessment of scaling processes. In addition to results-based management and as part of a ToC, we propose that articulating a specific *theory of scaling* (ToS) could complement current efforts to use theories of change to guide research and innovation programs toward impact at scale. In this chapter, we present a ToS framework to help decision-makers unpack what is involved in scaling processes in order to improve theories of change. Figure 25.2 illustrates part of what such unpacking involves and identifies the assumptions involved. It also illustrates the need to be realistic about what claims can reasonably be made about links between research, innovations, and aspired impact at scale (Leeuwis et al. 2018). Assumptions may relate to,

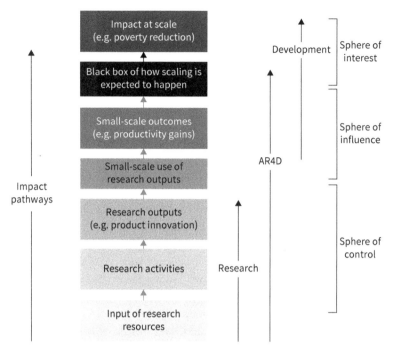

Figure 25.1 A simplified impact-pathway perspective on theories of change in relation to agricultural research and the often missing articulation of assumptions relating to dimensions and dynamics of scaling processes
Source: Adapted from CGIAR (2016); Thornton et al. (2017).

for example, roles that particular stakeholders need to play and required capacities involved, environmental and political conditions, and the motivation and ownership of primary stakeholders.

We focus here on theories of scaling in the context of research and innovation initiatives. Such initiatives take place within wider governance frameworks that require them to contribute to political agendas (such as the SDGs), for example by creating incentives and disincentives for particular scaling processes. We are not discussing governance dynamics here, but we will return to this topic in our discussion section in light of the multitude of scaling initiatives that somehow need to work together toward achieving shared societal goals.

The outputs and outcomes from articulating a ToS using a systematic process such as we suggest in our ToS framework can perform two key functions in support of decision-making. First, it can provide a shared reference framework regarding scaling processes among stakeholders, involving a) a shared vision for the scaling initiative and shared scaling ambitions among stakeholders, and b) shared assumptions and plausibility structure about what would make for effective and responsible scaling. Second, it can support decision-making in scaling initiatives by a) helping

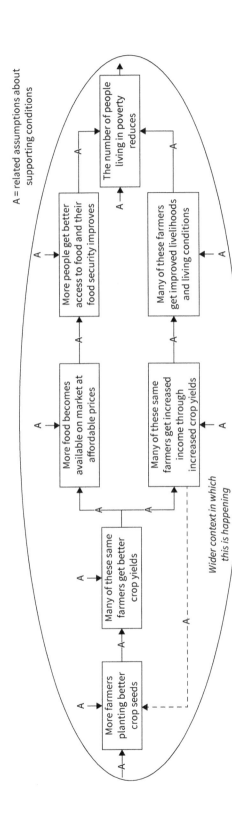

Figure 25.2 A simplified illustration of sequential scaling processes involved in impact pathways and related causal relationships

Source: the authors

to consider what is important to take into account in the design and implementation of the scaling initiative, b) raising awareness about different strategic options for engaging with scaling processes, c) raising awareness about specific needs for capacities and conditions in scaling initiatives, and d) addressing scaling-specific monitoring and evaluation (MandE) needs.

We address the following questions: 1) Which elements of a ToS framework would enable the systematic unpacking of scaling processes, serving as a structured way to articulate a ToS? and 2) How can such a framework be used to assess scaling initiatives and be part of a wider ToC? In section 2, we present the suggested ToS framework. In section 3, we discuss the six dimensions of the framework in more detail and translate it into a decision-making process for partners and stakeholders. In section 4, we discuss broader implications in terms of options for using theories of scaling as part of theories of change. We also reflect on our research questions, and draw conclusions on the potential for articulating theories of scaling to enhance preparedness for effective and responsible scaling, and on the need for further research and development.

2. The Theory of Scaling Framework

Scaling in the context of agricultural research and innovation has been the subject of a large body of research (e.g., Garb and Friedlander 2014; Hermans et al. 2013; Johansson et al. 2015; Millar and Connell 2010; Rogers 2003; Wigboldus et al. 2016). Scaling relates to both actively promoted processes (e.g., Pachico and Fujisaka 2004) and processes that are not steered by human actors (e.g., diseases or climate change, as in West 1999). Scaling may involve human agency but still not be the result of promotion by policy or research agencies (e.g., use of mobile phones in Africa, or urbanization in many countries, as in Bettencourt et al. 2007). Scaling may also be catalyzed but then develop a dynamic of its own (e.g., Chambers 1992). In commerce, scaling relates to such things as sales numbers, expanding production (capacity), and franchising (e.g., Galitopoulou and Noya 2016; Gradl and Jenkins 2011).

To build a framework to guide the development and use of theories of scaling, we identified key elements of typical theories of change (see, e.g., Douthwaite et al. 2007; Mayne and Johnson 2015; Vogel 2012) and used these to identify what should inform the development and use of a ToS (Figure 25.3).

3. Informing the Development of a Theory of Scaling

We discuss six dimensions of scaling in a particular order in the following, but as they partly overlap they need to be considered interactively and iteratively.

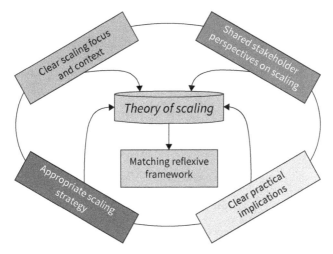

Figure 25.3 Dimensions of a reflection framework to inform the development of a theory of scaling
Source: the authors

3.1 Creating Clarity About the Scaling Focus and Context

This dimension of scaling is about using a scaling-specific theory of change as a "canvas" for outlining innovation characteristics, relevant context characteristics, and their implications for scaling.

3.1.1. Facilitating Development of Rich Perspectives on How Scaling Happens

Scaling processes involve more complexities than are generally taken into account in scaling initiatives (Wigboldus et al. 2016). Developing a ToS therefore requires the use of analytical tools that help to create rich perspectives on what may affect scaling processes and on what scaling processes may have an effect, including potentially undesired effects.

The literature provides a rich basis from which to draw in developing initial perspectives on what needs to be considered in a particular scaling initiative (see Table 25.1). Wigboldus et al. (2016) reviewed several conceptual frameworks that help to create an integrated systemic perspective on options for, implications of, and potential complications in scaling initiatives. They argue that a multilevel perspective (MLP) (Blesh and Wolf 2014; Elzen et al. 2012; Geels 2002; Hinrichs 2014) helps create a perspective on the interaction between novelties (innovations) that emerge under specific conditions (niches) and dominant institutional conditions (regimes) in, for example, key institutions.

Table 25.1 Dimensions of scalability of a particular innovation

---

The chance of an innovation going to scale increases if the innovation:
- Is feasible and can in principle be used more widely.
- Is credible, based on sound evidence, or espoused by respected persons or institutions.
- Is observable, potential users can see the result in practice; this may involve trialling (on a limited basis).
- Is easy to transfer and adopt, relating to simplicity and ease of use.
- Can be tested without committing the potential user to complete adoption when results have not yet been seen.
- Is suitable for reinvention in terms of modification/adaptation to create ownership and fit-for-purpose.
- Is relevant for addressing persistent or sharply felt problems.
- Has a relative advantage over existing practices.
- Is compatible with existing users' established values, norms, and facilities, not requiring big changes in existing practices.
- Is enabled by conducive communication processes (networks, peer-to-peer).

Source: Adapted from Cooley and Kohl (2006); Holcombe (2012); and Rogers (2003).

In general, analytical tools need to be able to help develop an appropriate understanding about a number of things, including: 1) the relevant context (potentially multiple situations) in which the scaling initiative takes place both in terms of origins (e.g., where piloted) and target (the contexts in which scaling is envisaged to happen); 2) characteristics of products or processes (innovations) involved; and 3) relevant stakeholder dynamics, in terms both of those affecting conditions for scaling and of those being affected by the scaling processes in positive or negative ways.

3.1.2 Considering Relevant Innovation Characteristics and Their Implications

The ease of scaling an innovation depends on the nature of the innovation itself, conditions under which it was tested, and conditions under which it is meant to be applied at scale (Table 25.1). The possibility of wider use can only be assessed if we know what exactly is meant to be used or applied more widely. Scaling will often involve adaptation and translation (Coe et al. 2014; Garb and Friedlander 2014). Key questions, therefore, include: scaling of exactly what (e.g., use of a specific technology), on what scale (e.g., area or number of farmers), and in what contexts (e.g., geographic).

From a perspective of *responsible* scaling, scalability may be assessed more specifically along the lines of economic feasibility, social acceptability, cultural appropriateness, ethical propriety, geographical determinants, political preferences, and ideological purposes (Wigboldus and Leeuwis 2013). We will return to the topic of responsible scaling later in this chapter.

3.1.3 Considering Relevant Context Conditions and Their Implications

The concept of scaling spaces allows for the creation of an integrated perspective on conditions for scaling. Table 25.2 lists a number of such spaces. Scaling initiatives

Table 25.2 Spaces for scaling

Space (conditions) for scaling	Description
Natural resource/ environmental space	The extent to which the impact of the scaling initiative on natural resources and the environment must be considered, harmful effects mitigated, or beneficial impacts promoted.
Political space	The extent to which political support for a scaling initiative can be ensured. This may require alignment with political agendas, including such things as the SDGs.
Cultural space	The extent to which there are cultural obstacles and the extent to which the scaling initiative can be suitably adapted to support responsible scaling in culturally diverse environments.
Analytical space	The extent to which appropriate analysis informs decision-making regarding the scaling initiative.
Social space	The extent to which the scaling initiative is embedded in conducive (multistakeholder) relationships and interactions, and the extent to which appropriate leadership and facilitation can support this.
Partnership space	The extent to which partners can be mobilized to coordinate efforts relevant for the scaling up of the initiative effort.
Legitimacy space	The extent to which the scaling initiative has a recognized mandate from relevant stakeholders to guide collaborative efforts (e.g., mandate for multistakeholder partnership).
Capacity/ competency space	The extent to which appropriate capacities and competences can carry the scaling initiative forward.
Management space	The extent to which there is a match between the scale of management (institutions) and the scale/s of the social, economic, and ecological processes being targeted through the scaling initiative.
Facilitation space	The extent to which multistakeholder processes relating to the scaling initiative can be facilitated through agents such as brokers and intermediaries, and whether conducive functions can be put in place such as innovation and scaling platforms, hubs, labs, networks, and alliances.
Fiscal/financial space	The extent to which fiscal and financial resources can be mobilized to support the scaling initiative and/or the extent to which the costs of the initiative can be adapted to fit into the available fiscal/financial space.
Learning space	The extent to which knowledge about what does and does not work in scaling can be harnessed through monitoring and evaluation, knowledge sharing, and training, and the extent to which the scaling strategy is dynamic and adapts to an evolving process (no blueprints involved).

Source: Based on Wigboldus (2016), who adapted material from Cumming et al. (2006); Gillespie (2004); Jonasova and Cooke (2012); IFAD (2011).

need to consider the extent to which such conditions are conducive or not and what that means for scaling strategy options (Hounkonnou et al. 2018).

3.2 Getting to a Shared Stakeholder Approach

Scaling initiatives invariably involve a range of partners and stakeholders. A shared perspective on the role and nature of scaling processes, as well as on the way in which they can effectively and responsibly contribute to shared objectives, is critical.

3.2.1 Clarifying Conceptual Understanding

The term scaling is used in many different ways (Fixsen 2009; Wigboldus et al. 2016). It is important for partners who contribute to a scaling initiative to have a shared conceptual understanding about the scaling process.

Different concepts such as scaling up (e.g., increasing production volumes), scaling out (e.g., geographical spreading of the use of an innovation), horizontal scaling (often understood in the same way as scaling out), and vertical scaling (improving institutional embedding) are commonly used but sometimes lack clear definitions. Almost always, multiple types of scaling (relating to different scales) will be involved, each potentially having different implications for the appropriate scaling strategy, such as in the case of integrated pest management (IPM), which often comprises a number of different practices.

The question "At what scale?" can be asked in two ways: 1) What type of scale (e.g., numeric, spatial) applies? and 2) What level on that particular scale (e.g., few or many) applies? Often a variety of different types of scaling processes will be involved, such as faster transactions, more people involved, and more intensive use of resources. A combination of technical and institutional innovations may also be involved, such as when increased use of a particular technology requires new policies, legislation, or social arrangements (Sartas et al. 2017).

The adoption of a new technology usually involves a variety of interrelated scaling processes for different products and practices. The particular scale to which they relate (numeric, spatial, etc.) will often be different as well (Wigboldus et al. 2016). This means that impact will relate to a variety of processes on different scales. Furthermore, it may involve processes of both scaling up and scaling down. For example, scaling up the practice of using herbicides may lead to scaling down the practice of mechanical weeding. Some scaling processes may be unintentionally triggered; for example, scaling up the practice of using chemical fertilizers may trigger the scaling down of plant biodiversity.

Creating conceptual clarity among stakeholders enhances opportunities for informed dialogue on options and their implications, and for developing shared perspectives on what the scaling initiative needs to take into account.

3.2.2 Considering Guiding Principles and Orientations

Theories of change involve fundamental ideas on what makes for progress and development. These involve worldviews and subjective preferences and therefore are not value-neutral (Stirling 2011; Sumberg et al. 2013). Theories of scaling involve that same potential for contestation. This leads to questions such as: Who drives this scaling agenda? What interests are at stake? What historical processes leading up to the current situation need to inform our decision-making because they affect people's perspectives and preferences? What consequences can be foreseen? (e.g., Aggestam et al. 2017; Johnson et al. 2016).

Scaling is commonly understood as a process of finding out what works and doing more of the same. This relates to so-called proven innovations and solutions that are suitable to be used more widely, by other actors, in new places, and often at larger scale. In this rhetoric, scaling leads to impact at scale. However, what works at one scale level and/or in a particular context does not necessarily work the same way at other scale levels and in other contexts (Cumming et al. 2006; Wigboldus et al. 2016). In the absence of an integrated systems perspective, situations may be created in which positive impact in one sphere of life and for one particular group (e.g., income of large corporations) goes hand in hand with negative impact in another sphere of life (e.g., reduced land security for smallholders). The idea of responsible scaling relates to such considerations (Wigboldus and Leeuwis 2013). Responsibility in that context includes awareness about potentially undesired consequences of scaling processes that might have been anticipated and then could have informed different decision-making (Wigboldus et al. 2016).

Stilgoe et al. (2013) have suggested four key dimensions of responsible innovation: anticipation, inclusion, responsiveness, and reflexivity. These four dimensions of responsible innovation translate well to the context of scaling processes (Wigboldus et al. 2016). This leads to four key design questions to direct scaling strategy considerations:

1. What are important things for the scaling initiative to anticipate in relation to the consequences of successful scaling, the target situations for scaling, and the future context dynamics?
2. To what does the scaling initiative need to respond in terms of both societal needs and societal concerns expressed by different stakeholders?
3. What does the scaling initiative need to include in processes of analysis and sense-making, who does it need to involve in decision-making processes and in collaborative effort, and who is meant to benefit in exactly what way?
4. What does the scaling initiative need to include in analysis and strategic guidance to inform reflexive and adaptive management in light of the defined purpose?

Table 25.3 Example ranges of options in focusing an appropriate strategy

Focus on adoption of solutions through scaling	←→	Focus on system change supported by matching scaling processes
Focus on direct intervention (control/influence)	←→	Focus on indirect intervention (catalysis)
Focus on engaging as individual organization	←→	Focus on engaging as broad collaborative effort
Focus on one grand scaling initiative with central leadership	←→	Focus on network of multiple interactive scaling efforts related to common goal
Focus on blueprint for scaling (roll-out) i.e., fixed selected innovations to be scaled up	←→	Focus on flexible scaling (adaptive/organic/coevolutionary process guided by reflexive monitoring)
Focus on achieving scale rapidly	←→	Focus on more biological or organic growth involving gradual absorption
Focus on how to make scaling of particular innovations happen (effectiveness focus)	←→	Focus on how scaling can align with wider societal processes and goals (responsibility focus)

Source: the authors.

Other principles may, of course, be used as well, such as how scaling affects resilience and/or sustainability (e.g., de Bruijn et al. 2017). In articulating a ToS, those involved in a scaling initiative need to consider which design principles should be guiding their efforts.

3.3 Deciding on an Appropriate Scaling Strategy

Articulating a scaling strategy will involve considering trade-offs in light of implications of different strategy choices. Partners and stakeholders may view such implications differently. The choice of strategy will need to connect to relevant required levels of complexity, uncertainty, ambiguity involved, levels of actor capability and knowledge available, and levels of connectivity (between partners, stakeholders) (Wigboldus and Leeuwis 2013).

3.3.1 Considering the General Strategy

There are many conceivable scaling strategies. Strategies always need to be context-specific rather than following standard processes. In relation to multistakeholder processes and monitoring and evaluation processes, many experts have been trained over the past few decades to support strategy development. For some reason, no scaling experts have been trained, even though scaling features as an important dimension of theories of change. This is a gap to be addressed as scaling experts could help stakeholders to think creatively about strategic options. Table 25.3 explores a number of options to consider in designing an appropriate scaling strategy.

A more direct, solutions-driven strategy will take as its point of departure a technology or practice that needs to go to scale to see its benefits multiplied (e.g., Bozeman et al. 2015), often involving a pilot and followed by a roll-out program (van de Fliert et al. 2010), commonly called dissemination and extension. A more indirect, vision-driven strategy will focus on creating an environment (e.g., achieved through subsidies or legislation) that attracts scaling processes that support the realization of a vision (e.g., food and nutrition security). What exactly will go to scale will then still be rather open. Potters and de Wolf (2014) discuss the case of scaling the IPM application, undertaken through new policies and legislation that created conditions favorable for IPM, rather than by pushing particular innovations.

Different strategy options can be mutually supportive where, for example, policymakers may focus more on enhancing institutional conditions and other actors more on generating options for scaling. Such combinations of scaling strategies happen in larger multistakeholder scaling initiatives such as SUN (SUN Movement 2019) and GAIN (GAIN 2019). This involves forging multistakeholder partnerships and platforms in acknowledgment of the multiplicity of interacting scaling processes (Leeuwis and Wigboldus 2017; Schut et al. 2015).

3.3.2 Considering Scaling Methods and Their Implications

Scaling methods are essentially about the question of how to get from one/few to many, from small to large. Extension services, farmer field schools, and innovation platforms (e.g., Adekunle et al. 2016; Millar and Connell 2010; Muilerman and Vellema 2017) are examples of such scaling methods. Marketing, subsidies, and taxation are other possible scaling methods. Different types of scaling methods involve different types of roles to be played by different actors in a scaling initiative (Hermans et al. 2013; Wigboldus 2016). Table 25.4 illustrates roles to be played in relation to the choice of different scaling methods.

3.3.3 Considering Scaling Scenarios and Their Implications

Scaling scenarios in agriculture are essentially about foresight analysis (Lehtonen et al. 2007; Struif Bontkes and van Keulen 2003). Scaling initiatives will interact with wider trends and developments, meaning that scaling processes will be part of a complex interaction of a host of scaling processes (Leeuwis and Wigboldus 2017). Scaling up production capacity, for example, may not make sense if world market prices are expected to drop (scale down) or if negative effects on the environment are expected. Furthermore, scaling the production of one particular crop may increase vulnerability to potential outbreaks of diseases or falling demand in the market.

Table 25.4 Possible roles to be played in scaling initiatives

Variety of possible roles	Description
Marketing	Through market studies understand target group preferences and communicate in ways connecting to those preferences so they will buy into the advertised benefits of the innovation at scale. The focus will be on dissemination and transferability (also see Little 2011).
Selling	Through promotion, publicity, or even propaganda, entice people to start making use of certain products or services at scale. The focus may be on branding.
Sharing/proposing	Generating options, informing people about them, and waiting to see what happens and whether this eventually leads to scaling of innovations. The approach may include open sourcing.
Facilitating/enabling	Creating capacities and conditions that make it easier for known innovations to go to scale. The approach may include cooperation and participation.
Aggregating	Connecting and taking up a full or intermediary role as part of a network or alliance to work on multiple scaling processes with multiple actors in relation to a common (scaling) goal. The approach may include collaborative networks.
Catalysing	Through policies and legislation creating conditions for scaling of yet unknown innovations that align with the system/sector/ societal aspirations to which those policies and legislation relate. The approach may include institutional change.

Source: Based on Little (2011); Tayabali (2014; Westley et al. (2014); Wigboldus (2016).

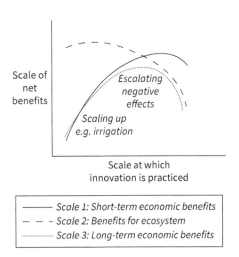

Figure 25.4 Possible scaling scenarios regarding the relationship between scale and net benefit

Source: Adapted from Wigboldus and Jochemsen (2018).

Finally, there is the big question of "What if this goes to scale?" Scaling the use of groundwater for crop irrigation may lead to dramatic hydrological effects (Hossain 2006). Decision-makers need to develop a sense of a "return on scaling": up to what scale level will scaling keep adding shared value and from what level can diminishing returns on scaling be expected? This also relates to developing a sense of how the net benefit of, for example, using a particular innovation ("net" meaning, in light of all relevant interests) would vary at different scales (Figure 25.4).

Constructing scaling scenarios and related foresight analysis may be one of the most important contributions of a ToS to the enhanced capability of decision-makers to engage effectively and responsibly with scaling processes.

3.4 Creating Clarity About Practical Implications

The organization and implementation of scaling initiatives may look very different, depending on the adopted principles for scaling, the relevant context, and the chosen scaling strategy. The focus of scaling may range from a simple project to a multistakeholder alliance or network (Ubels and Jacobs 2016). Central roles may reside more with the private sector, with the public sector, with research organizations, or with civil society organizations. This translates into different requirements regarding partnership development and facilitation, governance and organizational arrangements, and management processes (Gillespie 2004; Middleton et al. 2002). This may, for example, require interdonor coordination and capacity development in relation to responsiveness to institutional arrangements (Gillespie 2004). Related transdisciplinary collaboration and multistakeholder processes may also require different ways of working for researchers (Hoffmann et al. 2017; König et al. 2013; Wigboldus et al. 2016 2017). Rather than being located in the scaling phase of a program, this needs to be considered as early in research planning processes as possible (Ghiron et al. 2014).

These considerations translate into potential implications, including the need for appropriate competencies and capacities to deal effectively with pertinent scaling conditions and requirements; the need for appropriate collaborative arrangements (e.g., partnerships, alliances) with relevant actors who may significantly affect, or be significantly affected by, the scaling initiative; the need for the provision of appropriate programmatic arrangements and incentives, including realistic expectations about what single actors can achieve in terms of impact at scale; and the need for a "navigation plan" that allows for adaptive response to the realities encountered as the initiative unfolds.

As noted earlier, scaling processes as a topic is, unfortunately, not yet a specific field of expertise for which training and education are available. As a result, scaling initiatives are often managed by people who may be experts in the field of research and innovation, even in piloting options, but lack knowledge and expertise in the field of the complexities involved in scaling processes.

3.5 Consolidating and Articulating the Theory of Scaling

The consolidated ToS is not a scaling model, but rather an articulated shared perspective on how the scaling initiative would plausibly achieve its objectives (Douthwaite et al. 2003 2013; Mayne and Johnson 2015; Springer-Heinze et al. 2003; Thornton et al. 2017). It may be rendered in various kinds of visual formats, just as theories of change are (van Es et al. 2015; Vogel 2012). A ToS will include at least the following three components: 1) a timeframe showing interaction between processes, actors, and dimensions, as well as their anticipated sequencing over time, which represents a view on assumed causal relationships; 2) key assumptions underpinning the ToS in terms of how scaling is expected to happen, including assumptions about relevant partnerships, alliances, and network arrangements, and about roles to be played and capacities needed (e.g., Archibald et al. 2016; Christiaensen 2017; Shortall 2013; Ton et al. 2015); and 3) critical uncertainties about causal relationships, actors' activities, and the way in which various processes may play out.

If donors and other stakeholders explicitly accept the plausibility of the ToS, this creates a basis for "being in it together"; this can help prevent having to prove effectiveness through mere achievement of predefined scaling targets that may both be unrealistic and create a culture of mere target setting (Douthwaite and Hoffecker 2017; MacCormack 2014). This also relates to roles and responsibilities along impact pathways and the need for appropriate expectations about related contributions to impact (Leeuwis et al. 2018). As is the case with theories of change in general, a ToS, rather than being a fixed guidance instrument, will need to be revisited and, if needed, revised over time as it becomes clear how the scaling initiative is faring in reality.

3.6 Defining a Theory-of-Scaling Reflexive Framework

To understand how a scaling initiative is faring, we suggest four points of reference: the extent to which 1) understanding about the scaling focus and context and related assumptions are found to be correct and valid; 2) the scaling strategy (strategies) and related assumptions are found to be appropriate and effective; 3) the consolidated ToS is found to be a sound basis for strategic and operational management of the scaling initiative; and 4) the strategic and operational management of the scaling initiative is found to be appropriate and effective. This needs to be complemented by processes that assess emerging effects of scaling in light of both intended benefits and of possible unintended effects.

A well-articulated ToC, including a ToS, will spell out key assumptions, critical uncertainties, and (causal) connections between actors and factors that are expected to lead to aspired impact at scale. In sound monitoring and evaluation practice, this translates into strategic questions: How will we know our assumptions turn out to be valid, that uncertainties are not turning out to place major obstructions in the impact pathway, and that change is coming about as envisaged (Kusters et al. 2017)? If questions are clear and relevant, the program knows what it needs to know and can define its information needs at different points in time.

Scaling processes are usually influenced by different actors, and related effects often occur over a timespan of many years. This complicates attribution claims; many actors may claim the same impact, and impact assessment may therefore make little sense if it is done only in relation to separate initiatives (Ton et al. 2011 2014; Maru et al. in press). Contribution analysis (e.g., Befani and Mayne 2014; Delahais and Toulemonde 2012; Mayne 2001) is one of the ways to address this situation.

Negative or positive impacts may occur long after a particular initiative has ended (Sabiha et al. 2015; Urruty et al. 2016). This may cause complications in multiactor settings and long-term impact of scaling processes: Who is responsible for keeping track of what? This underscores the importance of informing policymaking and governance processes with big-picture and long-term perspectives on the way in which a multitude of scaling initiatives work out in complementary or conflicting ways in light of societal goals (Gee et al. 2013; Padt et al. 2014; Stilgoe et al. 2013).

3.7 From Theory of Scaling to Decision-Making Processes

Scaling "generalizable solutions," emerging through STI, is widely considered to be the blueprint for achieving the SDGs. The SDGs were designed as an integral perspective on *the world we want*, but in reality, scaling initiatives will focus on particular targets related to particular goals. When related innovations go to scale, this tends to have unanticipated and potentially undesired effects on other spheres of life and thus on (targets related to) other goals. This relates to similar principles as those underpinning total cost accounting.

Following the articulation of a ToS, decision-making will need to address a range of questions at three different levels.

Level 1: How to scale things in the right way? This relates to the operational side of a scaling initiative, as we discussed in relation to the ToS dimensions of clear practical implications, and the reflexive framework. Even if an envisioned scaling initiative is considered appropriate (see level 2 and 3), it still matters how the scaling is implemented.

Level 2: How to scale the right things? This relates to the strategic aspect of scaling, such as a clear scaling focus and context, stakeholder perspectives on scaling, and scaling strategy. This is not about finding generalizable solutions to be rolled out, but rather about finding generalizable principles of design and purpose to be applied in diverse ways to connect appropriately to variable contexts (e.g., through "baskets of options").

Level 3: How to engage responsibly with scaling processes? This connects to an integrated perspective on the SDGs. It relates to questions which we briefly referred to at the start of this chapter, but have not dealt with extensively in the ToS framework since it is about fundamental ideas on what makes for progress and development (Wigboldus and Jochemsen 2018).

Addressing particular SDG targets may undermine achievements in relation to other SDG targets. For example, efforts to address SDG targets 2.1, 2.2, and 2.3 (which are about food and nutrition security, and productivity) may conflict with target 2.4 (which includes maintaining ecosystems). Targets under SDG15 are even more specific about environmental conditions than target 2.4. A focus on SDG2 may only address environmental concerns in a very general way; it may prevent further encroaching into pristine forests, but leaving alone existing degradation of forests. The lesson is that these types of trade-offs need to be identified and carefully considered in decision-making processes.

4. Discussion and Conclusion

The issue of scaling tends not to be adequately addressed in theories of change. As a result, the capabilities of decision-makers and managers to deal effectively and responsibly with scaling are limited. We have introduced a framework to encourage and facilitate the adoption of a systematic approach to the articulation of a theory of scaling, to be used as a tool for reflection and decision-making by researchers and managers of scaling initiatives, policymakers, and donors concerned about the effective use of their donations.

The process of articulating a ToS and the resulting product can help inform decision-makers when they are making choices on how to connect to and engage with scaling processes (Table 25.5).

A ToS should be an integral part of a ToC, not a separate product. Scaling-inclusive ToCs can highlight the importance of a collaborative perspective in scaling initiatives, the need to anticipate scaling processes and define related implications for research and innovation design, and the need for interactions with partners and stakeholders early on in research and innovation in anticipation of envisioned scaling.

The process of articulating a ToS may enrich existing diagnostic and planning approaches such as the rapid assessment of agricultural innovation systems (e.g., Schut et al. 2015), participatory impact pathway analysis (Alvarez et al. 2010; Douthwaite et al. 2007), and the practice of results-based management, which is increasingly being applied in relation to agricultural research efforts (Schuetz et al. 2014). Quick-scan explorations, which may be facilitated through soft systems methodology (e.g., rich pictures) and interviews with selected focus groups and key informants, can help prevent the articulation of the ToS from turning into a research project on its own (Wigboldus et al. 2017).

Table 25.5 Potential contribution of the articulation of theories of scaling to the design and implementation of scaling initiatives

Limiting practice in scaling initiatives	Potential offered by ToS-based scaling initiatives
Mere rhetoric on scaling in proposals	Carefully thought through scaling approaches that include considerations about who drives scaling and why
Wishful thinking about anticipated scaling regarding how it would happen as well as regarding its wider effects and implications	Transparency about ambitions, ideas about how scaling is expected to happen, and would benefit the right people and about related assumptions
Considering scaling always to be a good idea if a related innovation is considered to have its merits (e.g., seen as a solution)	Awareness that the quality and impact of an innovation is codetermined by its original context; alertness to the fact that at scale and in other contexts (ecology, institutional, social, etc.) performance and effects of innovations may work out quite differently
Scaling processes considered only after initial research and innovation efforts	Scaling processes anticipated and taken into account in at early stages in research and innovation design and implementation
Narrow, silver-bullet-focused scaling strategies	Well-considered, contextualized, complexity-aware, and creative scaling strategies that are also informed through strategic foresight analysis
Organizations trying to make things go to scale mainly through their own efforts	Timely development of effective networking, alliances, and partnerships as a basis for a collaborative approach to scaling
Lack of articulated scaling narratives in proposals that include assumptions about how scaling is thought to work out	Insightful scaling narratives creating shared perspectives and a sense of shared direction in multistakeholder partnerships in scaling
Trial-and-error scaling initiatives	Scaling initiatives ready to engage effectively and responsibly with scaling processes through anticipatory, responsive, inclusive, and reflexive decision-making
Being oblivious to potential negative impacts of achieving scale	Strategic foresight supports scaling initiatives that have considered potential implications of, and trade-offs involved in, scaling, including considering potential effects across scales and social, economic, and environmental system boundaries

Source: the authors.

In this chapter, we focused on scaling in the context of agricultural research and innovation. These issues of scaling for agriculture are also highly relevant to the SDGs. Scaling initiatives that contribute to one of the SDGs may work out negatively for another SDG. For example, growing crops for biofuels may contribute to increased access to renewable energies (SDG7), but also go hand in hand with land grabbing

(SDG1) and reduced food security for certain groups (SDG2). Also, what appears to be an attractive innovation in small-scale and/or particular contexts may work out differently at larger scale, in other contexts, and in interaction with other conditions or innovations, including other innovations at scale (Raworth 2017; Rockström et al. 2009).

This perspective underscores the role of scaling-sensitive policymaking that defines the policy space for scaling processes. Policymakers need to consider sector-level and society-level theories of scaling when defining, for example, incentives and disincentives (such as subsidies or penalties) for agricultural and industrial development, as they are actively stimulating (or stopping) scaling of innovations through these measures. The focus of policymaking tends to be on considering what innovations match policy perspectives, without considering whether such innovations at scale and in different contexts would still achieve the relevant goals (Kanie and Biermann 2017; Padt et al. 2014).

The reflection framework suggested for articulating theories of scaling is a first step toward making the idea of theories of scaling more concrete. It will be important to develop further guidance on the role of policymaking and governance in view of many different scaling initiatives which have compound (therefore possibly crossing the optimal point for scale levels) and possibly conflicting impact in relation to the SDGs. Also, more field-testing in relation to different types of scaling initiatives, a description of process facilitation, and further development of guiding frameworks, are needed. As discussed in relation to scaling strategy, there is also a need to develop specific expertise in the field of guiding scaling initiatives and in the field of policymaking that recognizes and understands scaling, much in the same way that expertise in the field of monitoring and evaluation and in the field of multistakeholder partnership has been developed over the past few decades. As far as we know, scaling processes have thus far not been considered a particular field of expertise, as few if any training workshops or other educational efforts appear to be advertised. We would argue that this has limited both the effectiveness and the appropriateness of scaling processes. The ToS framework discussed in this chapter may also be considered as a tentative outline of a curriculum for training experts in responsible scaling.

Acknowledgments

We acknowledge the CGIAR Research Programs on Integrated Systems for the Humid Tropics (Humidtropics) and on Roots, Tubers and Bananas (RTB), and the CGIAR Fund Donors, for their provision of core funding to support this research. We also thank the reviewers of an earlier draft of this chapter for their valuable comments and suggestions.

References

Adekunle, A., Fatunbi, A. O., and Kefasi, N. (2016). The theory of change underlying the efficiency of agricultural innovation platforms (IPs): the case of the Thyolo Vegetable IP in Malawi. In: Francis, J., Mytelka, L., van Huis, A., and Röling, N. (Eds.), *Innovation Systems: Towards Effective Strategies in support of Smallholder Farmers*, pp. 143–155. CTA and WUR (Covergence of Sciences-Strengthening Innovation Systems (CoS-SIS)), Wageningen.

Africa Union Commission (2014). *Science, Technology and Innovation Strategy for Africa 2024*. Africa Union Commission, Addis Ababa.

Aggestam, V., Fleiß, E., and Posch, A. (2017). Scaling-up short food supply chains? a survey study on the drivers behind the intention of food producers. *Journal of Rural Studies* 51, 64–72.

Alvarez, S., Douthwaite, B., Thiele, G., Mackay, R., Córdoba, D., and Tehelen, K. (2010). Participatory impact pathways analysis: a practical method for project planning and evaluation. *Development in Practice* 20, 946–958.

Apgar, J. M., Allen, W., Albert, J., Douthwaite, B., Ybarnegaray, R. P., and Lunda, J. (2016). Getting beneath the surface in program planning, monitoring and evaluation: learning from use of participatory action research and theory of change in the CGIAR Research Program on Aquatic Agricultural Systems. *Action Research* 15(1), 1–20.

Archibald, T., Sharrock, G., Buckley, J., and Cook, N. (2016). Assumptions, conjectures, and other miracles: the application of evaluative thinking to theory of change models in community development. *Evaluation and Program Planning* 59, 119–127.

Balmann, N. A., and Valentinov, V. (2016). Towards a theory of structural change in agriculture: just economics? Paper prepared for presentation at the 149th EAAE Seminar "Structural Change in Agri-Food Chains: New Relations between Farm Sector, Food Industry and Retail Sector," Rennes, France, October 27–28, 2016. Leibniz Institute of Agricultural Development in Transition Economies (IAMO), Halle, Germany.

Befani, B., and Mayne, J. (2014). Process tracing and contribution analysis: a combined approach to generative causal inference for impact evaluation. *Institute of Development Studies (IDS) Bulletin* 45, 17–36.

Bettencourt, L. M. A., Lobo, J., Helbing, D., Kühnert, C., and West, G. B. (2007). Growth, innovation, scaling, and the pace of life in cities. *Proceedings of the National Academy of Sciences of the United States of America* (PNAS) 104, 7301–7306.

Blesh, J., and Wolf, S. (2014). Transitions to agroecological farming systems in the Mississippi River Basin: toward an integrated socioecological analysis. *Agriculture and Human Values* 31, 621–635.

Bozeman, B., Rimes, H., and Youtie, J. (2015). The evolving state-of-the-art in technology transfer research: revisiting the contingent effectiveness model. *Research Policy* 44, 34–49.

CGIAR (2012). Strategic Overview of CGIAR Research Programs. Part I. Theories of Change and Impact Pathways. Independent Science and Partnership Council, CGIAR. http://ispc.cgiar.org/sites/default/files/ISPC_WhitePaper_TOCsIPs.pdf (accessed March 8, 2019)

CGIAR (2016). Towards a Performance-Based Management System for CGIAR Research. http://www.cgiar.org/wp-content/uploads/2016/11/SC3-03_Towards-PerformanceMgmtSystem_17Nov2016.pdf (accessed March 8, 2019)

CGIAR (2017). Focusing Agricultural and Rural Development Research and Investment on Achieving SDGs 1 and 2 (Ag4SDGs). https://www.cgiar.org/wp-content/uploads/2017/06/2-JointInitiativeAG4SDGS.pdf (accessed March 8, 2019).

Chambers, R. (1992) Spreading and self-improving: a strategy for scaling up. In: Edwards, M., and Hulme, D. (Eds.), *Making a Difference: NGOs and Development in a Changing World*, pp. 40–47. Earthscan, London.

Christiaensen, L. (2017). Agriculture in Africa: telling myths from facts: a synthesis. *Food Policy* 67, 1–11.

Coe, R., Sinclair, F., and Barrios, E. (2014). Scaling up agroforestry requires research "in" rather than "for" development. *Current Opinion Environmental Sustainability* 6, 73–77.

Cooley, L., and Kohl, R. (2006). *Scaling up: From Vision to Large-Scale Change*. Management Systems International (MSI), Washington, DC.

Cumming, G. S., Cumming, D. H. M., and Redman, C. L. (2006). Scale mismatches in sociale-cological systems: causes, consequences, and solutions. *Ecology and Society* 11, 14.

Darbas, T., Maru, Y. T., Alford, A., Brown, P. R., and Dixon, J. (2015). Getting to impact: enriching logframes with theories of change. Paper presented at the Australasian Aid Conference ANU, February 2015.

de Bruijn, K., Buurman, J., Mens, M., Dahm, R., and Klijn, F. (2017). Resilience in practice: five principles to enable societies to cope with extreme weather events. *Environmental Science and Policy* 70, 21–30.

Delahais, T., and Toulemonde, J. (2012). Applying contribution analysis: lessons from five years of practice. *Evaluation* 18, 281–293.

Douthwaite, B., Alvarez, B. S., Cook, S., Davies, R., George, P., Howell, J., Mackay, R., and Rubiano, J. (2007). Participatory impact pathway analysis: a practical application of program theory in research-for-development. *Canadian Journal of Program Evaluation* 22, 127–159.

Douthwaite, B., and Hoffecker, E. (2017). Towards a complexity-aware theory of change for participatory research programs working within agricultural innovation systems. *Agricultural Systems* 155, 88–102.

Douthwaite, B., Kamp, K., Longley, C., Kruijssen, F., Puskur, R., Chiuta, T., Apgar, M., and Dugan, P. (2013). Using Theory of Change to Achieve Impact in AAS. AAS Working Paper. World Fish, Penang.

Douthwaite, B., Kuby, T., van de Fliert, E., and Schulz, S. (2003). Impact pathway evaluation: an approach for achieving and attributing impact in complex systems. *Agricultural Systems* 78, 243–265.

Ely, A., Van Zwanenberg, P., and Stirling, A. (2014). Broadening out and opening up technology assessment: approaches to enhance international development, co-ordination and democratisation. *Research Policy* 43, 505–518.

Elzen, B., Barbier, M., Cerf, M., and Grin, J. (2012). Stimulating transitions towards sustainable farming systems. In: Darnhofer, I., Gibbon, D., and Dedieu, B. (Eds.), *Stimulating Transitions Towards Sustainable Farming Systems: Farming Systems Research Into the 21st Century: The New Dynamic*, pp. 431–455. Springer, Dordrecht.

Fixsen, A. A. M. (2009). Defining scaling up across disciplines: an annotated bibliography. Portland State University, Portland, OR.

Galitopoulou, S., and Noya, A. (2016). Policy Brief on Scaling The Impact of Social Enterprises. Policies for Social Entrepreneurship. EU and OECD, Luxembourg.

Garb, Y., and Friedlander, L. (2014). From transfer to translation: using systemic understandings of technology to understand drip irrigation uptake. *Agricultural Systems* 128, 13–24.

Gaunand, A., Hocdé, A., Lemarié, S., Matt, M., and de Turckheim, E. (2015). How does public agricultural research impact society? a characterization of various patterns. *Research Policy* 44, 849–861.

Gee, D., Grandjean, P., Hansen, S.F., van den Hove, S., MacGarvin, M., Martin, J., Nielsen, G., Quist, D., and Stanners, D. (Eds.) (2013). *Late Lessons From Early Warnings: Science, Precaution, Innovation*. EEA report No1/ 2013. European Environment Agency, Copenhagen.

Geels, F. W. (2002). Technological transitions as evolutionary reconfiguration processes: a multi-level perspective and a case-study. *Research Policy* 31, 1257–1274.

Ghiron, L., Shillingi, L., Kabiswa, C., Ogonda, G., Omimo, A., Ntabona, A., Simmons, R., and Fajans, P. (2014) Beginning with sustainable scale up in mind: initial results from a population, health and environment project in East Africa. *Reproductive Health Matters* 22, 84–92.

Gillespie, S. (2004). Scaling Up Community-Driven Development: A Synthesis of Experience. International Food Policy Research Institute (IFPRI), Washington, DC.

Gillespie, S., Menon, P., and Kennedy, A. L. (2015). Scaling up impact on nutrition: what will it take? *Advances in Nutrition* 6, 440–451.

Global Alliance for Improved Nutrition (GAIN) (2019). GAIN. http://www.gainhealth.org/ (accessed January 21, 2019).

Gradl, C., and Jenkins, B. (2011). Tackling Barriers to Scale: From Inclusive Business Models to Inclusive Business Ecosystems. Harvard Kennedy School, Cambridge, MA.

Grygoruk, M., and Rannow, S. (2017). Mind the gap! lessons from science-based stakeholder dialogue in climate-adapted management of wetlands. *Journal of Environmental Management* 186, 108–119.

Hermans, F., Roep, D., and Klerkx, L. (2016). Scale dynamics of grassroots innovations through parallel pathways of transformative change. *Ecological Economics* 130, 285–295.

Hermans, F., Stuiver, M., Beers, P. J., and Kok, K. (2013). The distribution of roles and functions for upscaling and outscaling innovations in agricultural innovation systems. *Agricultural Systems* 115, 117–128.

Hinrichs, C. C. (2014). Transition to sustainability: a change in thinking about food systems change? *Agriculture and Human Values* 31, 143–155.

Hoffmann, S., Pohl, C., and Hering, J. G. (2017). Exploring transdisciplinary integration within a large research program: empirical lessons from four thematic synthesis processes. *Research Policy* 46, 678–692.

Holcombe, S. (2012). *Lessons From Practice: Assessing Scalability.* The World Bank, Washington, DC.

Hossain, M. F. (2006). Arsenic contamination in Bangladesh: an overview. *Agriculture, Ecosystems and Environment* 113, 1–16.

Hounkonnou, D., Brouwers, J., van Huis, A., Jiggins, J., Kossou, D., Röling, N., Sakyi-Dawson, O., and Traoré, M. (2018). Triggering regime change: a comparative analysis of the performance of innovation platforms that attempted to change the institutional context for nine agricultural domains in West Africa. *Agricultural* Systems 165, 296–309.

International Fund for Agricultural Development (IFAD) (2011). *Section XXI: Guidelines for Scaling Up. Updated Guidelines and Source Book for Preparation and Implementation of a Results-Based Country Strategic Opportunities Programme (RB-COSOP). Volume 1: Guidelines.* International Fund for Agricultural Development (IFAD), Rome.

Johansson, K. E., Axelsson, R., Kimanzu, Ng., Sassi, S. O., Bwana, E., and Otsyina, R. (2013). The pattern and process of adoption and scaling up: variation in project outcome reveals the importance of multilevel collaboration in agroforestry development. *Sustainability* 5, 5195–5224.

Johnson, N. L., Kovarik, C., Meinzen-Dick, R., Njuki, J., and Quisumbing, A. (2016). Gender, assets, and agricultural development: lessons from eight projects, *World Development* 83, 295–311.

Jonasova, M., and Cooke, S. (2012). *Thinking Systematically About Scaling Up: Developing Guidance for Scaling Up World Bank-Supported Agriculture and Rural Development Operations.* World Bank, Washington, DC.

Kanie, N., and Biermann, F. (Eds.) (2017). *Governing through Goals: Sustainable Development Goals as Governance Innovation.* MIT Press, Cambridge MA.

König, B., Diehl, K., Tscherning, K., and Helming, K. (2013). A framework for structuring interdisciplinary research management. *Research Policy* 42, 261–272.

Kusters, C., Batjes, K., Wigboldus, S., Brouwers, J., and Baguma, S. D. (2017). *Managing for Sustainable Development Impact: An Integrated Approach to Planning, Monitoring and Evaluation*. Practical Action Publishing, Rugby, UK.

Leeuwis, C., Klerkx, L., and Schut, M. (2018). Reforming the research policy and impact culture in the CGIAR: integrating science and systemic capacity development. *Global Food Security* 16, 17–21.

Leeuwis, C., and Wigboldus, S. (2017). What kinds of "systems" are we dealing with? implications for systems research and scaling. In: Oburn, I., Vanlauwe, B., Phillips, M., Thomas, R., Brooijmans, W., and Atta-Krah, K. (Eds.), *Sustainable Intensification in Smallholder Agriculture. An Integrated Systems Research Approach*, pp. 319–333. Routledge, New York.

Lehtonen, H., Bärlund, I., Tattari, S., and Hilden, M. (2007). Combining dynamic economic analysis and environmental impact modelling: addressing uncertainty and complexity of agricultural development. *Environmental Modelling and Software* 22, 710–718.

Little, M. (2011). *Achieving Lasting Impact at Scale. Part One: Behavior Change and the Spread of Family Health Innovations in Low-Income Countries*. Social Research Unit, Dartington, UK.

MacCormack, C. (2014). The road to scale runs through improved donor strategies. Less donor fragmentation can lead to impact at a transformative scale. Stanford Social Innovation Review, May 2014, Stanford University, Stanford.

Maredia, M. K., Shankar, B., Kelley, T. G., and Stevenson, J. R. (2014). Impact assessment of agricultural research, institutional innovation, and technology adoption: introduction to the special section. *Food Policy* 44, 214–217.

Maru Y., Sparrow, A., Stirzaker, R., and Davies, J. (2018). Integrated agricultural research for development (IAR4D) from a theory of change perspective. *Agricultural Systems* 165, 310–320.

Matt, M., Gaunand, A., Joly, P. B., and Colinet, L. (2017). Opening the black box of impact: ideal-type impact pathways in a public agricultural research organization. *Research Policy* 46, 207–218.

Mayne, J. (2001). Addressing attribution through contribution analysis: using performance measures sensibly. *Canadian Journal of Program Evaluation* 16, 1–24.

Mayne, J. (2007). Challenges and lessons in implementing results-based management. *Evaluation* 13, 87–109.

Mayne, J., and Johnson, N. (2015). Using theories of change in the CGIAR Research Program on Agriculture for Nutrition and Health. *Evaluation* 21, 407–428.

Middleton T., de la Fuente, T., and Ellis-Jones, J. (2005). Scaling up successful pilot experiences in natural resource management: lessons from Bolivia. In: Stocking, M., Helleman, H., and White, R. (Eds.), *Renewable Natural Resources Management for Mountain Communities*, pp. 221–238. International Centre for Integrated Mountain Development (ICIMOD), Kathmandu.

Middleton, T., Roman, M. A., Jones, J. E., Garforth, C., and Goldey, P. (2002). *Lessons Learnt on Scaling Up from Case Studies in Bolivia, Nepal, and Uganda*. Silsoe Research Institute (SRI), Silsoe, UK.

Millar, J., and Connell, J. (2010). Strategies for scaling out impacts from agricultural systems change: the case of forages and livestock production in Laos. *Agriculture and Human Values* 27, 213–225.

Muilerman, S., and Vellema, S. (2017). Scaling service delivery in a failed state: cocoa smallholders, farmer field schools, persistent bureaucrats and institutional work in Côte d'Ivoire. *International Journal of Agricultural Sustainability* 15, 83–98.

National Academies of Sciences, Engineering, and Medicine (2017). *Review of Science, Technology, Innovation, and Partnership (STIP) for Development and Implications for The Future of USAID*. National Academies Press, Washington, DC.

Neufeldt, H., Negra, C., Hancock, J., Foster, K., Nayak, D., and Singh, P. (2015). Scaling Up Climate-Smart Agriculture: Lessons Learned from South Asia and Pathways for Success. ICRAF Working Paper No. 209. Nairobi, World Agroforestry Centre.

Notenbaert, A., Pfeifer, C., Silvestri, S., and Herrero, M. (2017). Targeting, out-scaling and prioritising climate-smart interventions in agricultural systems: lessons from applying a generic framework to the livestock sector in sub-Saharan Africa. *Agricultural Systems* 151, 153–162.

Oddsdóttir, F. (2014). *Evaluations of Scaling Up*. GSDRC Helpdesk Research Report 1097. GSDRC, University of Birmingham.

Pachico, D., and Fujisaka, S. (Eds.) (2004). *Scaling Up and Out: Achieving Widespread Impact Through Agricultural Research*. International Centre for Tropical Agriculture (CIAT), Cali, Columbia.

Padt, F., Opdam, P., Polman, N., and Termeer, C. (Eds.) (2014). *Scale-Sensitive Governance of the Environment*. John Wiley and Sons, Chichester, UK.

Passioura, J. B. (2010). Scaling up: the essence of effective agricultural research. *Functional Plant Biology* 37, 585–591.

Potters, J., and de Wolf, P. (2014). It is never too soon to think about scaling: scaling integrated pest management in Denmark. In: van den Berg, J. (Ed.), *Blowing the Seeds of Innovation: How Scaling Unfolds in Innovation Processes Towards Food Security and Sustainable Agriculture*, pp. 8–9. Wageningen University & Research, Wageningen.

Raworth, K. (2017). *Doughnut Economics: Seven Ways to Think Like a 21st Century Economist*. Chelsea Green Publishing, White River Junction, VT.

Rockström, J., Steffen, W., Noone, K., Persson, Å., Chapin, F. S., Lambin, E. F., Lenton, T. M., Scheffer, S., Folke, C., Schellnhuber, H. J., Nykvist, B., de Wit, C. A., Hughes, T., van der Leeuw, S., Rodhe, H., Sörlin, S., Snyder, P. K., Costanza1, R., Svedin, U., Falkenmark, M., Karlberg, L., Corell, R. W., Fabry, V. J., Hansen, J., Walker, B., Liverman, D., Richardson, K., Crutzen, P., and Foley, J. A. (2009). A safe operating space for humanity. *Nature* 461, 472–475.

Rogers, E. M. (2003). *Diffusion of Innovations*. 5th ed. Free Press, New York.

Rogers, E. M., Medina, U. E., Rivera, M. A., and Wiley, C. J. (2005). Complex adaptive systems and the diffusion of innovations. *The Innovation Journal* 10, 30.

Sabiha, N. E., Salim, R., Rahman, S., and Rola-Rubzen, M. F. (2015). Measuring environmental sustainability in agriculture: a composite environmental impact index approach. *Journal of Environmental Management* 166, 84–93.

Sartas, M., Schut, M., and Leeuwis, C. (2017). Scaling Readiness for Agricultural Innovations Fundamentals and Metrics. Presentation at Scaling Readiness and Scaling Strategy Development Workshop for African Cassava Agronomy Initiative (ACAI), Ibadan, Nigeria, International Institute of Tropical Agriculture /Wageningen University. doi:10.13140/ RG.2.2.27993.52324

Schuetz, T., Förch, W., Schubert, C., Thornton, P., and Cramer, L. (2014). Lessons and Insights from CCAFS Results-Based Management Trial. Learning Brief No 12. CGIAR Research Program on Climate Change, Agriculture and Food Security (CCAFS), Copenhagen.

Schurman, R. (2018). Micro(soft) managing a "green revolution" for Africa: the new donor culture and international agricultural development. *World Development* 112, 180–192.

Schut, M., Klerkx, L., Rodenburg, J., Kayeke, J., Hinnou, L.C., Raboanarielina, C. M., Adegbola, P. Y., van Ast, A., and Bastiaans, L. (2015). RAAIS: Rapid Appraisal of Agricultural Innovation Systems (Part I): a diagnostic tool for integrated analysis of complex problems and innovation capacity. *Agricultural Systems* 132, 1–11.

Shortall, O. K. (2013). "Marginal land" for energy crops: exploring definitions and embedded assumptions. *Energy Policy* 62, 19–27.

Springer-Heinze, A., Hartwich, F., Henderson, J. S., Horton, D., and Minde, I. (2003). Impact pathway analysis: an approach to strengthening the impact orientation of agricultural research. *Agricultural Systems* 78 (2), 267–285.

Stilgoe, J., Owen, R., and McNaghten, P. (2013). Developing a framework for responsible innovation. *Research Policy* 42, 1568–1580.

Stirling, A. (2011). Pluralising progress: from integrative transitions to transformative diversity. *Environmental Innovation and Societal Transitions* 1, 82–88.

Struif Bontkes, T., and van Keulen, H. (2003). Modelling the dynamics of agricultural development at farm and regional level. *Agricultural Systems* 76, 379–396.

Sumberg, J., Thompson, J., and Woodhouse, P. (2013). Why agronomy in the developing world has become contentious. *Agriculture and Human Values* 30, 71–83.

SUN Movement (2019). Scaling Up Nutrition. http://scalingupnutrition.org/ (accessed April 8, 2019).

Task Force on Impact Assessment and Evaluation European Initiative for Agricultural Research for Development (EIARD) (2003). Impact assessment and evaluation in agricultural research for development. *Agricultural Systems* 78, 329–336.

Tavella, E. (2016). How to make participatory technology assessment in agriculture more "participatory": the case of genetically modified plants, *Technological Forecasting and Social Change* 103, 119–126.

Tayabali, R. (2014). PATRI framework for scaling social impact. https://www.inclusivebusiness.net/sites/default/files/wp/PATRI-Framework.pdf (accessed March 8, 2019).

Thornton, P. K., Schuetz, T., Förch, W., Cramer, L. Abreu, D., Vermeulen, S., and Campbell, B. M. (2017). Responding to global change: a theory of change approach to making agricultural research for development outcome-based. *Agricultural Systems* 152, 145–153.

Ton, G., Klerkx, L., de Grip, K. and Rau, M. L. (2015). Innovation grants to smallholder farmers: revisiting the key assumptions in the impact pathways. *Food Policy* 51, 9–23.

Ton, G., Vellema, S., and De Ruyter DeWildt, M. (2011). Development impacts of value chain interventions: how to collect credible evidence and draw valid conclusions in impact evaluations? *Journal on Chain and Network Science* 11, 69–84.

Ton, G., Vellema, S., and Ge, L. (2014). The triviality of measuring ultimate outcomes: acknowledging the span of direct influence. *Institute of Development Studies (IDS) Bulletin* 45, 37–48.

Ubels, J., and Jacobs, F. (2016). Scaling: From Simple Models to Rich Strategies. PPPLab Explorations 04. https://ppplab.org/2016/11/explorations-04-scaling-from-simple-models-to-rich-strategies/ (accessed April 6, 2019).

United Nations Department of Economic and Social Affairs (UNDESA), United Nations Conference on Trade and Development (UNCTAD), and United Nations Office for Sustainable Development (UNOSD) (2017). Workshop on Science, Technology and Innovation for SDGs. Report of the Meeting. Songdo Convensia, Incheon, Republic of Korea, November 29–December 1, 2017.

United Nations Industrial Development Organization. (2017). *Industrial Development Report 2018*. Demand for Manufacturing: Driving Inclusive and Sustainable Industrial Development. Vienna.

Urruty, N., Deveaud, T., Guyomard, H., and Boiffin, J. (2016). Impacts of agricultural land use changes on pesticide use in French agriculture. *European Journal of Agronomy* 80, 113–123.

UTZ (2017). UTZ' Theory of Change. https://utz.org/?attachment_id=13887 (accessed March 8, 2019).

van de Fliert, E., Christiana, B., Hendayana, R., and Murray-Prior, R. (2010). Pilot roll-out: adaptive research in farmers' worlds. *Journal of Agricultural Education and Extension* 6, 63–71.

van Es, M., Guijt, I., and Vogel, I. (2015). *Theory of Change Thinking in Practice: A Stepwise Approach*. HIVOS, The Hague.

West, G. B. (1999). The origin of universal scaling laws in biology. *Physica A: Statistical Mechanics and its Applications* 263, 104–113.

Westley, F., Antadze, N., Riddell, D. J., Robinson, K., and Geobey, S. (2014). Five configurations for scaling up social innovation: case examples of nonprofit organizations from Canada. *Journal of Applied Behavioral Science* 50, 234–260.

Wigboldus, S. (2016). Using a Theory of Scaling to Guide Decision Making: Towards a Structured Approach to Support Responsible Scaling of Innovations in the Context of Agrifood Systems. Wageningen University and Research, Wageningen.

Wigboldus, S., Hammond, J., Xu, J., Yi, Z. F., He, J., Klerkx, L., and Leeuwis, C. (2017). Scaling green rubber cultivation in Southwest China: an integrative analysis of stakeholder perspectives. *Science of the Total Environment* 580, 1475–1482.

Wigboldus, S., and Jochemsen, H. (2018). Scaling under scrutiny: a critical assessment of the idea of scaling innovations for development and progress. In: Wigboldus, S. (Ed.), *To Scale, or Not to Scale—That is Not the Only Question: Rethinking the Idea and Practice of Scaling Innovations for Development and Progress*, pp. 35–85. Wageningen University and Research, Wageningen.

Wigboldus, S., Klerkx, L., Leeuwis, C., Schut, M., Muilerman, S., and Jochemsen, H. (2016). Systemic perspectives on scaling agricultural innovations: a review. *Agronomy for Sustainable Development* 36, 46.

Wigboldus, S., and Leeuwis, C. (2013). *Toward Responsible Scaling Up and Out in Agricultural Development: An Exploration of Concepts and Principles*. Centre for Development Innovation, and Knowledge, Technology and Innovation group. Wageningen University and Research Centre, Wageningen.

CONCLUSION

26
Conclusions and Future Policies for Meeting the Sustainable Development Goals

Ademola A. Adenle, Marian R. Chertow, Ellen H.M. Moors, and David J. Pannell

1. Introduction

Science, technology, and innovation (STI) policy interventions are key to addressing the 17 sustainable development goals (SDGs) leading to increased productivity, sustainable growth, and prosperity around the world, and especially in developing regions. The SDGs are more tightly and explicitly related to STI solutions than were the millennium development goals (MDGs). An illustration of this is the establishment in 2015 of the technology facilitation mechanism (TFM), which aims to facilitate technology transfer to address the gap between developed and developing countries in STI capacity and to coordinate a multistakeholder forum on sustainable development, including SDGs. Beyond technology transfer, delivering STI solutions will require the bolstering of legislative support, business commitment, coordination with governments and civil society organizations of all types, and the strengthening of STI strategies at the country level. There is a consensus within the international community that developing strong STI capacity at the national level is critical to building thriving economies and societies and finding long-term solutions to national and global problems (UN 2016).

This book has provided insight into actual and potential STI contributions for the achievement of SDGs, covering a wide range of STI initiatives across numerous countries and sectors. It has explored the interconnections between the SDGs and STI capacity, often with a focus on the implementation of SDG projects. This book has emphasized intersectoral collaboration and interdisciplinary approaches to addressing the challenges of SDGs across the themes of environment and energy, health, and agriculture. The content of the book includes case studies, experiences, and empirical analyses, identifying some of the most significant roles that STI solutions can play in meeting the socioeconomic and environmental challenges spelled out in the SDG targets.

The chapters have covered a broad range of emerging issues and factors influencing the development of effective STI policies, and how such policies may contribute to or undermine the achievement of SDGs. In this conclusion we review contributions from all of the chapters, drawing out common themes, differences, and key lessons. Conclusions are identified for each theme and we finish with a set of recommendations based on the insights provided in the earlier chapters.

Previous studies indicate the potential role of STI in tackling global challenges including poverty, food insecurity, diseases, and climate change, yet in many developing countries, little attention is paid to harnessing STI in addressing these problems (Adenle et al. 2015; UN 2014; World Bank 2008). The global development agenda, including MDGs, often underemphasized the potential for STI contributions, resulting in impacts that fall short of their potential. Contributors to this underperformance have included limited investment in research and development (R&D), poor protection of intellectual property rights (IPRs), insufficient knowledge infrastructure, lack of institutional support for development and diffusion of STI solutions, and lack of adequate regulatory instruments that could create incentives for investment in STI solutions. The experiences and evidence presented in this book illustrate how a failure to provide the institutions and resources needed to build STI capacity, and a failure of key actors to engage synergistically, can be serious impediments to development.

Prior to this book, there had been relatively little interdisciplinary writing about STI and the SDGs. The interdisciplinary approach used here has cast light on a number of questions, including: How can STI capacity be enhanced and harnessed at the national level to meet specific SDG targets? What can be done to improve international STI frameworks that can enhance the implementation of SDGs? How can we reliably test new ideas for harnessing STI to deliver SDGs?

The aims of this book are to examine the roles that STI can play in contributing to the SDGs, to understand factors aiding or impeding successful application of STI, taking into account the interlinkages among SDGs, and to investigate the interaction of STI and SDGs in a broad variety of settings and geographies.

2. Key Messages

Theme I of the book (environment and energy) offers the following set of findings to assist with achieving SDGs: 1) With respect to the four chapters on energy, renewable technologies present strong opportunities for inclusive social and economic development and can also contribute to addressing environmental problems, particularly those related to climate change and the implementation of the 2015 Paris Agreement; 2) Chapters 3 to 6, which focus on solar energy, laid out numerous conclusions that have emerged from research over the last 15 years, emphasizing specific sociological and governance conditions that slow the pace of diffusion, thereby inhibiting effective responses to climate change (SDG13) and biodiversity loss (SDG15). Solar energy is a gateway not only to increased human development, but also to significant changes in the current system of production and consumption across different sectors and industries, offering a sustainable development path based on a system of green innovation; 3) environment and energy outcomes can benefit from policy measures that target low-carbon innovation and market-development mechanisms while supporting national STI capacity building (e.g., research institutes, universities, R&D investment, and IPR systems); and 4) assessment tools such as input-output analysis (IOA) and life-cycle assessment (LCA) enable comparisons of different ways to achieve similar goals in environmental terms so that less harmful solutions can be identified and selected.

With respect to barriers to energy development, chapter 3 (Timilsima and Shah) highlights a wide range of problems facing the renewable energy sector, characterized by the presence of market, capacity, and regulatory barriers in the developing world. The authors make the case that these barriers must be overcome so that SDG7 and other relevant SDGs can benefit from the advantages of renewable energy technologies. Chapter 4 (Adenle) underscores the need to develop robust renewable energy policy that encourages investment in R&D for solar energy programs in view of limited institutional capacity across African countries. Overcoming barriers that limit the abilities of developing countries to introduce renewable energy technologies successfully enables participation in competitive market-based economies.

As described in chapter 5 (Schmidt et al.), regulatory challenges in Cambodia, Laos, and Indonesia remain important problems limiting the design, installation, and operation of renewable-energy micro-grids, which could undermine achievements within SDG7. Similarly, chapter 6 (Kemp et al.) contrasts the enormous growth of installed solar capacity in India based on auctions and other market-oriented approaches with China's reliance on the governance role of state institutions and programs for successfully harnessing STI to meet SDG7.

With respect to the environmental aspects of Theme 1, environment and energy, we report findings at the human/environment interface based on policies and programs in the natural environment (chapters 2 and 7) and on tools and technologies in the built environment (chapters 8 to 10). As with renewable energy, the environmental policy themes reflect that national and international level policies can be effective, but they require significant supporting resources.

Chapter 2 (Stevens) examines a target of SDG15 to halt biodiversity loss. By examining the role of innovation in biodiversity conservation, the author finds that the management pathway developed through the use of action plans advanced under the Convention on Biological Diversity can be used effectively at the international level but depends on a great deal of planning and coordination now and to the SDG deadline of 2030. Chapter 7 (Machado and Young) scrutinize the R&D needed for environmental conservation and sustainable use of natural resources in Brazil, but rather than placing as much attention on the interplay of institutions as Stevens did, the chapter investigates the financial resources available to meet the country's SDGs; a large shortfall is found by 2030. This leads the authors to discuss improvements needed for STI funding policy and to consider making SDG targets mandatory for determining resource allocation.

Chapter 8 (Chertow et al.) introduces several tools and strategies from the field of industrial ecology for determining how SDG-related activities and technologies can be assessed to determine their environmental and social performance. The authors report how these tools can identify trade-offs of different alternatives such as impacts of public transport versus private vehicles applicable to SDG9 on cities or where in a food supply chain the highest amounts of food waste are generated useful for SDG2 on hunger.

Delving further into technology, the case of automated vehicles discussed in chapter 9 (Wang and Oster) revolves around the call in SDG12 to make cities inclusive, safe, resilient, and sustainable. The authors show, for example, how providing low-speed driverless shuttles for city residents can increase safety and inclusivity by coordinating with public mass transit rather than competing with it. But of course technology also has its limits, which can lessen the impact of SDGs through rebound effects described in chapter 10 (Font Vivanco and Makov). The irony of rebound effects is that an environmentally sound technology may encourage consumers to use more of a given product than when the same function was fulfilled by the product when it had less desirable characteristics. The authors find, therefore, that rebound effects can easily serve as hidden barriers for achieving SDGs by offsetting some of the positive effects.

Theme II (health) describes scientific advances and the role of innovative technologies in addressing global health problems. Here, the authors emphasize the following key points toward achieving SDG3 (healthy lives) and other relevant SDGs: 1) harnessing STI is critical for the delivery of effective and low-cost health technologies; 2) removing regulatory barriers (including the logjam around international property rights) can facilitate access to vaccines against emerging and neglected infectious diseases in low-income regions; 3) increased availability, affordability, accessibility, and acceptability of anti-malaria drug development can facilitate production, diffusion, and deployment of new drugs in affected regions; 4) deployment of analytical tools such as development-focused health technology assessment (HTA) can help decision-makers to prioritize and make new health technologies and innovations more accessible both in developing and developed countries; 5) recent progress of digital health can be important in improving the quality of healthcare; 6) support for long-term transformations of urban sanitation services is important in low- and middle-income countries (LMICs) to reach the health-related SDGs; and 7) health innovations needed to deliver SDGs should be pursued in responsible and responsive ways.

The chapters in the health theme describe how STI are conceptualized in the SDGs to ensure healthy lives and to promote well-being for all (SDG3), and also how they are assumed to tackle health inequalities (SDG10) and to assure availability of water and sanitation (SDG6) (e.g., chapters 11, 13, 15, 16, and 17). By exploring cases in the field of vaccine innovations, anti-malarial drug developments, point-of-care diagnostics, cookstoves, urban sanitation projects, and animal-source foods, these chapters show that global health issues should be defined as multidimensional, multiactor problems. In contrast to the MDGs, the SDGs entail a global and equitable approach to these global health issues.

Chapter 11 (Possas et al.) shows that a number of the SDGs are related to vaccination problems. They elaborate on the specific technological and regulatory gaps in developing countries that inhibit the delivery of health-related SDGs, such as limited commercial interest in neglected diseases and limited access to patented vaccines. Adequate STI governance strategies can overcome many of these challenges and increase access to essential medicines and vaccines. An example is the Decade of Vaccines initiative, described in chapter 11. Vaccines developed by novel STI strategies require adaptations at a systems level, such as accompanying immunization services to reach the poorest areas with mobile units and adequate capacity-building and infrastructure.

Chapter 13 (de Haan and Moors) shows the importance of using an availability, affordability, accessibility, and acceptability framework to analyze the development and diffusion of anti-malarial medicines in light of the SDG goals, particularly the SDG3 targets. This chapter shows that the malaria burden in LMICs has been strongly reduced due to improved access to effective interventions for malaria prevention, diagnosis, and treatment. According to the authors, resistance to artemisinin and associated drugs has developed and is currently spreading. Therefore, innovative R&D approaches are required to restore anti-malarial efficacy and to circumvent the spread of resistance to other areas, which could have a huge public-health impact.

Chapter 14 (Poon et al.) shows the interesting relationship between emerging digitalization in health and its relation to various SDGs. The authors use a four-dimensional perspective, zooming in on translation, education, transformation, and technology to examine the progress of digital health in China and its potential to improve the quality and delivery of healthcare to enhance health at all ages (SDG3) and the way health information is shared through new value chains in the health system. Part of the authors' emphasis is that digital health literacy and expansion of complementary skills among health professionals and patients will become more important to ensure inclusive and equitable quality education and to promote lifelong learning, thereby contributing to the achievement of targets for SDG4 (equitable quality education). Additionally, transforming health service supply in LMICs through digital technology will enable better access to services, including in rural areas, which often have poorer access than do urban areas. Also, progress in digital health can also reduce inequality, thereby contributing to SDG10 targets.

Chapter 16 (van Welie and Truffer) shows the potential of STI to address urban sanitation problems, an important issue as progress toward the SDG on water and sanitation (SDG6) is very slow and problems are especially persistent for cities in the Global South. Such a complex problem demands a transformation of urban sanitation services and infrastructure in view of challenges identified by the authors. Some of these challenges include limited participation of local communities, lack of adequate public services, and lack of awareness. The authors argue that STI efforts should focus on improving alignment among various sanitation-service regimes in the city through making utility services work better in informal settlements and improving collaboration and participation of relevant stakeholders to scale on-site sanitation innovations.

Chapter 17 (de Bruyn et al.) focuses on STI to promote the value of animal-source foods for meeting health-related SDGs in resource-poor regions and other settings. A holistic approach is needed, targeting stakeholder engagement for promotion of animal-source foods to meet health-related SDGs. According to these authors, the use of mobile technologies, for example, can support important information for livestock production. The authors also mention that gender-sensitive and culturally sensitive communication (incorporating traditional beliefs) can encourage participation in nutrition programs in resource-poor settings. These insights may contribute to achievement of SDG3 targets.

Chapter 12 (Bouttell et al.) gives an overview of the role of development-focused HTA to achieve the health-related SDGs. LMICs are now beginning to develop HTA and apply it in their healthcare decision-making. This chapter shows how a development-focused HTA analysis of low-cost point-of-care diagnostics can potentially improve the return on investment in new technologies by improving the efficiency of research prioritization and development processes while ensuring that the needs of vulnerable populations are met. Furthermore, chapter 15 (Engel et al.) shows that taking a responsible and responsive lens on STI is very important for inclusive SDG policies taking a multiactor perspective of different stakeholder settings involved, and their perceptions and needs.

The point-of-care approach highlighted in chapters 12 and 15 underscores the importance of frugal innovation that can provide access to affordable and cost-effective diagnosis and treatment for low-income and poor populations. Frugal innovation is increasingly seen as a potential solution to overcome challenges such as access to water, sanitation, and energy security for those who live in resource-constrained communities around the world. In chapter 16, van Welie and Truffer mention that a pro-poor initiative for on-site sanitation services has been developed by social enterprises in Kenya to address sanitation problems. This is consistent with ideas laid out by Adenle in chapter 4 (Theme I) that government policy should support R&D efforts that support production and redesign of solar products, thereby making services and systems more affordable for the rural poor in Africa. With these measures being put in place, a targeted policy approach to support frugal innovation can help deliver some SDG targets around the world.

Taken together, the chapters in this theme highlight that health and STI policymakers should demand SDG-inclusive innovative health technologies. National STI policy should focus on responsible health technology development in context and should create STIs that are responsive and flexible for a range of situations. Rather than relying on health technology transfer from developed to developing countries, however, the SDGs require us to go deeper as described in this theme. More holistic approaches, given the high potential for unintended consequences, also address the impacts of STI with respect to reducing injustice and inequality and increasing health and well-being on a global basis.

Key messages emerging from Theme III (agriculture) include the following: 1) agricultural R&D is critical to ensure a flow of new technologies and innovations that allow farmers and others in agricultural supply chains to address SDGs; 2) there are concerns about the concentration into a few hands of intellectual property rights for the seeds of improved agricultural crop varieties, raising the risk of monopolistic pricing strategies; 3) sophisticated packages of technologies and practices designed and promoted to farmers can help advance SDGs, provided that the packages are well designed to meet farmers local needs and to fit with their context and challenges; 4) there is an increasing need for R&D to strengthen the emphasis on sustainability, not just productivity, when addressing the needs for farmers in developing countries; 5) well-designed agricultural innovations and technologies can contribute simultaneously to multiple SDGs, commonly including SDG1 (end poverty), SDG2 (end hunger), SDG8 (economic growth, employment, and decent work), and SDG13 (combat climate change) and sometimes others; 6) policy settings regulating technologies in developing countries should reflect needs and risks in those countries rather than in developed countries; 7) effective systems for agricultural extension (including awareness-raising measures and technical support) are needed to accelerate the spread of improved agricultural practices and technologies in developing countries; and 8) leverage points for delivery of SDGs may be along the food supply chain, rather than on farms.

Agricultural extension is one approach to scaling up the adoption of agricultural innovations that could contribute to delivery of SDGs. Others include farmer field schools, innovation platforms, subsidies, taxes, and marketing. Chapter 25 (Wigboldus et al.) focuses on the issue of scaling up, identifying it as a key issue for policymakers and development agencies to address. They provide a framework for thinking about and planning the scaling process, encompassing four key issues, in brief: the consequences of scaling, both societal needs and societal concerns, the needs for analysis and collaboration, and reflexive and adaptive management. These elements can contribute to defining a theory of scaling that the authors propose should form part of any theory of change. The approach provided in this chapter is relevant across all agricultural development issues.

Several chapters focus on particular agricultural technologies, practices, or packages. Chapter 18 (Harpankar) explored a range of technologies related to the nitrogen fertility of crops, and discussed their potential to contribute to various SDGs. A particular challenge with nitrogen fertility is providing sufficient nitrogen to achieve high crop yields (supporting SDGs 1, 2, and 8) while avoiding nitrogen pollution in water bodies (affecting SDG6 and SDG14). The technological options range from traditional (legume rotations) to high-tech (biotechnology) and all are seen as having potential. However, the barriers facing smallholder farmers in developing countries when considering adoption of these practices are substantial, including issues of affordability, technical expertise, infrastructure, and market access.

Chapter 19 (Rola-Rubzen et al.) and chapter 21 (Mwongera et al.) both dealt with complex packages of agricultural practices and technologies in developing counties. Both considered developing regions where traditional agricultural practices with low productivity are still widely used. Both packages are intended to contribute to the resilience of farmers, particularly in the face of climate change. Rola-Robzen et al. emphasized the importance of using effective strategies for agricultural extension, awareness raising, capacity-building, and technical advice. Mwongera et al. prioritized various practices in different agricultural regions based on farmers' preferences, and found that the top priorities contributed to food security and livelihoods. Understanding farmers' preferences and needs is an important element of designing effective approaches to agricultural extension.

The important role of policy and governance is highlighted in several chapters in the agriculture section. Adenle et al. (in chapter 20) observes that policy constraints on the use of genetically modified organisms (GMOs) or genetic modification (GM) technology in various developing countries are inhibiting the achievement of various SDGs. They question the application of a highly risk-averse policy approach, similar to that applied in Europe, in countries where the cost in lost development opportunities is much higher than in the wealthy countries of Europe. Wedig (in chapter 23) is also concerned with existing policy and governance settings, this time in the context of small-scale fishers on Lake Victoria in Africa. Lake Victoria supports large numbers of small-scale fishers, as well as larger commercial fishing companies. Wedig argues that existing governance favors the larger companies, with potential negative consequences for SDG1 (end poverty), SDG2 (end hunger), and SDG14 (sustainable marine resources). She argues for a new transformation governance based on a threefold structure: increasing eco-efficiency, redistributing access to natural resources, and recognizing eco-sufficiency as a guiding principle.

Chapter 22 (Flocco) examined the soybean production complex in Brazil, exploring various ways to make the system contribute better toward various SDGs. STI are central to many of the approaches discussed, including contributions from the private sector. Reinforcing the message of Adenle et al. in chapter 20, genetically modified crops have played major roles in advancing the development of Brazilian agriculture, including making it more sustainable through reductions in use of pesticides and use of less toxic herbicides. Flocco notes that there have been concerns about GM technology in Brazil too, although they tend to be more about monopoly power and mono-cropping rather than unsubstantiated concerns about food safety.

As in other sectors, there can be trade-offs between SDGs in agriculture. An example is that the success of GM soybeans in Brazil led to pressure for deforestation to increase the area of land used for soybean production (chapter 22). An interesting initiative to counter this was the Soy Moratorium, which substantially reduced clearing, in part by facilitating market access for soybeans not associated with deforestation (or slave labor or threats to indigenous lands).

Flocco also highlighted that there are many stakeholders of many different types who impinge on the performance and impacts of an agricultural production complex. For Brazilian soybeans she identified farmers, local communities, private sector, commerce, transport, trading institutions, banks and financing institutions, producer's associations and commercial chambers, government, certifying institutions, multistakeholder dialogue tables, nongovernmental organizations, academic and research institutions, and consumers. This highlights the importance of a broad-based multipronged approach to the delivery of SDGs, rather than a narrow focus on one group. STI has contributions to make in all parts of this complex system.

Finally, Dentoni et al. (chapter 24) present value network analysis (VNA) as a diagnostic tool for analyzing business models that are being reorganized. They illustrate its application to the agricultural commodity exchange (ACE) in Malawi, a business model with complex objectives: to increase value-chain efficiency while fostering food security and reducing rural poverty and marginalization. In this case study, the approach helps to identify options for building cross-sector partnerships and coordinating actions (including STI actions) to tackle relevant SDG targets.

3. The Importance of Systems Approaches

Just as the SDGs have evolved to be cross-cutting and interdisciplinary, STI proposals, too, are seen in many of the chapters to have embraced a broader, systems approach. We believe that this perspective can facilitate implementation of the 2030 Agenda, including the SDGs. An example from each theme illustrates this point.

In the environment and energy theme, chapter 9 on automated vehicles analyzes this new technology from the perspective of being embedded in a pre-existing city. As game-changing as the technology appears to be, the importance of numerous other social, political, and infrastructural considerations are critical aspects of whether this is a viable solution for meeting SDGs and how these issues affect the overall adoption timeline for this innovation.

From the health theme, chapter 16 presents a systemic perspective on the global sanitation challenge. This chapter highlights the potential of STI to solve urban sanitation challenges, which are serious and persistent in many cities of LMICs. Using a sociotechnical systems perspective, the authors analyze the interplay between technologies, infrastructure and associated actor networks, regulations, sanitation providers, and user practices regarding sanitation. The authors specifically mention that various system weaknesses related to actors, networks, and institutions represent significant challenges to the implementation of improved sanitation systems. Therefore, STI efforts should focus on improving alignment and collaboration with the key stakeholders to scale on-site sanitation innovations to contribute to the achievement of SDG6 targets.

Chapter 19, from the agriculture theme, analyzes integrated crop management (ICM) in the context of the multifaceted constraints faced by farmers in Timor Leste. ICM is a production system that combines multiple components and delivers to multiple SDGs. Despite evidence of significant benefits from ICM, evidence shows that adoption of the system by farmers has been partial and biased. One important factor is that technical advice and training has been targeted to male farmers; not surprisingly, adoption of the system by female farmers is substantially lower. Another observation, one that is common in studies of adoption of agricultural innovations, is that adoption is more likely on larger farms. This is explicable in terms of the broader gains that larger farmers stand to make, and their greater capacity to invest the resources needed to learn about and implement the new innovation. Understanding this social, technical, and economic system provides insights into the sorts of initiatives needed to deliver development outcomes more effectively and more equitably.

In summary, the benefits of systems approaches are readily apparent. National STI policies can help to foster systems and networks that include governments, the private sector, community leaders, and international organizations.

4. Policy Recommendations

Since the SDGs were adopted in 2015, donors have provided billions of dollars for various related initiatives, and citizens show increasing awareness of the global goals. From a policy perspective, the most pressing issues discussed in this book relate to the need to improve the international STI framework and to encourage countries to increase their STI capacity so that they can better address sustainable development challenges. With higher recognition and visibility of SDGs, intergovernmental organizations, national governments, and donors (from the public and private sectors, charitable organizations, multilateral agencies, and others) can direct more resources and attention to solving institutional, market, and political issues that are impeding STI progress for sustainable development.

Such an STI approach resonates with the European Union's increasingly popular mission-oriented approach to policies. Mazzucato (2018) stresses that innovation programs should be bold, inspirational, and of wide societal relevance to engage the public and attract investment. These programs should be targeted, measurable, and time-bound enough to allow effective monitoring and evaluation. Further, Mazzucato argues that missions should be formulated ambitiously (high-risk but realistically feasible, centered on STI innovation activities) (Mazzucato 2018). In this light, we offer the following suggestions and recommendations:

1. That national STI policies adopt strategies that are market-driven (where relevant), stimulate socioeconomic growth and well-being, promote cooperation, and provide protection of property rights but also access by the poor to beneficial IP.
2. That country-level governments advance national innovation systems specifically for developing strong STI capacity that can foster sustainable development. There are a range of issues to be addressed affecting the capacity of STI to contribute to SDGs including: educational curricula; educational and

research infrastructure; public funding for R&D; and translation, outreach, and extension to deliver the benefits of STI to business and the community.

3. That broad and equitable community benefits be seen as foundational to STI programs and that the governance systems aim to support efficient institutions, strong legal frameworks, transparent decision-making, and provision of incentives, including business incentives, that acknowledge and encourage STI for SDGs.

4. That STI policy foster initiatives designed to deliver frugal innovation—the process of reducing the cost and complexity of a technology to make it more accessible. Particularly in developing countries, frugal innovation can stimulate social entrepreneurship, socioeconomic empowerment, and sustainable development by addressing food insecurity, energy problems, and health and well-being at the grassroots level.

5. That STI policy bodies take into account the dimensions of availability, affordability, accessibility, and acceptability when analyzing emerging STI solutions to health and sustainability problems.

6. That R&D around lower cost renewable energy technologies continue to be a mainstay of international development, contributing to energy access and achievement of inclusive social and economic development. Such R&D can be complemented by local training and support designed to increase employment and human development.

7. That there be encouragement for digital health STI policies to transform health services in LMICs and to increase digital health literacy and skills of health professionals and patients. This would help to achieve better access to health services, reduce education inequality, and increase lifelong learning, contributing to universal health, sustainable education, and national equality.

8. That STI strategies be better contextualized by taking an inclusive multiactor perspective into account when considering the pathways to SDGs. This implies proactively addressing in STI strategies the specific needs, responsibilities, and perceptions of involved stakeholders and the particular settings in which they are operating, by applying a sociotechnical systems perspective that enables analysis of the interplay between technologies, infrastructures, and their associated actor networks, institutions, and user practices.

9. That the role of development-focused HTA to achieve the health-related SDGs be emphasized in STI policy, especially among LMICs. These countries should further apply this form of HTA in their healthcare decision-making, as it shows the efficiency of research prioritization and development processes, while ensuring that the needs of vulnerable populations are met.

10. That when making decisions about investments in STI initiatives, governments and other organizations explicitly consider likely trade-offs between SDGs, the distribution of benefits and costs, and the potential for unintended adverse effects. Prior to implementation, each initiative would be carefully assessed so that the full range of consequences is identified and evaluated.

11. That considering how the desired scale of impact from STI initiatives for delivering SDG-related benefits is important. Strategies such as development of a theory of scaling can help in this by emphasizing the need for a clear focus on scaling, developing perspectives on how scaling occurs, considering how innovation characteristics affect scaling, and creating shared perspectives about scaling outcomes.

12. That STI strategies and policies focus on improving coordination, alignment, and collaboration with key STI stakeholders to identify challenges impeding development of an international STI framework that includes governments and businesses. There are technical, legal, trade, and institutional barriers that need attention to improve knowledge transfer, leverage of expertise and skills, and access to finance, at all stages of R&D programs.

13. That to guide national investments, and to improve the likelihood of external funding support, STI initiatives be systematically prioritized, broken into smaller development packages, and presented to external donors with whose targets they align. Honing and clarifying these processes is important for building trust, assessing feasibility, establishing goals and timelines, and meeting milestones.

14. That STI policy in pursuit of sustainable agricultural production in developing regions address issues such as the balance between productivity and sustainability; the needs, constraints, and knowledge of resource-poor farmers; access to resources, markets, and infrastructure; the exercise of market power by input suppliers and output purchasers; effective agricultural extension that recognizes local conditions and farmer preferences; and efficient management of the supply chain beyond the farm gate.

References

Adenle, A. A., Azadi, H., and Arbiol, J. (2015). Global assessment of technological innovation for change adaptation and mitigation in developing world, *Journal of Environmental Management* 161, 261–275.

Mazzucato, M. (2018). Mission-oriented innovation policies: challenges and opportunities, *Industrial and Corporate Change* 27(5), 803–815.

United Nations (2014). Science, technology and innovation for the post-2015 development agenda, Economic and Social Council. Report of the Secretary-General. Seventeenth session Geneva, May 12–16.

United Nations (2016). Science, technology, innovation and capacity-building: Chapter II.G. Addis Ababa Action Agenda-Monitoring commitments and actions. http://www.un.org/esa/ffd/wp-content/uploads/2016/03/2016-IATF-Chapter2G.pdf (accessed March 31, 2019).

World Bank (2008). Science, Technology, and Innovation: Capacity Building for Sustainable Growth and Poverty Reduction. The World Bank, Washington, DC.

Index

Tables, figures, and boxes are indicated by *t, f,* and *b* following the page number

For the benefit of digital users, indexed terms that span two pages (e.g., 52–53) may, on occasion, appear on only one of those pages.